Radiation Effects in Solids

NATO Science Series

A Series presenting the results of scientific meetings supported under the NATO Science Programme.

The Series is published by IOS Press, Amsterdam, and Springer in conjunction with the NATO Public Diplomacy Division

Sub-Series

I. **Life and Behavioural Sciences**	IOS Press
II. **Mathematics, Physics and Chemistry**	Springer
III. **Computer and Systems Science**	IOS Press
IV. **Earth and Environmental Sciences**	Springer

The NATO Science Series continues the series of books published formerly as the NATO ASI Series.

The NATO Science Programme offers support for collaboration in civil science between scientists of countries of the Euro-Atlantic Partnership Council. The types of scientific meeting generally supported are "Advanced Study Institutes" and "Advanced Research Workshops", and the NATO Science Series collects together the results of these meetings. The meetings are co-organized bij scientists from NATO countries and scientists from NATO's Partner countries – countries of the CIS and Central and Eastern Europe.

Advanced Study Institutes are high-level tutorial courses offering in-depth study of latest advances in a field.
Advanced Research Workshops are expert meetings aimed at critical assessment of a field, and identification of directions for future action.

As a consequence of the restructuring of the NATO Science Programme in 1999, the NATO Science Series was re-organised to the four sub-series noted above. Please consult the following web sites for information on previous volumes published in the Series.

http://www.nato.int/science
http://www.springer.com
http://www.iospress.nl

Radiation Effects in Solids

edited by

Kurt E. Sickafus
Los Alamos National Laboratory, Materials Science & Technology Division,
Los Alamos, NM, U.S.A.

Eugene A. Kotomin
European Commission, Joint Research Center, Institute for Transuranium Elements,
Karlsruhe, Germany

and

Blas P. Uberuaga
Los Alamos National Laboratory, Materials Science & Technology Division,
Los Alamos, NM, U.S.A.

 Springer

Published in cooperation with NATO Public Diplomacy Division

Proceedings of the NATO Advanced Study Institute on
Radiation Effects in Solids
Erice, Sicily, Italy
17– 29 July 2004

A C.I.P. Catalogue record for this book is available from the Library of Congress.

ISBN 978-1-4020-5294-1
ISBN 978-1-4020-5295-8 (eBook)

Published by Springer,
P.O. Box 17, 3300 AA Dordrecht, The Netherlands.

www.springer.com

Printed on acid-free paper

Additional material to this book can be downloaded from http://extras.springer.com

TABLE OF CONTENTS

PREFACE

This book contains proceedings of the NATO Advanced Study Institute (ASI): The 32nd Course of the International School of Solid State Physics entitled *Radiation Effects in Solids*, held in Erice, Sicily, Italy, July 17-29, 2004, at the Ettore Majorana Centre for Scientific Culture (EMCSC). The Course had 83 participants (68 students and 15 instructors) representing 23 countries.

The purpose of this Course was to provide ASI students with a comprehensive overview of fundamental principles and relevant technical issues associated with the behavior of solids exposed to high-energy radiation. These issues are important to the development of materials for existing fission reactors or future fusion and advanced reactors for energy production; to the development of electronic devices such as high-energy detectors; and to the development of novel materials for electronic and photonic applications (particularly on the nanoscale).

The Course covered a broad range of topics, falling into three general categories:

Radiation Damage Fundamentals
> Energetic particles and energy dissipation
> Atomic displacements and cascades
> Damage evolution
>> Defect aggregation
>> Microstructural evolution

Material Dependent Radiation Damage Phenomena
(metals, alloys, semiconductors, intermetallics, ceramics, polymers, biomaterials)
> Atomic and microstructural effects (e.g., point defects, color centers, extended defects, dislocations, voids, bubbles, colloids, phase transformations, amorphization)
> Macroscopic phenomena
> (e.g., swelling, embrittlement, cracking, thermal conductivity degradation)

Special Topics
> Swift ion irradiation effects
> Ion beam modification of materials
> Nanostructure design via irradiation
> Nuclear fuels and waste forms
> Radiation detectors, dosimeters, phosphors, luminescent materials, etc.
> Solar and galactic cosmic particles
> Irradiation effects in bone

The Course served to demonstrate the crucial interplay between experimental and theoretical investigations of radiation damage phenomena. The Course explored computer simulation methods for the examination of radiation effects, ranging from molecular dynamics (MD) simulations of events occurring on short timescales (ps – ns), to methods such as kinetic Monte Carlo and kinetic rate theory, which consider damage evolution over times ranging from μs to hours beyond the initial damage event. The Course also examined the plethora of experimental techniques used to assess radiation damage accumulation in solids, including transmission electron microscopy, ion channeling, nanoindentation, and positron annihilation, to name only a few techniques.

Two international schools on radiation damage effects preceded this one: (1) *Radiation Damage in Solids* – The XVIII Course of the International School of Physics << Enrico Fermi >>, September 5 – 24, 1960, Centro di Studi Nucleari di Ispra del Comitato Nazionale per L'Energia Nucleare, Ispra (Varese) ITALY[1]; and (2) Fundamentals of Radiation Damage – an International Summer School on the Fundamentals of Radiation Damage, August 1 – 12, 1993, University of Illinois, Urbana, Illinois, USA.[2]

These proceedings are organized loosely as follows: state-of-the-art computational procedures for simulating radiation damage effects in solids are reviewed in Chapters 1 & 2; in-situ transmission electron microscopy (TEM) observations of radiation damage effects

---

[1] D. S. Billington and S. i. d. fisica, Eds., Radiation Damage in Solids: International School of Physics "Enrico Fermi" (1960; Varenna, Italy, (Academic Press, New York, 1962).

[2] Fundamentals of Radiation Damage, edited by I. M. A. Robertson, R. S., Tappin, D. K. Rehn, L. E., published in Journal of Nuclear Materials, Vol. 216, pp. 1 – 368 (1994).

are described in Chapter 3; different particle types and their effects are described in Chapter 4; radiation effects in metals are described in Chapters 5 & 6; radiation effects in ceramics and ionic compounds are discussed in Chapters 7 – 9; phase transformations induced by radiation are described in Chapters 10 – 12; and finally, selected special topics on radiation damage effects are covered in Chapters 13 – 19.

We, the Editors of this proceedings, wish to acknowledge the NATO Science Committee who helped to make this Course a success. Committee members included: Kurt Sickafus (Los Alamos National Laboratory); Eugene Kotomin (formerly University of Latvia; presently Institute for Transuranium Elements (ITU)); Steven Zinkle (Oak Ridge National Laboratory); Christina Trautmann Gesellschaft für Schwerionenforschung (GSI); and Giorgio Benedek (Università di Milano-Bicocca). [G. Benedek is also the Director of the International School of Solid State Physics.] We'd especially like to thank all of the authors for the many hours they spent preparing their contributions to these proceedings. We are also indebted to Susan Rhyne, Ishwari Sollohub, and James Valdez for their editing assistance in finalizing the manuscripts for this publication. We hope you enjoy this volume.

Kurt E. Sickafus
Eugene A. Kotomin
Blas P. Uberuaga
May, 2006

Chapter 1

INTRODUCTION TO THE KINETIC MONTE CARLO METHOD

Arthur F. Voter
Los Alamos National Laboratory, Los Alamos, NM 87545 USA

1   INTRODUCTION

Monte Carlo refers to a broad class of algorithms that solve problems through the use of random numbers. They first emerged in the late 1940's and 1950's as electronic computers came into use [1], and the name means just what it sounds like, whimsically referring to the random nature of the gambling at Monte Carlo, Monaco. The most famous of the Monte Carlo methods is the Metropolis algorithm [2], invented just over 50 years ago at Los Alamos National Laboratory. Metropolis Monte Carlo (which is *not* the subject of this chapter) offers an elegant and powerful way to generate a sampling of geometries appropriate for a desired physical ensemble, such as a thermal ensemble. This is accomplished through surprisingly simple rules, involving almost nothing more than moving one atom at a time by a small random displacement. The Metropolis algorithm and the numerous methods built on it are at the heart of many, if not most, of the simulations studies of equilibrium properties of physical systems.

In the 1960's researchers began to develop a different kind of Monte Carlo algorithm for evolving systems *dynamically* from state to state. The earliest application of this approach for an atomistic system may have been Beeler's 1966 simulation of radiation damage annealing [3]. Over the next 20 years, there were developments and applications in this area (e.g., see [3–7]), as well as in surface adsorption, diffusion and growth (e.g., see [8–17]), in statistical physics (e.g., see [18–20]), and likely other areas, too. In the 1990's the terminology for this approach settled in as *kinetic Monte Carlo*, though the early papers typically don't use this term [21]. The popularity and range of applications of kinetic Monte Carlo (KMC) has continued to grow and KMC is now a common tool for studying materials subject to irradiation, the topic of this book. The purpose of this chapter is to provide an introduction to this KMC method, by taking the reader through the basic concepts underpinning KMC and how it is typically implemented, assuming no prior knowledge of these kinds of simulations. An appealing property of KMC is that it can, in principle, give the *exact* dynamical evolution of a system. Although this ideal is virtually

1

K.E. Sickafus et al. (eds.), Radiation Effects in Solids, 1–23.
© 2007 *Springer.*

never achieved, and usually not even attempted, the KMC method is presented here from this point of view because it makes a good framework for understanding what is possible with KMC, what the approximations are in a typical implementation, and how they might be improved. Near the end, we discuss a recently developed approach that comes close to this ideal. No attempt is made to fully survey the literature of KMC or applications to radiation damage modeling, although some of the key papers are noted to give a sense of the historical development and some references are given for the reader who wants a deeper understanding of the concepts involved. The hope is that this introductory chapter will put the reader in a position to understand (and assess) papers that use KMC, whether for simulations of radiation damage evolution or any other application, and allow him/her to write a basic KMC program of their own if they so desire.

2   MOTIVATION: THE TIME-SCALE PROBLEM

Our focus is on simulating the dynamical evolution of systems of atoms. The premiere tool in this class of atomistic simulation methods is molecular dynamics (MD), in which one propagates the classical equations of motion forward in time. This requires first choosing an interatomic potential for the atoms and a set of boundary conditions. For example, for a cascade simulation, the system might consist of a few thousand or million atoms in a periodic box and a high velocity for one atom at time zero. Integrating the classical equations of motion forward in time, the behavior of the system emerges naturally, requiring no intuition or further input from the user. Complicated and surprising events may occur, but this is the correct dynamical evolution of the system for this potential and these boundary conditions. If the potential gives an accurate description of the atomic forces for the material being modeled, and assuming both that quantum dynamical effects are not important (which they can be, but typically only for light elements such as hydrogen at temperatures below T=300K) and that electron-phonon coupling (non-Born-Oppenheimer) effects are negligible (which they will be unless atoms are moving extremely fast), then the dynamical evolution will be a very accurate representation of the real physical system. This is extremely appealing, and explains the popularity of the MD method. A serious limitation, however, is that accurate integration requires time steps short enough ($\sim 10^{-15}$ s) to resolve the atomic vibrations. Consequently, the total simulation time is typically limited to less than one microsecond, while processes we wish to study (e.g., diffusion and annihilation of defects after a cascade event) often take place on much longer time scales. This is the "time-scale problem."

Kinetic Monte Carlo attempts to overcome this limitation by exploiting the fact that the long-time dynamics of this kind of system typically consists of diffusive jumps from state to state. Rather than following the trajectory through every vibrational period, these state-to-state transitions are treated directly, as we explain in the following sections. The result is that KMC can reach vastly longer time scales, typically seconds and often well beyond.

3   INFREQUENT-EVENT SYSTEMS, STATE-TO-STATE DYNAMICS, AND THE KMC CONCEPT

An infrequent-event system is one in which the dynamics is characterized by occasional transitions from one state to another, with long periods of relative inactivity between these transitions. Although the infrequent-event designation is fairly general (and hence also the possible applications of KMC), for simplicity we will restrict our discussion to the case where each state corresponds to a single energy basin, and the long time between transitions arises because the system must surmount an energy barrier to get from one basin to another, as indicated schematically in Fig. 1. This is an appropriate description for most solid-state atomistic systems. For a system that has just experienced a knock-on event causing a cascade, this infrequent-event designation does not apply until the excess initial energy has dissipated and the system has thermally equilibrated. This usually takes a few ps or a few tens of ps.

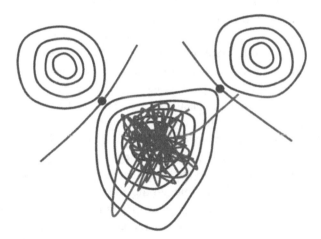

Figure 1. Contour plot of the potential energy surface for an energy-barrier-limited infrequent-event system. After many vibrational periods, the trajectory finds a way out of the initial basin, passing a ridgetop into a new state. The dots indicate saddle points.

To be a bit more concrete about the definition of a state, consider a 256-atom system in a perfect fcc crystal geometry with periodic boundary conditions. Remove one of the atoms and put it back into the crystal somewhere else to create an interstitial. Now, using a steepest descent or conjugate gradient algorithm, we can "relax" the system: we minimize the energy to obtain the geometry at the bottom of the energy basin where the forces on every atom are zero. This defines a particular state i of the system and the geometry at the minimum is R_i. If we heat the system up a bit, e.g., by giving each atom some momentum in a random direction and then performing MD, the system will vibrate about this minimum. As it vibrates, we still say it is in state i (assuming it has not escaped over a barrier yet) because if we stop the MD and minimize the energy again, the system will fall back to the exact same geometry R_i. Adjacent to state i there are other potential basins, each separated from state i by an energy barrier. The lowest barriers will correspond to moving the interstitial (perhaps through an interstitialcy mechanism) or moving an atom into the vacancy. Even though only one or a few atoms move in these cases, the entire system has been taken to a new state. This is an important point – we don't move atoms to new states, we move the entire system from state to state.

The key property of an infrequent-event system caught in a particular basin is that because it stays there for a long time (relative to the time of one vibrational period), it forgets how it got there. Then, for each possible escape pathway to an adjacent basin, there is a *rate constant k_{ij}* that characterizes the probability, per unit time, that it escapes to that state j, and these rate constants are independent of what state preceded state i. As we will discuss below, each rate constant k_{ij} is purely a property of the shape of the potential basin i, the shape of the ridge-top connecting i and j, and (usually to a much lesser extent) the shape of the potential basin j. This characteristic, that the transition probabilities for exiting state i have nothing to do with the history prior to entering state i, is the defining property of a Markov chain [22, 23]. The state-to-state dynamics in this type of system correspond to a Markov walk. The study of Markov walks is a rich field in itself, but for our present purposes we care only about the following property: because the transition out of state i depends only on the rate constants $\{k_{ij}\}$, we can design a simple stochastic procedure to propagate the system correctly from state to state. If we know these rate constants exactly for every state we enter, this state-to-state trajectory will be indistinguishable from a (much more expensive) trajectory generated from a full molecular dynamics simulation, in the sense that the probability that we see a given sequence of states and transition times in the KMC simulation is the same as the probability for seeing that same sequence and transition times in the MD. (Note that here we are assuming that at the beginning of an MD simulation, we assign a random momentum to each atom using a fresh random number seed, so that each time we perform the MD simulation again, the state-to-state trajectory will in general be different.)

4   THE RATE CONSTANT AND FIRST-ORDER PROCESSES

Because the system loses its memory of how it entered state i on a time scale that is short compared to the time it takes to escape, as it wanders around vibrationally in the state it will similarly lose its memory repeatedly about just where it has wandered before. Thus, during each short increment of time, it has the same probability of finding an escape path as it had in the previous increment of time. This gives rise to a first-order process with exponential decay statistics (i.e., analagous to nuclear decay). The probability the system has not yet escaped from state i is given by

$$p_{survival}(t) = \exp(-k_{tot}t), \tag{1}$$

where k_{tot} is the total escape rate for escape from the state. We are particularly interested in the probability distribution function $p(t)$ for the time of first escape from the state, which we can obtain from this survival probability function. The integral of $p(t)$ to some time t' gives the probability that the system has escaped by time t', which must equate to $1 - p_{survival}(t')$. Thus, taking the negative of the time derivative of $p_{survival}$ gives the probablity distribution function for the time of first escape,

$$p(t) = k_{tot}\exp(-k_{tot}t). \tag{2}$$

We will use this first-passage-time distribution in the KMC procedure. The average time for escape τ is just the first moment of this distribution,

$$\tau = \int_0^\infty t\, p(t)dt = \frac{1}{k_{tot}}. \tag{3}$$

Because escape can occur along any of a number of pathways, we can make the same statement as above about each of these pathways – the system has a fixed probability per unit time of finding it. Each of these pathways thus has its own rate constant k_{ij}, and the total escape rate must be the sum of these rates:

$$k_{tot} = \sum_j k_{ij}. \tag{4}$$

Moreover, for each pathway there is again an exponential first-escape time distribution,

$$p_{ij}(t) = k_{ij}\exp(-k_{ij}t), \tag{5}$$

although only one event can be the first to happen. For more discussion on the theory of rate processes in the context of stochastic simulations, see [24,25].

We are now almost ready to present the KMC algorithm. Not surprisingly, given the above equations, we will need to be able to generate exponentially distributed random numbers, which we quickly describe.

4.1 Drawing an Exponentially Distributed Random Number

Generating an exponentially distributed random number, i.e., a time t_{draw} drawn from the distribution $p(t) = k \exp(-kt)$, is straightforward. We first draw a random number r on the interval (0,1), and then form

$$t_{draw} = -(1/k)\ln(r). \qquad (6)$$

A time drawn in this way is an appropriate realization for the time of first escape for a first-order process with rate constant k. Note that the usual definition of the uniform deviate r is either $0 < r < 1$ or $0 < r \leq 1$; a random number generator implemented for either of these ranges will give indistinguishable results in practice. However, some random number generators also include $r = 0$ in the bottom of the range, which is problematic (causing an ill-defined ln(0) operation), so zero values for r must be avoided.

5 THE KMC PROCEDURE

Having laid the conceptual foundation, it is now straightforward to design a stochastic algorithm that will propagate the system from state to state correctly. For now, we are assuming all the rate constants are known for each state; in later sections we will discuss how they are calculated and tabulated.

Before discussing the procedure usually used by KMC practitioners, it is perhaps instructive to first present a more transparent approach, one that is less efficient, though perfectly valid. (This method is presented as the "first-reaction" method in an excellent early paper by Gillespie [24].) Our system is currently in state i, and we have a set of pathways and associated rate constants $\{k_{ij}\}$. For each of these pathways, we know the probability distribution for the first escape time is given by Eq. 5. Using the procedure in Sect. 4.1, we can draw an exponentially distributed time t_j from that distribution for each pathway j. Of course, the actual escape can only take place along one of these pathways, so we find the pathway j_{min} which has the lowest value of t_j, discard the drawn time for all the other pathways, and advance our overall system clock by $t_{j min}$. We then move the system to state j_{min}, and begin again from this new state. That is all there is to it. This is less than ideally efficient because we are drawing a random number for each possible escape path, whereas it will turn out that we can advance the system to the next state with just two random numbers.

We now describe the KMC algorithm that is in common use. The pathway selection procedure is indicated schematically in Fig. 2a. We imagine that for each of the M escape pathways we have an object with a length equal to the rate constant k_{ij} for that pathway. We put these objects end to end, giving a total length k_{tot}. We then choose a single random position along the length of this stack of objects.

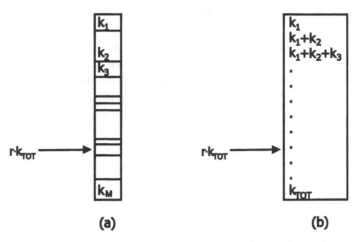

(a)   **(b)**

Figure 2. Schematic illustration of the procedure for picking the reaction pathway to advance the system to the next state in the standard KMC algorithm. (a) Objects (boxes for this illustration), each with a length proportional to the rate constant for its pathway, are placed end to end. A random number r on (0,1), multiplied by k_{tot}, points to one box with the correct probability. (b) In a computer code, this is achieved by comparing rk_{tot} to elements in an array of partial sums.

This random position will "touch" one of the objects, and this is the pathway that we choose for the system to follow. This procedure gives a probability of choosing a particular pathway that is proportional to the rate constant for that pathway, as it should. To advance the clock, we draw a random time from the exponential distribution for the rate constant k_{tot} (see Sect. 4.1). Note that the time advance has nothing to do with *which* event is chosen. The time to escape depends only on the total escape rate. Once the system is in the new state, the list of pathways and rates is updated (more on this below), and the procedure is repeated.

In a computer implementation of this procedure, we make an array of partial sums. Let array element $s(j)$ represents the length of all the objects up to and including object j,

$$s(j) = \sum_{q}^{j} k_{iq}, \qquad (7)$$

as shown in Fig. 2b. One then draws a random number r, distributed on (0,1), multiplies it by k_{tot}, and steps through the array s, stopping at the first element for which $s(j) > rk_{tot}$. This is the selected pathway.

This rejection-free "residence-time" procedure is often referred to as the BKL algorithm (or the "n-fold way" algorithm), due to the 1975 paper by Bortz, Kalos and Lebowitz [18], in which it was proposed for Monte Carlo simulation of Ising spin systems. It is also presented as the "direct" method in Gillespie's 1976 paper [24].

6   DETERMINING THE RATES

Assuming we know about the possible pathways, we can use transition state theory (TST) [26–28], to compute the rate constant for each pathway. Although TST is approximate, it tends to be a very good approximation for solid-state diffusive events. Moreover, if desired, the rate computed from TST can be corrected for recrossing effects to give the exact rate. By underpinning the KMC in this way, using high-quality TST rates that can be extended to exact rates if desired, the state-to-state dynamics of the KMC simulations can, in principle, be made as accurate as real molecular dynamics on the underlying potential. This concept was first proposed in [17].

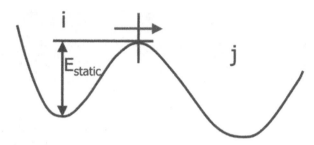

Figure 3. Illustration of the transition state theory rate constant. The unimolecular rate constant for escape from state i to state j, k_{ij}, is given by the equilibrium outgoing flux through the dividing surface separating the states.

6.1 Transition State Theory

Transition state theory (TST), first proposed in 1915 [26], offers a conceptually straightforward approximation to a rate constant. The rate constant for escape from state i to state j is taken to be the equilibrium flux through a dividing surface separating the two states, as indicated in Fig. 3. We can imagine having a large number of two-state systems, each allowed to evolve long enough that many transitions between these states have occurred, so that they represent an equilibrium ensemble. Then, looking in detail at each of the trajectories in this ensemble, if we count the number of forward crossings of the dividing surface per unit time, and divide this by the number of trajectories, on average, that are in state i at any time, we obtain the TST rate constant, k_{ij}^{TST}. The beauty of TST is that, because it is an equilibrium property of the system, we can also calculate k_{ij}^{TST} without ever looking at dynamical trajectories. For a thermal ensemble (the only kind we are considering in this chapter), k_{ij}^{TST} is simply proportional to the Boltzmann probability of being at the dividing surface relative to the probability of being anywhere in state i. Specifically, for a one-dimensional system with a dividing surface at $x = 0$,

$$k_{ij}^{\text{TST}} = \langle |dx/dt|\, \delta(x) \rangle_i, \tag{8}$$

where the angular brackets indicate a canonical ensemble average over the position coordinate x and momentum p, the subscript i indicates evaluation over the phase space belonging to state i ($x \leq 0$ in this case), and $\delta(x)$ is the Dirac delta function. Extension to many dimensions is straightforward [29], but the point is that once the dividing surface has been specified, k_{ij}^{TST} can be evaluated using, for example, Metropolis Monte Carlo methods [29, 30].

The implicit assumption in TST is that successive crossings of the dividing surface are uncorrelated; i.e., each forward crossing of the dividing surface corresponds to a full reactive event that takes the system from residing in state i to residing in state j. However, in reality, there is the possibility that the trajectory may recross the dividing surface one or more times before either falling into state j or falling back into state i. If this happens, the TST rate constant overestimates the exact rate, because some reactive events use up more than a single outgoing crossing. As stated above, the exact rate can be recovered using a dynamical corrections formalism [31–33], in which trajectories are initiated at the dividing surface and integrated for a short time to allow the recrossing events to occur. While the best choice of dividing surface is the one that minimizes the equilibrium flux passing through it (the best surface usually follows the ridgetop), this dynamical corrections algorithm recovers the exact rate constant even for a poor choice of dividing surface. This dynamical corrections formalism can also be extended to correctly account for the possibility of multiple-jump events, in which case there can be nonzero rate constants k_{ij} between states i and j that are not adjacent in configuration space [34].

In principle, then, classically exact rates can be computed for each of the pathways in the system. In practice, however, this is never done, in part because the TST approximation is fairly good for solid-state diffusive processes. In fact, most KMC studies are performed using a further approximation to TST, which we describe next.

6.2 Harmonic Transition State Theory

The harmonic approximation to TST, and further simplifications to it, are often used to calculate KMC rate constants. Harmonic TST (HTST) is often referred to as Vineyard theory [35], although equivalent or very similar expressions were derived earlier by others [36]. In HTST, we require that the transition pathway is characterized by a saddle point on the potential energy surface (e.g., the dots in Fig. 1). The reaction coordinate is defined as the direction of the normal mode at the saddle point that has an imaginary frequency, and the the dividing surface is taken to be the saddle plane (the hyperplane perpendicular to the reaction coordinate at the saddle). One assumes that the potential energy near the basin minimum is well described (out to displacements sampled thermally) with a second-order energy expansion – i.e., that the vibrational modes are harmonic – and that the same is true for the modes perpendicular to the reaction coordinate at the saddle point. Evaluation of the ensemble average in Eq. 8 for a system with N moving atoms then gives the simple form

$$k^{HTST} = \frac{\prod_i^{3N} \nu_i^{min}}{\prod_i^{3N-1} \nu_i^{sad}} \exp(-E_{static}/k_B T). \qquad (9)$$

Here E_{static} is the static barrier height (energy difference between the saddle point and the minimum) and k_B is the Boltzmann constant. In the preexponential factor (or prefactor), $\{\nu_i^{min}\}$ are the $3N$ normal mode frequencies at the minimum and $\{\nu_i^{sad}\}$ are the $3N - 1$ nonimaginary normal mode frequencies at the saddle [37]. The computation of k^{HTST} thus requires information only about the minimum and the saddle point for a given pathway. The HTST rate tends to be a very good approximation to the exact rate (e.g., within 10-20%) up to at least half the melting point for diffusion events in most solid materials (e.g., see [38, 39]), although there can be exceptions [40]. Further, since prefactors are often in the range of 10^{12} s^{-1} - 10^{13} s^{-1} (though they can be higher; e.g., see Fig. 4 in [41]), a common approximation is to choose a fixed value in this range to save the computational work of computing the normal modes for every saddle point.

The form of the Vineyard approximation merits further comment. Note that the only temperature dependence is in the exponential, and depends only on the static (i.e., $T=0$) barrier height [42]. No correction is needed, say, to account for the extra potential energy that the system has as it passes over the saddle region at a finite temperature. This, and all the entropy effects, cancel out in the integration over the normal modes, leaving the simple form of Eq. (9). Also note that Planck's constant h does not appear in Eq. (9). The kT/h preexponential sometimes found in TST expressions is an artifact of incomplete evaluation of the partition functions involved, or a dubious approximation made along the way. TST is a classical theory, so h cannot remain when the integrals are all evaluated properly. Confusion about this expression, which introduces the wrong temperature dependence and an inappropriate physical constant, has persisted because $k_B T/h$ at T=300K (6.2×10^{12} Hz) is coincidentally similar to a typical preexponential factor.

7 THE LATTICE ASSUMPTION AND THE RATE CATALOG

Typically in KMC simulations, the atoms in the system are mapped onto a lattice. An event may move one atom or many atoms, perhaps in a complicated way, but in the final state, each atom will again map onto a unique lattice point. Note that if, for example, harmonic TST is used to compute the rates, it requires that the system be relaxed to find the energy and frequencies at the minimum. After relaxation, the atoms will in general no longer be positioned on the lattice points, especially for atoms near defects. However, if each atom is much closer to one lattice point than any other lattice point, and if the mapping of the atoms onto the lattice points does not change during the relaxation, then it is safe to map the system onto a lattice in this way to simplify the KMC and the generation of the rate constants.

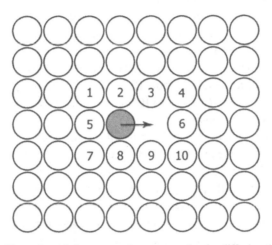

Figure 4. Schematic illustration of the rate catalog concept for the diffusional jump of a vacancy. Atoms in the lattice sites labelled 1-10 can affect the rate constant significantly, so TST rate constants are computed for all possible occupations (differing atom types or vacancies) of these sites. This list of rates makes up a *rate catalog*, which can be accessed during the KMC simulation to determine the rate constant for the jump of a vacancy in any direction for any environment.

Lattice mapping also makes it easy to exploit locality in determining rates. We assume that only the atoms near a defect affect the rate constant for any change or migration of that defect. An example of this is shown in Fig. 4 for a pathway schematically corresponding to the jump of a lattice vacancy. For each of the local environments of the jumping atom (i.e., in which each of the numbered sites in Fig. 4 is either vacant or filled with one atom type or another), we can compute a TST rate constant [17]. The number of possible rates, ignoring symmetry, is

$$n_{rate} = (n_{type} + 1)^{n_{site}}, \tag{10}$$

where n_{site} is the number of sites explicitly considered ($n_{site} = 10$ in Fig. 4) and n_{type} is the number of possible atom types that can be at each of those sites. Equation 10 results from the fact that each site, independently, may either be vacant, or have an atom of one of the n_{type} types. (For the purposes of this formal counting, we are overlooking the fact that some of the pathways involving multiple vacancies in the environment may be ill defined if the adatom has no neighbors.)

The set of rates computed in this way comprise a "rate catalog" [17], which we can then use to look up the rates we need for every state the system visits. By making the local environment larger, we can make the rates in it more accurate, and in principle we can make the environment as large as we need to achieve the accuracy we desire. In practice, however, the fact that the number of rates that will have to be computed grows as a strong power law in n_{site} means that we may settle for less than ideal accuracy. For example, for a vacancy moving in an fcc metal, including just nearest-neighbor sites of the jumping atom, n_{site}=18 and n_{type}=1, giving 2^{18} = 262,144 rates to be computed (many equivalent by symmetry). For a classical interatomic potential, this is feasible, using an automated procedure in which a nudged elastic band calculation [43,44] or some other saddle-finding algorithm (e.g., Newton-Raphson) is applied to each configuration. However, just increasing this to include second nearest neighbors (2^{28} = 2.7x10^8 rates, ignoring symmetry) or to consider a binary alloy (3^{18} = 3.8x10^8 rates, ignoring symmetry) increases the computational work enormously.

This work can be reduced somewhat by splitting the neighborhood into two sets of sites [45], one set of sites that most influence the active atom in the initial state (e.g., sites 1,2,3,5,7,8, and 9 in Fig. 4) and another set of sites that most influence the active atom at the saddle point (e.g., sites 2,3,8, and 9 in Fig. 4). Two catalogs are then generated, one for the minima and one for the saddle points. Each catalog entry gives the energy required to remove the active atom (the one involved in the jump) from the system. Subtracting these special vacancy formation energies for a given minimum-saddle pair gives the energy barrier for that process.

Another way to reduce the work is to create the rate catalog as the KMC simulation proceeds, so that rate constants are computed only for those environments encountered during the KMC.

While achieving convergence with respect to the size of the local environment is formally appealing, we will see in Sect. 9 that it is usually more important to make sure that all *types* of pathways are considered, as missing pathways often cause larger errors in the final KMC dynamics.

Finally, we note that the locality imposed by this rate-catalog approach has the benefit that in the residence-time procedure described in Sect. 5, updating the list of rates after a move has been accepted requires only fixed amount of work, rather than work scaling as the number of atoms in the entire system.

7.1 Assuming Additive Interactions

As discussed above (see Eq. 10), computing every rate necessary to fill the rate catalog may be undesirably expensive if n_{site} and/or n_{type} are large, or if a computationally expensive electronic structure calculation is employed to describe the system. Within the HTST framework, where the rate is specified by a barrier height and a preexponential factor, an easy simplification is to assume that the barrier height can be approximated by additive interactions. For example, beginning from the example shown in Fig. 4, the neighboring atoms can be categorized as class $m1$ (nearest neighbors to the jumping atom when the system is at the minimum, sites 2, 5 and 8), class $m2$ (second nearest neighbors to the jumping atom when the system is at the minimum, sites 1, 3, 7 and 9), class $s1$ (first neighbors to the jumping atom when the system is at the saddle point, sites 2, 3, 8, and 9), and so forth. This example is shown in Fig. 5. The barrier energy is then approximated by

$$E_{static} = E_{sad} - E_{min}, \tag{11}$$

where the energy of the minimum (E_{min}) and the energy of the saddle (E_{sad}) are given by

$$E_{min} = E_{min}^0 + n_{m1} E_{m1} + n_{m2} E_{m2} \tag{12}$$

$$E_{sad} = E_{sad}^0 + n_{s1} E_{s2} + n_{s2} E_{s2}. \tag{13}$$

Here, n_{m1} is the number of atoms in $m1$ positions, and similarly for n_{m2}, n_{s1}, and n_{s2}. In this way, the rate catalog is replaced by a small number of additive interaction energies. The energies $E_{min}^0, E_{m1}, E_{m2}, E_{sad}^0, E_{s1}$, and E_{s2} can be simply specified ad hoc or adjusted to give simulation results that match experiment (this is the way almost all KMC simulations were done until the mid 1980's, and many still are), or they can be obtained from a best fit to accurately calculated rate constants (e.g., see [46]). For the prefactor, $10^{12} - 10^{13} s^{-1}$ is a good estimate for many systems.

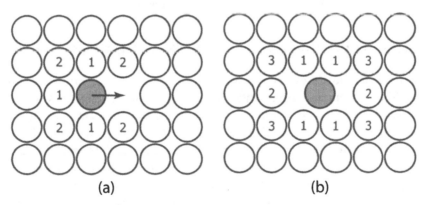

(a) (b)

Figure 5. Schematic illustration of the additive rate catalog for the diffusional jump of a vacancy. Sites are labeled by class for (a) the minimum and (b) the saddle point.

7.2 Obeying Detailed Balance

In any chemical system, we can make general statements about the behavior of the system when it is in equilibrium that are useful for understanding the dynamical evolution when the system is out of equilibrium (as it typically is). Formally, exact equilibrium properties can be obtained by gathering statistics on a very large number of systems, each of which has run for an extremely long time before the measurements are made. At equilibrium, the fractional population of state i, χ_i, is proportional to $\exp(-G_i/k_BT)$, where G_i is the free energy of state i. For every pair of connected states i and j, the number of transitions per unit time (on average) from i to j must equal the number of transitions per unit time from j to i. Because the number of escapes per time from i to j is proportional to the population of state i times the rate constant for escape from i to j, we have

$$\chi_i k_{ij} = \chi_j k_{ji}, \tag{14}$$

and the system is said to "obey detailed balance." Because the equilibrium populations and the rate constants are constants for the system, this detailed balance equation, which must hold even when the system is not in equilibrium, places requirements on the rate constants. If a rate catalog is constructed that violates detailed balance, then the dynamical evolution will not correspond to a physical system. This ill-advised situation can occur, for example, if a rate constant is set to zero, but the reverse rate is not, as might happen in a sensitivity analysis or a model study. It can also arise if there is an asymmetry in the procedure for calculating the rates (e.g., in the way that saddle points are found) that gives forward and reverse rates for a connected pair of states that are not compatible.

8   COMPUTATIONAL SCALING WITH SYSTEM SIZE

For a system with M escape pathways, the residence-time algorithm described in Sect. 5, in its simplest implementation, would require searching through a list of M rates to find the pathway that is selected by the random number. The computational work to choose each KMC step would thus scale as M. Over the years, papers [47–49] have appeared discussing how to implement the residence-time procedure with improved efficiency. Blue, Beichl and Sullivan [48] pointed out that by subdividing the list of rates into hierarchical sublists, the work can be reduced to that of searching a binary tree, scaling as $\log(M)$. Recently, Schulze [49] demonstrated that for a system in which there are equivalent rates that can be grouped (e.g., the rate for a vacancy hop in one part of the system is equivalent to the rate for a vacancy hop in another part of the system), the work can be reduced further, becoming independent of M.

After the pathway is selected and the system is moved to the new state, the rate list must be updated. In general, for this step the locality of the rate constants can be exploited, as discussed in Sect. 7, so that only a fixed amount of work is required, independent of M.

The overall computational scaling of KMC also depends on how far the system advances in time with each KMC step. In general, for a system with N atoms, the number of pathways M will be proportional to the number of atoms N [50]. If we increase the size of a system in a self-similar way, e.g., doubling N by placing two equivalent systems side by side, then the total escape rate k_{tot} will be proportional to N (see Eq. 4). Since the average time the system advances is inversely proportional to k_{tot} (see Eq. 3), this means that the overall work required to propagate a system of N atoms forward for a certain amount of time is proportional to N (within the Schulze assumption that there is a fixed number of unique rate constants) or at worst $N\log N$.

9 SURPRISES – THE REAL REASON KMC IS NOT EXACT

As claimed in the introduction, KMC can, in principle, give the exact state-to-state dynamics for a system. This assumes that a complete rate catalog has been generated, containing an accurate rate constant for every escape pathway for every state that will be encountered in the dynamics. We have discussed above the fact that the TST rate is not exact (unless augmented with dynamical corrections) and the difficulty in fully converging the environment size (see Eq. 10). However, for a typical system, neither of these effects are the major limitation in the accuracy of the KMC dynamics. Rather, it is the fact that the real dynamical evolution of a system will often surprise us with unexpected and complex reaction pathways. Because these pathways (before we have seen them) are outside our intuition, they will typically not be included in the rate catalog, and hence cannot occur during the KMC simulation.

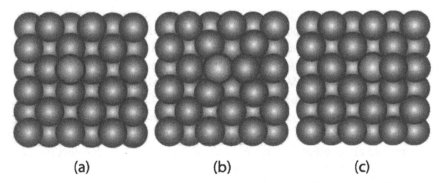

(a)   (b)   (c)

Figure 6. Exchange mechanism for adatom on fcc(100) surface. (a) initial state; (b) saddle point; (c) final state. This mechanism, unknown until 1990 [51], is the dominant diffusion pathway for some fcc metals, including Al, Pt, and Ir.

The field of surface diffusion provides a classic example of such a surprising pathway. Until 15 years ago, diffusion of an adatom on the simple fcc(100) surface was assumed to occur by the adatom hopping from one four-fold site to the next. Using density functional theory calculations, Feibelman discovered in 1990 that the primary diffusion pathway on Al(100) actually involves the exchange event shown in Fig. 6, in which the adatom plunges into the surface, pushing a substrate atom up into the second nearest neighbor binding site [51]. In field-ion microscope experiments, this exchange mechanism was shown to be the dominant pathway for Pt/Pt(100) [52] and Ir/Ir(100) [53]. For Pt/Pt(100), the barrier for the hop mechanism is roughly 0.5 eV higher than for the exchange mechanism. Thus, a KMC simulation of Pt adatoms on a Pt(100) surface, using a rate catalog built assuming hop events only (which was standard practice for KMC on the fcc(100) surface until recently), would give a seriously flawed description of the diffusion dynamics.

More recently, there have been many examples of unexpected surface and bulk diffusion mechanisms [54–58]. In some cases, the discovered mechanisms are so complex that it would not be easy to incorporate them into a KMC rate catalog, even after the existence of the pathway is known. This issue was the primary motivation for the development of the accelerated molecular dynamics methods [59] described in the next chapter.

10 SIMULATION TIME ACHIEVABLE WITH KMC

The total simulation time that can be achieved in a KMC simulation is strongly system dependent. Each KMC step advances the system by a time (on average) no greater than the inverse of the fastest rate for escape from the current state. This rate depends exponentially on the barrier height divided by the temperature, and the size of the lowest barrier can change, perhaps dramatically, as the system evolves. However, to get some sense of what is possible, we observe that on present-day computers, one can take roughly 10^{10} steps in a few hours of computer time (the exact value, of course, depends on the type and size of the system). If we assume that for every state there is one fast escape pathway with a fixed lowest barrier E_a and a prefactor of 10^{13}, then we can achieve a simulation time of $10^{10}/(10^{13}\exp(-E_a/k_BT))$. For $E_a = 0.5$ eV, this gives a total simulation time of 2.5×10^5 s at T=300K, 16 s at T=600K and 0.33 s at T=1000K. For a very low barrier, times are even shorter but the temperature dependence is much weaker. For example, $E_a = 0.1$ eV gives 50 ms at T=300K, 10 ms at T=600K, and 3 ms at T=1000K. These times are all significantly longer than one can achieve with direct molecular dynamics simulation (typically between 1 ns and 1 μs).

11 THE LOW-BARRIER PROBLEM

It is interesting to note how important the lowest barrier is. A persistent low barrier can significantly decrease the total accessible simulation time, and many systems exhibit persistent low barriers. For example, in metallic surface diffusion, adatoms that diffuse along the edge of a two-dimensional cluster or a step edge usually do so with a much lower barrier than for diffusion on an open terrace [17]. In bulk fcc materials, interstitials typically diffuse with very low barriers in the range of 0.1 eV or less. In glassy materials, low-barriers abound. This is a common and long-standing problem with KMC simulations.

One approximate approach to the problem is to raise the lowest barriers artificially to slow down the fastest rates. This will give accurate dynamics if the fast processes are reasonably well equilibrated under the conditions of interest, and if they are still able to reach equilibration when they are slowed down. In general, though, it may be hard to know for sure if this is corrupting the dynamics.

Often the structure of the underlying potential energy surface is such that the system repeatedly visits a subset of states. Among the states in this *superbasin*, an equilibrium may be achieved on a much shorter time scale than the time it takes the system to escape from the superbasin. In this situation, if all the substates are known and a list of all processes that take the system out of the superbasin can be enumerated, then one of these processes can be selected with an appropriate Boltzmann probability. An example of this kind of approach has been presented recently by Samant and Vlachos [60]. Two difficulties typically arise: 1) efficiently identifying and recognizing all the substates, the number of which may be very large; and 2) being sure the system is truly equilibrated in the superbasin.

Regarding the first problem, Mason et al.[61] have recently pointed out that a hashing procedure based on the Zobrist key [62], developed for recognizing previously stored configurations in chess, can be used to efficiently identify revisited states in a KMC simulation. In lattice-based KMC, as in chess, the number of possible states is typically astronomical, vastly exceeding the number of indices that can be stored in computer memory. The Zobrist approach maps a lattice-based configuration onto a non-unique index (key) that has a low probability of colliding with other configurations. Mason et al showed that retrieving previously visited states based on their Zobrist key saved substantial time for a KMC simulation in which rates were calculated from scratch for each new configuration. This type of indexing could also be powerful for enumerating and recognizing states in a superbasin.

Novotny has presented a general method [63] that circumvents the second problem (that of establishing equilibration in the superbasin). The implementation requires setting up and diagonalizing a transition matrix over the revisited states. While this probably becomes too costly for superbasins with large numbers of substates, the generality of the method is very appealing. It yields the time of the first transition out of the set of revisited states and the state the system goes to, while making no requirement of a local equilibrium or even that the complete set of states in the superbasin be known.

A recent review by Vlachos [64] discusses other methods relevant to these low barrier problems. Finally, we note that the low-barrier problem is also an issue for the accelerated molecular dynamics methods discussed in the next chapter, as well as for the on-the-fly KMC discussed in Section 13.

12   OBJECT KINETIC MONTE CARLO

A higher level of simulation, which is still in the KMC class, can be created by constructing state definitions and appropriate rate constants for multi-atom entities such as interstitial clusters, vacancy clusters, etc. This type of simulation is becoming more common in radiation damage annealing studies. As in basic atomic-level KMC, this *object KMC* approach is usually performed on a lattice using the residence time algorithm, perhaps augmented by additional rules. By treating the diffusive motion of a cluster, for example, as proceeding by simple KMC steps of the center of mass, rather than as a cumulative result of many individual basin-to-basin moves that move atoms around in complex (and often unproductive) ways, object KMC can reach much longer time and length scales than pure atom-based KMC. A good example of this approach, with references to earlier work, can be found in [65].

The tradeoff in this approach, however, is that important pathways may go missing from the rate catalog as atomistic details are eliminated. For example, the diffusion rate as a function of cluster size must be specifed directly, as well as rules and rates for coalescense or annihilation two clusters that encounter one another. When the real dynamics are explored, these dependencies sometimes turn out to be surprisingly complicated. For example, in the case of intersitial clusters in MgO, the diffusion constants are strongly non-monotonic with cluster size, and, worse, the cluster resulting from coalescence of two smaller clusters can sometimes form in a long-lived metastable state with dramatically different diffusion properties [58]. So, while object KMC is an effective way to reach even greater time and size scales than standard KMC, it is perhaps even more important to keep in mind the dangers of missing pathways.

13   ON-THE-FLY KINETIC MONTE CARLO

As discussed above, while ideally KMC simulations can be carried out in a way that is faithful to the real dynamics for the underlying interatomic potential, this is virtually never the case in real applications due to the fact that reaction pathways are invariably missing from the rate catalog. In part, this deficiency arises from the fact that keeping the system on lattice precludes certain types of diffusive events, but the far more dominant reason is simply that we usually make up rate catalogs based on our intuition about how the system will behave, and the real dynamics is almost always more complicated. This situation can be improved by gaining more experience on the system to be simulated, e.g., by observing the types of events that occur during extensive direct MD simulations. However, even this approach is usually inadequate for finding all the reactive events that could occur during the evolution of the system.

It is this situation that has motivated research in recent years to develop alternative methods that can reach long time scales while maintaining (or coming close to) the accuracy of direct MD. The next chapter in this book describes one powerful approach to this problem, accelerated molecular dynamics, in which the classical trajectory is retained (rather than collapsing the description to a set of states, as in KMC), and this classical trajectory is coaxed into finding each escape path more quickly. We finish the present chapter with a brief description of another approach, one which retains the flavor of KMC.

Recently, Henkelman and Jónsson [68] have proposed a variation on the KMC method, in which one builds a rate catalog on the fly for each state. The key to this approach is having an efficient way to search for saddle points that are connected to the current state of the system. For this, they use the "dimer" method [57]. Given a random starting position within the energy basin, the dimer algorithm climbs uphill along the lowest eigenvector of the Hessian matrix, reaching the saddle point at the top. Because it only requires first derivatives of the potential [69], it is computationally efficient. In principle, if *all* the bounding saddle points can be found (and hence all the pathways for escape from the state), the rate for each of these pathways can be supplied to the KMC procedure described in Section 5, propagating the system in a dynamically correct way to the next state, where the procedure is begun again. In practice, it is hard (probably impossible) to demonstrate that all saddles have been found, especially considering that the number of saddle points bounding a state grows exponentially with the dimensionality of the system. However, with a large number of randomly initiated searches, most of the low-lying barriers can be found, and this approach looks very promising [70]. Examining the pathways that these systems follow from state to state, often involving complicated multiple-atom moves, it is immediately obvious that the quality of the predicted dynamical evolution is substantially better than one could hope to obtain with pre-cataloged KMC. This approach can be parallelized efficiently, as each dimer search can be performed on a separate processor. Also, for large systems, each dimer search can be localized to a subset of the system (if appropriate). On the other hand, this type of on-the-fly KMC is substantially more expensive than standard KMC, so the user must decide whether the increased quality is worth the cost.

14 CONCLUSIONS

Kinetic Monte Carlo is a very powerful and general method. Given a set of rate constants connecting states of a system, KMC offers a way to propagate dynamically correct trajectories through the state space. The type of system, as well as the definition of a state, is fairly arbitrary, provided it is appropriate to assume that the system will make first-order transitions among these states. In this chapter, we have focused on atomistic systems, due to their relevance to radiation damage problems, which are the subject of this book. In this case, the states correspond to basins in the potential energy surface.

We have emphasized that, if the rate catalog is constructed properly, the easily implemented KMC dynamics can give exact state-to-state evolution of the system, in the sense that it will be statistically indistinguishable from a long molecular dynamics simulation. We have also pointed out, however, that this ideal is virtually never realizable, due primarily to the fact that there are usually reaction pathways in the system that we don't expect in advance. Thus, if the goal of a KMC study is to obtain accurate, predictive dynamics, it is advisable to perform companion investigations of the system using molecular dynamics, on-the-fly kinetic Monte Carlo (see Section 13), or accelerated molecular dynamics (see next chapter).

Despite these limitations, however, KMC remains the most powerful approach available for making dynamical predictions at the meso scale without resorting to more dubious model assumptions. It can also be used to provide input to and/or verification for higher-level treatments such as rate theory models or finite-element simulations. Moreover, even in situations where a more accurate simulation would be feasible (e.g., using accelerated molecular dynamics or on-the-fly kinetic Monte Carlo), the extreme efficiency of KMC makes it ideal for rapid scans over different conditions, for example, and for model studies.

15 ACKNOWLEDGMENTS

The author thanks H.L. Heinisch and P.A. Rikvold for supplying information about early KMC work, and B.P. Uberuaga, L.N. Kantorovich, and D.G. Vlachos for a critical reading of the manuscript and helpful suggestions. This work was supported by the United States Department of Energy (DOE), Office of Science, Office of Basic Energy Sciences, Division of Materials Science.

REFERENCES

[1] N. Metropolis, *Los Alamos Science*, **12**, 125 (1987).

[2] N. Metropolis, A.W. Rosenbluth, M.N. Rosenbluth, A.H. Teller, and E. Teller, J. Chem. Phys. **21**, 1087 (1953).

[3] J.R. Beeler, Jr., Phys. Rev. **150**, 470 (1966).

[4] D.G. Doran, Radiat. Eff. **2**, 249 (1970).

[5] J.-M. Lanore, Rad. Eff. **22** 153 (1974).

[6] H.L. Heinisch, D.G. Doran, and D.M. Schwartz, ASTM Special Technical Publication **725**, 191 (1981).

[7] H.L. Heinisch, J. Nucl. Mater. **117** 46 (1983).

[8] R. Gordon, J. Chem. Phys. **48**, 1408 (1968).

[9] F.F. Abraham and G.W. White, J. Appl. Phys. **41**, 1841 (1970).

[10] C.S. Kohli and M.B. Ives, J. Crystal Growth **16**, 123 (1972).

[11] G.H. Gilmer, J. Crystal Growth **35**, 15 (1976).

[12] M. Bowker and D.A. King, Surf. Sci. **71**, 583 (1978).

[13] D.A. Reed and G. Ehrlich, Surf. Sci. **105**, 603 (1981).

[14] P.A. Rikvold, Phys. Rev. A **26**, 647 (1982).

[15] E.S. Hood, B.H. Toby, and W.H. Weinberg, Phys. Rev. Lett. **55**, 2437 (1985).

[16] S.V. Ghaisas and A. Madhukar, J. Vac. Sci. Technol. B **3**, 540 (1985).

[17] A.F. Voter, Phys. Rev. B **34**, 6819 (1986).

[18] A.B. Bortz, M.H. Kalos, and J.L. Lebowitz, J. Comp. Phys. **17**, 10 (1975).

[19] K. Binder, in *Monte Carlo Methods in Statistical Physics* (Springer Topics in Current Physics, Vol. 7) edited by K. Binder (Springer, Berlin 1979) p. 1.

[20] K. Binder and M.H. Kalos, in *Monte Carlo Methods in Statistical Physics* (Springer Topics in Current Physics, Vol. 7) edited by K. Binder (Springer, Berlin 1979) p. 225.

[21] Early (and even some recent) KMC work can be found under various names, including "dynamic Monte Carlo," "time-dependent Monte Carlo," and simply "Monte Carlo."

[22] R. Norris, *Markov Chains* (Cambridge University Press, Cambridge, UK, 1997).

[23] W. Feller, *An Introduction to Probability Theory and its Applications, Vol. 1*, Wiley, New York (1966).

[24] D.T. Gillespie, J. Comp. Phys. **22**, 403 (1976).

[25] K.A. Fichthorn and W.H. Weinberg, J. Chem. Phys. **95**, 1090 (1991).

[26] R. Marcelin, Ann. Physique **3**, 120 (1915).

[27] E. Wigner, Z. Phys. Chem. **B 19**, 203 (1932).

[28] H. Eyring, J. Chem. Phys. **3**, 107 (1935).

[29] A.F. Voter and J.D. Doll, J. Chem. Phys. **80**, 5832 (1984).

[30] A.F. Voter, J. Chem. Phys. 82, 1890 (1985).

[31] J.C. Keck, Discuss. Faraday Soc. **33**, 173 (1962).

[32] C.H. Bennett, in *Algorithms for Chemical Computation*, edited by R.E. Christofferson (American Chemical Society, Washington, DC, 1977), p. 63.

[33] D. Chandler, J. Chem. Phys. **68**, 2959 (1978).

[34] A.F. Voter and J.D. Doll, J. Chem. Phys. **82**, 80 (1985).

[35] G.H. Vineyard, J. Phys. Chem. Solids **3**, 121 (1957).

[36] P. Hanggi, P. Talkner, and M. Borkovec, Rev. Mod. Phys. **62**, 251 (1990).

[37] For a three-dimensional periodic system with all atoms moving, discarding the translational modes leaves $3N$-3 and $3N$-4 real normal mode frequencies at the minimum and saddle, respectively. For a system that is free to rotate, there are $3N$-6 and $3N$-7 relevant modes.

[38] G. DeLorenzi, C.P. Flynn, and G. Jacucci, Phys. Rev. B **30**, 5430 (1984).

[39] M.R. Sørensen and A.F. Voter, J. Chem. Phys. **112**, 9599 (2000).

[40] G. Boisvert and L.J. Lewis, Phys. Rev. B **54**, 2880 (1996).

[41] F. Montalenti and A.F. Voter, Phys. Stat. Sol. (b) **226**, 21 (2001).

[42] An additional T dependence is introduced if the quasiharmonic method is employed to give a different lattice constant (and hence different barrier height and preexponential) at each temperature,but here we are assuming a fixed lattice constant.

[43] H. Jónsson, G. Mills, and K.W. Jacobsen, in *Classical and Quantum Dynamics in Condensed Phase Simulations*, edited by B.J. Berne, G. Ciccotti and D.F Coker (World Scientific, 1998), chapter 16.

[44] G. Henkelman, B.P. Uberuaga, and H. Jónsson, J. Chem. Phys. **113**, 9901 (2000).

[45] A.F. Voter, in *Modeling of Optical Thin Films*, M.R. Jacobson, Ed., Proc. SPIE **821**, 214 (1987).

[46] H. Mehl, O. Biham, K. Furman, and M. Karimi, Phys. Rev. B **60**, 2106 (1999).

[47] P.A. Maksym, Semicond. Sci. Technol. **3**, 594 (1988).

[48] J.L. Blue, I. Beichl and F. Sullivan, Phys. Rev. E **51**, R867, (1994).

[49] T.P. Schulze, Phys. Rev. E **65**, 036704 (2002).

[50] In fact, the number of saddle points accessible to a state may scale more strongly than linear in N if complicated, high-barrier mechanisms are considered, but in almost all KMC implementations it will be proportional to N.

[51] P.J. Feibelman, Phys. Rev. Lett. **65**, 729 (1990).

[52] G.L. Kellogg and P.J. Feibelman, Phys. Rev. Lett. **64**, 3143 (1990).

[53] C. Chen and T.T. Tsong, Phys. Rev. Lett. **64**, 3147 (1990).

[54] C.L. Liu and J.B. Adams, Surf. Sci. **268**, 73 (1992).

[55] R. Wang and K.A. Fichthorn, Molec. Sim. **11**, 105 (1993).

[56] J.C. Hamilton, M.S. Daw, and S.M. Foiles, Phys. Rev. Lett. **74**, 2760 (1995).

[57] G. Henkelman and H. Jónsson, J. Chem. Phys. **111**, 7010 (1999).

[58] B.P. Uberuaga, R. Smith, A.R. Cleave, F. Montalenti, G. Henkelman, R.W. Grimes, A.F. Voter, and K.E. Sickafus, Phys. Rev. Lett. **92**, 115505 (2004).

[59] A.F. Voter, F. Montalenti and T.C. Germann, Annu. Rev. Mater. Res., **32**, 321 (2002).

[60] A. Samant and D.G. Vlachos, J. Chem. Phys. **123**, 144114 (2005).

[61] D.R. Mason, T.S. Hudson, and A.P. Sutton, Comp. Phys. Comm. **165**, 37 (2005).

[62] A.L. Zobrist, Technical report 88, Computer Science Department, University of Wisconsin, Madison, 1970; Reprinted in: ICCA J. **13**, 69 (1990).

[63] M.A. Novotny, Phys. Rev. Lett. **74**, 1 (1994); Erratum **75**, 1424 (1995).

[64] D.G. Vlachos, Adv. Chem. Eng. **30**, 1 (2005).

[65] C. Domain, C.S. Becquart, and L. Malerba, J. Nucl. Mater. **335**, 121 (2004).

[66] S. Liu, Z. Zhang, J. Norskov, and H. Metiu, Surf. Sci. **321**, 161 (1994).

[67] Z.-P. Shi, Z. Zhang, A.K. Swan, and J.F. Wendelken, Phys. Rev. Lett. **76**, 4927 (1996).

[68] G. Henkelman and H. Jónsson, J. Chem. Phys. **115**, 9657 (2001).

[69] A.F. Voter, Phys. Rev. Lett. **78**, 3908 (1997).

[70] G. Henkelman and H. Jonsson, Phys. Rev. Lett. **90**, 116101-1 (2003).

Chapter 2

ACCELERATED MOLECULAR DYNAMICS METHODS

Blas P. Uberuaga and Arthur F. Voter
Los Alamos National Laboratory, Los Alamos, New Mexico 87545 USA

1 INTRODUCTION

The evolution of radiation damage in materials is a classic example of a problem that spans many time and length scales. The initial production of damage occurs on the atomic scale via collision cascades that take place on the picosecond time scale. However, this damage ultimately manifests itself macroscopically in the form of swelling or cracking which can take years to develop. There is a wide range of phenomena that bridge these two extremes, including defect diffusion, annihilation and aggregation, the formation of interstitial loops and voids, and the development of more complex microstructure. As a result, no one simulation method can be employed to study the problem of radiation damage on all relevant time and length scales. Rather, a combination of many techniques must be used to address this problem.

Molecular dynamics (MD) simulation, in which atom positions are evolved by integrating the classical equations of motion in time, is ideally suited to studying the collision cascade. MD simulations can probe timescales of ps to ns, which is the time typically required to initiate a cascade and to allow it to evolve until the thermal spike of the collision has dissipated. Thus, the damage produced in the collision cascade can be directly simulated using MD.

However, once that damage has been formed, diffusion and subsequent annihilation or aggregation of those defects can occur on much longer time scales, perhaps even seconds or beyond, depending on the conditions (temperature, pressure, etc.). Such phenomena must be accounted for in order to accurately predict larger scale features – including interstitial dislocation loops and vacancy voids – that comprise the overall microstructure of the system and lead to macroscopic response to radiation damage such as swelling and cracking. However, MD will not allow us to study defect behavior on these longer timescales.

Recently, methods based on a new concept have been developed for circumventing this time scale problem. For systems in which the long-time dynamical evolution is characterized by a sequence of activated events – typically the case for defect

25

K.E. Sickafus et al. (eds.), Radiation Effects in Solids, 25–43.

diffusion – these "accelerated molecular dynamics" methods [1] can extend the accessible time scale by orders of magnitude relative to direct MD, while retaining full atomistic detail. These methods – hyperdynamics, parallel replica dynamics, and temperature accelerated dynamics (TAD) – have already been demonstrated on problems in surface and bulk diffusion, surface growth, and molecular problems. With more development they will become useful for a broad range of key materials problems, including grain growth, dislocation climb and dislocation kink nucleation. Here we give an introduction to these methods, discuss their current strengths and limitations, and demonstrate their use in problems involving radiation damage.

While the treatment given here is similar to that presented previously [2], the examples used to illustrate the accelerated dynamics methods were chosen because of their relevance to studies of radiation damage.

2   BACKGROUND

2.1 Infrequent Event Systems

We begin by defining an "infrequent-event" system, as this is the type of system for which the accelerated dynamics methods are ideal. The dynamical evolution of such a system is characterized by the occasional activated event that takes the system from basin to basin, events that are separated by possibly millions of thermal vibrations within one basin. A simple example of an infrequent-event system is an adatom on a metal surface at a temperature that is low relative to the diffusive jump barrier. We will exclusively consider thermal systems, characterized by a temperature T, a fixed number of atoms N, and a fixed volume V; i.e., the canonical ensemble. Typically, there is a large number of possible paths for escape from any given basin. As a trajectory in the $3N$-dimensional coordinate space in which the system resides passes from one basin to another, it crosses a $(3N$-$1)$-dimensional "dividing surface" at the ridgetop separating the two basins. While on average these crossings are infrequent, successive crossings can sometimes occur within just a few vibrational periods; these are termed "correlated dynamical events" (e.g., see [3–5]). An example would be a double jump of the adatom on the surface. For this discussion it is sufficient, but important, to realize that such events can occur. In two of the methods presented below, we will assume that these correlated events do not occur. This is the primary assumption of transition state theory, which is actually a very good approximation for many solid-state diffusive processes. We define the "correlation time" (τ_{corr}) of the system as the duration of the system memory. A trajectory that has resided in a particular basin for longer than τ_{corr} has no memory of its history and, consequently, how it got to that basin, in the sense that when it later escapes from the basin, the probability for escape is independent of how it entered the state. The relative probability for escape to a given adjacent state is proportional to the rate constant for that escape path, which we will define below.

An infrequent event system, then, is one in which the residence time in a state (τ_{rxn}) is much longer than the correlation time (τ_{corr}). We will focus here on systems with energetic barriers to escape, but the infrequent-event concept applies equally well to entropic bottlenecks.[1] The key to the accelerated dynamics methods described here is recognizing that to obtain the right sequence of state-to-state transitions, we need not evolve the vibrational dynamics perfectly, as long as the relative probability of finding each of the possible escape paths is preserved.

2.2 Transition State Theory

Transition state theory (TST) [6–10] is the formalism underpinning all of the accelerated dynamics methods, directly or indirectly. In the TST approximation, the classical rate constant for escape from state A to some adjacent state B is taken to be the equilibrium flux through the dividing surface between A and B (Figure 1). If there are no correlated dynamical events, the TST rate is the exact rate constant for the system to move from state A to state B.

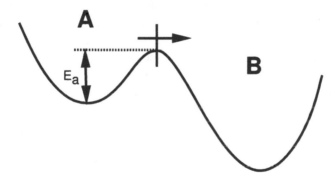

Figure 1. A two-state system illustrating the definition of the transition state theory rate constant as the outgoing flux through the dividing surface bounding state A.

The power of TST comes from the fact that this flux is an *equilibrium* property of the system. Thus, we can compute the TST rate without ever propagating a trajectory. The appropriate ensemble average for the rate constant for escape from A, $k_{A\rightarrow}^{TST}$, is

$$k_{A\rightarrow}^{TST} = \langle |dx/dt| \, \delta(x - q)\rangle_A, \tag{1}$$

where x is the position along the reaction coordinate and $x = q$ is the dividing surface bounding state A. The angular brackets indicate the ratio of Boltzmann-weighted integrals over $6N$-dimensional phase space (configuration space r and

---

[1] For systems with entropic bottlenecks, the parallel-replica dynamics method can be applied very effectively [1].

momentum space **p**). That is, for some property $P(\mathbf{r}, \mathbf{p})$,

$$\langle P \rangle = \frac{\int \int P(\mathbf{r}, \mathbf{p}) \exp[-H(\mathbf{r}, \mathbf{p})/k_B T] d\mathbf{r} d\mathbf{p}}{\int \int \exp[-H(\mathbf{r}, \mathbf{p})/k_B T] d\mathbf{r} d\mathbf{p}}, \tag{2}$$

where k_B is the Boltzmann constant and $H(\mathbf{r}, \mathbf{p})$ is the total energy of the system, kinetic plus potential. The subscript A in (1) indicates the configuration space integrals are restricted to the space belonging to state A. If the effective mass (m) of the reaction coordinate is constant over the dividing surface, (1) reduces to a simpler ensemble average over configuration space only [11],

$$k_{A\rightarrow}^{\text{TST}} = \sqrt{2k_B T/\pi m} \, \langle \delta(\mathbf{x} - \mathbf{q}) \rangle_A . \tag{3}$$

The essence of this expression, and of TST, is that the Dirac delta function picks out the probability of the system being at the dividing surface, relative to everywhere else it can be in state A. Note that there is no dependence on the nature of the final state B.

In a system with correlated events, not every dividing surface crossing corresponds to a reactive event, so that, in general, the TST rate is an upper bound on the exact rate. For diffusive events in materials at moderate temperatures, these correlated dynamical events typically do not cause a large change in the rate constants, so TST is often an excellent approximation. This is a key point; this behavior is markedly different than in some chemical systems, such as molecular reactions in solution or the gas phase, where TST is just a starting point and dynamical corrections can lower the rate significantly (e.g. see [12]).

While in the traditional use of TST, rate constants are computed after the dividing surface is specified, in the accelerated dynamics methods we exploit the TST formalism to design approaches that do not require knowing in advance where the dividing surfaces will be, or even what product states might exist.

2.3 Harmonic Transition State Theory

If we have identified a saddle point on the potential energy surface for the reaction pathway between A and B, we can use a further approximation to TST. We assume that the potential energy near the basin minimum is well described, out to displacements sampled thermally, with a second-order energy expansion – i.e., that the vibrational modes are harmonic – and that the same is true for the modes perpendicular to the reaction coordinate at the saddle point. Under these conditions, the TST rate constant becomes simply

$$k_{A\rightarrow B}^{HTST} = \nu_0 e^{-E_a/k_B T}, \tag{4}$$

where

$$\nu_0 = \frac{\prod_{i}^{3N} \nu_i^{min}}{\prod_{i}^{3N-1} \nu_i^{sad}}. \tag{5}$$

Here E_a is the static barrier height (the difference in energy between the saddle point and the minimum of state A (Figure 1)) $\{\nu_i^{min}\}$ are the normal mode frequencies at the minimum of A, and $\{\nu_i^{sad}\}$ are the non-imaginary normal mode frequencies at the saddle separating A from B [2]. This is often referred to as the Vineyard [13] equation. The analytic integration of (1) over the whole phase space thus leaves a very simple Arrhenius temperature dependence.[3] To the extent that there are no recrossings and the modes are truly harmonic, this is an exact expression for the rate. This harmonic TST expression is employed in the temperature accelerated dynamics method (without requiring calculation of the prefactor ν_0).

2.4 Complex Infrequent Event Systems

The motivation for developing accelerated molecular dynamics methods becomes particularly clear when we try to understand the dynamical evolution of what we will term complex infrequent event systems. In these systems, we simply cannot guess where the state-to-state evolution might lead. The underlying mechanisms may be too numerous, too complicated, and/or have an interplay whose consequences cannot be predicted by considering them individually. In very simple systems we can raise the temperature to make diffusive transitions occur on an MD-accessible time scale. However, as systems become more complex, changing the temperature causes corresponding changes in the relative probability of competing mechanisms. Thus, this strategy will cause the system to select a different sequence of state-to-state dynamics, ultimately leading to a completely different evolution of the system, and making it impossible to address the questions that the simulation was attempting to answer.

Many, if not most, materials problems are characterized by such complex infrequent events. We may want to know what happens on the time scale of milliseconds, seconds or longer, while with MD we can barely reach one microsecond. Running at higher T or trying to guess what the underlying atomic processes are can mislead us about how the system really behaves. Often for these systems, if we could get a glimpse of what happens at these longer times, even if we could only afford to run a single trajectory for that long, our understanding of the system would improve substantially. This, in essence, is the primary motivation for the development of the methods described here.

---

[2] Discarding translational modes for a periodic system, there are $3N$-3 and $3N$-4 real normal mode frequencies at the minimum and saddle, respectively. For a system that is free to rotate, there are $3N$-6 and $3N$-7.

[3] Note that although the exponent in (4) depends only on the static barrier height E_a, in this HTST approximation there is no assumption that trajectory passes exactly through the saddle point.

2.5 Dividing Surfaces and Transition Detection

We have implied that the ridge tops between basins are the appropriate dividing surfaces in these systems. For a system that obeys TST, these ridgetops are the optimal dividing surfaces; recrossings will occur for any other choice of dividing surface. A ridgetop can be defined in terms of steepest-descent paths – it is the $3N-1$-dimensional boundary surface that separates those points connected by steepest descent paths to the minimum of one basin from those that are connected to the minimum of an adjacent basin. This definition also leads to a simple way to detect transitions as a simulation proceeds, a requirement of parallel replica dynamics and temperature accelerated dynamics. Intermittently, the trajectory is interrupted and the potential energy is minimized via steepest descent. If this minimization leads to a basin minimum that is distinguishable from the minimum of the previous basin, a transition has occurred. An appealing feature of this approach is that it requires virtually no knowledge of the type of transition that might occur. Often only a few steepest descent steps are required to determine that no transition has occurred. While this is a fairly robust detection algorithm, more efficient approaches can be tailored to the system being studied. For example, in the simulations of carbon nanotubes presented below, transitions were defined as changes in atomic coordination.

In what follows, we describe the accelerated dynamics methods. There are currently three such methods that have been developed: parallel replica dynamics, hyperdynamics, and temperature accelerated dynamics. Two of them – parallel replica and temperature accelerated dynamics – have been applied to problems involving radiation damage and, as a result, are discussed in detail. Hyperdynamics has not been applied to problems specifically related to radiation damage and thus is only introduced here. The reader is referred to other publications for more details about hyperdynamics.

3   PARALLEL-REPLICA DYNAMICS

The parallel replica method [14] is the simplest and most accurate of the accelerated dynamics techniques, with the only assumption being that the infrequent events obey first-order kinetics (exponential decay); i.e., for any trajectory that has been in a state long enough to have lost its memory of how it entered the state (longer than τ_{corr}), the probability distribution function for the time of the next escape from that state is given by

$$p(t) = k_{tot}e^{-k_{tot}t} \qquad (6)$$

where k_{tot} is the rate constant for escape from the state. For example, (6) arises naturally for ergodic, chaotic exploration of an energy basin. Parallel replica allows for the parallelization of the state-to-state dynamics of such a system on M processors. We sketch the derivation here for equal-speed processors. For a state in which the rate to escape is k_{tot}, on M processors the effective escape rate will be Mk_{tot}, as the state is being explored M times faster. Also, if the time accumulated on one processor is t_1, on the M processors a total time of $t_{sum} = Mt_1$ will be accumulated. Thus, we find that

$$p(t_1)dt_1 = Mk_{tot}e^{-Mk_{tot}t_1}dt_1 \tag{7}$$

$$= k_{tot}e^{-k_{tot}t_{sum}}dt_{sum} \tag{8}$$

$$= p(t_{sum})dt_{sum} \tag{9}$$

and the probability to leave the state per unit time, expressed in t_{sum} units, is the same whether it is run on one or M processors. A variation on this derivation shows that the M processors need not run at the same speed, allowing the method to be used on a heterogeneous or distributed computer; see [14].

The algorithm is shown schematically in Figure 2. Starting with an N-atom system in a particular state (basin), the entire system is replicated on each of M available parallel or distributed processors. After a short dephasing stage during which each replica is evolved forward with independent noise for a time $\Delta t_{deph} \geq \tau_{corr}$ to eliminate correlations between replicas, each processor carries out an independent constant-temperature MD trajectory for the entire N-atom system, thus exploring phase space within the particular basin M times faster than a single trajectory would. Whenever a transition is detected on any processor, all processors are alerted to stop. The simulation clock is advanced by t_{sum}, the accumulated trajectory time summed over all replicas until the transition occurred.

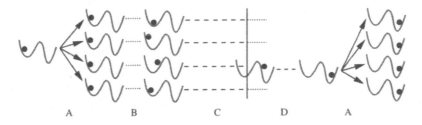

Figure 2. Schematic illustration of the parallel replica method (after [1]). The four steps, described in the text, are (A) replication of the system into M copies, (B) dephasing of the replicas, (C) propagation of independent trajectories until a transition is detected in any of the replicas, and (D) brief continuation of the transitioning trajectory to allow for correlated events such as recrossings or follow-on transitions to other states. The resulting configuration is then replicated, beginning the process again.

The parallel replica method also correctly accounts for correlated dynamical events (i.e., there is no requirement that the system obeys TST), unlike the other accelerated dynamics methods. This is accomplished by allowing the trajectory that made the transition to continue on its processor for a further amount of time

$\Delta t_{corr} \geq \tau_{corr}$, during which recrossings or follow-on events may occur. The simulation clock is then advanced by Δt_{corr}, the final state is replicated on all processors, and the whole process is repeated. Parallel replica dynamics then gives exact state-to-state dynamical evolution, because the escape times obey the correct probability distribution, nothing about the procedure corrupts the relative probabilities of the possible escape paths, and the correlated dynamical events are properly accounted for.

The efficiency of the method is limited by both the dephasing stage, which does not advance the system clock, and the correlated event stage, during which only one processor accumulates time. (This is illustrated schematically in Figure 2, where dashed line trajectories advance the simulation clock but dotted line trajectories do not.) Thus, the overall efficiency will be high when

$$\tau_{rxn}/M \gg \Delta t_{deph} + \Delta t_{corr}. \tag{10}$$

Some tricks can further reduce this requirement. For example, whenever the system revisits a state, on all but one processor the interrupted trajectory from the previous visit can be immediately restarted, eliminating the dephasing stage. Also, the correlation stage (which only involves one processor) can be overlapped with the subsequent dephasing stage for the new state on the other processors, in the hope that there are no correlated crossings that lead to a different state.

Parallel replica dynamics has the advantage of being fairly simple to program, with very few "knobs" to adjust – Δt_{deph} and Δt_{corr}, which can be conservatively set at a few ps for most systems.

Recently, parallel replica dynamics has been extended to driven systems [15], such as systems with some externally applied strain rate. The requirement here is that the drive rate is slow enough that at any given time the rates for the pathways in the system depend only on the instantaneous configuration of the system.

Figure 3 shows examples of driven parallel-replica simulations; carbon nanotubes (CNTs) with a pre-existing vacancy are stretched at different rates (10^6 and 5×10^8 s^{-1}) and temperatures (300 and 2000 K). The conditions used here are qualitatively similar to experiments in which CNTs, pinned on each end to a substrate, are irradiated with electrons. This introduces both defects (such as vacancies) and strain (via the loss of material) into the system [16]. The simulations find that the mobility of the vacancy is qualitatively different for different conditions. For low temperatures (300 K), the mobility of the vacancy is so low that the nanotube behaves almost as if there were no vacancy in the system, exhibiting essentially the same energy-vs-strain behavior as a T=0K system. As the temperature is raised, the vacancy behavior depends on the applied strain rate. For a high strain rate (5×10^8 s^{-1}) at a high temperature (2000 K) (Fig. 3b), the mobility of the vacancy is still low compared to the time scale of failure, and the vacancy does not move before failure occurs. However, when the strain rate is reduced to 10^6 s^{-1}, the time scale for vacancy diffusion is faster than that of mechanical failure (Fig. 3c).

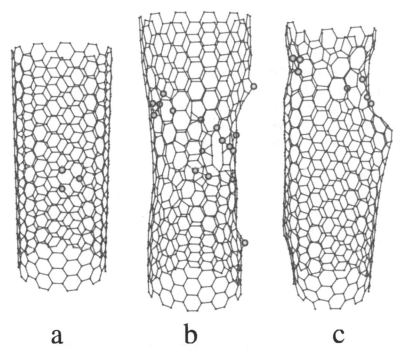

a   b   c

Figure 3. Structure of nanotubes after being evolved at different strain rates. (*a*) shows the nanotube before strain is applied, with a vacancy surrounded by three undercoordinated atoms (larger circles colored light/red). (*b*) and (*c*) show the nanotube just after yield when strained at rates of 5×10^8 s^{-1} and 10^6, respectively. Overcoordinated atoms are colored dark/green.

We observe that the vacancy is able to diffuse around the circumference of the nanotube. As the strain in the system is increased, the vacancy creates topological defects in the carbon network. These defects both reduce the strain in the nanotube as well as introduce morphological features consistent with experimental observation. Thus, we find that the vacancy is capable of inducing local rearrangements in the carbon bonding network that relieve strain before failure is reached.

Parallel-replica dynamics has been successfully applied to a number of varied problems, including the diffusion of H_2 in crystalline C_{60} [17], the pyrolysis of hexa-decane [18], the diffusion of defects in plutonium [19], the annealing of islands on Ag surfaces [1], and the folding dynamics of small proteins [20]. As parallel-computing environments become more common, parallel-replica will become an increasingly important tool for the exploration of complex systems.

4   HYPERDYNAMICS

As mentioned, because hyperdynamics has not been applied to radiation damage problems, we briefly introduce the method here and refrain from giving an example of the application of the method. Further details can be found in [1, 21].

Hyperdynamics builds on the basic concept of importance sampling [22, 23], extending it into the time domain. In the hyperdynamics approach [21], the potential surface $V(\mathbf{r})$ of the system is modified by adding to it a nonnegative *bias* potential $\Delta V_b(\mathbf{r})$. The dynamics of the system is then evolved on this biased potential surface, $V(\mathbf{r}) + \Delta V_b(\mathbf{r})$. A schematic illustration is shown in Figure 4. The derivation of the method requires that the system obeys TST – that there are no correlated events. There are also important requirements on the form of the bias potential. It must be zero at all the dividing surfaces, and the system must still obey TST for dynamics on the modified potential surface. If such a bias potential can be constructed, a challenging task in itself, we can substitute the modified potential $V(\mathbf{r}) + \Delta V_b(\mathbf{r})$ into (1) to find

$$k_{A\rightarrow}^{TST} = \frac{\langle |v_A|\, \delta(\mathbf{r})\rangle_{A_b}}{\langle e^{\beta \Delta V_b(\mathbf{r})}\rangle_{A_b}}, \tag{11}$$

where $\beta = 1/k_B T$ and the state A_b is the same as state A but with the bias potential ΔV_b applied. This leads to a very appealing result: the relative rates of events leaving A are preserved:

$$\frac{k_{A_b\rightarrow B}^{TST}}{k_{A_b\rightarrow C}^{TST}} = \frac{k_{A\rightarrow B}^{TST}}{k_{A\rightarrow C}^{TST}}. \tag{12}$$

This is because these relative probabilities depend only on the numerator of (11) which is unchanged by the introduction of ΔV_b since, by construction, $\Delta V_b = 0$ at the dividing surface.

Furthermore, it can be shown that a dynamical trajectory evolved on the modified surface, while relatively meaningless on vibrational time scales, evolves *correctly* from state to state at an accelerated pace. Moreover, the accelerated time is easily estimated as the simulation proceeds. For a regular MD trajectory, the time advances at each integration step by Δt_{MD}, the MD time step (often on the order of 1 fs). In hyperdynamics, the time advance at each step is Δt_{MD} multiplied by an instantaneous boost factor, the inverse Boltzmann factor for the bias potential at that point, so that the total time after n integration steps is

$$t_{hyper} = \sum_{j=1}^{n} \Delta t_{MD}\, e^{\Delta V(\mathbf{r}(t_j))/k_B T}. \tag{13}$$

Time thus takes on a statistical nature, advancing monotonically but nonlinearly. In the long-time limit, it converges on the correct value for the accelerated time

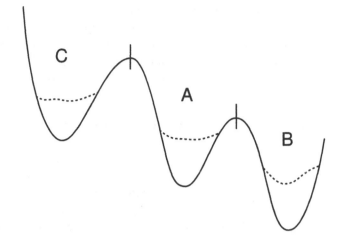

Figure 4. Schematic illustration of the hyperdynamics method. A bias potential ($\Delta V(\mathbf{r})$), is added to the original potential ($V(\mathbf{r})$, solid line). Provided that $\Delta V(\mathbf{r})$ meets certain conditions, primarily that it be zero at the dividing surfaces between states, a trajectory on the biased potential surface ($V(\mathbf{r}) + \Delta V(\mathbf{r})$, dashed line) escapes more rapidly from each state without corrupting the relative escape probabilities. The accelerated time is estimated as the simulation proceeds.

with vanishing relative error. The total computational boost is the raw boost factor corrected for the extra work required to compute the forces for the bias potential. The raw boost factor is simply given by

$$\text{boost(hyperdynamics)} = t_{hyper}/t_{MD} = \frac{1}{n}\sum_{j=1}^{n} e^{\Delta V(\mathbf{r}(t_j))/k_B T}, \qquad (14)$$

If all the visited states are equivalent (e.g., this is common in calculations to test or demonstrate a particular bias potential), (14) becomes an ensemble average on the biased potential within the state,

$$\text{boost(hyperdynamics)} = \langle e^{\Delta V(\mathbf{r})/k_B T} \rangle_{A_b}. \qquad (15)$$

The rate at which the trajectory escapes from a state is enhanced because the positive bias potential within the well lowers the effective barrier. Note, however, that the shape of the bottom of the well after biasing is irrelevant; no assumption of harmonicity is made.

The ideal bias potential should give a large boost factor, have low computational overhead (though more overhead is acceptable if the boost factor is very high), and, to a good approximation, meet the requirements stated above. This is very challenging, since we want, as much as possible, to avoid utilizing any prior knowledge of the dividing surfaces or the available escape paths. To date, proposed bias potentials typically have either been computationally intensive, have been tailored to very specific systems, have assumed localized transitions, or have been limited to low-dimensional systems. But the potential boost factor available from hyper-dynamics is tantalizing, so developing bias potentials capable of treating realistic many-dimensional systems remains a subject of ongoing research. See [1] for a detailed discussion on bias potentials and results generated using various forms, as well as [24] for an example of a recently developed form that looks promising.

5   TEMPERATURE ACCELERATED DYNAMICS

In the temperature accelerated dynamics (TAD) method [25], the idea is to speed up the transitions by increasing the temperature, while filtering out the transitions that should not have occurred at the original temperature. This filtering is critical, since without it the state-to-state dynamics will be inappropriately guided by entropically favored higher-barrier transitions. The TAD method is more approximate than the previous two methods, as it relies on harmonic TST, but for many applications this additional approximation is acceptable, and the TAD method often gives substantially more boost than hyperdynamics or parallel replica dynamics. Consistent with the accelerated dynamics concept, the trajectory in TAD is allowed to wander on its own to find each escape path, so that no prior information is required about the nature of the reaction mechanisms.

In each basin, the system is evolved at a high temperature T_{high} (while the temperature of interest is some lower temperature T_{low}). Whenever a transition out of the basin is detected, the saddle point for the transition is found. The trajectory is then reflected back into the basin and continued. This "basin constrained molecular dynamics" (BCMD) procedure generates a list of escape paths and attempted escape times for the high-temperature system. Assuming that TST holds and that the system is chaotic and ergodic, the probability distribution for the first-escape time for each mechanism is an exponential (6). Because harmonic TST gives an Arrhenius dependence of the rate on temperature (4), depending only on the static barrier height, we can then extrapolate each escape time observed at T_{high} to obtain a corresponding escape time at T_{low} that is drawn correctly from the exponential distribution at T_{low}. This extrapolation, which requires knowledge of the saddle point energy, but not the preexponential factor, can be illustrated graphically in an Arrhenius-style plot ($ln(1/t)$ vs. $1/T$), as shown in Figure 5. The time for each event seen at T_{high} extrapolated to T_{low} is then

$$t_{low} = t_{high}e^{E_a(\beta_{low}-\beta_{high})}, \tag{16}$$

where, again, $\beta = 1/k_BT$ and E_a is the energy of the saddle point. The event with the shortest time at low temperature is the correct transition for escape from this basin.

Because the extrapolation can in general cause a reordering of the escape times, a new shorter-time event may be discovered as the BCMD is continued at T_{high}. Let

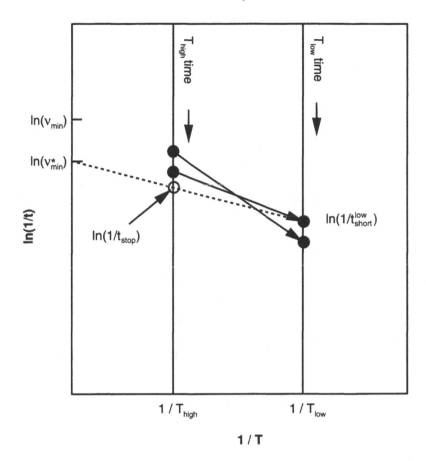

Figure 5. Schematic illustration of the temperature accelerated dynamics method. Progress of the high-temperature trajectory can be thought of as moving down the vertical time line at $1/T_{high}$. For each transition detected during the run, the trajectory is reflected back into the basin, the saddle point is found, and the time of the transition (solid dot on left time line) is transformed (arrow) into a time on the low-temperature time line. Plotted in this Arrhenius-like form, this transformation is a simple extrapolation along a line whose slope is the negative of the barrier height for the event. The dashed termination line connects the shortest-time transition recorded so far on the low temperature time line with the confidence-modified minimum preexponential ($\nu^{\star}_{min} = \nu_{min}/\ln(1/\delta)$) on the y axis. The intersection of this line with the high-T time line gives the time (t_{stop}, open circle) at which the trajectory can be terminated. With confidence 1-δ, we can say that any transition observed after t_{stop} could only extrapolate to a shorter time on the low-T time line if it had a preexponential factor lower than ν_{min}.

us make the additional assumption that there is a minimum preexponential factor, ν_{min}, which bounds from below all the preexponential factors in the system. We can then define a time at which the BCMD trajectory can be stopped, knowing that the probability that any transition observed after that time would replace the first transition at T_{low} is less than δ. This "stop" time is given by

$$t_{high,stop} \equiv \frac{\ln(1/\delta)}{\nu_{min}} \left(\frac{\nu_{min} t_{low,short}}{\ln(1/\delta)} \right)^{T_{low}/T_{high}}, \tag{17}$$

where $t_{low,short}$ is the shortest transition time at T_{low}. Once this stop time is reached, the system clock is advanced by $t_{low,short}$, the transition corresponding to $t_{low,short}$ is accepted, and the TAD procedure is started again in the new basin. Thus, in TAD, two parameters govern the accuracy of the simulation: δ and ν_{min}.

The average boost in TAD can be dramatic when barriers are high and T_{high}/T_{low} is large. However, any anharmonicity error at T_{high} transfers to T_{low}; a rate that is twice the Vineyard harmonic rate due to anharmonicity at T_{high} will cause the transition times at T_{high} for that pathway to be 50% shorter, which in turn extrapolate to transition times that are 50% shorter at T_{low}. If the Vineyard approximation is perfect at T_{low}, these events will occur at twice the rate they should. This anharmonicity error can be controlled by choosing a T_{high} that is not too high.

As in the other methods, the boost is limited by the lowest barrier, although this effect can be mitigated somewhat by treating repeated transitions in a "synthetic" mode [25]. This is in essence a kinetic Monte Carlo (KMC) treatment of the low-barrier transitions (see previous chapter for a discussion of KMC), in which the rate is estimated accurately from the observed transitions at T_{high}, and the subsequent low-barrier escapes observed during BCMD are excluded from the extrapolation analysis.

Recent advances, relying on knowledge of the minimum barrier to leave a state, have been made that increase the efficiency of TAD for certain kinds of systems. These depend upon the fact that if we know something about the minimum barrier to leave a given state, either because we have visited the state before and have a lower bound on this minimum barrier or because the minimum barrier is supplied *a priori*, we can accept a transition and leave the state earlier than the time given by (17). The first of these approaches improves the TAD efficiency for states that are revisited, with more improvement as the states are revisited more. See [26] for details.

However, this approach does not help for systems that rarely revisit previously seen states. This is often the case in radiation damage simulations, in which the system is evolving from some relatively high energy metastable state toward a lower energy configuration. By using some other procedure to find the minimum barrier for leaving a state, such as the dimer method [27], we can still exploit this information to leave the state earlier than the time prescribed by (17). This dimer-TAD approach does introduce new uncertainty, as one can never be completely confident that the dimer searches have found the minimum barrier to leave the state. However, if the dimer searches are employed conservatively, one can feel safe that large errors have not been introduced while increasing the efficiency of the TAD simulation. For systems with very large barriers relative to the temperature of interest, this gain in efficiency can be quite large.

Temperature accelerated dynamics has been demonstrated to be very effective for studying the long-time behavior of defects produced in collision cascades [28, 29]. Using pairwise Coulombic potentials for MgO, damage was produced via MD simulations of collision cascades. MD is ideal for this phase of the simulation, as the cascades typically settle down after a few picoseconds. However, while the defects that remain are still far from equilibrium, their diffusive behavior is governed by time scales much longer than those accessible via MD. Using dimer-TAD, we studied the room-temperature annealing of typical defects resulting from the MD simulations. We found that, given time, these defects will annihilate and aggregate and that the resulting aggregates can exhibit surprising behavior. Figure 6 illustrates a TAD simulation of two interstitial clusters – one containing two interstitials (I_2) and another containing four interstitials (I_4). The I_4 structure was found via a previous simulation in which two I_2 were allowed to aggregate. The I_2 is mobile on the time-scale of seconds at room temperature while the resulting I_4 is essentially immobile. In a simulation starting with I_2 and I_4, the I_2 cluster approaches the immobile I_4. On the time scale of seconds, the two clusters meet, forming a high-energy metastable complex containing six interstitials. This complex anneals further, exploring alternate I_6 geometries, until it eventually finds a form that is very mobile at room temperature. In fact, this I_6 structure is the most mobile species we have yet encountered in MgO, diffusing on the nanosecond time scale at room temperature, even faster than isolated interstitials. This structure is still not in the ground state for a six-interstitial cluster, but at room temperature it has a lifetime of years before it will decay to the ground state. In this particular TAD simulation, we obtained a boost factor of 10^9.

A similar MD/TAD procedure has also been applied to the simulation of thin film growth of Ag [30] and Cu [31] on Ag(100). Heteroepitaxial systems are especially hard to treat with techniques such as kinetic Monte Carlo due to the increased tendency for the system to go off lattice due to mismatch strain, and because the rate catalog needs to be considerably larger when neighboring atoms can have multiple types. Other applications for which TAD has proven effective include defect diffusion on oxide surfaces [32], the diffusion of interstitial clusters in Si [33] and defect diffusion in plutonium [19].

6   DISCUSSION AND CONCLUSION

As these accelerated dynamics methods become more widely used and further developed (including the possible emergence of new methods), their application

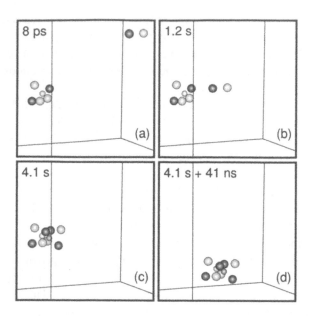

Figure 6. TAD simulation of the formation of I_6 at 300 K. Only defects in the lattice are shown (large spheres represent interstitials and small spheres represent vacancies). (a) An I_2 and I_4 begin about 1.2 nm apart. (b) By $t = 1.2$ s, the I_2 approaches the immobile I_4. (c) By $t = 4.1$ s, the combined cluster anneals to form the metastable I_6, (d) which diffuses on the ns time scale with a barrier of 0.24 eV (after [28, 29]).

to important problems in materials science will continue to grow. We conclude this article by comparing and contrasting the three methods presented here, with some guidelines for deciding which method may be most appropriate for a given problem. We point out some important limitations of the methods, areas in which further development may significantly increase their usefulness. Finally, we discuss the prospects for these methods in the immediate future.

The key feature of all of the accelerated dynamics methods is that they collapse the waiting time between successive transitions from its natural time (τ_{rxn}) to (at best) a small number of vibrational periods. Each method accomplishes this in a different way. TAD exploits the enhanced rate at higher temperature, hyperdynamics effectively lowers the barriers to escape by filling in the basin, and parallel-replica dynamics spreads the work across many processors.

The choice of which accelerated dynamics method to apply to a problem will typically depend on three factors. The first is the desired level of accuracy in following the exact dynamics of the system. As described above, parallel replica is the most exact of the three methods; the only assumption is that the kinetics are first order. Not even TST is assumed, as correlated dynamical events are treated correctly in the method. This is not true with hyperdynamics, which does rely upon the assumptions of TST. Finally, temperature accelerated dynamics makes the further assumptions inherent in the harmonic approximation to TST, and is thus the most approximate of the three methods. If complete accuracy is the main goal of the simulation, parallel replica is the superior choice.

The second consideration is the potential gain in accessible time scales that the accelerated dynamics method can achieve for the system. Typically, TAD is the method of choice when considering this factor. While in all three methods the boost for escaping from each state will be limited by the smallest barrier, if the barriers are high relative to the temperature of interest, TAD will typically achieve the largest boost factor. Recent advances in constructing bias potentials [24, 34, 35] have improved the efficiency of hyperdynamics, putting it on par with TAD. Finally, parallel replica dynamics usually offers the smallest boost given the typical access to parallel computing today (on the order of tens of processors per user for continuous use), since the maximum possible boost is exactly the number of processors. For some systems, the overhead of, for example, finding saddle points in TAD may be so great that parallel replica can give more overall boost. However, in general, the price of the increased accuracy of parallel replica dynamics will be shorter achievable time scales.

It should be emphasized that the limitations of parallel replica in terms of accessible time scales are not inherent in the method, but rather are a consequence of the limited computing power which is currently available. As massively parallel processing becomes commonplace for individual users so that a large number of processors can be used in the study of a given problem, parallel replica should become just as efficient as the other methods. This will be true when enough processors are available so that the amount of simulation time each processor has to do for each transition is on the order of ps. This analysis may be complicated by issues of communication between processors, but the future of parallel replica is very promising.

The last main factor determining which method is best suited to a problem is the shape of the potential energy surface (PES). Both TAD and hyperdynamics require that the PES be relatively smooth. In the case of TAD, this is because saddle points must be found and standard techniques for finding them often perform poorly for rough landscapes. The same is true for the hyperdynamics bias potentials that require information about the shape of the PES. Parallel replica, however, only requires a method for detecting transitions. No further analysis of the potential energy surface is needed. Thus, if the PES describing the system of interest is relatively rough, parallel replica dynamics may be the only method that can be applied effectively.

Currently, these methods are very efficient when applied to systems in which the barriers are much higher than the temperature of interest. This is often true for systems such as ordered solids, but there are many important systems that do not fall so cleanly into this class, a prime example being glasses. Such systems are characterized by either a continuum of barrier heights, or a set of low barriers that describe rapid events, such as conformational changes in a molecule. Low barriers typically degrade the boost of all of the accelerated dynamics methods, as well as the efficiency of standard kinetic Monte Carlo. However, even these systems will be amenable to study via accelerated dynamics methods as progress is made on this low-barrier problem.

The accelerated dynamics methods, as a whole, are still in their infancy. Even so, they are currently powerful enough to study a wide range of materials problems that were previously intractable. As these methods continue to mature, their applicability, and the physical insights gained by their use, can be expected to grow.

7   ACKNOWLEDGMENTS

We gratefully acknowledge vital discussions with Francesco Montalenti, Tim Germann, Graeme Henkelman, and Steve Stuart. This work was supported by the United States Department of Energy (DOE), Office of Science, Office of Basic Energy Sciences, Division of Materials Science.

REFERENCES

[1] A. F. Voter, F. Montalenti, and T. C. Germann, *Annu. Rev. Mater. Res.* **32**, 321 (2002).

[2] B. P. Uberuaga, F. Montalenti, T. C. Germann, and A. F. Voter, In S. Yip, editor, *Handbook of Materials Modeling*, page 629. Springer, The Netherlands, 2005.

[3] D. Chandler, *J. Chem. Phys.* **68**, 2959 (1978).

[4] A. F. Voter and J. D. Doll, *J. Chem. Phys.* **82**, 80 (1985).

[5] C. H. Bennett, in *Algorithms for Chemical Computation*, edited by R. E. Christofferson (American Chemical Society, Washington, DC, 1977), p. 63.

[6] R. Marcelin, *Ann. Physique* **3**, 120 (1915).

[7] E. P. Wigner, *Z. Phys. Chemie B* **19**, 203 (1932).

[8] H. Eyring, *J. Chem. Phys.* **3**, 107 (1935).

[9] P. Pechukas, *Annu. Rev. Phys. Chem.* **32**, 159 (1981).

[10] D. G. Truhlar, B. C. Garrett, and S. J. Klippenstein, *J. Phys. Chem.* **100**, 12771 (1996).

[11] A. F. Voter and J. D. Doll, *J. Phys. Chem.* **80**, 5832 (1984).

[12] B. J. Berne, M. Borkovec, and J. E. Straub, *J. Phys. Chem.* **92**, 3711 (1988).

[13] G. H. Vineyard, *J. Phys. Chem. Solids* **3**, 121 (1957).

[14] A. F. Voter. *Phys. Rev. B* **57**, 13985 (1998).

[15] B. P. Uberuaga, S. J. Stuart and A. F. Voter, submitted for publication.

[16] M. Terrones, F. Banhart, N. Grobert, J. C. Charlier, H. Terrones, and P. M. Ajayan, *Phys. Rev. Lett.* **89**, 075505 (2002).

[17] B. P. Uberuaga, A. E. Voter, K. K. Sieber, and D. S. Sholl, *Phys. Rev. Lett.* **91**, 105901 (2003).

[18] O. Kum, B. M. Dickson, S. J. Stuart, B. P. Uberuaga, and A. F. Voter, *J. Chem. Phys.* **121**, 9808 (2004).

[19] B. P. Uberuaga, S. M. Valone, M. I. Baskes, and A. F. Voter, *AIP Conference Proceedings* **673**, 213 (2003).

[20] M. Shirts and V. S. Pande, *Science*, **290**, 1903 (2000).

[21] A. F. Voter, *J. Chem. Phys.* **106**, 4665 (1997).

[22] J. P. Valleau and S. G. Whittington, In B. J. Berne, editor, *Statistical Mechanics. A. A Modern Theoretical Chemistry*, volume 5, pages 137–68. Plenum, New York, 1977.

[23] B. J. Berne, G. Ciccotti, and D. F. Coker, editors, *Classical and Quantum Dynamics in Condensed Phase Simulations*. World Scientific, Singapore, 1998.

[24] R. A. Miron and K. A. Fichthorn, *J. Chem. Phys.* **119**, 6210 (2003).

[25] M. R. Sørensen and A. F. Voter, *J. Chem. Phys.* **112**, 9599 (2000).

[26] F. Montalenti and A. F. Voter, *J. Chem. Phys.* **116**, 4828 (2002).

[27] G. Henkelman and H. Jónsson, *J. Chem. Phys.* **111**, 7010 (1999).

[28] B. P. Uberuaga, R. Smith, A. R. Cleave, F. Montalenti, G. Henkelman, R. W. Grimes, A. F. Voter, and K. E. Sickafus, *Phys. Rev. Lett.* **92**, 115505 (2004).

[29] B. P. Uberuaga, R. Smith, A. R. Cleave, G. Henkelman, R. W. Grimes, A. F. Voter, and K. E. Sickafus, *Phys. Rev. B* **71**, 104102 (2005).

[30] F. Montalenti, M. R. Sørensen, and A. F. Voter, *Phys. Rev. Lett.* **87**, 126101 (2001).

[31] J. A. Sprague, F. Montalenti, B. P. Uberuaga, J. D. Kress, and A. F. Voter, *Phys. Rev. B* **66**, 205415 (2002).

[32] D. J. Harris, M. Y. Lavrentiev, J. H. Harding, N. L. Allan, and J. A. Purton, *J. Phys.: Condens. Matter* **16**, L187 (2004).

[33] M. Cogoni, B. P. Uberuaga, A. F. Voter, and L. Colombo. *Phys. Rev. B* **71**, 121203 (2005).

[34] R. A. Miron and K. A. Fichthorn, *Phys. Rev. Lett.* **93**, 128301 (2004).

[35] R. A. Miron and K. A. Fichthorn, *Phys. Rev. B* **72**, 035415 (2005).

Chapter 3

RADIATION INDUCED STRUCTURAL CHANGES THROUGH IN-SITU TEM OBSERVATIONS

Chiken Kinoshita

Kyushu University, Fukuoka, JAPAN

1   INTRODUCTION

Electron microscopy provides structural information and chemical analysis of materials through elastic and inelastic interactions between electrons and materials [1]. The same scattering processes produce radiation effects in materials. Inorganic materials such as metals and ceramics undergo radiation effects via the processes of direct atomic displacement and/or electronic excitation. Electron microscopy, therefore, provides many experimental advantages for investigating radiation effects. The advantages include high electron flux, easy control of electron energy, easy alignment of crystallographic orientation, easy control of irradiation temperature and so on.   However, the most important advantage of electron microscopy is that it permits direct observations of phenomena during electron irradiation.

Stimulated mainly by those advantages, high voltage electron microscopy (HVEM), analytical electron microscopy, high resolution electron microscopy and HVEM combined with ion accelerators have been used to investigate not only radiation effects themselves, but also the kinetic processes associated with point defects. Systematic experiments have led to an understanding of fundamental aspects of radiation effects in metallic, ionic and covalent crystals. The purpose of this chapter is to introduce induced phenomena in alloys and ceramics under irradiation with electrons and/or ions [2]. Particular emphasis is placed on radiation-induced phenomena, such as the formation of defects clusters, chemical disordering, segregation and precipitation, spinodal decomposition, amorphization, re-crystallization and phase transformations near the surface.

K.E. Sickafus et al. (eds.), Radiation Effects in Solids, 45–63.
© 2007 *Springer.*

2   FORMATION OF DEFECT CLUSTERS [3-8]

The primary defects in alloys or metallic crystals created by energetic electron irradiation are isolated Frenkel pairs of interstitials and vacancies. Most of these point defects have a short life-time and annihilate via the interstitial-vacancy recombination process. Some of the point defects, on the other hand, aggregate to form interstitial or vacancy clusters, such as dislocation loops, stacking fault tetrahedra or voids. The micrographs in Figure 1 show typical examples of interstitial-type dislocation loops in metals, the majority of which nucleate at the beginning of irradiation and generally grow quickly into a well-defined shape [3].

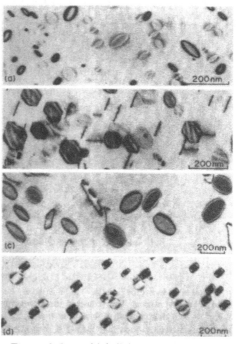

Figure 1. Interstitial dislocation loops formed in (a) Al, o(b) Cu, (c) Ni and (d) Fe irradiated with 2 MeV electrons. (After Yoshida and Kiritani [3])

The kinetic behavior of such clusters is directly related to the behavior of point defects.

On the basis of systematic experimental and theoretical works [3-8] on interstitial dislocation loops, the kinetics of loop nucleation and growth in foil specimens for electron microscopy is classified in terms of the mobility of interstitials M_I and of vacancies M_V. The density of interstitial loops saturates at the beginning of irradiation. The dependencies of the volume density C_L and the size D of interstitial dislocation loops on irradiation time t, electron flux ϕ and/or the mobility of point defects are

$$C_L \propto \phi^{1/2} M_I t^0 \quad \text{and} \quad D \propto t^{1/3}$$
$$\text{for} \quad M_I \gg M_V \lesssim 10^2 \text{ jumps/s,} \tag{1}$$

$$D \propto \phi^{1/2} M_V^{1/2} t$$
$$\text{for} \quad M_I \gg M_V \gtrsim 10^3 \text{ jumps/s.} \tag{2}$$

In the case of diatomic ionic crystals, electronic excitation produces Frenkel pairs only on a single sub-lattice (usually the anion one), while elastic collision produces Frenkel pairs on the both sub-lattices, though the production rates usually differ according to ion species. A dislocation loop in diatomic ionic crystals involves an equal number of cation and anion interstitials, while a void consists of an equal number of cation and anion vacancies. Clustering of vacancies on the cation- and anion-sublattices leads to gas bubbles and metal colloids, respectively, while interstitial clusters on the cation- and anion-sublattices respectively create metal colloids and gas bubbles. Only neutral vacancies or interstitials can form clusters on each sub-lattice. In view of these arguments, four methods of clustering of irradiation-induced Frenkel pairs in diatomic crystals are possible, as shown in Table 1. It should be emphasized that clustering of point defects provides the means for decomposing of ionic crystals into separate phases of its constituent atoms. Such decomposition occurs for irradiation at temperatures where both interstitials and vacancies are mobile and where the respective phases are stable against dissolution. Several examples have been shown to lead to precipitation of anions or cation as separate phases [9,10]. There is evidence that during electron irradiation in an HVEM [11], both void formation and Al precipitation can occur in thin foils of α-Al_2O_3 , though Al precipitation could also arise from the predominance of Al displacements in such experiments.

Table 1.   Possible types of defect clusters in diatomic ionic crystals formed by irradiation at high temperatures where both interstitial and vacancy are mobile.

Type	Interstitials	Vacancies
1	Dislocation loops involving N cation and N anion interstitials	Voids involving N cation and N anion vacancies
2	Dislocation loops involving N cation and N anion interstitials	Gas bubbles involving N cation vacancies and metal colloids involving N anion vacancies
3	Dislocation loops involving N cation and N anion interstitials	Gas bubbles involving $2N$ cation vacancies and N anti-site centers or metal colloids involving $2N$ anion vacancies and N anti-site centers
4	Gas bubbles involving N anion interstitials or metal colloids involving N cation interstitials	Gas bubbles involving N cation vacancies or metal colloids involving N anion vacancies

3   IRRADIATION-INDUCED CHEMICAL DISORDERING [12-14]

In atomic displacement processes, replacement collision sequences play an important role as a medium for transporting energy and mass through crystalline materials. Most of the displacements occur via replacement collision sequences, leaving a vacancy at the beginning of the sequence and depositing an excess atom as an interstitial at the end of the sequence. When super-lattice alloys such as FeAl and Ni_3Al are irradiated with high energy particles at low temperatures, where interstitials and vacancies are immobile, the displacement and/or replacement collisions are accompanied by chemical disordering. Two chemical disordering mechanisms involve: {1} replacement collisions along mixed atom rows, and {2} the recombination of displaced atoms with vacancies on the "wrong" sub-lattice.

The HVEM has been used for measuring changes in the long-range order parameter S of alloys during electron irradiation, and results have been discussed in terms of chemical ordering and/or disordering mechanisms. However, the super-lattice intensity method has mostly been used for measuring S, which is the electron microscopy analogue of the classical X-ray method. This method is very convenient to use, but dynamical electron diffraction effects produce serious ambiguities in experimental data. Kinoshita et al. [12], on the other hand, proposed a method for determining S from the measurement of thickness fringes in electron micrographs. They applied it to clarifying the dominant mechanism for electron irradiation-induced chemical disordering of Fe-48.6at%Al, which has a B2-type super-lattice and a high concentration of structural vacancies [13,14]. The extinction distance of the 100 super-lattice reflection of Fe-48.6at%Al at the exact 200 Bragg position depends more strongly on S as the electron energy increases. The relative positions of the thickness fringes in 100 and 200 dark-field micrographs taken at the exact 200 Bragg condition were related to S by the many-beam dynamical theory with an appropriate choice of atomic scattering factors based on the experimental results. The micrographs in Figure 2 are typical examples of pairs of 100 and 200 dark-field micrographs obtained from a wedge-shaped specimen under irradiation, showing respectively an increase and a decrease in the spacing of thickness fringes under electron irradiation. These phenomena correspond to the chemical disordering.

The variation under irradiation was measured for irradiation along directions near <100> and <110> with electrons of energy 250-1250 keV at room temperature. The results are summarized as a ln S *versus* ϕt plot in Figure 3. The values of S decrease with irradiation time t, and follow the equation

$$S = S_0 \exp(-K\phi t), \tag{3}$$

where S_0 is the initial value of S, K the chemical disordering cross section which can be expressed in units of barns, and ϕ the electron flux. The value of K increases with increasing energy of electrons and is independent of irradiation directions. Equation (3) is also supported by theoretical predictions based on the disordering mechanisms {1} and {2} described earlier. Using the electron-atom collision theory, the theoretical cross section K, based on mechanism {1}, is shown to depend strongly on irradiation direction, which is not shown by the experimental results. The successful application of the mechanism {2} to the experimental results suggests that the recombination of displaced atoms with vacancies on the "wrong" sub-lattice is the dominant process for disordering in the alloy Fe-48.6at%, which contains a high concentration of structural vacancies.

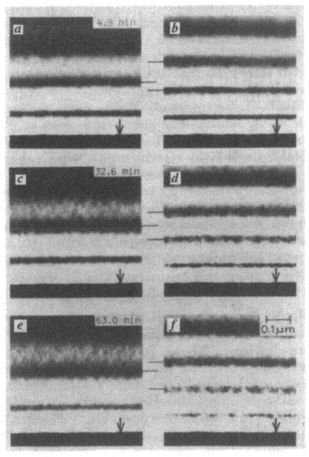

Figure 2.   Electron micrographs showing changes in spacings of thickness fringes during electron irradiaton. The pairs of dark-field micrographs for 100 (*a, c, e*) and 200 (*b, d, f*) reflections were taken at 294 s (*a,b*) 1956 s (*c,d*) 3780 s (*e,f*) during irradiation at 383 K with 500 keV electrons at a flux of 1.4×10^{22} e/m^{2}s. (After Mukai et al. [14])

4   IRRADIATION-INDUCED SEGREGATION AND PRECIPITATION [15-17]

The effects of electron irradiation on phase stability have been extensively studied in alloys, where displacements are solely due to direct displacements, not only because of scientific interest, but also for the assessment of nuclear materials. Two types of irradiation-induced precipitation have been found in many systems, and a number of mechanisms have been suggested to explain these phenomena. One is inhomogeneous precipitation around point defect sinks. This type of irradiation-induced precipitation has been successfully interpreted as resulting from the accumulation of solute atoms that have drifted toward defect sinks or the matrix [15]. The other is homogeneous in the form of coherent or incoherent precipitates [16,17]. One example of a homogeneous irradiation-induced phase transformation is presented in the next section along with its mechanism.

5   IRRADIATION-INDUCED SPINODAL DECOMPOSITION [18-22]

Electron irradiation induces a modulated structure in alloys, which show spinodal decomposition, even at temperatures higher than their spinodal temperatures. Figure 4 shows a typical dark-field micrograph for Au-70%Ni showing the modulated structure along the <100> direction and satellite spots along the same direction on the corresponding diffraction pattern. The wavelength of the modulated structure λ remains constant for increasing irradiation time after rapid formation of the modulated structure. Such kinetic behavior, according to the Cahn-Hilliard theory, suggests a modulated structure being induced by spinodal decomposition. The Cahn-Hilliard theory gives the wave number of modulated structure β_m ($\equiv 2\pi/\lambda$) in terms of the spinodal temperature T_s as

$$\beta_m^2 \propto (T - T_s), \qquad (4)$$

where T is the annealing temperature.

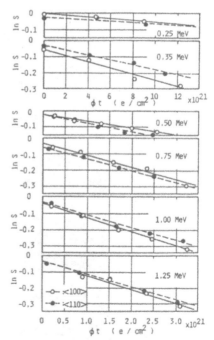

Figure 3. Variation of the order parameter S with electron dose ϕt during irradiation at room temperature with various energy electrons. The open circles and black dots correspond to irradiation along the <100> and the <110> directions, respectively. (After Mukai et al. [14])

If Equation (4) holds for spinodal decomposition under irradiation, a plot of observed β_m^2 *versus* T, where T is irradiation temperature in this case, should be linear and intersect the T-axis at T_s. The value of T_s under irradiation increases with increasing electron flux ϕ. The maximum spinodal temperatures with respect to ϕ for Au-Ni alloys with different compositions are plotted on the phase diagram (Figure 5), as an example of the alloy systems examined. The results shown in Figure 5 suggest the inducement of spinodal decomposition up to the chemical spinodal temperature. This tends to confirm that relaxation of the coherent strain associated with the modulated structure expands the spinodal region, and that electron irradiation relaxes the coherent strain.

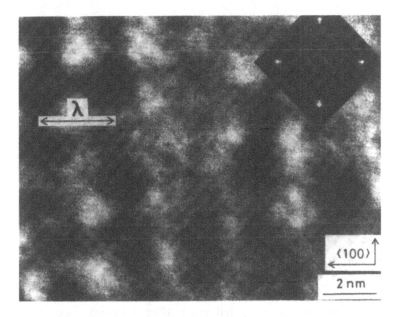

Figure 4.   Dark-field electron micrograph and the corresponding diffraction pattern showing modulated structure in Au-70at%Ni. (After Nakai and Kinoshita [19])

It is then instructive to examine the variation of lattice sites in the modulated structure caused by irradiation. Information obtained regarding lattice strain allows an investigation of the behavior of point defects. Lattice arrangements associated with the modulated structure during irradiation is studied through high resolution electron microscopy. Atom and/or lattice displacements are anticipated due to relaxation of the coherent strain under irradiation. Multi-beam lattice images were obtained from one area of a Au-70at%Ni specimen as a function of 400 keV electron irradiation dose at room temperature.

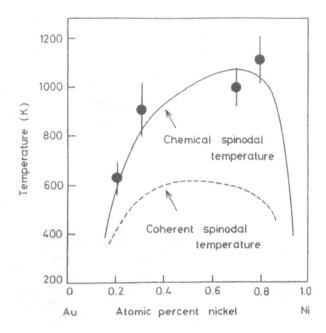

Figure 5.   The Au-Ni phase diagram showing the maximum values of spinodal temperatures under 1000 keV electron irradiation (•) at a flux of ~ 5 x 10^{23} electrons/m^{2}s. The values are in good agreement with the chemical spinodal temperatures within the error limits. (After Nakai and Kinoshita [20])

Figure 6 shows inter-planar spacings associated with (200) planes under 400 keV electron irradiation. The fluctuation of (200) inter-planar spacings as a function of distance within the observed sample area, clearly diminishes with increasing irradiation time, resulting in the inter-planar spacings approaching a constant. The striations corresponding to the composition modulation persisted even after disappearance of the phase modulation. These results suggest that the modulation of lattice sites is reduced despite of the existence of composition modulation under irradiation. The relaxation of the coherent strain associated with the modulated structure occurs without destruction of coherency between atoms through the alternate accumulation of interstitials and vacancies with a periodicity of λ_0 into expansive and compressive regions, respectively.

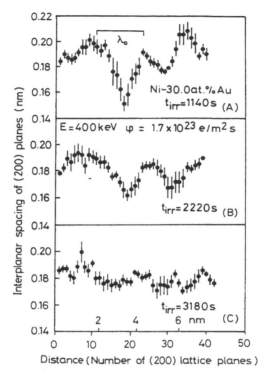

Figure 6. Variation of (200) inter-planar spacing with distance along the [100] direction under 400 keV electron irradiation at 298 K in Au-70at%Ni alloy. Figures (A) to (C) were obtained at t_{irr}=1140, 2220 and 3180 s, respectively. The inter-planar spacings were determined using multi-beam lattice image obtained from one specific area under irradiation. The value of λ_0 corresponds to the wavelength of the modulated structure existing before irradiation. (After Nakai and Kinoshita [22]).

6   IRRADIATION-INDUCED AMORPHIZATION [23-27]

It is well known that many non-metallic crystals undergo amorphization under heavy-ion irradiation. Naguib and Kelly [23] have shown that substances having ionicities less than 0.47 amorphize on ion impact, while substances with values larger than 0.59 are stable in crystalline form. Furthermore, Burnett and Page [24] have presented a sophisticated expression for the critical deposition-energy density as a function of ionicity, as shown in Figure 7. Irradiation with high-energy electrons easily induces an amophous phase transformation in covalent crystals such as graphite and SiC at low temperatures. In this section experimental results which provide valuable insights into the mechanisms of irradiation-induced amorphization will be shown [25,27].

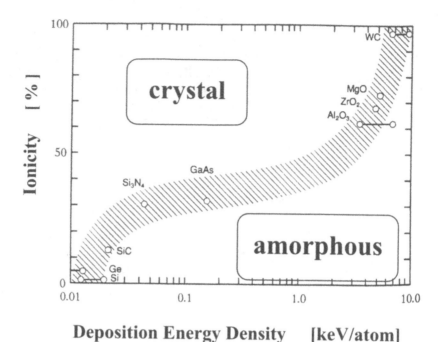

Figure 7.   Critical deposition energy density for amorphization of crystals having various ionicity. (After Burnett and Page [24])

Convergent beam electron diffration (CBED) was used for examining irradiation-induced microstructural evolution in highly-oriented pyrolytic graphite (HOPG) obtained from Union Carbide Co. Thin sections of graphite were irradiated with ions and 100-1000 keV electrons at temperatures from 110 to 1100 K. Amorphization was observed at temperatures below 500 K, where the diffraction contrast such as bend contours and thickness fringes disappeared and the corresponding diffraction pattern gradually changed from a spot pattern to a halo one during irradiation. Diffraction contrast consisting of black or white dots or dendrites were induced by interstitial clusters at temperatures higher than 500 K.

A dose required for full amorphization was determined through the halo pattern typical of amorphous, and is defined as the critical amorphization dose (D_c). Figure 8 shows temperature dependence of D_c for a variety of projectile masses, energies and dose rates [27]. The critical amorphization dose remains constant at a value of about 0.5 displacement per atom (dpa) at lower temperatures, and increases rapidly with increasing irradiation temperature. The amorphization dose was also determined at 300 K as a function of electron energy, and it decreases with increasing electron energy; that is, the cross section for amorphization increases with increasing electron energy, as does the elastic collision cross-section.

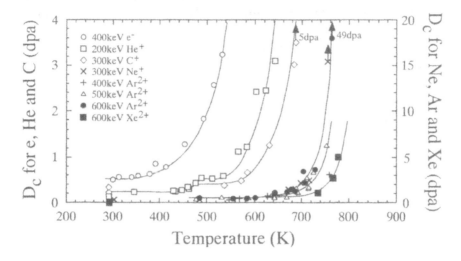

Figure 8. Temperature dependence of the critical amorphization dose for graphite irradiated with various kinds of projectiles. (After Abe et al. [27])

The amorphization process was more carefully examined during electron irradiation through observations using CBED, high resolution electron microscopy (HREM) and electron energy loss spectroscopy (EELS). CBED patterns were obtained from a small region irradiated with electrons, and they were converted to the lattice constant. The variation of the lattice constant along the *c*-axis is shown in Figure 9. The lattice constant increases in the early stage of irradiation and reaches a saturation value ($\Delta C/C{\sim}5 \times 10^{-3}$) at about 1×10^{24} e/m^2 ($\sim 1 \times 10^{-3}$ dpa), where $\Delta C/C$ is the fractional change in the lattice constant. Vacancies scarcely affect the lattice constant along the *c*-axis, but interstitials and di-interstitial carbon (C_2) molecules increase it. The fractional change $\Delta C/C$ is expected to be $3.4x$ for interstitials and $1.2y$ for C_2 molecules, where x and y are the concentrations of interstitials and C_2 molecules, respectively. Therefore, ~0.15% of isolated interstitials or ~0.11% of C_2 molecules are introduced into the matrix following irradiation to a dose of ~ 1×10^{-3} dpa [26].

Structure images were obtained using HREM, which revealed an array of periodic dots. This array became irregular and some parts were gradually chopped into small blocks with concomitant loss in periodicity within a fluence of 1.0×10 e/m^2 ($\sim 1 \times 10^{-2}$ dpa) [25] (though diffraction spots remained). Furthermore, variations in vacancy concentrations were detected, based on the structure of the EELS K-edge, or intensities of π^* and σ^* peaks. Irradiation-induced amorphization of graphite proceeds via the following processes: abrupt swelling along the c-axis within 1×10^{-3} dpa, followed by a gradual increase of the vacancy concentration up to 0.5 dpa, at which point amorphization is complete [25].

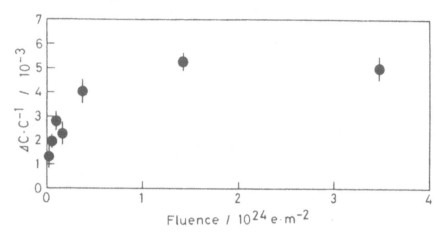

Figure 9. Fractional variation of the lattice constant along the c-axis of HGPO irradiated with a 200 keV electron flux at 110 K. (After Kinoshita et al. [25])

The primary defects in graphite created by electron irradiation are isolated Frenkel pairs which may induce amorphization. According to Iwata's systematic studies [28], however, C_2 molecules are formed near 100 K and they migrate to form their clusters above 160 K. Therefore, it is concluded that some displaced carbon atoms aggregate to form C_2 molecules and/or their clusters under irradiation at lower temperatures than 500 K and induce swelling along the c-axis. A high concentration of C_2 molecules and/or their clusters induce strains which cause fading of the periodic structure image at $\sim 1 \times 10^{-2}$ dpa. A variety of topological structures associated with configurations of C_2 molecules and vacancies presumably play an important role in relaxing the strain to induce amorphization.

7   IRRADIATION-INDUCED RECRYSTALLIZATION [29,30]

Thin foil specimens of HOPG were irradiated with 30 keV He$^+$, Ar$^+$ or Xe$^+$ ions at room temperature, 10° from the [0001] direction [30]. At the thinner part of the specimen, amorphization was followed by microcrystal evolution under Ar$^+$ or Xe$^+$ ion irradiation. Under He$^+$ ion irradiation only amorphization was detected.

In the thicker regions, on the other hand, a twin network was formed at the beginning of irradiation before the inducement of amorphization.

Furthermore, Ar^+ and Xe^+ ions evolved microcrystals in the amorphized region. An example of such sequential changes is shown in Figure 10. In summary, ion irradiation evolves peculiar structures with increasing fluence in the following order: a twin network; an amophization transformation; and microcrystal formation within the amorphous regions. Each of the phenomena is induced at a lower fluence with increasing mass of projectiles. The results suggest that the twin network is induced with the stress field caused by the difference of defect concentrations between the regions within and outside the projected range of ions. The accumulation of defects alters the lattice parameters as mentioned before, and produces the critical stress field for twinning through lattice coherency. The nucleation of micro-crystals in the amorphous phase evolves dominantly with increasing projectile mass. This can be explained in terms of the deposition energy density within the cascade.

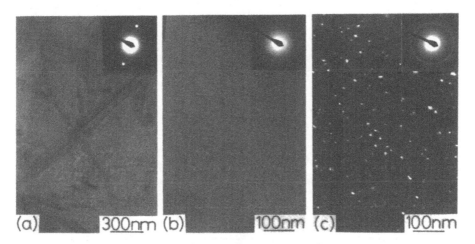

Figure 10. A series of electron micrographs showing twinning, amorphization and re-crystallization of HOPG exposed to 30 keV Ar^+ irradiation at 300 K. (After Nakai et al. [30])

8 CHARACTERISTIC MICROSTRUCTURAL EVOLUTION UNDER CONCURRENT IRRADIATION CONDITIONS WITH ELECTRONS AND IONS [29,31]

An HVEM-accelerator facility was also used for investigating concurrent effects of electrons and ions [29,31]. Figure 11 shows the accumulation process of cascades in Ge for various ratios of electron to ion flux. We observed decreasing cascade contrast accumulation with increasing electron flux; that is, there seems to be a concurrent effect of electron irradiation on the accumulation process of cascades.

The micro-structural evolution of cascades depends strongly on the spatial distribution of defects in individual displacement cascades. A cascade is thought of as having a vacancy-rich core surrounded by a high concentration of interstitials. The stability of cascades is, therefore, controlled by the behavior of interstitials around the vacancy-rich core. The importance of interstitials and of cascade overlap increases with increasing irradiation time and slows down the formation rate of cascade contrasts, eventually leading to saturation. The concurrent effect of 1000 keV electrons contributes to the annihilation of cascades. In the case of covalent crystals irradiated with heavy ions, a part of the cascade sometimes consists of an amorphous phase. The electron irradiation also affects the stability of the amorphous phase.

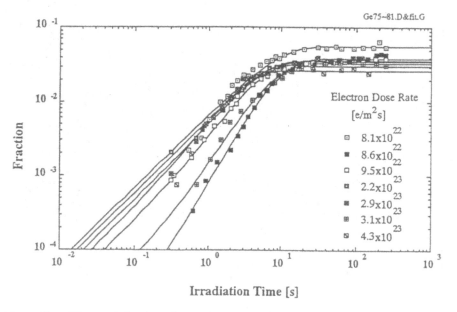

Figure 11.   The areal density of cascade contrasts in Ge under irradiation with various ratios of the 1000 keV electron fluxes to a 60 keV Ar^+ ion flux of 2.2 x 10^{14} ions/m²s. (After Abe et al. [31])

Figure 12 is a phase diagram for Si exposed to concurrent irradiation with various kinds of ions and electrons, where the critical electron flux for preventing amorphization is shown as a function of ion flux [31]. The retardation of ion-induced amorphization is enhanced by electrons with lower energy through irradiation induced migration of point defects or electronic excitation [31].

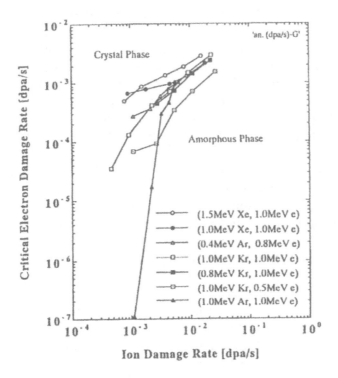

Figure 12. A phase diagram for Si showing crystalline and amorphous phases as functions of the electron damage rate and the ion damage rate with various combinations of ions and electrons. The damage rates were calculated by TRIM-90 with a displacement threshold energy of 16 eV. (After Abe et al. [31])

9   IRRADIATION-INDUCED PHASE TRANSFORMATIONS NEAR SURFACES [32-34]

Irradiation-induced phase transformations are very sensitive to the atmospheric conditions in electron microscopes. Oxidation and reduction are frequently observed in conventional electron microscopes. Clean high-vacuum also provides new types of phase transformations under electron irradiation. An example is a structural change induced by electron irradiation at or near the surface of α-Al_2O_3 specimens in clean-vacuum electron microscopes [33,34].

When thin foils of α-Al_2O_3 are observed in a JEM-4000EX, mottled contrast appears within several minutes. The contrast becomes wavy, or like a "patchwork quilt," with increasing observation time. After this, holes and dark-line contrast appear. Anisotropic hole drilling was observed in MgO even when circular electron probes were employed [35]. The mottled or patchwork-quilt-like contrast can be seen at 200 keV also, suggesting that the roughening and hole drilling originate from desorption induced by electronic transitions at surfaces. During observation along the [110] direction at 400 keV, dark-line contrast appears, as shown in Figure 13. The number density of dark-lines increases with increasing irradiation time. The dark-line contrast appears in thick regions as well as a thin areas, though the contrast is not always dark. The lines are parallel to the (001) basal plane, showing streaks along the [001] direction in the corresponding diffraction pattern. The streaks are observable when many dark- or bright-lines appear and suggest the presence of planar defects [33]. These dark- or bright-lines were also produced at 200 keV. Bursill et al. [36] observed similar results even at 100 keV, which is much lower than the displacement energy for aluminum or oxygen ions. On the other hand, no line contrast appears even after 3600s of electron irradiation at 200 keV in a JEM-2000FX, where the vacuum around the specimen is estimated to be about one order magnitude worse than in the JEM-4000EX. These results indicate that the formation of planar defects is very sensitive to the quality of vacuum around the specimen.

The planar defect is a result of rearrangements of Al ions and Al vacancies in the (001) basal plane. A model of the rearrangement is shown in Figure 14(a). In the [110] projection of the planar defect, columns of Al vacancies alone are lost and those consisting of 50% of Al ions and Al vacancies are present, whereas columns of Al-vacancies (open circles) are arranged with every other two columns of Al ions (closed circles) in the perfect structure. This means that the ratio of the number of vacancies to the number of Al ions in the defect plane changes from 1/2 to 1/3 in a perfect plane (Figure 14 (c)). An atomic arrangement in the defect plane is shown in Figure 14(b) along with the one in the perfect plane. A row of Al ions alone appears along the [110] direction and the identical directions with every other row which consists of 50% Al ions and 50% Al vacancies. Al ions and Al vacancies are arranged so the structure may have six-fold symmetry and a double period compared with the original structure. The model of the planar defect shown here is qualitatively consistent with the streaking and the extra spots in the corresponding diffraction pattern. It should be noted that the ratio of the number of Al vacancies to the atomic sites in the defect Al plane is equal to one fourth, which corresponds to the composition of Al_3O_2, while the ratio in the perfect Al plane in α-Al_2O_3 is one third. The rearranged structure shown in Figure 14(b) is actually the same as the one of [111] layers of γ-Al_3O_4 (spinel-type). It is noteworthy that Bursill et al. [36] and Lin [37] interpreted the dark lines as being due to facets terminated with a monolayer α-Al_3O_4 structure, based on computer simulation of [110] HREM images. These results suggest the local reduction of α-Al_2O_3 on a mono-layer scale. We have, however, not detected any sign in HREM images and diffraction patterns to indicate the presence of an α-Al₃O₄ phase or a metallic Al phase. On the other hand, Bonevich and Marks [34] observed small crystallites of Al at surface of α-Al₂O₃ instead of the dark-lines when the specimen was irradiated under ultra-high vacuum conditions. This result also indicates that the damage process is very sensitive to the degree of vacuum around the specimen.

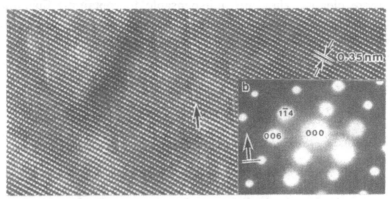

Figure 13. HREM image of α-Al₂O₃ irradiated with 400 keV electrons along the [110] direction and the corresponding diffraction pattern, showing striations parallel to the (001) basal plane. (After Tomokiyo et al. [33])

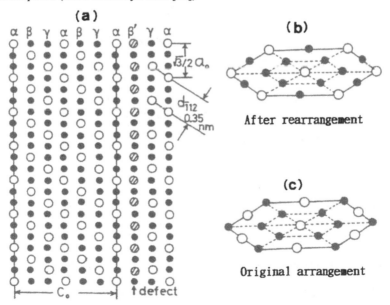

Figure 14. Model of the planar defect in α-Al₂O₃. (a) The [110] projection: (●) columns consisting of Al ions alone, shaded circles indicate (◐) columns consisting of Al vacancies alone, (○) columns of 50% Al ions and 50% Al vacancies. (b) The rearranged structure in the defect plane. (c) The original arrangement in the perfect plane; (●) Al ions, (○) Al vacancies. (After Tomokiyo et al. [33])

10   SUMMARY

Energetic particles induce point defect clustering and chemical disordering. In some cases, they also induce phase transformations, such as precipitation, spinodal decomposition and amorphization. High-resolution and high-vacuum transmission electron microscopes reveal such radiation effects more clearly. They are indispensable for obtaining quantitative structural and chemical information about materials, and they also facilitate our ability to establish mechanisms of radiation damage.

REFERENCES

[1]   C. Kinoshita, *J. Electron Microsco.*, **34** (1985), 299.

[2]   C. Kinoshita, *J. Electron Microsco.*, **40** (1991), 301.

[3]   N. Yoshida and M. Kiritani, *J. Phys. Soc. Jpn.*, **35**(1973), 1418.

[4]   N.Yoshida, M. Kiritani and F. E. Fujita, *J. Phys. Soc. Jpn.*, **39**(1975), 170.

[5]   M. Kiritani, N. Yoshida, H. Takaka and Y. Maehara, *J. Phys. Soc. Jpn.*, **38**(1975),1677.

[6]   M. Kiritani, *J. Phys. Soc .Jpn.*, **40**(1976), 1035.

[7]   M. Kiritani and H. Takata, *J. Nucl. Mat.*, **69&70**(1978), 277.

[8]   M. Kiritani, *Ultramicroscopy*, **39**(1991), 135.

[9]   L. W. Hobbs, in Introduction to Analytical Electron Microscopy (J. Hren, J. I. Goldstein and D. C. Joy, ed. Plenum Press, New York, 1979).

[10]   L. W. Hobbs, *J. Phys.*, **37**(1976), 3.

[11]   T. Shikama and G. P. Pells, *Phil. Mag.*, **47**(1983), 381.

[12]   C. Kinoshita, T. Mukai and S. Kitajima, *Acta. Cryst.*, **A33** (1977), 605.

[13]   T. Mukai, C. Kinoshita and S. Kitajima, *J. Phys. Soc. Jpn.*, **45** (1978), 1676.

[14]   T. Mukai, C. Kinoshita and S. Kitajima, *Phil. Mag.*, **A47** (1983), 255.

[15]   P.R. Okamoto and H. Wiedersich, *J. Nucl. Mat.*, **53** (1974), 336.

[16]   C. Kinoshita, L.W. Hobbs and T.E. Mitchell, *The Met. Soc. of AIME*, (1981) 561.

[17] K. Nakai, C. Kinoshita, S. Kitajima and T.E. Mitchell, *J. Nucl. Mat.*, **133-134** (1985) 694.

[18] C. Kinoshita, K. Nakai and S. Kitajima, *Mat. Sci. Forum*, **15-18** (1987), 1403.

[19] K. Nakai, C. Kinoshita and Y. Asai, *Suppl. to Trans JIM.*, **29** (1988), 167.

[20] K. Nakai and C. Kinoshita, *J. Nucl. Mat.*, **169** (1989), 116.

[21] Y. Asai, Y. Isobe, K. Nakai, C. Kinoshita and K. Shinohara, *J. Nucl. Mat.*, **179** (1991) 1050.

[22] K. Nakai, C. Kinoshita and N. Nishimura, *J. Nucl. Mat.*, **179-181** (1991), 1046.

[23] H.M. Naguib and R. Kelly, *Rad. Effects*, **25** (1975), 1.

[24] P. J. Burnett and T. F. Page, *Rad. Effects*, **97**(1986), 283.

[25] C. Kinoshita, K. Nakai, A. Matsunaga and K. Shinohara, *Proceedings Jpn. Academy*, **65B** (1989), 182.

[26] A. Matsunaga, C. Kinoshita, K. Nakai and Y. Tomokiyo, *J. Nucl. Mat.*, **179-181** (1991), 457.

[27] H. Abe, H. Naramoto, A. Iwase and C. Kinoshita, *Nucl. Inst. Meth. in Phys. Res.*, **B127**(1977), 681.

[28] H. Maeta, T. Iwata and S. Okuda, *J. Phys. Soc. Jpn.*, **38** (1975), 1538.

[29] C. Kinoshita, Y. Isobe, H. Abe, Y. Denda and T. Sonoda, *J. Nucl. Mater.*, **206** (1993), 341.

[30] K. Nakai, C. Kinoshita and A. Matsunaga, *Ultramicroscopy*, **39** (1991) 361.

[31] H. Abe, C. Kinoshita and K. Nakai, *J. Nucl. Mater.*, **179-181**, (1991) 917.

[32] C. Kinoshita, Y. Tomokiyo and K. Nakai, *Ultramicroscopy*, **56** (1994), 216.

[33] Y. Tomokiyo, T. Kuroiwa and C. Kinoshita, *Ultramicroscopy*, **39** (1991), 213.

[34] J.E. Bonevich and L.D. Marks, *Ultramicroscopy*, **35** (1991), 161.

[35] P.S. Turner, T.J. Bullough, R.W. Devenish, D.M. Maher and C.J. Humphreys, *Phil. Mag. Lett.*, **61** (1990), 181.

[36] L.A. Bursill, Peng Ju Lin and D.J. Smith, *Ultramicroscopy*, **23** (1987), 223.

[37] L.A. Bursill and Peng Ju Lin, *Phil. Mag.*, **A60** (1989) 307.

Chapter 4

RADIATION DAMAGE FROM DIFFERENT PARTICLE TYPES

Gary S. Was[1] and Todd R. Allen[2]
[1]*University of Michigan, Ann Arbor, MI 48109*
[2]*Engineering Physics, University of Wisconsin, Madison, WI 53706*

1. INTRODUCTION

In the 1960s and 70s, heavy ion irradiation was being developed for the purpose of simulating neutron damage in support of the fast breeder reactor program. [1-3] Ion irradiation and simultaneous He injection have also been used to simulate the effects of 14 MeV neutron damage in conjunction with the fusion reactor engineering program. Lately, the application of ion irradiation (defined here as any charged particle, including electrons) to the study of neutron irradiation damage is being revisited by the light water reactor community in an effort to solve the irradiation assisted stress corrosion cracking (IASCC) problem. [4-6], as additionally, ion irradiation is being used to understand performance in reactor pressure vessel steels, Zircaloy fuel cladding, materials for advanced reactor concepts in the GenIV program and materials for the advanced fuel cycle initiative, AFCI. While the environment plays a role in IASCC, the appearance of a "threshold" fluence demonstrates that the observed behavior is strongly related to irradiation-induced changes in the alloy that may involve microstructural changes, microchemical changes, transmutation, or some synergistic combination of these effects.

Clearly there is significant incentive to use ion irradiation to study neutron damage as this technique has the potential for yielding answers on basic process in addition to the potential for enormous savings in time and money. The nature of performing neutron irradiation experiments is not amenable to studies involving a wide range of conditions, which is precisely what is required for investigations of the basic damage processes. Simulation by ions allows easy variation of the irradiation conditions.

K.E. Sickafus et al. (eds.), Radiation Effects in Solids, 65–98.
© 2007 Springer.

Regarding cost and time, typical neutron irradiation experiments in test reactors require 1-2 years of exposure in core. However, this is accompanied by at least another year of capsule design and preparation as well as disassembly and cooling. Analysis of microchemical changes by Auger electron spectroscopy (AES) or microstructural changes by energy dispersive spectroscopy via scanning transmission electron microscopy (STEM) and mechanical property or stress corrosion cracking (SCC) evaluation can take several additional years because of the precautions and special facilities and instrumentation required for handling radioactive samples. The result is that a single cycle from irradiation through microanalysis and mechanical property/SCC testing may take between 4 and 6 years. Such a long cycle length does not permit for iteration on irradiation or material conditions that is a critical element in any experimental research program. It also requires the stability of financial support for periods of at least 5 years, which is often difficult to guarantee. Because of the long cycle time, the requirement of special facilities and special sample handling, the costs for neutron irradiation experiments are very high.

In contrast to neutron irradiation, ion (heavy ions, light ions or electrons) irradiation enjoys considerable advantages in both cycle length and cost. Ion irradiations of any type rarely require more than a few tens of hours to reach 1-5 dpa levels. Irradiation produces little or no residual radioactivity allowing handling of samples without the need for special precautions. These features translate into significantly reduced cycle length. For instance, samples of 304 SS have been irradiated with protons to 1 dpa, the grain boundary composition was characterized via AES, the microstructure was characterized via STEM-EDS and constant extension rate tensile (CERT) experiments were conducted in 288°C water to determine IASCC susceptibility, all in a time period of 3-4 months! Analysis of results indicated the need for another experiment in which the alloy composition was slightly modified in order to isolate the effect of impurities, which could not have been foreseen. This next experiment (iteration) was conducted over the course of the next 3-4 months. The rapidity of the technique allows for several such iterations in the course of a year! In terms of cost, on a per sample basis, ion irradiation is about 1/100 the cost of neutron irradiation and requires less than 1/10 the time. If one considers the cost of obtaining the same amount of information over a common time period, then the cost per data point by ion irradiation is approximately 1/100 x 1/10 or 1/1000 of that for neutron irradiation. The magnitude of the savings is something that can't be ignored. The challenge is then to verify the equivalency of the results of neutron and ion irradiation.

A question that needs to be answered is how do results from neutron and charged particle irradiation experiments compare? How, for example is one to compare the results of a component irradiated in core at 288°C to a fluence of 1 x 10^{21} n/cm^2 (E>1MeV) over a period of 8.5 months, with an ion irradiation experiment using 3 MeV protons at 400°C to 1 dpa (displacements per atom) at a dose rate of 10^{-5} dpa/s, or 5 MeV Ni^{++} at 500°C to 10 dpa at a dose rate of 10^{-3} dpa/s? The first question to resolve is the *measure* of radiation effect. In the IASCC problem, concern has centered on two effects of irradiation: segregation of major alloying elements or impurities to grain boundaries which then cause embrittlement or enhance the intergranular stress corrosion cracking (IGSCC) process, and hardening of the matrix which results in restricted deformation modes and embrittlement. The appropriate *measure* of the radiation effect in the former case would then be the alloy concentration at the grain boundary or the amount of impurity segregated to the grain boundary. This is a *measurable* quantity, either by AES or by STEM-EDS. For the latter case, the *measure* of the radiation effect would be the nature, size, density and distribution of dislocation loops, black dots and the total dislocation network. Hence, specific and measurable effects of irradiation can be determined for both neutron and ion irradiation experiments.

The next concern is determining how ion irradiation translates into the environment describing neutron irradiation. That is, what are the irradiation conditions required for ion irradiation to yield the same *measure* of radiation effect as that for neutron irradiation? *This is the key question, for in a post-irradiation test program, it is only the <u>final state</u> of the material that is important in the determination of equivalence, and not the path taken.* Therefore, if one could devise ion irradiation experiments that yielded the same measures of irradiation effects as observed in neutron irradiation experiments, then the data obtained in post-irradiation experiments will be equivalent. In such a case, ion irradiation experiments can provide a direct substitute for neutron irradiation.

2. RADIATION DAMAGE

The first problem in determining the equivalence between the measure of radiation effect in charged particle and neutron irradiation is the use of a common dose unit. The basic (measurable) dose unit for neutron irradiation is n/cm^2 above some energy threshold (E>x MeV), where x is the energy threshold. For charged particles it is the integrated current or charge, Q/cm^2. The particle beam community is accustomed to reporting dose in units of dpa and dose rate as dpa/s using one of several models for the determination of dpa. Although somewhat more complicated, due to the existence of an energy spectrum rather than a monoenergetic ion beam, the same conversion can be made from n/cm^2 (E>x MeV), although it is not always done. A fundamental difference between ion and neutron irradiation effects is the particle energy spectrum that arises due to the difference in how the particles are produced. Ions are produced in accelerators and emerge in monoenergetic beams with vary narrow energy widths, Fig. 1a. However, the neutron energy spectrum extends over several orders of magnitude in energy, thus presenting a much more complicated source term for radiation damage, Fig. 1b.

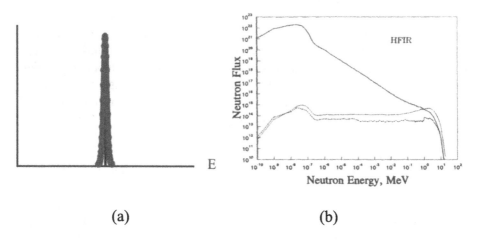

(a)   (b)

Figure 1. Energy spectrum of a) incident particles in a monoenergetic ion beam and b) in the High Flux Isotope Reactor.

Another major difference in the characteristics of ions and neutrons is their depth of penetration. As shown in Fig. 2, ions lose energy quickly because of high electronic energy loss, and result in a spatially non-uniform energy deposition due to the varying importance of electronic and nuclear energy loss during slowing down. Their penetration distances range between 0.1 and 100 μm for ion energies that can practically be achieved by laboratory-scale accelerators or implanters. By virtue of their electrical neutrality, neutrons can penetrate very large distances and produce spatially flat damage profiles over many mm of material.

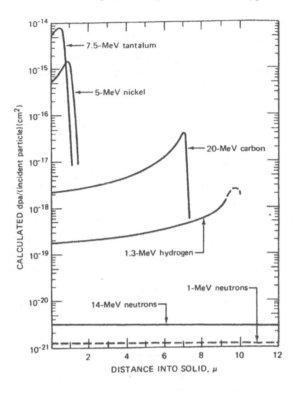

Figure 2. Damage depth profile for neutrons and ions of varying energies. [7]

2.1 Damage Function

The parameter commonly used to correlate the damage produced by different irradiation environments is the total number of displacements per atom. Kinchin and Pease [8] were the first to attempt to determine the number of displacements occurring during irradiation and a modified version of their model known as the NRT model [9] is generally accepted as the international standard for quantifying the number of atomic displacements in irradiated materials. According to the NRT model, the number of Frenkel pairs $\nu(T)$ generated by a primary knock-on atom (PKA) of energy T is given by:

$$\nu(T) = \frac{\kappa E_D(T)}{2E_d}, \tag{1}$$

where $E_D(T)$ is the damage energy (energy of the PKA less the energy lost to electron excitation), E_d is the displacement energy, that is, the energy needed to displace the struck atom from its lattice position, and κ is the displacement efficiency (usually taken as 0.8). Integration of the NRT damage function over recoil spectrum and time gives the atom concentration of displacements known as the NRT displacements per atom (dpa):

$$dpa = \phi\sigma_{FP}t = \phi t \frac{dT(d\sigma(E < T)}{dT} v(T). \qquad (2)$$

The displacement damage is accepted as a measure of the amount of change to the solid due to irradiation and is a much better measure of irradiation effect than is the particle fluence. As shown in Fig. 3, seemingly different effects of irradiation on low temperature yield strength for the same fluence level (Fig. 3a) disappear when the particle energy spectrum is accounted for (Fig. 3b). However, there is more to the description of radiation damage than just the number of dpa. There is the issue of the spatial distribution of damage production, which can influence the microchemistry and microstructure, particularly at temperatures where diffusional processes are important to microstructural development. In fact, the "ballistically" determined value of dpa calculated using such a displacement model is not the appropriate unit to be used for dose comparisons between particle types. The reason is the difference in the primary damage state among different particle types.

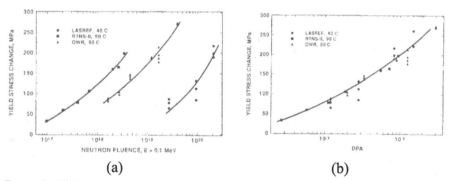

(a)   (b)

Figure 3. Yield stress vs. neutron fluence produces different amounts of hardening for the same fluence level (a). But after accounting for the particle energy spectra through the use of displacement damage theory (b), the yield stress is shown to depend on the damage level irrespective of the facility in which the experiment was conducted. [10]

2.2   Primary and Weighted Recoil Spectra

A description of irradiation damage is not complete without considering the distribution of recoils in energy and space. The primary recoil spectrum describes the relative number of collisions in which an energy between T and T + dT is transferred from the primary recoil atom to other target atoms. The fraction of recoils between the displacement energy E_d, and T is:

$$P(E,T) = \frac{1}{N} dT \frac{d\sigma(E,T)}{dT}, \qquad (3)$$

where N is the total number of primary recoils and $d\sigma(E,T)$ is the differential cross section for a particle of energy E to create a recoil of energy T. Figure 4 shows the difference in the types of damage that is produced by different types of particles. Light ions such as electrons and protons will produce damage as isolated Frenkel pairs or in small clusters while heavy ions and neutrons produce damage in large clusters. For 1 MeV particle irradiation of copper, half the recoils for protons are produced with energies less than ~60 eV while the same number for Kr occurs at about 150 eV. Recoils are weighted toward lower energies because of the screened Coulomb potential that controls the interactions of charged particles. For an unscreened Coulomb interaction, the probability of creating a recoil of energy T varies as $1/T^2$. However, neutrons interact as hard spheres and the probability of a creating a recoil of energy T is independent of recoil energy. In fact, a more important parameter describing the distribution of damage over the energy range is a combination of the fraction of defects of a particular energy and the damage energy. This is the "weighted average" recoil spectra, W(E,T) which weights the primary recoil spectra by the number of defects or the damage energy produced in each recoil:

$$W(E,T) = \frac{1}{E_D(E)} \frac{dT(d\sigma(E,T)}{dT} E_D(T), \qquad (4)$$

where E_d is the threshold displacement energy, T_{max} is the maximum recoil energy given by $T_{max} = 4E(m_1 m_2)/(m_1 + m_2)^2$ and $d\sigma/dT$ is the differential cross section for energy transfer from a particle of energy E to a target atom in the energy range T to T+dT. For the extremes of Coulomb and hard-sphere interactions, the differential cross sections are:

$$\frac{d\sigma_{coul}}{dT} dT = \frac{\pi m_1 (Z_1 Z_2 e^2)^2}{E} \frac{dT}{T^2}, \qquad (5a)$$

and $\qquad \frac{d\sigma_{HS}}{dT} dT = A \frac{dT}{E}. \qquad (5b)$

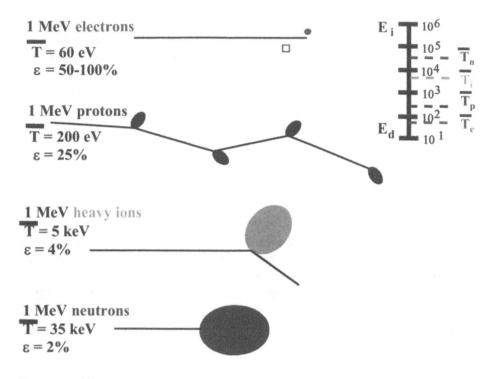

1 MeV electrons
T = 60 eV
ε = 50-100%

1 MeV protons
T = 200 eV
ε = 25%

1 MeV heavy ions
T = 5 keV
ε = 4%

1 MeV neutrons
T = 35 keV
ε = 2%

Figure 4. Difference in damage morphology, displacement efficiency and average recoil energy for 1 MeV particles of different type incident on nickel.

Ignoring electron excitations and allowing $E_{D_D}(T) = T$, then the weighted average recoil spectra, eqn. (4), are:

$$W_{coul}(T) = \frac{\ln T - \ln T_{min}}{\ln T_{max} - \ln T_{min}}, \tag{6a}$$

and

$$W_{HS}(T) = \frac{T^2 - T_{min}^2}{T_{max}^2}. \tag{6b}$$

Equations 6a and 6b are graphed in Fig. 5 for 1 MeV particle irradiations of copper. The Coulomb potential is a good approximation for proton irradiation while the hard sphere potential is a good approximation for neutron irradiation. The Coulomb forces extend to infinity and slowly increase as the particle approaches the target. In a hard sphere interaction, the particles and target do not "feel" each other until their separation reaches the hard sphere radius at which point the repulsive force goes to infinity. A screened Coulomb is most appropriate for heavy ion irradiation. Note the large difference in W(T) between the various types of irradiations. While heavy ions come closer to reproducing the energy distribution of recoils of neutrons than do light ions, neither is accurate in the "tails" of the distribution. This does not mean that ions are poor simulations of radiation damage, but it does mean that damage is produced differently and that this needs to be considered when assessing the microchemical and microstructural changes due to irradiation.

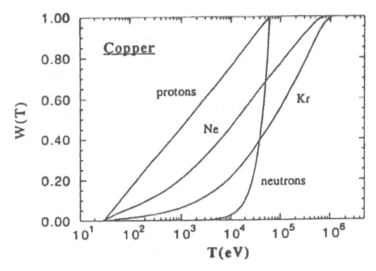

Figure 5. Fraction of recoils with energy above E_d and below T, where T is the energy transferred in the collision. [11]

2.3 Radiation Damage Morphology

The actual number of defects that survive the displacement cascade and their spatial distribution in the solid will determine their effect on the irradiated microstructure. The displacement efficiency is defined as the fraction of the "ballistically" produced Frenkel pairs (FPs) which survive the cascade quench and are available for long range migration. These are referred to variously as "freely migrating" or "available migrating" defects. [12] The freely migrating defects are the only defects that will affect the amount of grain boundary segregation, which is one measure of radiation effect. The fraction of the total number of defects produced that are "freely migrating" is termed the displacement efficiency, ε. This number can be very small, approaching a few percent at high temperatures. The displacement cascade efficiency, ε can be broken into three components:

$\gamma_{i,v}$: the isolated point defect fraction,

$\delta_{i,v}$: clustered fraction including mobile defect clusters such as di-interstitials, and

ζ: fraction initially in isolated or clustered form after the cascade quench that is annihilated during subsequent short term ($>10^{-11}$ s) intracascade thermal diffusion. They are related as:

$$\varepsilon = \delta_i + \gamma_i + \zeta_i = \delta_v + \gamma_v + \zeta_v$$

Figure 6 shows the history of defects born as vacancies and interstitials as described by the NRT model. Due to significant recombination in the cascade, Fig. 7, a significant fraction of the defects recombine, leaving a small minority that are fee to migrate from the displacement zone [30]. These defects can recombine outside of the cascade region, be absorbed at sinks in the matrix (voids, loops) or be absorbed at the grain boundaries, providing for the possibility of radiation-induced segregation.

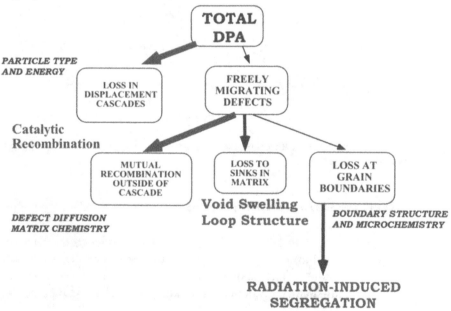

Figure 6. History of point defects after creation in the displacement cascade.

Despite the equivalence in energy among the four particle types described in Fig. 4, the average energy transferred and the defect production efficiencies vary by almost two orders of magnitude! This is explained by the differences in the cascade morphology among the different particle types. Neutrons and heavy ions produce dense cascades that result in substantial recombination during the cooling or quenching phase. However, electrons are just capable of producing a few widely spaced Frenkel pairs that have a low probability of recombination. Protons produce small widely spaced cascades and many isolated FPs due to the Coulomb interaction and therefore, fall between the extremes in displacement efficiency defined by electrons and neutrons.

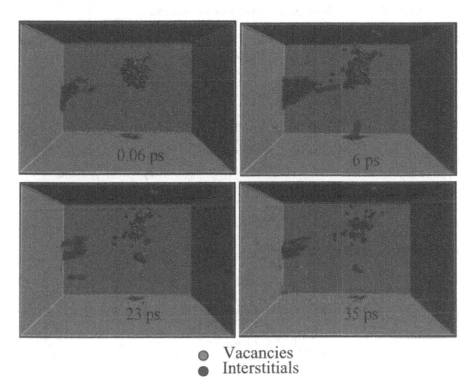

● Vacancies
● Interstitials

Figure 7. Loss of vacancies and interstitials in the cascade due to mutual recombination in the first 100 ps after cascade creation following 20 keV Cu self-ion irradiation.

We will focus on the comparison between four types of particle irradiation in order to outline a methodology for establishing equivalence between neutron and charged particle irradiation. The four types are given in Table 1 and are taken from experiments conducted to study the IASCC problem. Each experiment is characterized by the particle type and energy, irradiation temperature, reported dose rate and reported total dose. The displacement efficiency is calculated using Naundorf's model [13], which is based on two factors. The first is that energy transfer to atoms is only sufficient to create a single Frenkel pair. The second is that the Frenkel pair lie outside a recombination (interaction) radius so that the nearby FPs neither recombine nor cluster. The model follows each generation of the collision and calculates the fraction of all defects produced that remain free. According to Naundorf, the free single FPs are classified according to the generation i in which they were produced, i.e. the relative amount η_1 is that amount that is produced by primary collisions (first generation), while η_2 is the relative amount produced by secondary collisions (second generation). Thus the total number of free single FPs produced is:

Table 1. Comparison of different particle irradiation experiments.

Particle type	Energy (MeV)	Temp. (°C)	Eff.	Displ. Rate (dpa/s) Rept'd	Real*	Total Dose (dpa) Rept'd Real	Dose	Dose to steady state**(dpa) Rept'd Real
Electrons Kato [15]	1.0	450	1.0	2x10⁻³	2x10⁻³	10	10	28 28
Protons Was [16]	3.4	400	0.2	7x10⁻⁶	1x10⁻⁶	1	0.2	7 3
N++ ions Bruemmer [17]	5.0	500	0.04	5x10⁻³	2x10⁻⁴	10	0.4	25 7
Neturons Jacobs[18]	Fission reactor	288	0.02	~5x10⁻⁸	9x10⁻¹⁰	1	0.02	4 1.4

* efficiency corrected value
** as calculated by the Perks model for RIS

$$\eta = \sum_i \eta_i, \tag{7}$$

where that produced by primary collisions is:

$$\eta_1 = \left(\frac{\beta_p}{\sigma_d}\right) \int_{E_d}^{\alpha E} dT K_{I,A}(E,T), \tag{8}$$

and that produced by secondary collisions is:

$$\eta_2 = \left(\frac{1}{\sigma_d}\right) \int_{E_d}^{\alpha E} dT K_{I,A}(E,T)[Z(T)\beta_A(T)/\sigma_A(T)] \int_{E_d}^{2.5E_d} dT' K_{A,A}(T,T'). \tag{9}$$

The primary displacement cross section for the incident ion is:

$$\sigma_p = \int_{E_d}^{\alpha E} dT K_{I,A}(E,T), \tag{10}$$

and the total displacement cross section σ_d is given in the Kinchen-Pease model by:

$$\sigma_d = \int_{E_d}^{\alpha E} dT K_{I,A}(E,T)\nu(T). \tag{11}$$

$K_{I,A}(E,T)$ is defined as the differential cross section of an incident ion (I) of energy E which transfers the energy $T \leq T_{max}$ to an atom (A) of the crystal, E_d is the displacement energy, the maximum energy transferred is αE ($\alpha = 4M_I M_A/(M_I + M_A)^2$) and $v(T)$ is the K-P displacement function defined earlier. $Z(T)$ is the total number of secondary collisions produced above E_d by a primary of energy T along its path. The distance λ between two primary collisions is distributed according to an exponential law:

$$W(\lambda) = \tfrac{1}{\lambda_p} \exp(-\tfrac{\lambda}{\lambda_p}), \qquad (12)$$

with the mean distance

$$\lambda_p = \tfrac{\Omega}{\sigma_p}: \qquad (13)$$

where Ω is the atomic volume. The condition that the distance between two consecutive collisions must be larger than an appropriate interaction radius, r_{iv} (so that FPs produced near each other neither recombine nor cluster) reduces the amount of all possible free single FPs by:

$$\beta_p = \exp(-\tfrac{r_{iv}}{\lambda_p}), \qquad (14)$$

and is illustrated in Fig. 8. The model provides the efficiency for the production of freely migrating defects. This efficiency is shown in Fig. 9 for various ions

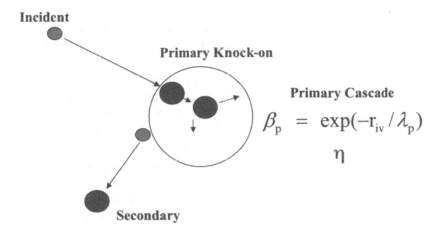

Figure 8. Illustration of the effect of the interaction radius on single Frenkel pair production.

Results of the model are applied to the four particle types in table 1. The "corrected" displacement rate and "corrected" total dose for each particle type are determined by multiplying the reported (uncorrected) values times the efficiency factor.

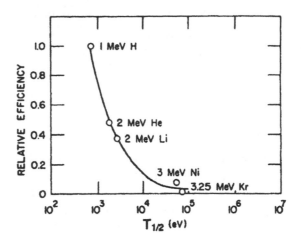

Figure 9. Relative efficiency of freely migrating defect production for ions of various mass and energy. [14]

3.   DEPENDENCE OF RIS ON PARTICLE TYPE

The quantity of interest in RIS modeling is the amount of chromium depleted from the grain boundary, or the area inside the Cr concentration profile, Fig. 10. The appropriate measure of depletion is somewhat questionable. One could use the grain boundary chromium value as the measure of the extent of chromium depletion. Alternatively, the FWHM of the depletion profile has been used. In fact both of these quantities are useful and can be obtained from measured depletion profiles. However, the area inside the Cr concentration profile represents changes to a volume of material and as will become evident later, is more sensitive to changes in the profile shape than either the grain boundary value or the FWHM alone. The amount of Cr depletion is determined by integrating the concentration profile for that element with distance from the grain boundary,

$$M = \int_0^{l(t)} \left[C_A^o - C_A(x,t) \right] dx, \qquad (15)$$

where M is the segregated area, C_A^o is the bulk atom concentration, $C_A(x,t)$ is the atom concentration near the surface, and l(t) is the half-width of the depleted zone. The amount of Cr depletion for "corrected" and "reported" dose rates is given in table 1 for each experiment. Also given are the values of Cr depletion at steady state and the doses required to reach steady state. Figure 11 shows the calculated amount of Cr depletion as a function of temperature and displacement rate at *steady state*. Steady state is reached at different dose levels for each experiment. At a given displacement rate, the segregated area peaks at some intermediate temperature and falls off at both higher and lower temperatures. This is due to the dominance of recombination at low temperatures and back diffusion at high temperatures. [19] Also note that the effect of a decreasing displacement rate is to shift the curves to higher maxima at lower temperatures. For a given dose, a lower displacement rate yields lower steady state defect concentrations, reducing the number of defects lost to recombination, and shifting the curve to lower temperatures while increasing the degree of segregation.

Note the change in the calculated values for the amount of Cr depletion in the four experiments shown in Fig. 11. Since electrons are assumed to be 100% efficient in producing defects available to affect segregation, there is no change in the segregated area after accounting for efficiency. However, there is a difference with protons, heavy ions and neutrons. The difference is largest for neutrons and smallest for protons. The difference is a function of not only the displacement efficiency, but also the slope of the dose rate curves. Nevertheless, substantial differences result in the expected amounts of grain boundary segregation when the displacement efficiency is taken into account.

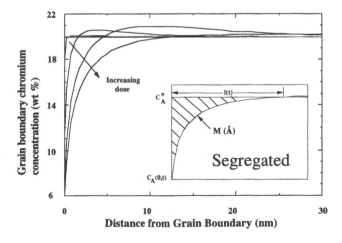

Figure 10. Definition of the segregated area, M, in a quantitative assessment of RIS.

Figure 11 shows the effect of three of the four parameters defining an experiment: particle type, temperature and dose rate. It does not show the effect of dose since this is a steady state result that is achieved at different doses for each of the experiments described in table 1. Figure 12 shows the dose required to reach steady state as function of temperature and dose rate. Each of the experiments is plotted for both the reported and the corrected displacement rates. Note the large difference in the dose to reach steady state between electrons and neutrons. In general, irradiation at a lower dpa rate will result in a lower dose to reach steady state and the difference is greatest for this comparison. Correspondingly, proton and heavy ion irradiation fall between neutrons and electrons for the experiments described in table 1. This can be understood by considering the chemical rate equations;

$$dC/dt = K_o - K_{iv}C_iC_v - K_{vs}C_vC_s, \qquad (16)$$

where the first term is the production rate, the second is the loss by mutual recombination and the third by annihilation of defects at sinks. At steady state, $C_{i,v}$ a $K_o^{1/2}$ at low temperature and $C_{i,v}$ a K_o at high temperature (see the lesson on Radiation Enhanced Diffusion and Radiation-induced Segregation). So the resulting point defect concentrations are strong functions of the production rate.

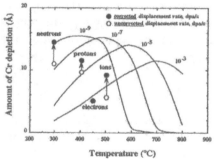

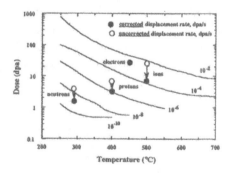

Figure 11. Chromium depletion at steady state as a function of temperature, dose rate and particle efficiency. [19]

Figure 12. Dose to reach steady state as a function of temperature, dose rate and particle efficiency. [19]

Figure 13 shows a plot of segregation as a function of temperature for particles with displacement rates characteristic of their sources. Note that the temperature at which segregation is a maximum (concentration is a minimum) shifts to higher temperature with increasing dose rate. This is due to the trade-off between temperature and dose rate. Figure 14 shows the temperature – dose rate inter-dependence for stainless steel over a wide range of temperatures and dose rates. Also noted are the positions where reactor irradiations by neutrons occur, and where proton and Ni ion irradiations occur. This graph explains why the experiments conducted at the highest dose rates are also conducted at the highest temperatures.

A simple method for examining the tradeoff between dose and temperature in comparing irradiation effects from different particle types is found in the invariance requirements. For a given change in dose rate, we would like to know what change in dose (at the same temperature) is required to cause the same number of defects to be absorbed at sinks. Alternatively, for a given change in dose rate, we would like to know what change in temperature (at the same dose) is required to cause the same number of defects to be absorbed at sinks. The number of defects per unit volume that have recombined up to time τ is given by [21]:

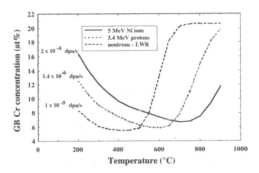

Figure 13. Dependence of grain boundary chromium concentration on temperature for particles with various dose rates.

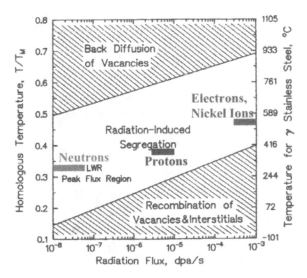

Figure 14. Relation between temperature and dose rate in the context of radiation induced segregation, and the locations of neutron, proton and Ni ion irradiations.

$$N_R = R_{iv} \int_o^\tau C_i C_v dt, \tag{17}$$

where R_{iv} = vacancy-interstitial recombination coefficient. Similarly, the number of defects per unit volume that are lost to sinks up to time τ is:

$$N_{sj} = \int_o^\tau K_{sj} C_j dt. \tag{18}$$

The ratio of vacancy loss to interstitial loss is:

$$R_s = \frac{N_{sv}}{N_{si}}, \tag{19}$$

where j = v or i, and K_s is the sink strength. The quantity N_s is important in describing microstructural development involving total point defect flux to sinks (e.g., RIS), while R_s is the relevant quantity for the growth of defect aggregates such as voids that requires partitioning of point defects to allow growth. In the *steady-state recombination dominant regime*, for N_s to be invariant at a fixed dose, the following relationship between **dose rate and temperature** must hold:

$$T_2 - T_1 = \frac{\left(\dfrac{kT_1^2}{E_{vm}}\right) \ln\left(\dfrac{K_2}{K_1}\right)}{1 - \left(\dfrac{kT_1}{E_{vm}}\right) \ln\left(\dfrac{K_2}{K_1}\right)}. \tag{20}$$

In the *steady-state recombination dominant regime*, for the R_s to be invariant at a fixed dose, the following relationship between **dose rate and temperature** must hold:

$$T_2 - T_1 = \frac{\left(\dfrac{kT_1^2}{E_{vm} + 2E_{vf}}\right) \ln\left(\dfrac{K_2}{K_1}\right)}{1 - \left(\dfrac{kT_1}{E_{vm} + 2E_{vf}}\right) \ln\left(\dfrac{K_2}{K_1}\right)}. \tag{21}$$

In the *steady-state recombination dominant regime*, for N_s to be invariant at a fixed temperature, the following relationship between **dose and dose rate** must hold:

$$\frac{\Phi_2}{\Phi_1} = \left(\frac{K_2}{K_1}\right)^{\frac{1}{2}}. \tag{22}$$

Finally, in the *steady-state recombination dominant regime*, for N_s to be invariant at a fixed dose rate, the following relationship between **dose and temperature** must hold:

$$T_2 - T_1 = \frac{\left(\frac{-2kT_1^2}{E_{vm}}\right)\ln\left(\frac{\Phi_2}{\Phi_1}\right)}{1 - \left(\frac{kT_1}{E_{vm}}\right)\ln\left(\frac{\Phi_2}{\Phi_1}\right)}. \tag{23}$$

The invariance requirements predict that in an experiment to simulate neutron radiation in stainless steel at a dose rate of 4.5×10^{-8} dpa/s at 288°C (typical core conditions in a BWR) with protons at 7.0×10^{-6} dpa/s (typical accelerator-generated proton flux), the temperature should be 416°C for invariant N_s (e.g. RIS), and 316°C for invariant R_s (e.g. swelling or loop growth). In other words, the temperature "shift" due to the higher dose rate is dependent on the microstructural property of interest.

4   ADVANTAGES AND DISADVANTAGES OF IRRADIATIONS USING VARIOUS PARTICLE TYPES

Each particle has its advantages and disadvantages for use in the study of radiation effects. Table 2 lists the advantages and disadvantages for each of three particle types; electrons, heavy ions and light ions (protons), and they are discussed in detail in the following sections.

Table 2. Advantages and disadvantages of irradiations with various particle types

---

Electrons

Advantages
  Relatively "simple" source - TEM
  Uses standard TEM sample
  High dose rate - short irradiation times

Disadvantages
  Energy limited to ~1 MeV
  no cascades
  Very high beam current (high dpa rate)
  requires high temperature
  Poor control of sample temperature
  Strong "Gaussian" shape (nonuniform
  intensity profile) to beam
  No transmutation

Heavy Ions

Advantages
  High dose rate - short irradiation times
  High T_{avg}
  Cascade production

Disadvantages
  Very limited depth of penetration
  Strongly peaked damage profile
  Very high beam current (high dpa rate)
  requires high temperature
  No transmutation
  Potential for composition changes at
  high dose via implanted ion

Protons

Advantages
  Accelerated dose rate - moderate
  irradiation times
  Modest ΔT required
  Good depth of penetration
  Flat damage profile over tens of μm

Disadvantages
  Minor sample activation
  Smaller, widely separated cascades
  No transmutation

4.1   Electrons

Electron irradiation is easily carried out in a transmission electron microscope and as such, it uses a rather simple ion source, that being either a hot filament or a field emission gun. An advantage is that the same instrument used for irradiation damage can be used to image the damage. Another advantage is that the high dose rate requires very short irradiation time, but this will also require a large temperature shift as explained in the previous section.

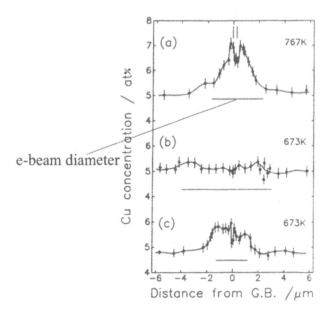

e-beam diameter

Figure 15. Enrichment of copper surrounding a local depletion at the grain boundary. The enrichment is caused by the high defect flux away from the irradiated region defined by the horizontal line. [22]

There are several disadvantages to electron irradiation using a TEM. First, energies are generally limited to 1 MeV. This energy is sufficient to produce an isolated Frenkel pair in transition metals, but not cascades. The high dose rate requires high temperatures and that must be closely monitored and controlled, which is difficult to do precisely in a typical TEM sample stage. Another drawback is that since irradiations are often conducted on thinned foils, defects are created in close proximity to the surface and their behavior may be affected by the presence of the surface. Perhaps the most serious drawback is the Gaussian beam shape that can give rise to strong dose rate gradients across the irradiated region. Figure 15 shows the composition profile of copper around a grain boundary in Ni-39Cu following electron irradiation. Note that while there is local depletion at the grain boundary, the region adjacent to the minimum is strongly enriched in copper due to the strong defect flux out of the irradiated zone defined by the horizontal line below the spectra. Another often observed artifact in electron irradiation is very broad grain boundary enrichment and depletion profiles. Figure 16 shows that the enrichment profile for Ni and the depletion profiles for Fe and Cr have widths on the order of 75-100 nm, which is much greater than the 5-10 nm widths observed following neutron or proton irradiation under similar conditions. A similar effect was noted by Wakai in electron and D^+ irradiation of the same alloy in which it was observed that the segregation was much greater and narrower around the grain boundary in the deuteron-irradiated sample as compared to the electron irradiation, Fig. 17.

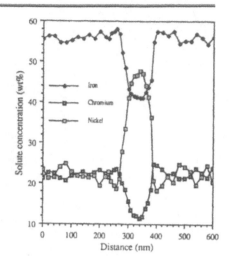

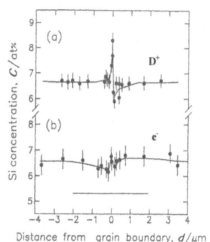

Figure 16. Broad grain boundary enrichment and depletion profiles in Fe-20Cr-25Ni-0.75Nb-0.5Si following irradiation with electrons at 420°C to 7.2 dpa. [23]

Figure 17. Comparison of electron and deuteron irradiation showing the greater amount of segregation and the narrower profile for the deuteron irradiation. [24]

4.2   Heavy Ions

Heavy ions enjoy the benefit of being produced at high dose rates resulting in accumulation of high doses in short times. Also, because they are typically produced in the energy range of a few MeV, they are very efficient at producing dense cascades, similar to those produced by neutrons. The disadvantage is that as with electrons, the high dose rates require large temperature shifts so that irradiations must be conducted at temperatures of ~500°C in order to create similar effects as neutron irradiation at ~300°C. Clearly, there is not much margin for studying neutron irradiations at high temperature since the required ion irradiation temperature would cause excessive annealing. Another drawback is the short penetration depth and the continuously varying dose rate over that penetration depth. Figure 18 shows the damage profile for several heavy ions incident on nickel. Note that the damage rate varies continuously and peaks sharply at only 2 μm depth below the surface. As a result, the challenge is to reproducibly study a region of known depth so that the varying dose and dose rate do not affect the observations.

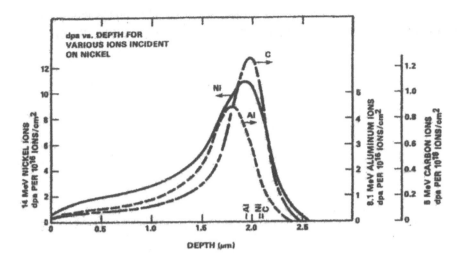

Figure 18. Damage profiles for C, Al and Ni in a nickel target. [25]

4.3   Protons

In many ways, proton irradiation overcomes the negatives with electron and neutron irradiation. With only a few MeV beam, the penetration depth can reach in excess 40 μm and results in a fairly flat dpa depth profile in which the dose rate varies by only a factor of two over several tens of μm. Further, the depth of penetration is sufficient to assess such properties as irradiation hardening through microhardness measurements, and stress corrosion cracking through crack initiation tests such as the slow strain rate test. Figure 19 shows a schematic of a 5 MeV Ni++ damage profile and a 3.2 MeV proton damage profile in stainless steel. Superimposed on the depth scale is a grain structure with a grain size of 10 μm. Note that with this grain size, there are numerous grain boundaries and a significant irradiated volume over which the damage rate is flat. The dose rate for proton irradiations is 2-3 orders of magnitude smaller than that for electrons or ions, thus requiring only a modest temperature shift, but since it is still 100-1000 times higher than neutron irradiation, modest doses can be achieved in reasonably short irradiation time.

The disadvantages are that because of the small mass of the proton compared to heavy ions, the recoil energy is smaller and the resulting damage morphology is characterized by smaller, more widely spaced cascades than with ions or neutrons. Also, since only a few MeV are required to surmount the Coulomb barrier for light ions, there is also a minor amount of sample activation that increases with proton energy.

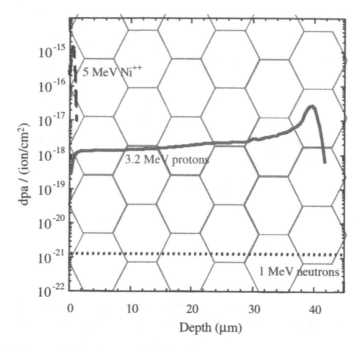

Figure 19. Damage profile for 5 MeV Ni^{++} and 3.2 MeV protons in stainless steel.

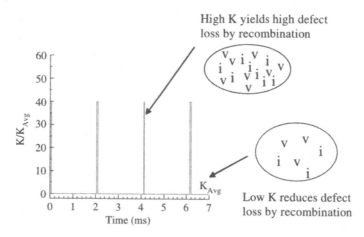

Figure 20. The effect of a scanned beam on the instantaneous production rate of point defects with the same time averaged rate as a continuous source.

Two additional considerations for all particle irradiations are the lack of transmutation reactions and the effect of a raster-scanned beam. With the exception of some minor transmutation reactions that can occur with light ion irradiation, charged particles do not reproduce the types of transmutation reactions that can occur in reactor cores due to interactions with neutrons. The most important of these is the production of He by neutron reactions with either Ni or B. But a second consideration is that of a raster-scanned beam in which any volume element of the target is "seeing" the beam for only a fraction of the raster-scan cycle. For proton irradiation in the Tandetron accelerator at the University of Michigan, that fraction of time the beam is on any particular volume element is 1/40. Thus, the instantaneous dose rate during the "beam-on" portion of the cycle is 40 times that of the average, Fig. 20. The result is that the defect production rate is very high and defects can anneal out in the remaining 39/40 portion of the cycle before the beam again passes through the volume element. As such, the effective defect production rate in raster-scanned systems will be less and must be accounted for.

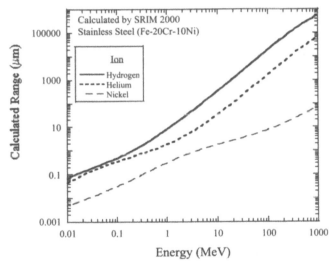

Figure 21. Range of protons, helium ions and nickel ions in stainless steel as a function of ion energy.

5 CHOOSING BEAM PARAMETERS FOR PARTICLE IRRADIATIONS

In the process of setting up an ion irradiation experiment, a number of parameters that involve beam characteristics (energy, current/dose) and

Radiation Effects in Solids

beam-target interaction need to be considered. One of the most important considerations is the depth of penetration. Figure 21 shows the range vs. particle energy for protons, helium ions and nickel ions in stainless steel as calculated by SRIM. Figure 22 shows how several other parameters describing the particle-target interaction vary with energy; dose rate, time to reach 1 dpa, deposited energy and the maximum permissible beam current (which will determine the dose rate and total dose) given a temperature limitation of either 360°C or 400°C. Finally, Fig. 23 shows how competing features of the irradiation vary with beam energy, creating tradeoffs in the beam parameters. For example, while greater depth is generally favored in order to increase the volume of irradiated materials, it is achieved by irradiating at higher energies, which leads to lower dose rates near the surface and higher residual radioactivity. For proton irradiation, the optimum energy range is achieved by balancing these factors and will lie between 2 and 6 MeV.

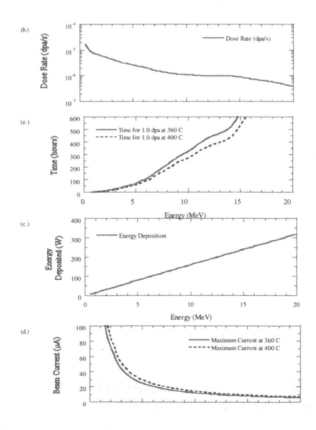

Figure 22. Behavior of beam-target interaction parameters as a function of beam energy for protons and helium ions.

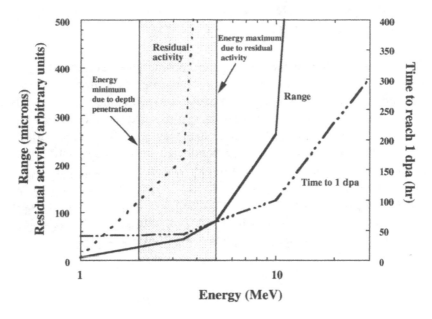

Figure 23. Variation of ion range, residual activity and time to reach 1 dpa for protons as a function of particle energy.

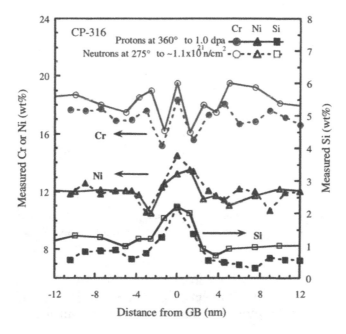

Figure 24. Comparison of grain boundary segregation of Cr, Ni and Si in commercial purity 316 stainless steel following irradiation with either protons or neutrons to similar doses. [26]

6   PROTON IRRADIATION AS A RADIATION DAMAGE EMULATION TOOL

Proton irradiation has undergone considerable refinement over the past decade as a radiation damage tool. Numerous experiments have been conducted and compared to equivalent neutron irradiation experiments in order to determine if proton irradiation can capture the effects of neutron irradiation on microstructure, microchemistry and hardening. In some cases, benchmarking exercises were conducted on the same native heat as neutron irradiation in order to eliminate heat-to-heat variations that may obscure comparison of the effects of the two types of irradiating particles. The following examples cover a number of irradiation effects on several alloys in an effort to demonstrate the capability of charged particle irradiation to capture the critical effects of neutron irradiation.

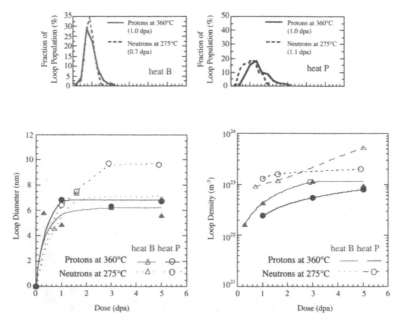

Figure 25. Comparison of a) loop size distributions and b) loop diameter and number density for commercial purity 304 and 316 stainless steels irradiated with neutrons or protons to similar doses, [26]

Figures 24 to 27 show direct comparisons of the same irradiation feature on the same heat of starting material following either neutron irradiation at 275°C or proton irradiation at 360°C to similar doses. Figure 24 compares the RIS behavior of Cr, Ni and Si in a 316 stainless steel alloy following irradiation to approximately 1 dpa. Neutron irradiation results are in open symbols and proton irradiation results are in solid symbols. This dose range was chosen as an extreme test of proton irradiation to capture the "W" shaped chromium depletion profile caused by irradiation of a microstructure containing grain boundary chromium enrichment prior to irradiation. Note that the two profiles trace each other extremely closely both in magnitude and spatial extent. The agreement extends across all three elements. Figure 25 shows the agreement in the dislocation microstructure as measured by the dislocation loop size distribution (Fig. 25a) and the size and number density of dislocation loops for 304 (heat B) and 316 (heat P) alloys. Note that the main features of the loop size distributions are captured by both irradiations; a sharply peaked distribution in the case of 304 SS and a flatter distribution with a tail for the case of 316 SS. The agreement on loop size is good for the 304 SS alloy, while loops are smaller for the proton-irradiated 316 alloy. The loop density is about a factor of three less for the proton-irradiated case than for the neutron irradiated case, which is expected since the proton irradiation temperature was optimized to track RIS (higher temperature) than the dislocation loop microstructure (see the result at the end of section 3). Figure 26 shows the comparison in irradiation hardening between the two types of irradiation. Not surprisingly, the results are close, with proton irradiation resulting in slightly lower hardness, since hardening is determined by the square root of the product of the dislocation loop size and number density. Figure 27 shows the IASCC susceptibility as measured by the %IG on the fracture surface following constant load testing (neutron-irradiated samples) and constant extension rate testing (proton-irradiated samples) in BWR normal water chemistry. Despite the significantly different testing mode, the results are in excellent agreement in that both irradiations result in the onset of IGSCC at about 1 dpa, which agrees with the literature.

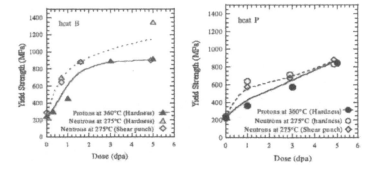

Figure 26. Comparison of irradiation hardening in commercial purity 304 and 316 stainless steel irradiated with neutrons or protons. protons to similar doses. [26]

Radiation Effects in Solids

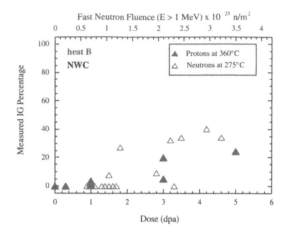

Figure 27. Comparison of the extent of intergranular stress corrosion cracking in commercial purity 304 stainless steel following similar SCC tests of either neutron or proton irradiated samples from the same heat. [26]

Figure 28 contains results on the behavior of swelling as a function of alloy nickel content for proton, Ni ion and neutron irradiation. While these experiments were conducted on different sets of alloys, and highly disparate irradiation conditions, they all show the same dependence of swelling on nickel content. In the last example on stainless steel alloys, Fig. 29 shows the relaxation of residual stress by neutron and proton irradiation. Here again, results are on different alloys and are from different types of tests, but both show the same dependence of stress relaxation with dose.

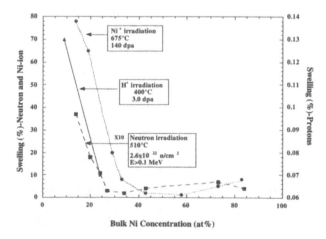

Figure 28. Comparison of swelling dependence on alloy nickel content in neutron, nickel ion or proton irradiated stainless steels.

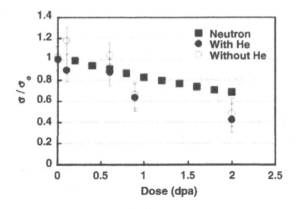

Figure 29. Comparison of neutron based prediction with proton-induced creep after removing the effect of thermally induced relaxation. [27]

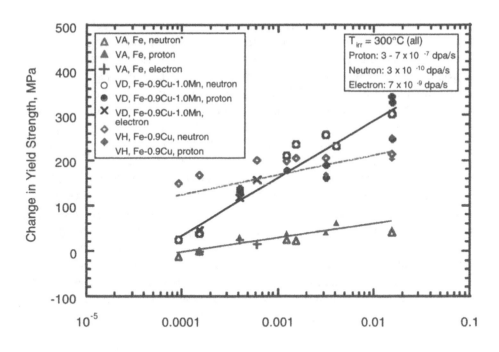

Figure 30. Irradiation hardening in model reactor pressure vessel steels following neutron, proton and electron irradiation at about 300°C. [28]

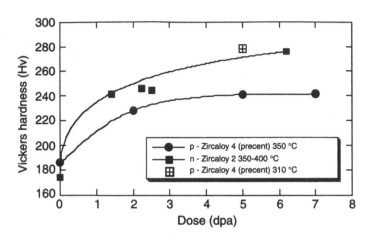

Figure 31. Hardening of Zircaloy 4 irradiated with 3 MeV protons at 310°C and comparison to neutron-irradiated Zircaloy 2 and Zircaloy 4. [29]

The next examples are from reactor pressure vessel steel and Zircaloy. Figure 30 shows an experiment on model reactor pressure vessel alloys in which the same heats were irradiated with neutrons, electrons or protons around 300°C to doses spanning two orders of magnitude. The alloys include an Fe heat (VA) that hardens very little under irradiation, an Fe-0.9Cu (VH) heat in which the initial hardening is rapid, followed by a slower rate, and a Fe-0.9Ce-1.0Mn alloy (VD) in which the hardening rate is greatest over the dose range studied. Despite the very different compositions and hardening rates, the results of the three types of irradiation agree extremely well. Figure 31 shows hardening for Zircaloy 2 and Zircaloy 4 irradiated with either neutrons or protons. Although the irradiations were not conducted on the same heats of material, nor using similar irradiation parameters, there is good agreement in the magnitude and dose dependence of hardening. These examples represent a comprehensive collection of comparison data between proton and neutron irradiation and taken together, serve as a good example of the capability for charged particles to emulate the effect of neutron irradiation on the alloy microstructure.

7. SUMMARY

Interest in the use of charged particles for irradiation effects studies has increased due to several factors; 1) the decline in the number and availability of test and research reactors, 2) the improvement in methods to improve the relevance of charged particle irradiation to neutron irradiation, and 3) the low cost and time relative to reactor irradiations. Radiation effects research has been conducted with a variety of energetic particles; neutrons, electrons, protons, He ions and numerous heavy ions. The damage state and microstructure resulting from irradiation depend upon the particle type. The degree to which irradiation by ions emulates neutron irradiation is also a function of the damage state and irradiated microstructure. As a result, each type of charged particle irradiation has both positive and negative factors. Overall, proton irradiation combines the advantages of relatively rapid irradiation, the accessibility of moderate doses and deep enough penetration to allow for some mechanical and stress corrosion property measurements. Extensive and recent research shows that light ion irradiation can effectively emulate most of the critical effects of irradiation (RIS, microstructure, hardening, SCC susceptibility, phase stability and stress relaxation) in a variety of relevant alloy systems; austenitic stainless steels and nickel base alloys, pressure vessel steels and Zircaloy. Consequently, charged particle irradiation provides a cost effective, rapid alternative to reactor irradiations to study the effects of neutron irradiation on materials.

8   REFERENCES

[1]   F. A. Garner, JNM 117 (1983) 177.

[2]   D. J. Mazey, JNM 174 (1990) 196.

[3]   "Standard Practice for Neutron Irradiation Damage Simulation by Charged Particle Irradiation," Designation E521-89, American Standards for Testing and Materials, Phila, 1989, p. D-9.

[4]   G. S. Was and P. L. Andresen, JOM 44 #4 (1992) 8.

[5]   P. L. Andresen, F. P. Ford, S. M. Murphy and J. M. Perks, Proc. Fourth Int'l Symp. on Environmental Degradation of Materials in Nuclear Power Systems - Water Reactors, National Association of Corrosion Engineers, Houston, 1990, p. 1-83.

[6]   P. L. Andresen, in *Stress Corrosion Cracking, Materials Performance and Evaluation*, R. H. Jones, ed., ASM International, Meals Park, OH, 1992, p. 181.

[7]   G. L. Kulcinski, J. L. Brimhall and H. E. Kissinger, in Radiation-Induced Voids in Metals, J. W. Corbett and L. C. Ianniello, eds, AEC Symposium Series, No. 26 (CONF-710601), 1972, p. 449.

[8]   G. H. Kinchin and R. S. Pease, Prog. Phys. 18 (1955) 1.

[9]   M. J. Norgett, M. T. Robinson and I. M. Torrens, Nucl. Eng. Des. 33 (1974) 50.

[10] H. L. Heinisch, M. L. Hamliton, W. F. Sommer, and P. Ferguson, J. Nucl. Mater. 191-194 (1992) 1177.

[11] R. S. Averback, JNM 216 (1994) 49.

[12] S. J. Zinkle and B. N. Singh, JNM 199 (1993) 173.

[13] V. Naundorf, J. Nucl. Mater. 182 (1991) 254.

[14] L. E. Rehn and H. Wiedersich in Surface Alloying by Ion, Electron and Laser Beams, L. E. Rehn, S. T. Picraux and H. Wiedersich, eds., American Society for Metals, Metals Park, OH, 1986, p. 137.

[15] T. Kato, H. Takahashi and M. Izumiya, JNM 189 (1992) 167.

[16] D. Damcott, D. Carter, J. Cookson, J. Martin, M. Atzmon and G. S. Was, Rad. Eff. Def. Sol. 118 (1991) 383.

[17] S. Bruemmer, JNM 186 (1991) 13.

[18] A. J. Jacobs and G. P. Wozadlo, Corrosion 91, National Association of Corrosion Engineers, Houston, TX, 1991, paper 41.

[19] G.S. Was and T. Allen, J. Nucl. Mater, 205 (1993) 332-338.

[20] P. R. Okamoto and L. E. Rehn, JNM 83 (1979) 2.

[21] L. K. Mansur, J. Nucl. Mater. 216 (1994) 97.

[22] T. Ezawa and E. Wakai, Ultramicroscopy 39 (1991) 187.

[23] J. A. Ashworth, D. I. R. Norris and I. P. Jones, JNM 189 (1992) 289.

[24] Wakai, Trans. JIM 33 (10) (1992) 884.

[25] J. B. Whitley, Thesis for Doctor of Philosophy - Nuclear Engineering, University of Wisconsin-Madison (1978).

[26] G. S. Was, J. T. Busby, T. Allen, E. A. Kenik, A. Jenssen, S. M. Bruemmer, J. Gan, A. D. Edwards, P. Scott and P. L. Andresen, J. Nucl. Mater., 300 (2002) 198-216.

[27] B.H. Sencer, G.S. Was, H. Yuya, Y. Isobe, M. Sagisaka and F.A. Garner, J. Nucl. Mater., in press.

[28] G. S. Was, M. Hash and G. R. Odette, Phil. Mag. A, in press.

[29] M. Atzmon, L.M. Wang and G. S. Was, Phil. Mag. A, in press.

[30] B. Wirth, private communication, July, 2004.

Chapter 5

HIGH DOSE RADIATION EFFECTS IN STEELS

Todd R. Allen
University of Wisconsin, Madison, WI 53706-1687 USA

1   INTRODUCTION

In this chapter, two examples of high-dose radiation effects in steels are described, specifically void swelling, and stress relaxation. For each example, the importance of the phenomena to reactor materials is presented and the effect is explained in reference to other lectures chapters in this series.

2   VOID SWELLING

Void swelling is the isotropic volume increase that occurs due to the accumulation of radiation-induced vacancies into stable cavities. Void swelling can cause unacceptable dimensional changes as well as unacceptable embrittlement in reactor structural materials. After void swelling was discovered in irradiated austenitic stainless steels [1], many studies were undertaken to determine not only the magnitude of the swelling problem but also the variables that influenced swelling. A comprehensive review of void swelling data is contained in reference [2]. A detailed description of the theory of void swelling is found in reference [3]. A classic example of void swelling is shown in figure 1.

Void swelling occurs because the point defects created by radiation are absorbed preferentially at different sinks (see figure 2). Specifically, the stress fields at dislocations lead to preferential absorption of interstitials. This leaves an excess of vacancies that cluster together to form voids. These voids can lead to two deleterious effects. First, they lead to a net increase in volume. For reactor components that are required to have a fixed geometry (e.g., heat transfer channels) or that are meant to be retrievable (e.g., fuel rods), excessive volumetric increase will be unacceptable. Second, at high density of voids, typically that corresponding

K.E. Sickafus et al. (eds.), Radiation Effects in Solids, 99–121.
© 2007 *Springer.*

to roughly 10% volumetric swelling, materials can become brittle with the void network providing a weak path for crack propagation.

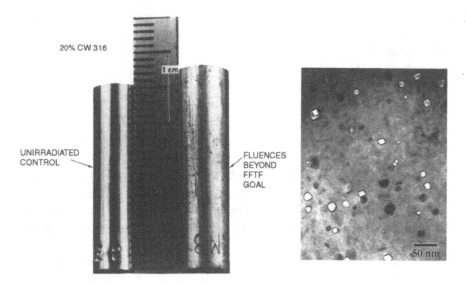

Figure 1. Easily observed void swelling in 20% cold-worked AISI 316 stainless steel (left) [2] and example of voids in 316 stainless steel (right).

2.1 Dependence on Irradiation and Material Parameters

Swelling is of concern at intermediate temperatures (roughly 0.3-0.5 of the melting temperature) (figure 3). At low temperatures, defects are not sufficiently mobile to diffuse and cluster into cavities. At high temperatures, cavities are not likely to form as the metal can thermodynamically support large concentrations of vacancies.

The typical dose dependence of swelling is shown in figure 4. The swelling rate slowly increases with increasing total dose until reaching a terminal swelling rate. The terminal swelling rate in austenitic stainless steels is roughly 1%/dpa. For ferritic-martensitic steels, the terminal swelling rate is roughly 0.2%/dpa. Swelling is largest for a particular material at a specific dose when the transient period is the shortest. As will be explained later, the development of swelling resistant materials all aimed at lengthening this transient period.

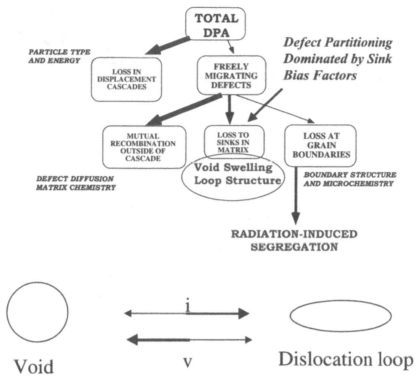

Figure 2. Partitioning of defects to sinks, with preferential absorption of interstitials at dislocations and vacancies at voids leads to swelling (defect partitioning flow chart courtesy of S. Bruemmer, Pacific Northwest National Laboratory).

The affect of temperature on the transient period can be seen by examining the swelling data from cold-worked 316 stainless steel (figure 5). This particular steel was irradiated at temperatures from 370-444°C to maximum dose of around 55 dpa. The swelling data can be fit to a power law description ($\frac{\Delta V}{V} = A\Phi^n$), where Φ is the total dose in dpa and n is an exponent that describes the rate of increase of the swelling. A larger n describes a shorter transient period). For the higher temperature data in figure 5, the exponent is around 3.3 while for the lower temperature data, the exponent is 1.6. At higher temperatures the swelling develops faster.

Swelling is also influenced by the rate at which the radiation damage occurs. Figure 6 shows the swelling data for 304 stainless steel irradiated at temperatures from 370-450°C [12]. The data came from material irradiated in different rows of the EBR-II reactor. The outer rows (higher row number) experienced radiation at a smaller rate. Those samples irradiated at a lower rate have a shorter transient and thus a greater swelling at any specific dose.

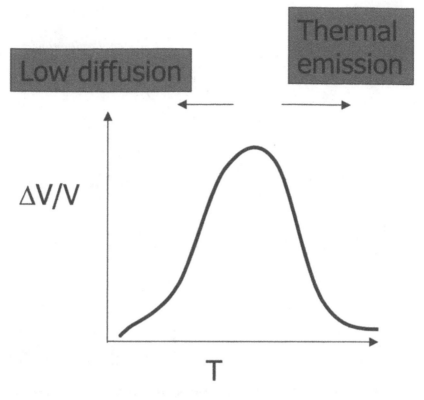

Most metals swell in temperature range of $0.3\ T_m < T < 0.55\ T_m$

Figure 3. Temperature dependence of swelling.

The swelling response is also a function of the specific alloy composition. Since early fast reactor cladding was made of austenitic stainless steels, most understanding of swelling derives from studies of austenitic stainless steels. An example of the affect of alloy composition on swelling can be seen in figure 7, which shows that increasing the bulk nickel composition in an alloy significantly decreases the swelling. Details of the effect of alloying elements on swelling can be found in [2]. As noted above, the terminal swelling rate for ferritic-martensitic steels is smaller than for austenitic stainless steels. For this reason, ferritic-martensitic steels are now the principal choice for high dose components.

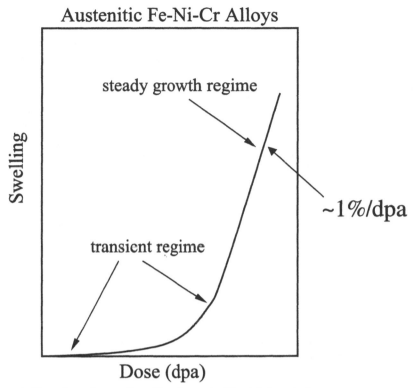

Figure 4. Dose dependence of swelling for austenitic stainless steels.

Figure 8 demonstrates the historical steps made to improve swelling resistance. 304 stainless steel was replaced by 316 stainless steel as a compositional modification that increased the period of transient swelling. The main difference between the two steels was the increased nickel and the addition of molybdenum in 316. The next step was to cold-work the 316. The increased number of point defect sinks reduced the point defects available to an individual void and thus reduced swelling. The shift from cold-worked 316 to HT9 was a shift from austenitic stainless steels to ferritic-martensitic steels. Steels with a BCC structure have longer transients and a smaller terminal swelling rate. Not shown on figure 8 was the development of D9, an optimized swelling resistant austenitic stainless steel that was used between CW316 and HT9. D9 was optimized by composition to minimize swelling. The details of the effect of all the minor alloying elements that went into the establishment of D9 can be found in reference [2].

Radiation Effects in Solids

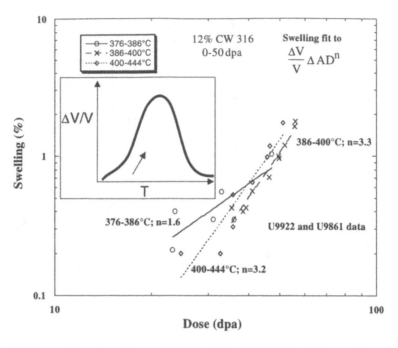

Figure 5. Temperature dependence of 12% cold-worked 316 stainless steel. At lower temperatures, the transient period is longer as described by the lower power law exponent.

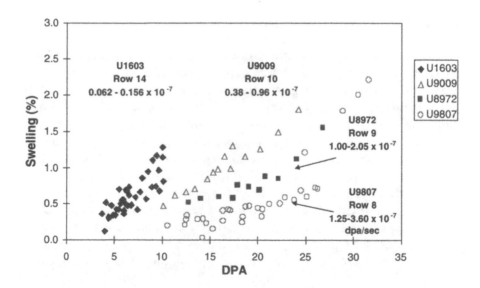

Figure 6. The affect of dose rate on swelling (figure courtesy F. Garner, Pacific Northwest National Laboratory). For temperatures in the range of 370-450°C, the transient period for swelling is shorter for lower dose rate irradiation.

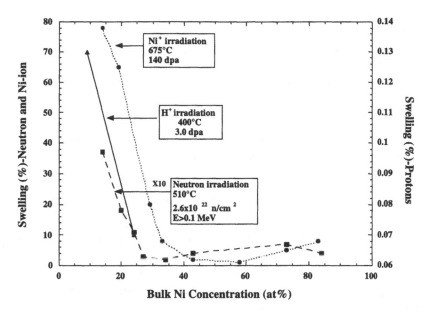

Figure 7. The effect of bulk nickel concentration on austenitic steels with Cr=16-20 at%.

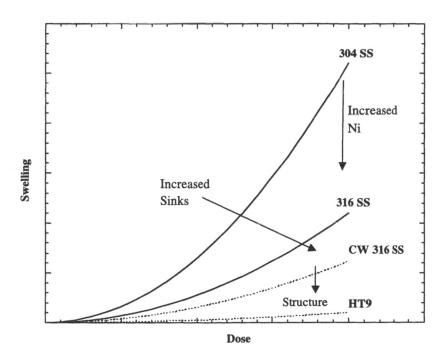

Figure 8. Techniques for improving swelling resistance.

2.2   Swelling Theory

The growth rate of a void is given by eq. (1)

$$\frac{dr_c}{dt} = \frac{\Omega}{r_c}\left[Z_v^c D_v C_v - Z_i^c D_i C_i - Z_v^c D_v C_v^e(r_c)\right] \qquad (1)$$

where r_c is the cavity (void) radius, Ω is the atomic volume, Z is the cavity bias for vacancies Z_v^c or interstitials Z_i^c, D_v is the vacancy diffusion coefficient, D_i is the interstitial diffusion coefficient, C_v is the vacancy concentration, C_i is the interstitial concentration, and C_v^e is the thermal equilibrium vacancy concentration [3]. For a void to grow, the arrival of vacancies to the void must exceed the arrival rate of interstitials plus the rate of emission of vacancies from the void. Using the steady-state vacancy and interstitial concentrations calculated from the point defect kinetics equations (see the chapter on RIS and RED in this series), the void growth rate can be expressed as follows.

$$\frac{dr_c}{dt} = \frac{\Omega K_0 Q_i Q_v \left(Z_i^d Z_v^c - Z_v^d Z_i^c\right)}{r_c Z_i^d Z_v^d \rho_d \left(1+Q_i\right)\left(1+Q_v\right)} \qquad (2)$$

for the case where sinks dominate point defect loss or

$$\frac{dr_c}{dt} = \frac{\Omega}{r_c}\left(\frac{D_i D_v K_0}{Z_i^d Z_v^d R}\right)^{\frac{1}{2}} \frac{Q_i^{\frac{1}{2}} Q_v^{\frac{1}{2}}\left(Z_i^d Z_v^c - Z_v^d Z_i^c\right)}{\left(1+Q_i\right)\left(1+Q_v\right)} \qquad (3)$$

for the case where recombination dominates point defect loss. The Q terms are given by

$$Q_{i,v} = \frac{Z_{i,v}^d \rho_d}{4\pi r_c N_c Z_{i,v}^c}$$

K_0 is the production rate of point defects, D_i and D_v are the diffusion coefficients of vacancies and interstitials, Ω is the atomic volume, R is the vacancy-interstitial recombination coefficient, N_c is the cavity density, r_c is the cavity radius, and r_d is the dislocation density.

In either case, for swelling to occur,

$$\frac{Z_v^c}{Z_i^c} > \frac{Z_v^d}{Z_i^d} \tag{4}$$

the void must have a net bias for vacancies as compared to other defect sinks (d denotes dislocation in eq. (4)). The key features of eq. (2) and (3) are to note that the void growth rate depends on

1. Sink strength r_d
2. The point defect creation rate K_0
3. Point defect diffusion coefficients D_i and D_v
4. Point defect bias Z

In the following section, the experimental database will be examined to show that the point defect bias and dislocation density are the critical factors in driving swelling.

2.3 Influence of Material and Experimental Variables on Swelling

2.3.1 Sink Strength

As mentioned earlier, one of the methods used to limit swelling was to cold-work 316 stainless steel. The addition of a large number of dislocations from the cold-working reduced the swelling rate. This effect can be explored by looking at eq. (2) for the case of a very large sink density ($Q \gg 1$). In this case, eq. (2) simplifies to

$$\frac{dr_c}{dt} \approx \frac{\Omega K}{r_c} \frac{\left(Z_i^d Z_v^c - Z_v^d Z_i^c\right)}{Z_i^d Z_v^d} \frac{1}{\rho_d}. \tag{5}$$

As the dislocation density increases, the void growth rate decreases.

The development of sinks over the course of the radiation also affects void growth. For very low sink density with $Q_{i,v} \gg 1$ (corresponding to an annealed material in the recombination dominant region with dislocations but no voids), equation (3) can be simplified to show that

$$\frac{dr_c}{dt} \propto \left(Z_i^d Z_v^c - Z_v^d Z_i^c \right) \frac{\rho_d}{\left(4\pi r_c N_c \right)^2} \qquad (6)$$

Okita et al. have recently preformed detailed microscopy on Fe-15Cr-16Ni samples irradiated in FFTF at temperatures from 408-444°C to doses of 70 dpa [4]. The samples in the Okita study were annealed prior to irradiation. Figure 9 shows $\dfrac{\rho_d}{\left(4\pi r_c N_c \right)^2}$ plotted as a function of dose. As the cavity density N_c increases relative to the dislocation density r_d with dose, the void growth rate should decrease. Each individual void competes for the fixed number of available vacancies with all the other developing voids and defect sinks in the system. Figure 10, adapted from [4], shows that the void growth rate does decrease over this same dose range.

Okita et al. also preformed detailed microscopy. Figure 11 plots the dislocation density as a function of dose from Okita's experiment, with the data points grouped by similar displacement rate. Those samples irradiated at lower displacement rates have a higher dislocation loop density for a given dose. Recall from the data in figure 6 that greater swelling occurs at lower displacement rate. The dependence of swelling on displacement rate correlates with the development of dislocations as predicted by eq. (6). Bias of dislocations for interstitials must exist for swelling to occur and changing the dislocation concentration can drive differences in swelling behavior.

2.3.2   Gas Loading

During irradiation, transmutation gasses such as helium can be formed. These gasses can diffuse to and be trapped in voids. The addition of gas to a void can stabilize the void, making void growth easier. The theory of gas loading on critical void size (the size above which a void nuclei is stable and will grow), is described by Mansur [3]. A figure from his work is shown in figure 12. The figure was derived by solving eq. (1) and including gas loading in the thermal vacancy emission term. The equation for the thermal emission is

$$C_v^e \left(r_c \right) = C_v^0 \left(r_c \right) \exp \left[\left(\frac{2\gamma}{r_c} - P \right) \frac{\Omega}{kT} \right] \qquad (7)$$

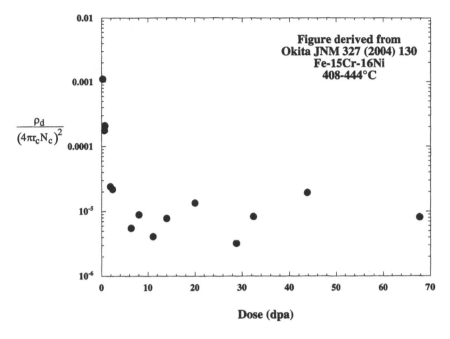

Figure 9. Changing sink density as a function of dose. The ratio decreases rapidly during the first few dpa. Adapted from [4] .

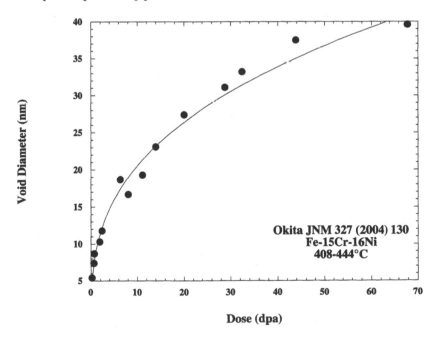

Figure 10. Void diameter as a function of dose. The void growth rate decreases rapidly during the first few dpa. Adapted from [4].

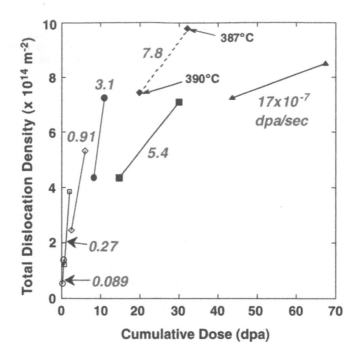

Figure 11. Dislocation density as a function of dose [4].

where γ is the void surface tension and P the gas pressure. The solution has two roots, as shown in Figure 12. The larger radius solution is the radius above which a void is stable. The addition of a gas pressure decreases the critical radius. With sufficient gas pressure, all voids are stable.

2.3.3   Point Defect Creation Rate

Both eq. (2) and (3) predict that increasing the point defect creation rate should increase the void growth rate. As shown in figure 6, the exact opposite happens. Those materials irradiated at lower dose have a greater swelling. The effect of point defect creation on void growth rate is not the primary driving force. As noted in discussing the Okita study, the dislocation concentration and bias are more important.

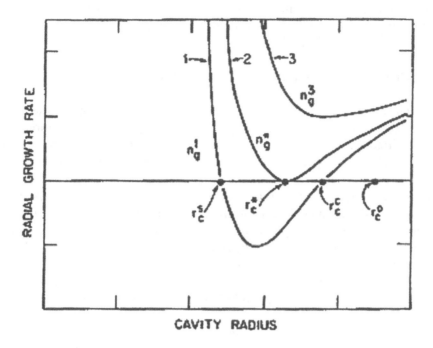

Figure 12. The predicted effect of gas loading on critical radius for void growth [3].

Point Defect Diffusion

Equations (2) and (3) predict that increasing the vacancy and interstitial diffusion coefficients should either have no effect (sink dominated case eq. (2)) or increase the void growth rate (recombination dominated case eq. (3)). Figure 7 showed that swelling decreased strongly at higher nickel concentration. For higher nickel concentration, the point defect diffusion coefficients increase. Equation (3) predicts greater void growth at higher diffusion coefficient (higher nickel concentration). This is not what happens. The effect of diffusion on void growth rate is not the primary driving force, the dislocation concentration and bias are more important.

The effect of concentration on swelling was also shown in an experiment at Oak Ridge National Laboratory. In the experiment, increasing the bulk nickel content from Fe-15Cr-15Ni to Fe-15Cr-35Ni decreased swelling by increasing the critical void radius above which void growth was energetically favorable [5]. The increase in critical radius corresponded with a change in dislocation microstructure, with higher bulk nickel concentration leading to a larger concentration of faulted loops, which are weaker sinks for vacancies than dislocation networks. The

ORNL study, consistent with the Okita study, showed that the dislocation structure and the bias are the important drivers for swelling.

2.3.4　Bias and Radiation-induced Segregation

Having established that bias is critical for swelling to occur, the next question to discuss is how bias changes for different alloy compositions or different irradiation conditions. A primary driver for changes in bias is radiation-induced segregation at void surfaces (radiation-induced segregation is described in a separate chapter). As noted when discussing figure 6, bulk composition can dramatically affect swelling, in this instance the influence of bulk nickel concentration. Figure 13 shows the ratio of the radiation-induced chromium depletion to nickel enrichment as a function of alloy composition. The trends in this ratio of segregation are very similar to those of swelling (figure 7). For alloys with a small chromium-to-nickel depletion ratio (large nickel enrichment compared to chromium depletion), swelling is smaller. A relationship between swelling and RIS is suggested by the data.

Figure 14 shows the radiation-induced segregation as a function of dose for alloys with similar bulk chromium concentration but varying bulk nickel. For those alloys with lower bulk nickel, the segregation develops at a much slower rate. These alloys with lower bulk nickel, which swell the most, develop changes in concentration near the void surface at much slower rates.

Wolfer and colleagues [6-9] predicted that segregation affects void growth through changes in elastic properties. Wolfer et al. showed that a compositional change which increases the shear modulus or lattice parameter locally around a void embryo causes the void to become a preferential sink for vacancies, thus increasing the rate at which embryos mature to voids. Wolfer's theory predicts that increasing the shear modulus or lattice parameter locally around a void embryo presents an energy barrier to interstitial motion toward the void. A discrete segregation shell is not required around a void, but the average values in the segregation gradient are adequate to describe the energetics of vacancy motion toward the void.

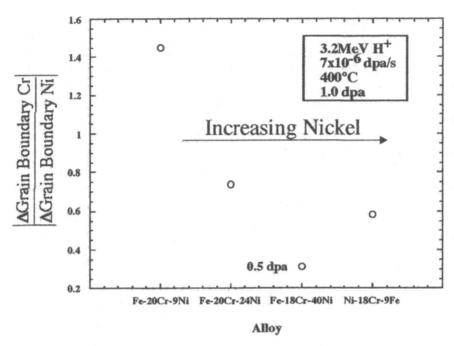

Figure 13. Ratio of the change in grain boundary chromium depletion to nickel enrichment as a function of bulk nickel concentration.

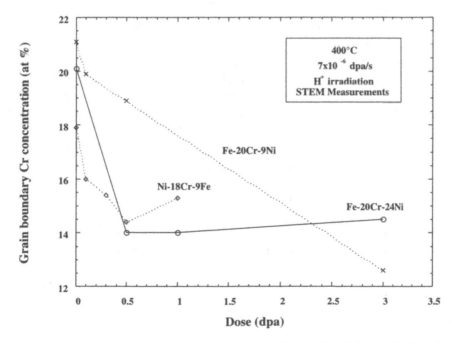

Figure 14. Segregation as a function of dose for alloys with similar bulk chromium concentration but varying bulk nickel concentration. Data from [26-27].

The segregation shell also affects the void nucleation rate. A segregation shell that increases the shear modulus or lattice parameter locally around a void embryo reduces the nucleation barrier significantly, increasing the probability that the cavity becomes a stable void. In fact, the reduction in the nucleation barrier for a given increase in the shear modulus or lattice parameter is larger in magnitude than the increase in nucleation barrier for a decrease in the shear modulus or lattice parameter of the same magnitude. The predicted effect of changing lattice parameter was more sensitive than the effect of changing shear modulus. An increase of only 0.2% in lattice parameter could significantly increase the void nucleation rate (by roughly six orders of magnitude) while a 1% increase in shear modulus only caused a predicted increase of three orders of magnitude.

For Fe-Cr-Ni alloys with compositions near 304/316 stainless steel, the lattice parameter and shear moduli increase with increasing Cr concentration and decrease with increasing Ni concentration [6]. For such alloys, RIS causes Cr to deplete and Ni to enrich around a void during irradiation. Figures 15 and 16 are ternary diagrams for the lattice parameter and shear modulus, respectively, as a function of composition for the Fe-Cr-Ni system. Superimposed on the figure are the shifts in concentration that occur near grain boundaries for an Fe-18Cr-8Ni alloy and an Fe-18Cr-40Ni alloy. For both alloys, the segregation shifts the grain boundary and void shell composition toward regions of smaller lattice parameter and lower shear modulus.

Figure 17 provides a schematic for a segregation profile at a void surface. To conserve mass, the chromium depletion and nickel enrichment at the void surface must be balanced by chromium enrichment and nickel depletion further into the bulk. Therefore, two regions surround the void. One region, at the void surface, has a composition that retards void formation and a second region, further into the bulk, which enhances void formation.

Swelling and segregation data were measured in three different alloy sets irradiated with high-energy protons at 400°C to doses of 0.5-1.0 dpa. The data can be examined to examine the interplay between segregation and swelling, in the context of Wolfer's theory. Figure 17 plots the swelling as a function of decrease in lattice parameter using all 8 alloys in the data set. The decrease in lattice parameter was extracted from figure 15 using the bulk composition to determine the starting lattice parameter.

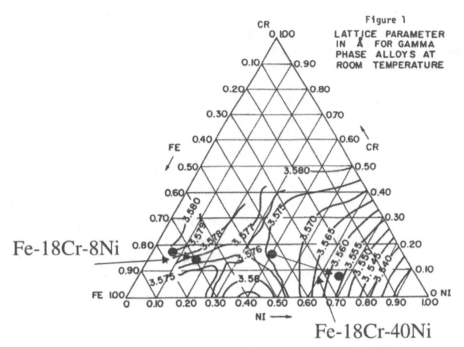

Figure 15. Lattice parameter as a function of composition in Fe-Cr-Ni alloys (taken from [9]).

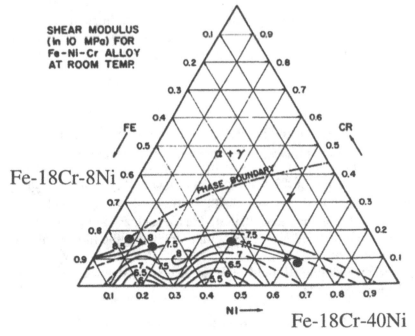

Figure 16. Shear modulus as a function of composition in Fe-Cr-Ni alloys (taken from [9]).

As can be seen in figure 18, those alloys with large decreases in lattice parameter near the boundary resist swelling. This is true regardless across alloys. A similar attempt to correlate swelling with change in shear modulus does not provide a recognizable correlation. As noted above, swelling is expected to be more sensitive to changes in lattice parameter as is apparently seen in this data set.

The second shell, further away from the boundary, is predicted to have a greater effect on swelling because of a greater sensitivity to increases in lattice parameter and shear modulus. The "second shell," although at times distinguishable, is not always experimentally evident. Although the second shell must exist to conserve mass, in many instances it is below the sensitivity of measurements. The small changes in void bias predicted for the large nickel enrichment and chromium depletion near the boundary appear to be more important than the large changes in void bias predicted for the nickel depletion and chromium enrichment in the second shell. This may be due to the "second shell" being broad with compositions close to the bulk.

3   STRESS RELAXATION

Radiation can cause creep to occur at temperatures lower than the thermal creep regime. Extensive reviews of radiation-induced creep can be found in [3, 10-11]. Although multiple mechanisms for radiation-induced creep are possible, each mechanism predicts that the creep rate is proportional to the radiation dose rate and the applied stress.

$$\dot{\varepsilon} = B\dot{\phi}\sigma \tag{8}$$

where $\dot{\varepsilon}$ is the effective strain rate, $\dot{\phi}$ is the displacement damage rate, σ is the effective stress, and B is the temperature- and material-dependent creep coefficient. Equation (8) is only valid at lower doses where swelling is negligible. At higher doses, the creep rate is affected by the swelling. A detailed explanation of the relationship between creep and swelling can be found in [2]. For components that are loaded such as springs and bolts, this creep can relax the load (stress relaxation), reducing the functionality of the loaded component.

When equation (8) holds, the residual stress decreases exponentially with displacement dose:

$$\frac{\sigma}{\sigma_o} = \exp\left(-BE\dot{\phi}t\right) \tag{9}$$

where E is the elastic modulus. A plot of $\ln\dfrac{\sigma}{\sigma_0}$ versus dose ($\dot{\phi}t$ gives a steady state slope equal to $-BE$. If the plot of $\ln\dfrac{\sigma}{\sigma_0}$ versus dose f is not linear at high dose, then the radiation-induced creep strain is no longer simply proportional to irradiation dose rate (an example is the effect of swelling on creep).

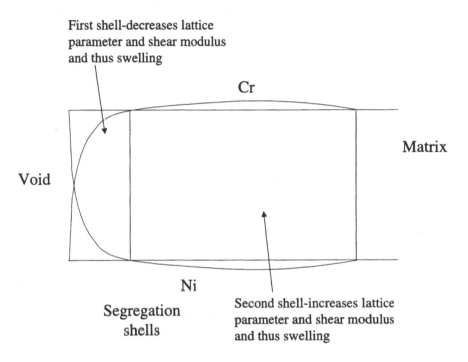

First shell-decreases lattice parameter and shear modulus and thus swelling

Cr

Matrix

Void

Ni

Segregation shells

Second shell-increases lattice parameter and shear modulus and thus swelling

Figure 17. General schematic of radiation-induced segregation near a void surface.

Springs, the design of which are shown in figure 19, were compressed by a sleeve and then irradiated at 375-415°C, at dose rates ranging from about 3×10^{-8} to 7×10^{-8} dpa/s, and to doses up to 20 dpa. Springs with three different nominal pre-loads were used, with magnitudes 10.7, 23, and 36.5 N. The length of the sleeve used to compress the spring determined the pre-load. The force versus deflection of each spring is measured. From the plot of force versus deflection, the slope (m) and intercept (b) are determined. The residual stress (s) for each spring is determined from the value of the force at zero deflection in the measured force versus deflection curve. The stress relaxation is determined by dividing the residual stress in the irradiated or aged samples by the residual stress in the unirradiated or unaged samples (σ_0). Figure 20 shows a plot of stress relaxation (defined as $\dfrac{\sigma}{\sigma_0}$) as a function of displacement dose.

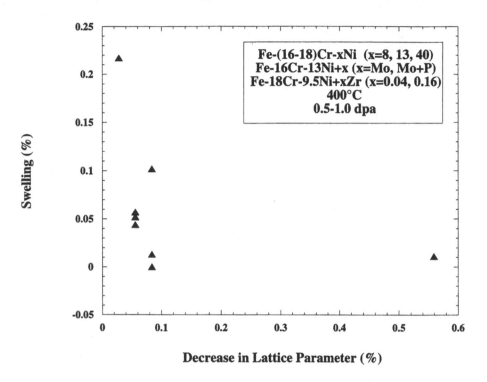

Figure 18. Decrease in swelling as a function of changing lattice parameter caused by radiation-induced segregation. A decrease in the lattice parameter indicates smaller lattice parameter at the boundary surface.

Although the scatter is larger in the springs irradiated to larger dose, the linear relationship between $\ln\dfrac{\sigma}{\sigma_0}$ and dose ϕ persists to the largest dose measured. Therefore, the proportional relationship between strain rate, stress, and dose rate of equation (8) remains true and any effect of swelling is negligible (measured swelling is <0.1% for X750 irradiated to 20 dpa at 375-415°C). Extrapolating the exponential fit of figure 20, the residual stress reaches 10% at around 20 dpa. The stress relaxation is irradiation-induced. A similar set of springs were thermally aged at 371°C for 6525 days (roughly 18 years) and no relaxation was noted, figure 21.

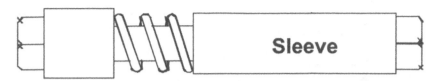

Figure 19. Inconel X750 spring design. The length of the sleeve determines the initial compression force.

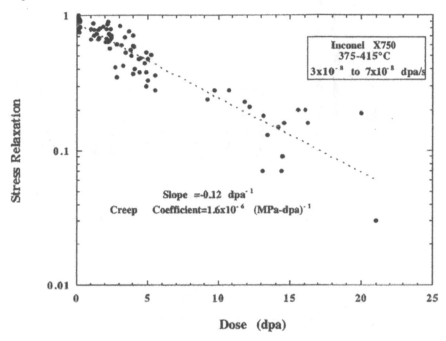

Figure 20. Stress relaxation as a function of irradiation dose for Inconel X750 springs.

4   SUMMARY

Void swelling is an irradiation-induced dimensional change in materials irradiated at high dose at temperatures near 30-50% of their melting temperature. For voids to grow, they must preferential absorb vacancies (simultaneously interstitials are preferentially absorbed at dislocations). Swelling is affects by temperature, dose, dose rate, and material composition. The major microstructural factors that determine a specific materials susceptibility to swelling are the total sink density and the net bias of voids for vacancies. This net bias is affected by radiation-induced segregation at void surfaces.

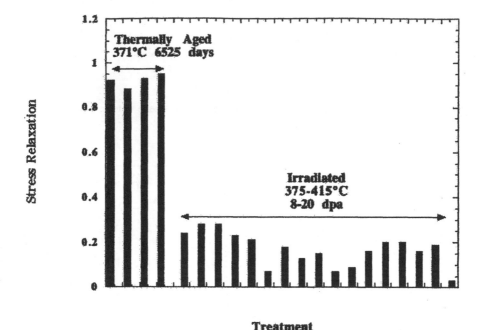

Figure 21. The stress relaxation is a function of the irradiation and not time at temperature.

Radiation-induced creep can caused loaded components to relax their stress. For lower doses where swelling is negligible, the stress relaxation follows exponential behavior. This irradiation-induced stress relaxation occurs at temperatures where thermal creep is negligible.

REFERENCES

[1]   Cawthorne, C. and Fulton, J. E., Nature, Vol. **216**, 1967 p. 515.

[2]   Garner, F. A. Irradiation Performance of Cladding and Structural Steels in Liquid Metal Reactors, in Materials Science and Technology, A Comprehensive Treatment, Vol. **10A** Nuclear Materials, Eds., R. W. Cahn, P. Haasen, and E. J. Kramer, (VCH Weinheim, 1994).

[3]   Mansur, L. K., J, Nucl. Mater. **216** (1994) 97.

[4]   Okita, T. and Wolfer, W. G., J. Nucl. Mater. **327** (2004) 130.

[5]   Lee, E. H. and Mansur, L. K., Phil. Mag., **52** (4) (1985) 493.

[6] Si-Ahmed, A. and Wolfer, W. G., Effects of Radiation on Materials: Eleventh Conference, ASTM STP 782, H. R. Brager and J.S. Perrin Eds., American Society for Testing and Materials, 1982, p.1008.

[7   Wolfer, W. G. and Mansur, L. K., J. Nucl. Mater. **91** (1980) 265.

[8] Sniegowski, J. J. and Wolfer, W. G, in: Proc. of Topical Conference on Ferritic Alloys for Use in Nuclear Energy Technologies, Snowbird, Utah, 19-23 June 1983.

[9] Wolfer, W. G, Garner, F. A, and Thomas, L. E., Effects of Radiation on Materials: Eleventh Conference, ASTM STP 782, H. R. Brager and J.S. Perrin Eds., American Society for Testing and Materials, 1982 p.1023.

[10] Matthews, J. R. and Finnis, M. W., J. Nucl. Mater. **159** (1988) 257.

[11] Simonen, E. P., Met. Trans. A **21A** (1990) 1053.

[12] Garner, F. A, private communication.

Chapter 6

RADIATION-ENHANCED DIFFUSION AND RADIATION-INDUCED SEGREGATION

Todd R. Allen [1] and Gary S. Was [2]
[1] *University of Wisconsin, Madison, WI 53706*
[2] *University of Michigan, Ann Arbor, MI 48109*

1   INTRODUCTION

In this chapter, the basic point defect kinetic equations and the solutions in different temperature and microstructural regimes are presented. The transient and steady-state solutions to the point defect kinetic equation are then described. Equations for radiation-induced segregation (RIS) are developed by adding diffusion terms to the point defect kinetic equations and allowing for multiple constituents in an alloy. The solution to the RIS equations are developed and segregation in austenitic Fe-Cr-Ni alloys is described in detail to provide insight into segregation behavior. Finally, a simple model that includes composition dependent diffusion parameters is described. This composition-dependent model improved the ability of RIS models to predict segregation in Fe-Cr-Ni alloys.

2   POINT DEFECT KINETIC EQUATIONS

The development of radiation-induced vacancy and interstitial concentrations occurs due to competing processes. Frenkel defects (isolated vacancy-interstitial pairs) are created from the collisions between high-energy particles and lattice atoms (details can be found in the chapter on radiation technique). These defects can be lost either through recombination of vacancies and interstitials or loss to a defect sink (void, dislocation, dislocation loop, or precipitate). These competing processes can be mathematically described in the manner shown in eq. (1) (the description

K.E. Sickafus et al. (eds.), Radiation Effects in Solids, 123–151.
© *2007 Springer.*

that follows is adapted from Ref [1] and is a simplification that does not consider point defect clusters).

$$\frac{dC_v}{dt} = K_0 - K_{iv}C_iC_v - K_{vs}C_vC_s \qquad (1a)$$

$$\frac{dC_i}{dt} = K_0 - K_{iv}C_iC_v - K_{is}C_iC_s \qquad (1b)$$

C_v=vacancy concentration
C_i=interstitial concentration
K_0= defect production rate
K_{iv}= vacancy-interstitial recombination rate coefficient
K_{vs}= vacancy-sink recombination rate coefficient
K_{is}= interstitial-sink recombination rate coefficient

Equation (1) assumes a homogeneous system with uniformly distributed sinks such that diffusion can be ignored (no net loss of point defects out of a specific volume due to diffusion). The equations also neglect thermal equilibrium vacancy concentrations. The rate coefficients are given by

$$K_{iv} = \frac{4\pi r_{iv}(D_i + D_v)}{\Omega} \approx \frac{4\pi r_{iv}D_i}{\Omega} \qquad (2a)$$

$$K_{is} = \frac{4\pi r_{is}D_i}{\Omega} \qquad (2b)$$

$$K_{vs} = \frac{4\pi r_{vs}Dv}{\Omega} \qquad (2c)$$

r_{iv}= vacancy-interstitial interaction radius
r_{vs}= vacancy-sink interaction radius
r_{is}= interstitial-sink interaction radius
D_i=interstitial diffusion coefficient
D_v=vacancy diffusion coefficient
Ω=atomic volume

D_i>>D_v because the interstitial migration energy is much less than the vacancy migration energy.
Equations (1a) and (1b), can be solved subject to the following limitations.

1. The material is a pure metal with a single D_i and single D_v.
2. The sink concentration and strength are time-independent
3. Other than point defect interactions with each other, defect-defect interactions are ignored (e.g., no void-void interactions are considered)
4. Bias factors for diffusion of defects to sinks are set to unity (no preferential absorption of specific point defects at specific sinks)
5. Diffusion terms in and out of a specific volume are not considered
6. Thermal equilibrium vacancy concentration is neglected

The solution to these equations and a description of the build-up of point defects is summarized in the next section.

2.1   Low Temperature, Low Sink Density

The approximate solutions to equations (1a), (1b) for low temperature and low sink density are given in figure 1. Initially, defect concentrations build up according to $dC/dt = K_o$ with $C_i \sim C_v$, so $C_i = C_v = C = K_o t$. The initial concentrations are too low for either recombination or sinks to have an effect. The buildup of point defects will start to level off when the production rate is compensated by recombination rate. Quasi-steady state concentrations are given by:

$$\frac{dC}{dt} = K_o - K_{iv}C^2 = 0 \qquad (C_i = C_v)$$

$$\therefore \ C = \left(\frac{K_o}{K_{iv}}\right)^{1/2}$$

Equating this concentration with that during buildup gives the time at which losses to recombination compensate for the production rate from radiation.

$$K_o t = \left(\frac{K_o}{K_{iv}}\right)^{1/2} \quad,$$

$$t = \tau_1 = (K_o K_{iv})^{-1/2}$$

where t is a time constant or characteristic time for the onset of mutual recombination.

Eventually, the vacancies and interstitials will start to be lost at sinks, with the sinks contributing to annihilation. Because $D_i > D_v$, more interstitials are lost to sinks than vacancies and vacancies and interstitials buildup and decay (respectively) according to:

$$C_v(t) = \left[\frac{K_0 K_{is} C_s t}{K_{iv}}\right]^{1/2}$$

$$C_i(t) = \left[\frac{K_0}{K_{iv} K_{is} C_s t}\right]^{1/2}.$$

The time at which this occurs is given by:

$$C_v = \left(\frac{K_0}{K_{iv}}\right)^{1/2} = \left[\frac{K_0 K_{is} C_s t}{K_{iv}}\right]^{1/2}$$

$$C_i = \left(\frac{K_0}{K_{iv}}\right)^{1/2} = \left[\frac{K_0}{K_{iv} K_{is} C_s t}\right]^{1/2}$$

$$t = \tau_2 = (K_{is} C_s)^{-1}$$

After a while, at time t_3, true steady state will be achieved. t_3 is the time constant of the slowest process, the interaction of vacancies with sinks. Solving for the steady state concentration of vacancies and interstitials by setting $dC_v/dt = dC_i/dt = 0$ gives:

$$C_v^{ss} = -\frac{K_{is} C_s}{2K_{iv}} + \left[\frac{K_0 K_{is}}{K_{iv} K_{vs}} + \frac{K_{is}^2 C_s^2}{4K_{iv}^2}\right]^{1/2}. \tag{3a}$$

$$C_i^{ss} = -\frac{K_{vs} C_s}{2K_{iv}} + \left[\frac{K_0 K_{vs}}{K_{iv} K_{is}} + \frac{K_{vs}^2 C_s^2}{4K_{iv}^2}\right]^{1/2} \tag{3b}$$

For the case of low temperature and low sink density, C_s is small

$$C_v^{ss} \cong \sqrt{\frac{K_0 K_{is}}{K_{iv} K_{vs}}} \quad ; \quad C_i^{ss} \cong \sqrt{\frac{K_0 K_{vs}}{K_{iv} K_{is}}}.$$

Equating these expressions to those from the previous region (buildup) gives the time constant for the buildup process.

$$C_v = \left[\frac{K_0 K_{is} C_s t}{K_{iv}}\right]^{1/2} = \left[\frac{K_0 K_{is}}{K_{iv} K_{iv}}\right]^{1/2}.$$

$$t = \tau_3 = (K_{vs} C_s)^{-1}$$

The buildup shown in figure 1 is really a schematic and not the actual buildup. The transitions between regimes are not so sudden. For example, if the sink density is assumed to be zero, the exact solution to eq. (1a) is:

$$C_v(t) = \sqrt{\frac{K_0}{K_{iv}}} \tanh\left(\sqrt{K_{iv} K_0}\, t\right) \tag{4}$$

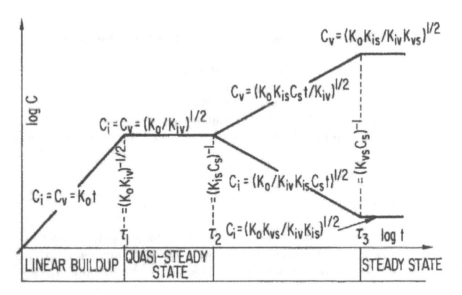

Figure 1. Buildup of radiation-induced vacancy and interstitial concentrations for low temperature, low sink density conditions.

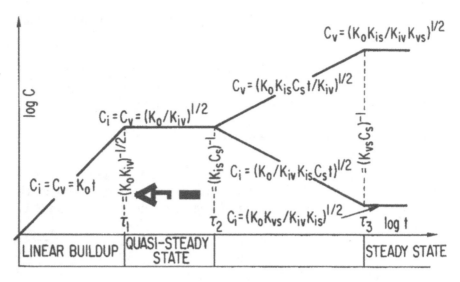

Figure 2. Buildup of radiation-induced vacancy and interstitial concentrations for low temperature, intermediate sink density conditions.

2.2   Low Temperature, Intermediate to High Sink Density

Increasing the sink density has the effect of bringing t_2 closer to t_1 (see figure 2). That is, the region of mutual recombination is shrunk at the expense of annihilation at sinks. In fact, when $t_1 = t_2$ or $(K_0 K_{iv})^{-1/2} = (K_{is} C_s)^{-1}$, the plateau disappears. The effect of a high sink density is that interstitials find the sinks before they find vacancies because $C_s \gg C_v$ in the early going (figure 3).

2.3   High Temperature

At high temperature, after a brief buildup time t_2, defects reach sinks quickly, keeping point defect concentrations low. In equations 1(a) and (1b), the recombination term is considered negligible. The point defect buildup follows the path shown in figure 4. The solution displayed in figure 4 ignores the presence of thermal vacancies, which may be significant at higher temperatures. The buildup of radiation-induced vacancies and interstitials at high temperature, including an initial presence of thermal equilibrium vacancies is shown in figure 5.

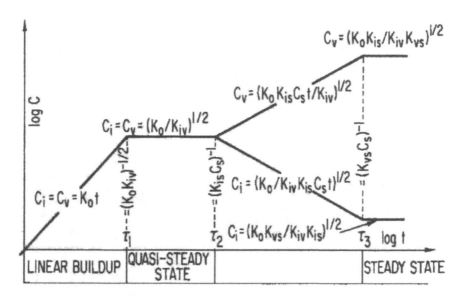

Figure 3. Buildup of radiation-induced vacancy and interstitial concentrations for low temperature, high sink density conditions.

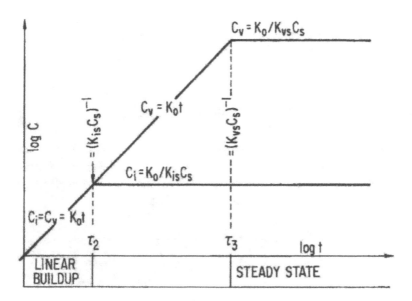

Figure 4. Buildup of radiation-induced vacancy and interstitial concentrations for high temperature conditions, ignoring the presence of thermal equilibrium vacancies.

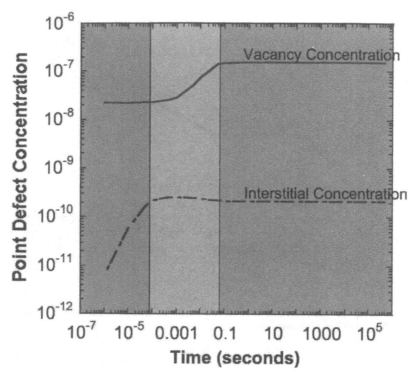

Figure 5. Buildup of radiation-induced vacancy and interstitial concentrations for high temperature conditions, including the presence of thermal equilibrium vacancies.

2.4   Properties of the Point Defect Balance Equations

By setting equations (1a) and (1b) equal to zero, we can solve for the steady-state vacancy and interstitial concentrations. If we assume that the sink strength is the same for both vacancies and interstitials (no preferential absorption of either type of defect at any sink), then:

$$K_0 = K_{iv}C_iC_v + K_{vs}C_vC_s$$
$$K_0 = K_{iv}C_iC_v + K_{is}C_iC_s$$

or

$$K_{vs}C_v = K_{is}C_i.$$

The absorption rate of interstitials and vacancies at sinks is equal. If the interstitial-sink capture radius and the vacancy-sink capture radius are equal, then

$$D_V C_V = D_i C_i.$$

Even though the steady-state concentration of interstitials is much lower than the steady-state concentration of vacancies, they each contribute equally to atom mobility because of the faster rate of diffusion of interstitials. For any particular sink to grow, it must have a net bias for either vacancies or interstitials. In real metals, K_{vs} and K_{is} are not equal. Specific sinks have a bias for certain point defects, allowing sinks to grow. This behavior is described in more detail in the chapter on high dose effects.

In a pure material, the diffusion coefficient under radiation is given by

$$D_{rad} = D_V C_V + D_i C_i.$$

Because the concentration of vacancies and interstitials is much greater under radiation than those produced thermally, the radiation-enhanced diffusion coefficients are much larger than thermal diffusion coefficients. The importance of radiation-induced point defects on diffusion is explained in detail in the chapter on ion beam mixing.

3   RADIATION-INDUCED SEGREGATION (RIS)

Radiation-induced segregation is the non-equilibrium segregation that occurs at grain boundaries or other defect sinks during irradiation of an alloy at intermediate temperature (30 to 50 percent of the melting temperature). Radiation produces quantities of point defects far in excess of equilibrium concentrations. At high temperatures, these defects are mobile and travel to low energy sites such as surfaces, grain boundaries, dislocations, and other defect sinks. Segregation occurs when a given alloying component has a preferential association with the defect flux. Enrichment or depletion of each element occurs according to the relative participation of each element in the defect flux. The segregation profiles have typical widths of 5-10 nm. The radiation-induced depletion of chromium and enrichment of nickel in a 304 stainless steel is shown in figure 6.

Radiation Effects in Solids

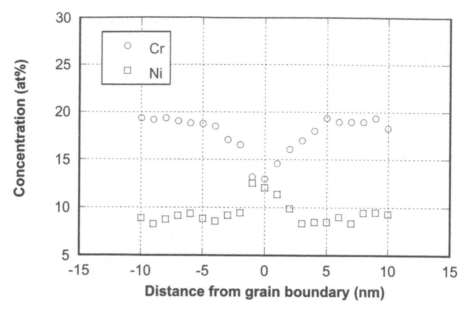

Figure 6. Radiation-induced segregation in 304 stainless steel irradiated with 3.2 MeV protons at 400°C to 1.0 dpa.

3.1   Theory of Radiation-induced Segregation

The following description is a condensation of that given by Lam et al. [2] for a ternary alloy whose components A, B and C are present in concentrations C_A, C_B and C_C (number of atoms per unit volume). The theory uses the point defect kinetic equations of equations (1a) and (1b) and add two new mechanisms: diffusion in and out of a specific volume and atom specific diffusion coefficients. Vacancy and interstitial concentrations during irradiation vary with time according to:

$$\frac{\partial C_v}{\partial t} = -\nabla \bullet J_v + K - R, \tag{5a}$$

$$\frac{\partial C_i}{\partial t} = -\nabla \bullet J_i + K - R, \tag{5b}$$

where $\nabla \bullet J_{v,i}$ are the divergences of the vacancy and interstitial fluxes at the defect sinks and K and R are the local total rates of production and mutual recombination, respectively, of point defects. As a simplification, equation (5) assumes a negligible sink density such that all point defect losses are due to recombination. The defect fluxes are partitioned into those occurring via A-, B- and C-atoms in the alloy according to:

$$J_i = J_i^A + J_i^B + J_i^C \tag{6a}$$

$$J_v = J_v^A + J_v^B + J_v^C, \tag{6b}$$

where the subscripts indicate the species of flux (interstitial or vacancy) and the superscripts indicate the complementary species by which the flux occurs (atom A, B. or C). The partial interstitial fluxes are in the same direction as the corresponding atom fluxes while the partial vacancy fluxes are in the opposite direction to the atom fluxes:

$$J_i^A = J_A^i; J_i^B = J_B^i; J_i^C = J_C^i \tag{7a}$$

$$J_v^A = J_A^v; J_v^B = J_B^v; J_v^C = J_C^v, \tag{7b}$$

and eq. (6) can be written as:

$$J_i = J_A^i + J_B^i + J_C^i \tag{8a}$$

$$J_v = -\left(J_A^v + J_B^v + J_C^v \right), \tag{8b}$$

These equations express the coupling between defect and atom fluxes across any fixed lattice plane. In general, the partitioning of the interstitial and vacancy fluxes via A-, B- and C-atoms is not in the same proportion as the atom fractions in the alloy. Interstitials may preferentially migrate by one atom type while vacancies preferentially exchange with another atom type. This preferential coupling of defect and atom fluxes is the physical origin of radiation-induced segregation.

As with the defect compositions, the alloy composition in time and space can be described by the conservation equations:

$$\frac{\partial C_A}{\partial t} = -\nabla \bullet J_A , \tag{9a}$$

$$\frac{\partial C_B}{\partial t} = -\nabla \bullet J_B, \tag{9b}$$

$$\frac{\partial C_C}{\partial t} = -\nabla \bullet J_C, \tag{9c}$$

where J_A, J_B and J_C are the total fluxes of the alloying elements, which can be partitioned into partial fluxes occurring by vacancies and interstitials:

$$J_A = J_A^i + J_A^i \tag{10a}$$

$$J_B = J_B^i + J_B^i \tag{10b}$$

$$J_B = J_B^i + J_B^i \tag{10c}$$

The defect and atom fluxes are expressed in terms of the concentration gradients of the different species;

$$J_k^i \ (\equiv J_i^k) = -D_k^i \alpha \nabla C_k - D_i^k \nabla C_i \tag{11a}$$

$$J_k^v \ (\equiv -J_k^k) = -D_k^v \alpha \nabla C_k + D_v^k \nabla C_v \tag{11b}$$

where k = A, B or C, a is the thermodynamic factor which relates the concentration gradient to the chemical potential gradient of atoms, and D_k^i D_k^v, D_i^k and D_v^k are the partial diffusion coefficients of atoms k by interstitials, and vacancies, and of interstitials and vacancies by atoms, respectively. The partial diffusion coefficients have the form:

$$D_k^i = d_{kj} N_j \quad \text{and} \quad D_j^k = d_{kj} N_k \tag{12}$$

where j = i or v, $N_j = \Omega C_j$ and $N_k = \Omega C_k$ are the atomic fractions of defects and of k-atoms, respectively, W is the average atomic volume in the alloy, and d_{kj} are the diffusivity coefficients for conjugate atom-defect pairs kj:

$$d_{kj} = \frac{1}{6} \lambda_k^2 z_k v_{kj}^{eff} \tag{13}$$

Here λ_k is the jump distance, z_k the coordination number, and $n^{eff}{}_{kj}$ the effective jump or exchange frequency of the pair. The total diffusion coefficients for interstitials and vacancies are defined as:

$$D_i = \sum_k d_{ki} N_k \tag{14a}$$

$$D_v = \sum_k d_{kv} N_k \tag{14b}$$

and for atoms:

$$D_k = d_{ki} N_i + d_{kv} N_v \tag{15}$$

From eq. (9), (11), (12), (13) and (15), the defect and atom fluxes with respect to a coordinate system fixed on the crystal lattice are:

$$J_i = -(d_{Ai} - d_{Ci}) \Omega C_i \alpha \nabla C_A - (d_{Bi} - d_{Ci}) \Omega C_i \alpha \nabla C_B - D_i \nabla C_i \tag{16a}$$

$$J_v = (d_{Av} - d_{Cv})\Omega C_v \alpha \nabla C_A + (d_{Bv} - d_{Cv})\Omega C_v \alpha \nabla C_B - D_v \nabla C_v \quad (16b)$$

$$J_A = -D_A \alpha \nabla C_A + d_{Av}\Omega C_A \nabla C_A - d_{Ai}\Omega C_A \nabla C_i \quad (16c)$$

$$J_B = -D_B \alpha \nabla C_B + d_{Bv}\Omega C_B \nabla C_B - d_{Bi}\Omega C_B \nabla C_i \quad (16d)$$

$$J_C = -D_C \alpha \nabla C_C + d_{Cv}\Omega C_C \nabla C_C - d_{Ci}\Omega C_C \nabla C_i \quad (16e)$$

Small perturbations arising from the presence of point defects are neglected so that $C_A + C_B + C_C = 1$ and $\nabla C_C = -(\nabla C_A + \nabla C_B)$. Of the five flux equations (16a-e), only four are independent because the defect and atom fluxes across a marker plane must balance:

$$J_A + J_B + J_C = J_i - J_v \quad (17)$$

A system of four coupled partial differential equations describing the space and time dependence of the atoms and defects in the solid is determined by substituting the defect and atom fluxes given by eq. (16) into eq. (6) and (10):

$$\frac{\partial C_v}{\partial t} = \nabla \bullet [-(d_{Av} - d_{Cv})\Omega C_v \alpha \nabla C_A - (d_{Bv} - d_{Cv})\Omega C_v \alpha \nabla C_B + D_v \nabla C_v] + K - R$$

$$\frac{\partial C_i}{\partial t} = \nabla \bullet [(d_{Ai} - d_{Ci})\Omega C_i \alpha \nabla C_A + (d_{Bi} - d_{Ci})\Omega C_i \alpha \nabla C_B + D_i \nabla C_i] + K - R$$

$$\frac{\partial C_A}{\partial t} = \nabla \bullet [D_A \alpha \nabla C_A + \Omega C_A (d_{Ai} \nabla C_i - d_{Av} \nabla C_v)]$$

$$\frac{\partial C_B}{\partial t} = \nabla \bullet [D_B \alpha \nabla C_B + \Omega C_B (d_{Bi} \nabla C_i - d_{Bv} \nabla C_v)] \quad (18)$$

Numerical solutions of eq. (18) are obtained for a planar sample under irradiation with the aid of computer codes designed to solve stiff differential equations (equations whose development time are significantly different) [3]. The grain boundary is equated to a free surface and the calculations are performed for only a single grain, taking advantage of the symmetry of the problem. The initial conditions are the thermodynamic equilibrium of the alloy. Conditions at the boundary are defined as follows. At the grain center, all concentration gradients are set equal to zero. At the grain boundary, the concentrations of interstitials and vacancies are fixed at their thermal equilibrium values. The grain boundary atom concentrations are determined by the conservation of the numbers of atoms in the specimen. Atom concentrations are assumed to be initially uniform. Parameters used in the calculation of segregation in Fe-Cr-Ni alloys are given in ref. [2] for the Lam model and in Ref. [4] for the Perks model as coded by Simonen [5].

Equations (18a-d) can be solved in steady-state to provide a relationship between gradients in vacancy and atom gradients. These relationships are known as determinants M and are a function of the concentrations and diffusivities. For the case of a binary alloy, the determinant is given by:

$$M_A = \frac{\nabla C_A}{\nabla C_v} = \frac{C_A C_B d_{Ai} d_{Bi}}{\alpha(d_{Bi} C_B D_A + d_{Ai} C_A D_B)}\left(\frac{d_{Av}}{d_{Ai}} - \frac{d_{Bv}}{d_{Bi}}\right). \quad (19)$$

If the determinant is positive, then the atom depletes at the sink during radiation. If the determinant is negative, then the atom enriches at the sink during radiation. The sign of the determinant is determined by the relative ratios of atom-vacancy and atom-interstitial diffusivities between the two atoms. Figure 7 is a schematic that demonstrates how these diffusivities $\left(\frac{d_{Av}}{d_{Ai}} - \frac{d_{Bv}}{d_{Bi}}\right)$ influence the direction of segregation.

The determinant for a three-component alloy is given by:

$$M_j = \frac{\nabla C_j}{\nabla C_v} = \frac{\dfrac{d_{jv} C_j}{D_j} \sum_{k \neq j} \dfrac{d_{ki} C_k}{D_k} - \dfrac{d_{ji} C_j}{D_j} \sum_{k \neq j} \dfrac{d_{kv} C_k}{D_k}}{\alpha \sum_k \dfrac{d_{ki} C_k}{D_k}}. \quad (20)$$

3.2   Trends in Radiation-induced Segregation in Fe-Cr-Ni Alloys

Because of a possible link between irradiation-assisted stress corrosion cracking (IASCC) and RIS in austenitic stainless steels (for details see the chapter on IASCC), a significant database of RIS measurements has been taken on irradiated Fe-Cr-Ni alloys. This database will be examined to explain the trends in RIS as a function of temperature, dose rate, dose, and material composition.

Figure 8 shows the trends in segregation (specifically the chromium depletion) as a function of radiation dose. These samples were irradiated with 3.2 MeV protons. The segregation approaches a terminal value over roughly the first few dpa, with the rate of approach determined by the specific alloy composition.

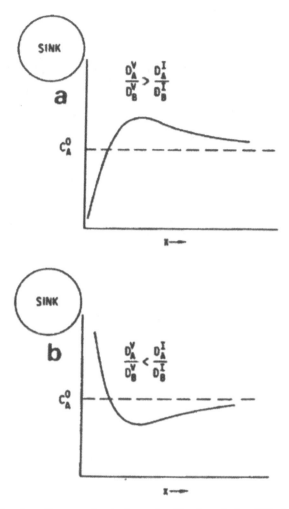

Figure 7. The direction of segregation is determined by the relative diffusivities.

Figure 9 shows the chromium depletion as a function of temperature. These samples were also irradiated with 3.2 MeV protons. For a specific alloy composition, the segregation reaches a maximum at an intermediate temperature. The temperature at which the peak segregation occurs differs for different alloy compositions.

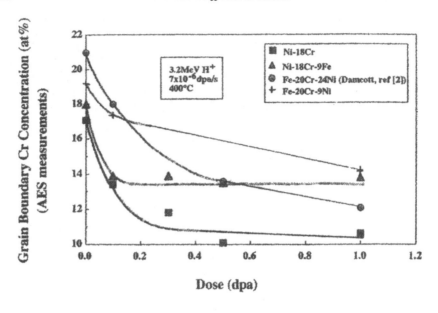

Figure 8. RIS in austenitic Fe-Cr-Ni alloys as a function of dose.

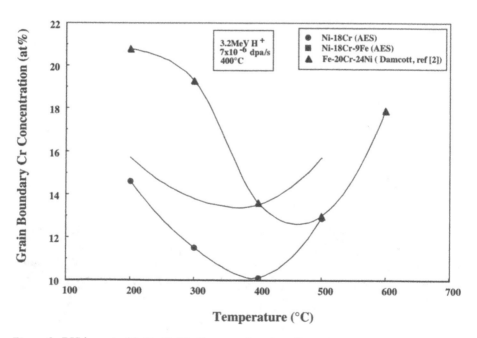

Figure 9. RIS in austenitic Fe-Cr-Ni alloys as a function of temperature.

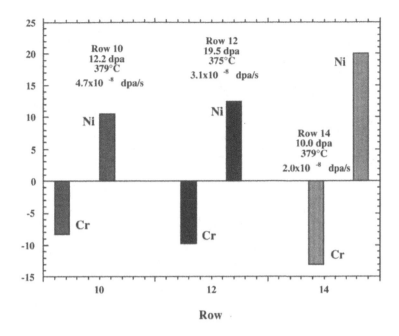

Figure 10. RIS in austenitic Fe-Cr-Ni alloys as a function of displacement rate.

Figure 10 shows the chromium depletion as a function of dose rate for 304 stainless steel samples irradiated in the EBR-II reactor. For the temperature of 370°C, as the dose rate decreases, the amount of grain boundary segregation increases.

Figure 11 summarizes the predicted temperature-dose rate response of RIS. For a given material, there is an intermediate temperature and dose rate regime where segregation occurs. At low temperatures, the diffusion rates are too low for segregation to occur. At high temperatures, thermal defect concentrations are large and the back diffusion of vacancies tends to retard the segregation process.

3.3   The Mechanism of Radiation-induced Segregation in Fe-Cr-Ni Alloys

In general, preferential interaction with vacancies and/or interstitials can contribute to RIS. Two different mechanisms have been proposed to describe RIS in austenitic Fe-Cr-Ni alloys. One segregation mechanism is inverse Kirkendall behavior, where preferential association with a defect gradient (either a vacancy gradient or an interstitial gradient) causes a solute flux (the description leading to equation 14 and figure 7 describes the inverse Kirkendall effect).

Radiation Effects in Solids

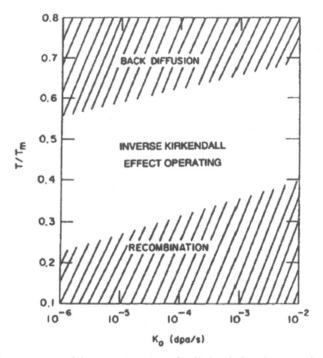

Figure 11. Temperature and dose rate response of radiation-induced segregation.

A second segregation mechanism is the formation of mobile defect-solute complexes (interstitial-solute or vacancy-solute), where the solute is dragged with the defect flux. If a solute binds with a defect to form a complex, and the complex undergoes significant diffusion before dissociating, then the solute can be dragged to the boundary. The diffusion of defect-solute complexes will enrich the solute near the sink. Mobile defect-solute complexes have been shown to be important in segregation in dilute alloys [6] but the concept of a defect-solute complex becomes ill-defined in the case where the nearest neighbors to a defect change significantly as the defect migrates. Although defect-solute complexes are not expected to form in concentrated alloys, the possibility has been raised that undersized elements may be more likely to exist as interstitials [7]. Greater participation in the interstitial flux would lead to an enrichment of an undersized element at the grain boundary.

The simplest form of an inverse Kirkendall driving force can be explained with the assistance of the data presented in figure 12 and was the concept supporting the RIS model proposed by Perks [4]. Figure 12 shows that diffusion in nickel via vacancy exchange is faster for chromium than for iron or nickel. Diffusion of iron is faster then nickel

$$d_{Cr}^{v} > d_{Fe}^{v} > d_{Ni}^{v}.$$

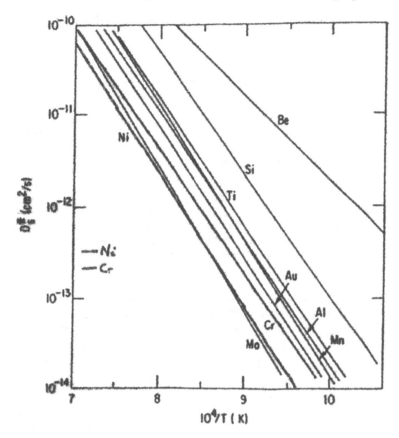

Figure 12. Tracer impurity diffusion coefficients for various solutes in nickel.

This relationship also holds true in Fe-base austenitic alloys [8]. Since relative diffusion coefficients for atom-interstitial diffusion are not known, Perks assumed that

$$d^i_{Cr} = d^i_{Fe} = d^i_{Ni}.$$

As can be seen by examining the equations for the determinants (eq.19-20), these conditions for vacancy and interstitial diffusion lead to chromium depletion and nickel enrichment. The data in Table 1 supports the vacancy driven segregation model as the trends in segregation can be obtained solely through the properties of atom-vacancy diffusion.. In this table, the determinants are calculated for a series of Fe-Cr-Ni alloys assuming interstitial diffusivities are equal for Fe, Cr, and Ni and that atom-vacancy diffusivities can be taken from thermal diffusion measurements. Using this simple approach correctly predicts the trends in the segregation of Fe, Cr, and Ni in every case.

Table 1. Determinants and segregation direction in Fe-Cr-Ni alloys

Alloy	M_{Cr}	M_{Fe}	M_{Ni}	Cr	Fe	Ni	Analysis Method
Ni-18Cr	3.9	------	-3.9	Depletes	----------	Enriches	AES
Ni-18Cr	3.9	------	-3.9	Depletes	----------	Enriches	STEM/EDS
Ni-18Cr-9Fe	5.0	0.4	-5.4	Depletes	Depletes	Enriches	AES
Ni-18Cr-9Fe	5.0	0.4	-5.4	Depletes	Depletes	Enriches	STEM/EDS
Fe-20Cr-9Ni	5.0	-3.0	-2.0	Depletes	Enriches	Enriches	AES

While the Perks approach correctly predicts the trends in segregation across a wide variety of compositions of Fe-Cr-Ni alloys, early attempts to model RIS using Perks' approach typically overpredicted the actual segregation (see figure 13), even when accounting for the resolution of the instrument used to measure the RIS [9]. Sensitivity studies showed that varying the atom-vacancy migration energy values could eliminate the model overprediction used in the calculation within the experimental uncertainty of the measured migration energies [9]. Thus, within the uncertainty of the measured atom-vacancy diffusion parameters, the Perks model, assuming all segregation is driven by preferential interaction with the vacancy flux, can predict the measured segregation. Additionally, to properly predict the trends in segregation across a set of alloys of varying bulk composition, the diffusivities used as model inputs needed to be chosen for the specific bulk composition [10]. Studies to date indicate that all the measured segregation in Fe-Cr-Ni alloys can be predicted using the Perks approach. This does not eliminate a possible contribution due to differences in interstitial diffusivity, but any such contribution is small enough to be masked by uncertainties in the vacancy parameters.

A study did show that defect-solute binding as proposed by Okamoto [7] is not a viable mechanism in austenitic Fe-Cr-Ni alloys [11]. The Okamoto model assumes that atom-interstitial diffusivities are not equal and that undersized atoms preferentially bind to the interstitial flux with a binding energy E_{Ai}^{b}. This increases the concentration of certain atoms as interstitials above the bulk concentration.

$$C_A^i = N_i \frac{C_A \exp\left(\dfrac{E_{Ai}^b}{kT}\right)}{C_A \exp\left(\dfrac{E_{Ai}^b}{kT}\right) + C_B + C_C} \tag{21}$$

$$d_{Ai} \neq d_{Bi} \neq d_{Ci}$$

For any reasonable set of input parameters, the interstitial binding model severely over predicted the nickel enrichment (Figure 13).

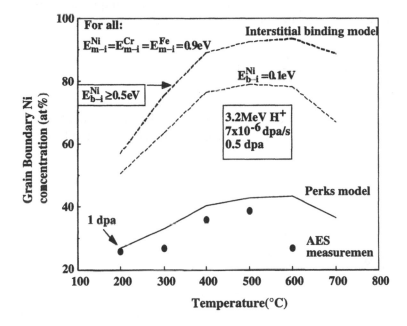

Figure 13. Comparison of Perks' inverse Kirkendall model and the interstitial binding model with RIS data. The Perks' approach fairly accurately predicts the trends in segregation while the interstitial binding model severely overpredicts the segregation.

For Fe-Cr-Ni alloys, comparisons of model calculations to measurements lead to the following conclusions.

1. The segregation can be explained primarily by an inverse Kirkendall mechanism, primarily driven by interactions of atoms with the vacancy flux
2. The diffusivity parameters used in the models must be functions of alloy composition
3. Short range order effects should be included in models

To model RIS in a way that includes composition diffusivities and short-range order, the Perks model can be modified such that at each location in the alloy, migration energies are calculated based on the local composition. To calculate the migration energies, a set of input parameters describing point defect properties in pure elements is required. These input parameters and the basis for there selection are described next.

3.4   Modified Inverse Kirkendall Model

The Perks model can be modified to include both alloy dependent diffusivities and short range ordering forces. In the Perks model, the rate of segregation of each element is described by a diffusivity of the form:

$$d^{Cr} = d_0^{Cr} \exp(\frac{-E_{vm}^{Cr}}{kT}) \ , \tag{22a}$$

$$d^{Ni} = d_0^{Ni} \exp(\frac{-E_{vm}^{Ni}}{kT}) \ , \tag{22b}$$

$$d^{Fe} = d_0^{Fe} \exp(\frac{-E_{vm}^{Fe}}{kT}), \tag{22c}$$

where d is the diffusivity of each element, d_0 is a pre-exponential factor, E_{vm} is the vacancy migration energy for the given element, k is the Boltzmann factor, and T is the temperature. In the majority of published calculations using the Perks model, the migration energies for each element are assumed to be equal [2,4, 5,10], with differences in segregation rates arising from differences in the pre-exponential factors. To include ordering effects, the migration energy term in the exponential must be described as a function of composition. Radiation-induced segregation calculations that use local compositions to calculate migration energies were first proposed by Grandjean et al. [12] for a binary Ni-Cu alloy. The method was extended to ternary alloys (see [13] and is described in the following paragraphs. Nastar developed a similar approach [14].

For an atom to migrate (Cr migrating in an Fe-Cr-Ni lattice will be used as an example), it must move from its equilibrium position in the lattice (with equilibrium energy E_{eq}^{Cr}) and travel through a position of maximum potential (known as the saddle point, with saddle point energy $ES_{Fe-Cr-Ni}^{Cr}$), moving to a new lattice site. This relationship between the energies is shown schematically in figure 14. The migration energy is the difference between the saddle point energy and the equilibrium energy:

$$E_{vm}^{Cr} = ES_{Fe-Cr-Ni}^{Cr} - E_{eq}^{Cr}. \tag{23}$$

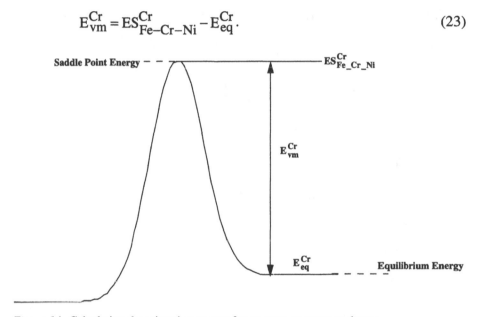

Figure 14. Calculating the migration energy for an atom-vacancy exchange.

In a simple model, the equilibrium energy can be described as the interaction energy between nearest neighbors:

$$E_{eq}^{Cr} = Z[C_{Cr}E_{CrCr} + C_{Ni}E_{NiCr} + C_{Fe}E_{FeCr} + C_v E_{Crv}], \tag{24}$$

where Z is the number of nearest neighbors, C is the atomic concentration of atoms and/or vacancies, and E_{xy} is the pair interaction energy between an atom x and an atom or vacancy y. Pair interaction energies between unlike neighbors are defined to be a linear average of the like atom pair energies minus any ordering energy:

$$E_{NiCr} = \frac{E_{NiNi} + E_{CrCr}}{2} - E_{NiCr}^{ord}. \tag{25}$$

Combining equations (23-25) and requiring that $C_{Fe} + C_{Cr} + C_{Ni} = 1$, the migration energy for Cr, Ni and Fe can be expressed:

$$E_{vm}^{Cr} = ES_{Fe-Cr-Ni}^{Cr} - Z\left[\frac{1}{2}(C_{Cr}+1)E_{CrCr} + \frac{C_{Ni}}{2}E_{NiNi} + \frac{C_{Fe}}{2}E_{FeFe} + C_v E_{Crv}\right]$$
$$+ ZC_{Ni}E_{NiCr}^{ord} + ZC_{Fe}E_{FeCr}^{ord} \tag{26a}$$

$$E_{vm}^{Ni} = ES_{Fe-Cr-Ni}^{Ni} - Z\left[\frac{1}{2}(C_{Ni}+1)E_{NiNi} + \frac{C_{Cr}}{2}E_{CrCr} + \frac{C_{Fe}}{2}E_{FeFe} + C_v E_{Niv}\right]$$
$$+ ZC_{Cr}E_{NiCr}^{ord} + ZC_{Fe}E_{FeNi}^{ord} \tag{26b}$$

$$E_{vm}^{Fe} = ES_{Fe-Cr-Ni}^{Fe} - Z\left[\frac{1}{2}(C_{Fe}+1)E_{FeFe} + \frac{C_{Cr}}{2}E_{CrCr} + \frac{C_{Ni}}{2}E_{NiNi} + C_v E_{Fev}\right]$$
$$+ ZC_{Cr}E_{FeCr}^{ord} + ZC_{Ni}E_{FeNi}^{ord} \tag{26c}$$

To determine the values of migration energies from eq. (26), pair interaction energies and saddle point energies must be calculated. To calculate the pair interaction energies for like atoms, the cohesive energy is divided by the number of bond pairs between nearest neighbors:

$$E_{CrCr} = E_{coh}^{Cr}/(Z/2), \tag{27a}$$

$$E_{FeFe} = E_{coh}^{Fe}/(Z/2), \tag{27b}$$

$$E_{NiNi} = E_{coh}^{Ni}/(Z/2). \tag{27c}$$

Assuming the calculation is for an FCC alloy, because both pure Fe and Cr are BCC, the energy required to convert Fe and Cr to FCC structure must be included in calculating E_{CrCr} and E_{FeFe} to properly describe the equilibrium energy.

$$E_{CrCr}^{FCC} = E_{CrCr}^{BCC} + \Delta G_{Cr}^{BCC \rightarrow FCC}, \tag{28a}$$

$$E_{FeFe}^{FCC} = E_{FeFe}^{BCC} + \Delta G_{Fe}^{BCC \rightarrow FCC}. \tag{28b}$$

Pair interaction energies between unlike neighbors are a linear average of the like atom pair energies minus any ordering energy.

$$E_{NiCr} = \frac{(E_{NiNi} + E_{CrCr})}{2} - E_{NiCr}^{ord}, \tag{29a}$$

$$E_{FeCr} = \frac{(E_{FeFe} + E_{CrCr})}{2} - E_{FeCr}^{ord}, \tag{29b}$$

$$E_{FeNi} = \frac{(E_{FeFe} + E_{NiNi})}{2} - E_{FeNi}^{ord}. \tag{29c}$$

The pair interaction energy for atoms and vacancies is fitted to the formation energy of the pure metal and is given by:

$$E_{Cr-v} = \left(\frac{E_{coh}^{Cr} + E_{vf}^{Cr}}{Z} \right), \tag{30a}$$

$$E_{Ni-v} = \left(\frac{E_{coh}^{Ni} + E_{vf}^{Ni}}{Z} \right), \tag{30b}$$

$$E_{Fe-v} = \left(\frac{E_{coh}^{Fe} + E_{vf}^{Fe}}{Z} \right). \tag{30c}$$

To calculate the migration energy, the saddle point energy must also be known. The saddle point energy in Fe-Cr-Ni alloys is calculated using the saddle-point energy of the pure Fe, Cr, and Ni. The saddle-point energy in the pure metal is calculated to reproduce the vacancy migration energy in the pure metal. For example, assume in pure Cr that both an atom and a vacancy are extracted and placed at a saddle point. The energy to extract the Cr atom is given by:

$$E_{Cr} = Z(C_{Cr}E_{CrCr} + C_v E_{Crv}). \tag{31}$$

The energy to extract the vacancy is:

$$E_v = Z(C_v E_{vv} + C_{Cr}E_{Crv}). \tag{32}$$

Since $C_{Cr} + C_v = 1$ in a pure metal and since $C_v \ll C_{Cr}$, the sum of $E_{Cr} + E_v$ becomes

$$E_{Cr} + E_v = Z(E_{CrCr} + E_{Crv}). \tag{33}$$

Equation 33 represents the equilibrium energy for pure Cr. We want to place the atom at a saddle point with an energy such that

$$E_{vm}^{Cr} = ES_{pure}^{Cr} - \left(E_{Cr} + E_v\right) = ES_{pure}^{Cr} - Z\left(E_{CrCr} + E_{Crv}\right). \quad (34)$$

The saddle point energy in the pure metal is then given as:

$$ES_{pure}^{Cr} = E_{vm}^{Cr} + Z\left(E_{CrCr} + E_{Crv}\right). \quad (35)$$

To calculate the saddle point energy for an element in an Fe-Cr-Ni alloy, a combination of pure element saddle point energies must be chosen. At one extreme, the saddle point could be chosen to be independent of the lattice with the saddle point for each element chosen as the pure element saddle point. This formulation makes the saddle point dependent only on the diffusing species. At the other extreme, the saddle point could be chosen as independent of the diffusing species, with the saddle point for all elements being defined by the average local bulk composition. This formulation makes the saddle point dependent only on the local composition. Each of these two formulations represents an extreme. The saddle point should be a function of both the local composition and the diffusing species.

One simple method is to choose the saddle point to combine features of the bulk lattice and the individual elements. The lattice average saddle point energy is defined as:

$$ES_{Fe-Cr-Ni}^{avg} = \sum_{i=Fe,Cr,Ni} ES_{pure}^i C_i. \quad (36)$$

The saddle point used for each element is the average of the pure element saddle point energy and the average lattice saddle point energy. For instance, the saddle point energy for Cr, Ni, and Fe in an Fe-Cr-Ni lattice is:

$$ES_{Fe-Cr-Ni}^{Cr} = \frac{ES_{pure}^{Cr} + ES_{Fe-Cr-Ni}^{avg}}{2}, \quad (37a)$$

$$ES_{Fe-Cr-Ni}^{Ni} = \frac{ES_{pure}^{Ni} + ES_{Fe-Cr-Ni}^{avg}}{2}, \quad (37b)$$

$$ES_{Fe-Cr-Ni}^{Fe} = \frac{ES_{pure}^{Fe} + ES_{Fe-Cr-Ni}^{avg}}{2}. \quad (37c)$$

This formulation is the simplest method that makes the saddle point energy dependent both on the local composition and the diffusing species. Two approximations are inherent in eq. 37. First, that the saddle point in the alloy is an average of the pure element saddle point and the lattice average saddle point. The second approximation is that the lattice average saddle point can be calculated by averaging the pure element saddle points. For this approach to be strictly correct, the equilibrium energies for Cr, Ni, and Fe would have to be equal and this is not necessarily the case.

The approach described in this section was used to model RIS and compared to measurements taken on a range of Fe-base and Ni-base austenitic alloys irradiated at temperatures from 200-600°C and doses to 3 dpa. Figures 15 and 16 clearly demonstrate the improvement of this approach over the simpler Perks approach.

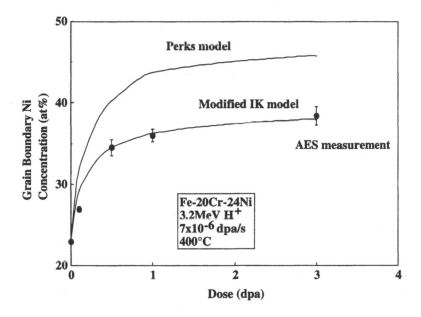

Figure 15. The Modified Inverse Kirkendall (MIK) model removes overpredictions inherent in the Perks model.

Radiation Effects in Solids

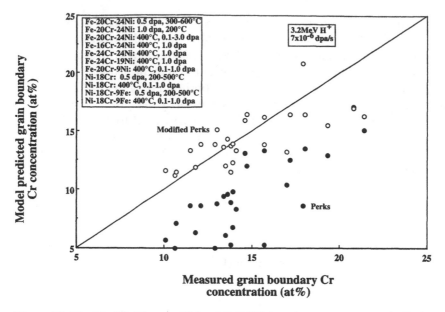

Figure 16. The Modified Inverse Kirkendall (MIK) is an improvement over the Perks model across a wide range of compositions.

4   SUMMARY

Radiation-enhanced diffusion and radiation-induced segregation contribute to chemical redistribution of constituent atoms at temperatures too low for thermal diffusion to be significant. Each mechanism relies on the build-up and diffusion of radiation created point defects. The basic rate equations describing each mechanism have been described and the trends in the experimental data explained. A detailed model for radiation-induced segregation in Fe-Cr-Ni alloys has been described and shown to be an advancement over previous generations of RIS models.

REFERENCES

[1]   Sizman, R., J. Nucl. Mater. **69&70**, 386 (1968)

[2]   Lam, N. Q. Kumar, A. and Wiedersich, H. *Effects of Radiation on Materials: Eleventh conference, ASTM STP 782*, ed by H. R. Brager, J. S. Perrin (American Society for Testing and Materials 1982) pp 985-1007

[3]   Hindmarsch, A. C. UCID-30001, Rev 3 (December 1974)

[4]   Perks, J.M., Marwick, A.D. and English, C.A. AERE R 12121 (June 1986)

[5]   Simonen, E. P, Bruemmer, S. M. *Proceedings of the 1994 MRS Fall Meeting: Symposium Y: Microstructure of Irradiated Materials*, ed by I. M. Robertson, L. E. Rehn, S. J. Zinkle, W. J. Phythian (Mater. Res. Soc. Proc. 373, Pittsburgh, PA 1995) p 95

[6]   Okamoto, P. R. and Rehn, L. E. J. Nucl. Mater. **83**, 2 (1979)

[7]   Okamoto, P. R. and Wiedersich, H, J. Nucl. Mater. **53**, 336 (1974)

[8]   Rothman, S.J., Nowicki, L.J., and Murch, G.E. J. of Physics F: Metal Physics **10**, 383 (1980)

[9]   Allen, T. R. and Was, G. S. *Proceedings of the 1994 MRS Fall Meeting: Symposium Y: Microstructure of Irradiated Materials.*, ed by I. M. Robertson, L. E. Rehn, S. J. Zinkle, W. J. Phythian (Mater. Res. Soc. Proc. 373, Pittsburgh, PA 1995) p 101

[10] Damcott, D. L., Allen, T. R., and Was, G. S., J. Nucl. Mater. **225**, 97 (1995)

[11] Allen, T. R., Busby, J. T., Was, G. S., Kenik, E. A. J. Nucl. Mater. **255**, 44 (1998)

[12] Grandjean, Y., Bellon, P., and Martin, G., Phys. Rev. B **50**, 4228 (1994)

[13] Allen, T. R. and Was, G. S., Acta Met. **46**, 3679 (1998)

[14] Nastar, M. and Martin, G., Mat. Sci. Forum **294-296**, 83 (1999)

Chapter 7

THE KINETICS OF RADIATION-INDUCED POINT DEFECT AGGREGATION AND METALLIC COLLOID FORMATION IN IONIC SOLIDS

Eugene A. Kotomin[a] and Anatoly I. Popov[b,c]
[a]European Commission, Joint Research Center, Institute for Transuranium Elements, 76125 Karlsruhe, Germany
[b]University of Latvia, Kengaraga 8, LV-1063 Riga, Latvia
[c]Institut Laue-Langevin, 6 rue Jules Horowitz, 38042 Grenoble Cedex 9, France

1   INTRODUCTION

By definition, a metallic colloid is "a particle, which size is sufficiently small that there is at least a possibility that its properties will differ from those of the bulk material" [1]. The existence of metallic colloids in ionic crystals has been known for a long time [2-6]. Metallic colloids can be produced in ionic solids by a number of methods. 1) *Additive coloration (thermochemical reduction)*, where a crystal is heated in the metal vapour at high temperature, so that the excess metal is introduced into the crystal. 2) *Electrolytic coloration* is carried out by applying an electric field (~100V/cm) at elevated temperatures. 3) *Ionizing radiation,* where conditions of colloid formation depend on both type of radiation and material. 4) *Ion implantation.*

The following experimental techniques are widely used in the study of the colloid structure, its manifestation, and the processes of colloid formation, transformation and destruction: optical spectroscopy, magnetic susceptibility and magnetic resonance, conduction electron spin resonance, Electron-Nuclear Double Resonance (ENDOR), electrical conductivity, dielectric constant measurements, thermal conductivity, electron and optical microscopy, X-ray and neutron diffraction, Small-Angle Neutron Scattering (SANS), and crystal hardening. Light absorption and scattering at spherical metallic particles is in particular quite characteristic and depends on the particle size (generally known as Mie theory), which is widely used for colloid identification [4-6]. In this Chapter we present a review of experimental and theoretical studies of metallic colloids in ionic materials.

K.E. Sickafus et al. (eds.), Radiation Effects in Solids, 153–192.
© 2007 Springer.

Table 1 contains the list of different types of ionic materials where colloids were observed and studied (discussed below). It is well known that the primary radiation defects in alkali halides and binary oxide solids, F centers (electron trapped by an anion vacancy) and H centers (interstitial halide/oxide atoms), aggregate under intensive irradiation at temperatures high enough to allow defect diffusion. This leads to the formation of gas bubbles and alkali metallic colloids. The process is generally believed to occur via diffusion-controlled aggregation of single F centers. Their smallest aggregate centers (F_2(M), F_3(R), F_4(N), i.e., two, three or four anion vacancies with trapped electrons) can be identified by characteristic optical absorption bands, whereas large aggregates transform into metal colloids with a broad optical extinction band (Table 2).

Table 1. Ionic crystals where colloids were detected experimentally

Type of materials	F center	Examples
Alkali halides	Yes	NaCl, KCl, KI, RbCl, LiF, CsBr
Alkaline earth fluorides	Yes	CaF_2 , BaF_2 and SrF_2
Silver halides	No	AgCl, AgBr
Lead halides	No	PbI_2
Hydrides	probably yes	LiH, AlH_3
Azides	probably yes	NaN_3, AgN_3, $Pb(N_3)_2$
Oxides	Yes	Li_2O, Al_2O_3, MgO

The two experimentally most investigated crystals with colloids are *alkali halides* -- LiF and NaCl. More than 40 years ago, Guinier and Lambert observed, in neutron-irradiated LiF after irradiation, the nucleation of lithium platelets of atomic thickness. In neutron-irradiated LiF, Li particles (were?) have been detected by x-ray diffraction [12], NMR [13], ESR [14,15] and thermal analysis [16]. Later, Van den Bosch performed magnetic susceptibility measurements of neutron-irradiated and γ-irradiated LiF crystals for the characterization of Li colloids [17,18]. He attributed the temperature-independent paramagnetic contribution to the presence of metallic lithium. Later, he showed that although the radiation damage in - irradiated samples is similar to that for the neutron-irradiated specimens, a quantitative observation reveals differences which give experimental evidence for the statement that the F-centers under these conditions occur in a more aggregated form in the γ- than in the neutron-irradiated crystals

[18]. Taupin observed by X-ray diffraction and ESR the presence of two kinds of colloids; those called '*platelets*' are particularly interesting because they show a quantum size effect [19].

Table 2. Positions of the absorption band maxima (in nm) for the F-type defects in some halides (after [3,4, 6-11])

Center		LiF	NaF	NaCl	CaF$_2$	MgF$_2$
F		250	341	458	375	250
F$_2$		444	505	725	366; 521	219; 317; 369; 403
F$_2^+$		645	740	1030	360; 551	
F$_2^-$		950				
F$_3$					678	
	R$_1$	313	394	545		
	R$_2$	380	435	596		
F$_3^+$		452	545.6(zpl)		734	
F$_3^-$		660 690	870			
F$_4$						
	N$_1$	523	587	826		
	N$_2$	541	623	861		
Intrinsic colloid		450	460	560	540-560	282
Extrinsic colloid		520(C$_{Na}$) 680(C$_K$)				

Quite recently, L'vov et al. [20] produced Li colloids in electron-irradiated LiF and studied them by means of ESR and ESR-imaging. Li colloids were also observed in the optical absorption studies on Ar- and Ne-ion implanted LiF samples [21] and on the surface of low-energy electron-irradiated LiF crystals [22].

Irradiated LiF crystals are of special interest since they provide excellent thermal stability of complex colour centers, and are used, e.g., for dosimetry [4] and laser applications [8,23]. Despite numerous experimental studies of the F center aggregates produced in LiF under different kinds of irradiation [8,23-28], the *kinetics* of the F center aggregation leading to the metal colloid formation is still not well understood. In particular, there is uncertainty in the experimental (indirect) estimates of the activation energies for the single F center activation energy for diffusion hops. There exist several indirect estimates of the diffusion energy published in the literature, e.g., 0.66 eV [29] and 0.85 eV [23]. Recently, the F center aggregation kinetics in LiF was studied with a focus on Li colloid growth during sample annealing after irradiation with low energy electrons at 300 K [30].

Figure 1 shows typical series of absorption spectra in LiF where at low temperatures F and M centers are predominant. This clearly demonstrates that, during heating, the F centers and dimer-, trimer-centers transform into a broad band at 2.9 eV associated with Li metal colloids [24]. The F-type center aggregation is accelerated above 360 K and is completed at 490 K. The observed shape of the 2.9 eV band is typical for the extinction due to

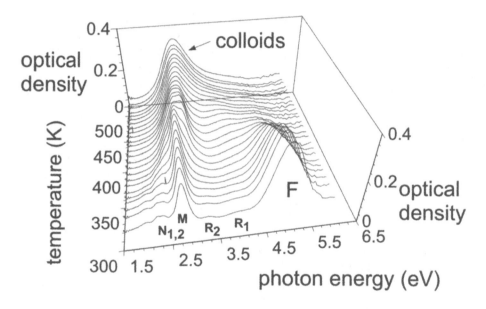

Figure 1. Development of optical colloid extinction spectra during heating of a LiF crystal after electron irradiation with low-energy (2.5 keV) electrons for 30 min. at 300 K [30]. Positions of isolated (F), dimer (M) and trimer (R) electron centers as well as 4-electon (N) centers are shown.

metal colloids but cannot be well fitted using simple Mie theory [31] due to the *non*-spherical shape of Li metal colloids, which is in agreement with earlier conclusions that these colloids have a *platelet* morphology [18,24,32]. (The great difference of LiF and, say, CaF_2 crystals is that in the latter the free Ca metal lattice almost perfectly matches its sublattice in CaF_2, whereas in LiF the Li sublattice constant is much smaller than for a free Li metal {the difference in a volume of the unit cell is as big as 25%} and also the crystalline structure is different.) The initial F center concentration could be as large as $n_0 = 6 \ 10^{20} \ cm^{-3}$ (which is a necessary step for the effective colloid nucleation) due to a small low-energy electron penetration depth ($\sim 10^3$ Å for 2.5 eV electrons [33,34]). We discuss below, in Section 3.2, the interpretation of these experimental data and the kinetics of metallic colloid formation.

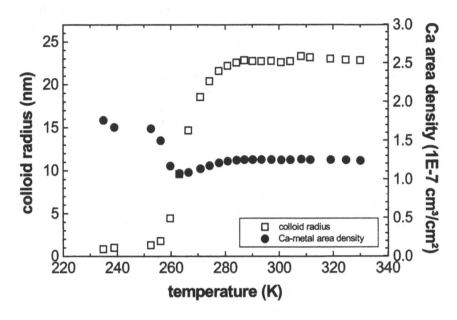

Figure 2. Mean colloid radius and areal metal density of calcium colloids in CaF_2 as a function of temperature derived using the Mie theory [34].

Long ago, Savostianova [35] found that colloid centers could be produced in additively colored NaCl under slow cooling and well characterized by the Mie theory [31]. Later on, the metal colloid formation was also intensively studied in detail in NaCl under different irradiation rates and temperatures [36-38]. Interest in the radiation damage of NaCl stems partly from the idea of using old salt mines for a long-time nuclear waste repository. The danger is, however, that Na colloids and Cl-gas accumulated under such prolonged irradiation could recombine as a result of an explosive back reaction, with possible radioactive contamination of the environment. Interesting metal–insulator transitions in Na colloids were observed recently in pure NaCl [39] and doped NaCl- KBF_4 single crystals after heavy neutron irradiation [40]. In these samples, up to about 10% of the NaCl molecules are transformed into extremely small metallic Na particles and Cl_2 precipitates. At high temperatures a single-line ESR signal, i.e., common mode due to a strong exchange interaction between conduction electrons and the *F*-aggregate centres was observed. Different dopants have resulted in quite different colloid densities [41,42] due to different nucleation efficiency.

Metallic colloids have been found in a series of *alkaline earth fluorides*: CaF_2, BaF_2 and SrF_2 Note that CaF_2 is the important material for optical components and nano-lithography, whereas BaF_2 is a good scintillator. Bulk colloid formation after electron irradiation has been investigated using the ESR [43], optical spectroscopy [34,44], and electron microscopy [45]. In ESR measurements, CaF_2 was irradiated with electrons of 75 keV energy. After irradiation, resonance lines typical for conduction band electrons were found, and the results indicate the formation of Ca clusters with a radius < 30 nm [43]. The low-energy electron irradiation has also been performed for CaF_2 crystals [34]. In this study no single *F* centers have been detected, even under irradiation at such low temperatures as 200 K, only a broad absorption peak was observed and attributed to colloids. Upon subsequent heating this peak undergoes a transformation in position and shape until it is stabilized at a certain temperature. It was found that Mie theory well describes this peak.

Figure 2 shows that the extracted colloid radius in CaF_2 rapidly grows up to 20-25 nm within a narrow temperature interval around 250 K where the *F* centers become mobile. The beginning of the *F* center mobility around 250 K is supported by several independent experiments. However, it is unclear why no single *F* centers are observed in the above-mentioned experiments at 200 K. Another question is why colloid growth due to diffusion-controlled *F* aggregation is not accompanied by a significant decrease of the metal area density as a result of the *F-H* center recombination. We address this problem below.

Among *oxides*, Li_2O is the best studied material, due to the fact that it was considered as a potential first-wall cladding material for fusion reactors [46]. At that time Noda and co-workers found an occasional extra line near *g*=2.003 in their EPR spectra, which was attributed to metallic Li clusters, mainly because of its isotropic character. Later Vajda and Beuneu [47] showed that this line is not due to Li colloids but most probably to small nonmetallic aggregates of the F^+ centers.

Careful observations by means of the electronic and optical microscopies, ESR, and microwave conductivity measurements have recently been conducted on electron-irradiated Li_2O. The observations demonstrated a simultaneous formation of two kinds of colloids of metallic lithium, one of which is associated with oxygen bubbles of a typical size of >10 μm, the other consisting of nanoclusters <10 nm [47-50]. They show different conduction electron spin resonance (CERS) signals and exhibit quite distinctive recovery behavior during thermal annealing.

Recently, the structural properties of the metallic Li precipitates were explored in more detail during several neutron scattering experiments. The structure of the large colloids was identified as b.c.c. in agreement with the structure of elemental lithium metal, and a specific orientational relation with respect to the oxide matrix was found [51]. Using SANS, the presence of small non-spherical precipitates with a typical size on the order of 10 nm was detected. Finally, an analysis of the elastic diffuse distortion scattering in the vicinity of the Bragg peaks [50] showed that the small colloids consist of a population of slightly anisotropic particles with a typical size of 5 nm representing Li clusters, which contain about 10^4 atoms. Later it was concluded that the concentration of oxygen nanoclusters, if they are present at all, must be very low in a comparison with the number of the small lithium colloids.

Until now, unambiguous observations of intrinsic metallic colloids in other oxides were reported only for Al in a-Al_2O_3 in a high-voltage electron microscope [52]. After long and unsuccessful efforts to produce colloids under irradiation in MgO [53], Li_2ZrO_3 [54], and $LiAlO_2$ [54,55], Mg colloids were finally produced under high temperature annealing of thermochemically reduced (additively colored) MgO with very high *F* center concentration [56,57].

Speaking of *hydrides*, lithium metal colloids were observed in β - [58] or γ-irradiated [59] LiH. Metallic lithium particles precipitated in LiH under the UV irradiation were recently investigated by ESR, ENDOR, and ESR at low temperatures, under conditions of the bistable Overhauser effect [60]. The ENDOR clearly shows the existence of the two well-defined populations of lithium particles with different crystallographic structures. Al-metal precipitation in irradiated AlH_3 has also been studied [61]. The short analysis of the metallic colloids presented in this section demonstrates a wide range of materials and conditions under which colloids could be produced. In the next section we discuss theoretical approaches for describing this process and the main results obtained for defect aggregation in several ionic materials under very different conditions (electron irradiation, thermochemical reduction, heavy ion irradiation).

2   THEORY

Existing theories of the radiation-induced defect aggregation and colloid formation could be classified, in terms of mathematical formalisms used, into three categories: macroscopic, mesoscopic and microscopic [62-64]. The first, *macroscopic* approach is based on *rate equations,* and contains many (typically about 20) parameters such as reaction rates, which should be found from independent experiments or calculations [36,39,65-67]. Defect interactions are incorporated only via the reation rates. Any spatial inhomogeinities in reactant distribution are neglected here. The latter two effects are taken into account in the *mesoscopic* theory [68-70], which is suited to study conditions (dose rate, defect mobilities and defect interactions) under which spatial local fluctuations in defect densities are no longer damped but increase in amplitude. This results in the breaking of the translational symmetry of the homogeneous defect distribution via the *spatial modulation* of the average defect concentrations. However, this formalism is unable to predict the time development (*kinetics*) of defect accumulation under irradiation, which is experimentally measured.

Lastly, the most refined and accurate microscopic formalism to be used in this paper treats basic processes on the atomic level and uses no *a priori* parameters except diffusion coefficients of defects and their interaction energies, which could be calculated quantum mechanically. Mathematically, the relevant formalism of many-particle densities is based on the coupled kinetic equations for *joint densities* of similar and dissimilar defects (vacancies and interstitials) treated in terms of a modified Kirkwood superposition approximation [62-64,71]. Combination of all three approaches is important for understanding a whole process of colloid formation. Critical comparison of mesoscopic and microscopic formalisms was presented in ref. [72,73].

A qualitatively new feature of our microscopic (atomistic) theory compared to several previous (macroscopic or mesoscopic) approaches is a direct incorporation of the effects of *relative spatial distribution* of similar (*F-F, H-H*) and dissimilar (*F-H*) defects, which is done in terms of three kinds of the *joint correlation functions*, (comma?)F_{ij} (r,t), i, j= *F, H* centers. The spatial distribution of the *F* and *H* centers turns out to be closely related and strongly affect reaction kinetics. For instance, the aggregation of *F* centers results in the decrease of the concentration of close *F, H* pairs and thus increases their average distance, as compared to the random defect distribution. In its turn, this strongly reduces the rate of the *F, H* recombination. Another new feature of theory is incorporation of *similar defect interactions* into diffusion-controlled kinetics, which qualitatively changes the defect average mobility due to similar defect aggregation, and thus the reaction kinetics under study. As a result, our kinetic equations, unlike previous simple theories, turn out to be strongly non-linear. In fact, we solve a set of 35 coupled equations for the macroscopic defect concentrations, joint correlation functions, effective interaction potentials, etc [64,71].

In our calculations dealing with the initial stages of colloid formation, we consider the creation of *F* and *H* centers with a given dose rate and their recombination when, during their random walks in a lattice with the diffusion coefficients D_F and D_H, defects approach each other to within the nearest neighbour (NN) distance. Defect interaction is incorporated in our model via NN attraction energies between similar defects, E_{FF} and E_{HH}. Thus, the input parameters for colloid growth simulations are activation energies for diffusion, E_F and E_H, and attraction energies between defects, as well as the temperature and dose rate p (which in present experiments is estimated as 10^{17} cm^{-3}s^{-1}). The activation energy for *H* center diffusion in LiF is 0.15 eV [74] whereas for the *F* centers the relevant energy is uncertain. This is why we have modelled the F aggregation kinetics varying the F center activation energy from 1 to 1.5 eV. In CaF_2 we use the diffusion energies for the H and F centers to be 0.46 eV and 0.7 eV [75]. The attraction energies determining F and H center attachment/detachment to/from similar-particle aggregates are even less well known. Calculations of the elastic interaction between the two nearest F or H centers in KBr yield an attraction energy of about 0.04 eV [76]. We also used this value in our calculations as an initial guess. For simplicity, we assume that similar particle interactions are equal for different defects, i.e., $E_{int}=E_{FF}=E_{HH}$.

We calculate the *time development* (kinetics) of both electron and hole defect *total* concentrations (which are densities of all defective lattice sites, i.e., the total concentration of F-type centers is nothing but a combination of single, dimer, trimer and higher aggregate concentrations), $n_F(t)=n_H(t)$, as well as concentrations of single (isolated) defects (no other defects in NN lattice sites) and *dimer* defects (two similar defects are NN). Larger aggregates are characterized by the integral values of the *number of particles* N_F, N_H therein and their *radii* R_F and R_H. Large aggregates of the F centers transform into metal colloids.

3   MAIN RESULTS FOR COLLOID FORMATION UNDER ELECTRON IRRADIATION

3.1   LiF

During LiF electron irradiation at 300 K the H centers are quite mobile. The temporal evolutions of the total H-type center concentration, as well as single H and dimer H_2 centers are plotted in Fig. 3 [30]. It is seen that, due to fast aggregation of the H centers at the very early stages of irradiation, the concentration of H_2 centers rapidly increases and approaches that of single H centers. This is why at a short time t_0 the H center concentration begins to decrease. Intensive H-aggregation leads to an almost simultaneous decrease of the density of H_2 centers due to a growth of larger H aggregates at time t_1. This process is greatly enhanced by sample heating after irradiation.

In its turn, up to 100 minutes all F centers are immobile and thus remain single. Their concentration grows linearly with irradiation time (Fig. 3) exceeding $3\,10^{20}$ cm^{-3}; i.e., several percent of the fluorine lattice sites are occupied by F centers. This is in agreement with experimental data [9]. F center aggregation begins only after sample heating after irradiation up to 350 K. In Fig. 4 we have plotted the decay of the total F-type defect concentration (A) and concentration of the single F centers (B) for three activation energies of diffusion, varying it from 1 eV to 1.5 eV and assuming the defect *interaction energy* of 0.04 eV. Calculations for 1 eV diffusion energy yield a very sharp *F* center decay clearly contradicting the experimental data in Fig. 2. For the 1.2 eV and 1.5 eV diffusion energies the *F* center concentration decays much more smoothly and the temperature range well agrees with experimental data.

Calculations of the aggregate radius and of the number of *F* centers inside show that for a diffusion energy of 1.5 eV both quantities grow intensively above 350 K, reaching a radius of 22 nm containing more than 10^5 defects at 425 K. Calculations for the *F* center activation energies of 1.2 eV and 1.5 eV (Fig. 4) predict a considerable decay of the total *F* center concentration, due to recombination of a considerable fraction of the isolated *F* centers with aggregates of the *H* centers, which happens when mobile *F* centers start random walks instead of joining other *F* centers or their aggregates. The experiments show a much smaller decay of concentrations and do not yield information on *H* aggregates whose optical absorption lie in UV region of the spectrum. (Probably, in reality, most of the *H* centers are not accumulated in loose aggregates in the bulk, as is assumed in our model, but disappear in dislocation loops, and a large portion of them are desorbed from the surface).

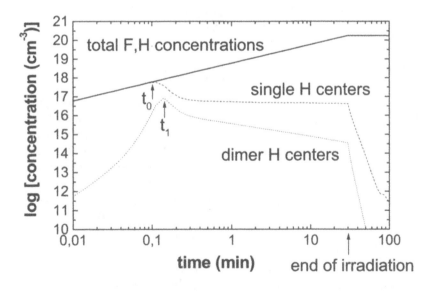

Figure 3. Prediction of the *F* and *H* center and H_2 aggregate center concentrations for a defect production rate of 10^{17} cm^{-3}s^{-1}.

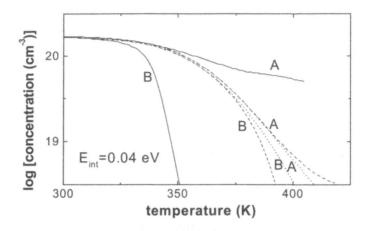

Figure 4. Calculated development of the total concentration of the *F*-type (A) and single *F* centers (B) during heating. The three activation energies used for *F* center diffusion are: 1 eV (solid lines), 1.2eV (dashed lines) and 1.5 eV (dotted lines). The attraction energy between defects was assumed to be 0.04 eV.

Figure 5 demonstrates how strongly the interaction energy between the defects affects the *F* center aggregation kinetics; an interaction energy increase by 25 to 30% (A-B-C) shifts the *F* center decay temperature up by 50 K. Obviously, this is a result of the *F* center recombination with mobile *H* centers escaped from their aggregates: the higher an attraction energy between defects inside the aggregate, the stronger *H* defects are bound to each other and, respectively, the higher the temperature at which a fraction of them can be released from the aggregates and recombine with the *F* centers.

3.2 CaF$_2$

We calculated the kinetics of defect concentration growth under CaF$_2$ irradiation at low temperatures [34,71] when the *F* centers are definitely immobile but the H centers are moving either slowly (150 K) or are already quite mobile (193 K). With the dose rate of p=10^{17} cm^{-3}s^{-1} corresponding to the experimental conditions, until the end of 30 minute irradiation, the defect concentration still grows almost linearly with time; the concentration saturation could be expected after 2 hours of irradiation only. The magnitude of the defect concentration achieved (10^{20} cm^{-3}) agrees well with the experimental value. Simulation of a very intensive irradiation by three order of magnitude exceeding used in the experiments [34], gives a theoretical prediction of the saturation concentration of 10^{22}cm^{-3}s^{-1}. Calculations clearly demonstrate that if one neglects defect interaction, this

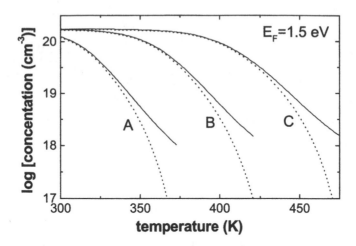

Figure 5. Calculated development of the total concentration of the *F*-type (full lines) and single *F* centers (dotted lines) for three attraction energies between defects: 0.03 eV (A), 0.04 eV (B), 0.05 eV (C). The activation energy for diffusion is 1.5 eV

leads to up to two orders of magnitude reduction in defect concentrations and a very fast recombination when irradiation is switched off.

The analysis of the time development of a total *H*-aggregate concentration, single and dimer centers shows that total concentrations of *H* centers monotonously grow and almost coincide at the two temperatures as $t_0 < 70$ min when irradiation is switched off and the sample is heated up by about 50 K. In contrast, as for LiF (Fig. 3), the concentration of single *H* centers starts to decrease rapidly due to growth of dimer concentration, and the latter also drops at a certain time t_0 due to growth of larger aggregates. The time t_0 decreases by three orders of magnitude when the irradiation temperature increases just by 43 K, from 150 K up to 193 K. The *F* centers at these temperatures are still immobile and their aggregation occurs only under sample heating after irradiation. For the defect interaction energy $E_{int}=0.02$ eV the size of *F* center aggregate (curve 1 in Fig. 6a) and the mean number of defects therein (curve 2 in Fig. 6b) start to grow at 250 K, just when the *F* centers become mobile. This temperature is in good agreement with experimental data discussed above as well as with the temperature of a rapid growth of M_A center concentration under pulsed electron irradiation of CaF_2 doped with Na; this process also occurs through *F* center diffusion [77]. An increase of the interaction energy by 0.01 eV considerably shifts the colloid growth process to higher temperatures (curve 2 in Fig. 6a).

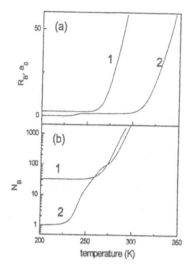

Figure 6. (a) Heating-induced growth of the mean radius R of *F* aggregates in CaF$_2$ (in unit of the F-F distance, 2.73 A). (b) Mean number of defects in each aggregate, N. Curves 1 and 2 in (a) are for defect attraction energies of 0.02 and 0.03 eV, respectively. In (b) curve 1 shows the effect of the radiation-enhanced *F* center diffusion, whereas in curve 2 it is neglected.

For the *F* center diffusion energy of 0.7 eV, no *F* center aggregation can take place below 250 K. However, the electron irradiation experiments reveal small (1-2 nm in radius) aggregates of *F* centers whose size remains constant until heating up to 250 K (Fig. 2). This indicates a process that takes place during irradiation and is characterized by a low activation energy (considerably less than that for the *F* center normal diffusion). These are two speculative *scenarios* of such *radiation-enhanced* processes. (i) Radiation-enhanced *F* center migration when the *F* center traps an unrelaxed hole, which are present in large concentrations under irradiation, converts into an anion vacancy which diffusion energy is about 0.4 eV, makes several hops and transforms again, sooner or later, into the *F* center by trapping an electron. (A similar mechanism was presented in the mid-60's in semiconductors, by Prof. J. Corbett). (ii) Self-trapped excitons possessing diffusion motion above 160 K could decay into a pair of the *F, H* centers not in regular lattice sites but preferentially nearby pre-existing *F* centers (e.g., due to drift in a lattice stress caused by *F* centers). High mobility of the self-trapped excitons in CaF$_2$ is known from an efficient exciton energy transfer from host lattice to impurities in CaF$_2$:Eu [78]. At room temperature its lifetime is estimated to be as long as 10^{-5} s [79]. Decay of excitons nearby defects was demonstrated for doped KCl crystals. In fact, in the second mechanism, *F* center aggregation induced by irradiation is again analogous to the *F* center diffusion but with a reduced activation energy.

To simulate such a radiation-enhanced process, we performed calculations for the *F* center effective activation energy of 0.4 eV under irradiation and with (the?) usual energy of 0.7 eV after irradiation. Curve 1 in Fig. 6b demonstrates formation of stable small *F* aggregates with a radius of several nm. These small aggregates transform into larger aggregates *only* at temperatures above the *F* center mobility edge (remember that the critical temperature depends also on the interaction energy). Unfortunately, due to computational difficulties we cannot follow the latest stage of the large aggregate growth but the reached typical size of the aggregate, R=100 a_0 = 28 nm, agrees qualitatively with the experimental value (25 nm).

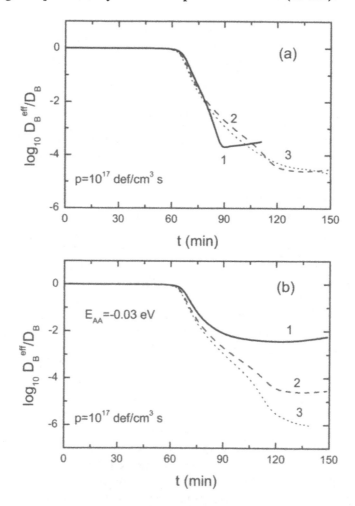

Figure 7. (a) Time development of the dimensionless effective diffusion coefficients for the *F* centers with the attraction energies of 0.02, 0.03 and 0.04 eV (curves 1, 2 and 3, respectively). (b) The effect of unequal interaction energies; E_{HH} = 0.03 eV is fixed, whereas E_{FF}= 0.02, 0.03, and 0.04 eV (curves 1-3, respectively).

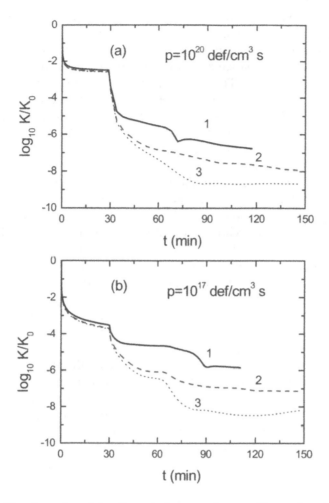

Figure 8. The dynamics of the dimensionless reaction rate K as a function of time, for two dose rates. Curves 1-3 correspond to the attraction energies 0.02, 0.03, and 0.04 eV, respectively.

Analysis of the calculated dynamics of the total F center concentration as a function of the temperature shows that it decreases insignificantly during heating up to 350 K, in agreement with the experiment, thus indicating that most F centers have now transformed to large aggregates.

Important information about the F center aggregation comes from calculations of the *effective diffusion coefficient*, which takes into account defect blocking in aggregates and defect-defect attraction (Fig. 7). As one can see, it decreases by three orders of magnitude at $t=80-90$ min which corresponds to the beginning of the F mobility at T=250 K. Stabilization of

D_{eff} means formation of stable aggregates where defects are bound by mutual attraction. This is in a dynamical equilibrium when the single F centers leave and join the aggregates. At high enough temperatures the aggregates must disappear since attraction energy becomes small compared to the thermal energy kT. The stronger the attraction between defects (curves 1 to 3 in Fig. 7), the lower the effective F center mobility.

Defect aggregation is accompanied with a great decrease in the F and H recombination rate K, as shown in Fig. 8. First, K drops rapidly in the very beginning of irradiation – by three orders of magnitude in about a minute. Its stabilized value does not depend on the attraction energy between defects. This is caused by the fast H center aggregation discussed above. As a result, such H aggregates are quite immobile, which prevents H and F recombination. The main contribution to the recombination comes from newly created centers whose concentration is proportional to the dose rate p. This is why, when irradiation is switched off, the reaction rate reduces additionally, by one-two orders of magnitude (cf. Fig. 8).

Lastly, the spatial correlations of similar and dissimilar defects can well be seen in the joint correlation functions shown in Fig. 9. The absence of the spatial correlation corresponds to the correlation function equal to unity, irrespective of the mutual defect distance. Large values of the correlation functions of similar defects F_{AA}, F_{BB} (A=H center, B=F center, note there logarithmic scale) at short relative distances r clearly demonstrate a strong aggregation of both H and F centers. At the end of irradiation shown in (a) a relative distance where F_{BB} (curve 2) approaches the unity is r=10 a_0. This gives an estimate of the aggregate radius. The effective radius of the H aggregates is larger, 30 a_0. After 25 minutes of heating shown in (b) the radii of the F and H aggregates increase to 30 a_0 and 70 a_0, respectively. The correlation function of dissimilar defects F_{AB} (curve 5) is anticorrelated to F_{AA} and F_{BB}. At the end of irradiation it increases from zero at r< a_0 up to unity at r= 30 a_0 , which gives us an estimate of the average distance between H and F aggregates. The analysis of the correlation functions of empty site with the H or F center (curves 3 and 4) shows that these aggregates have a small, dense core (there are almost no empty sites in their centers) but they are quite loose on their periphery, r>10 a_0.

3.3   Colloid Formation in Thermochemically Reduced MgO

As it was mentioned in the Introduction, annealing of the F-centers in additively colored (also known as thermochemically reduced) alkali halides gives rise to optical extinction bands due to intrinsic metallic colloids (Table 2). These bands have a bell-shaped dependence on the temperature and are usually analyzed using Mie theory. The first observation of this process in MgO single crystals subjected to a very severe thermochemical reduction (TCR) process has been recently reported [56]. Unlike irradiation with particles, such as neutrons and ions, TCR results in stoichiometric excess of substitutional Mg ions, without the presence of oxygen interstitials. The resulting concentration of F centers was extraordinarily large, 6×10^{18} cm^{-3}.

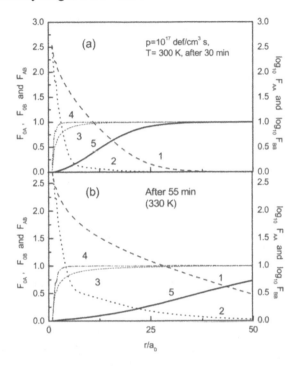

Figure 9. The joint correlation functions Fij (r), i, j= 0 (empty site), A defect (H center), or B defect (F center) vs the relative distance r between defects in CaF$_2$ (in units of F-F distance). (a) corresponds to the end of irradiation whereas (b) shows relative defect distribution subsequent to 25 min heating. Curves 1 and 2 demonstrate the H-H and F-F center spatial correlations; curves 3 and 4 empty site—F center, and empty site--H center correlations; curve 5 correlations of dissimilar (F-H) defects. Note semilogarithmic scale for F$_{HH}$ and F$_{FF}$.

The formation of an extinction band centered at ≈ 3.6 eV, which produces a brown coloration in the crystals, occurs after subsequent annealing in a reducing atmosphere in the temperature range of 1373-1673 K, and has a typical bell-shaped dependence on the temperature (Fig. 10). This extinction band was associated with nanocavities with their walls plated with magnesium with an average size of 3 nm, as imaged by transmission electron microscopy.

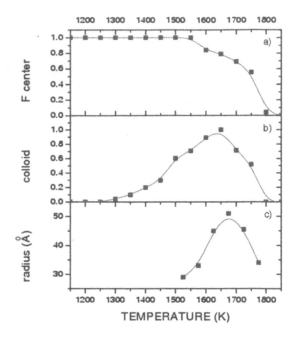

Figure 10. Experimental kinetics of single *F*-center concentration decay and Mg colloid growth [56]. Colloid radius estimate is based on Mie theory, $n_0 = 6 \times 10^{18} cm^{-3}$

The analysis of the *F*-center annealing kinetics in thermochemically reduced MgO crystals has demonstrated that the annealing rate strongly depends on the initial *F*-center concentration: the higher the concentration, the lower the rate [57]. We have proposed that for MgO crystals with low and intermediate *F*-center concentrations ($< 10^{17}$ cm^{-3}) their thermal destruction is due to the more mobile defects such as Mg vacancies or impurities. Only in samples with an extremely high *F*-center concentration ($> 10^{18}$ cm^{-3}), do the F-center *intrinsic* diffusion and aggregation result in unusual extended defects: magnesium-plated nanocavities. The experimentally estimated activation energy for the latter process is 3.4 eV [57], close to the theoretically predicted value of 3.1 eV [80,81] and much larger than generally believed. The present paper describes a model and first simulations the nanocavity formation process.

3.3.1 Defect Aggregation Model

In order to develop a model of metal colloid formation, we will first analyze the experimental data. Figure 10 clearly shows that the *F*-concentration decay and the growth of the Mg-colloid band start simultaneously at the temperature $T_1 \approx 1500$ K, whereas complete destruction of both single F centers and colloids takes place at $T_2 \approx 1800$ K. We assume that at temperature T_1 single F centers become mobile, make random walks in the lattice, and aggregate upon meeting each other in the nearest lattice (NN) sites, giving rise to F-aggregates (F_2, F_3, etc.), leading finally to metallic colloids. In the initial stage of the diffusion-controlled aggregation kinetics, the mean distance, l_0, between single F centers is $l_0 = (n_0)^{-1/3}$, where n_0 is the initial F-center concentration. To form an F_2 dimer center, two single F centers during their random walking for   sec have to diffuse the distance $l_0 = \bullet D(T)$; here $D = D_0 \exp (-E_a/kT)$ is the diffusion coefficient, E_a the activation energy, and D_0 the pre-exponential factor. Using the above-mentioned E_a estimate of 3.5 eV, a typical value of $D_0 = 10^3$ cm^2 s^{-1}, and $n_0 = 10^{18}$ cm^{-3}, we obtain $= 10$ minutes, in good agreement with the experimental data. It is generally believed that at temperature T_2 the F centers disappear at the external sample surfaces. However, under this assumption, defects have to move over a distance $d \approx 0.1$cm (sample thickness), which implies a diffusion time eight orders of magnitude longer than for aggregation! Indeed, increasing the temperature from T_1 to T_2 enhances the diffusion coefficient by only two orders of magnitude and the corresponding walking distance by one order of magnitude, whereas $l_0 = 10^{-6}$ cm and $d = 0.1$ cm differ by 5 orders of magnitude. Thus, *at these temperatures (1500-1800 K) mobile F centers do not have a chance to reach the external sample surfaces and can only annihilate at internal sinks.*

    An additional strong argument in favor of the key role played by the internal sinks in the annihilation of oxygen vacancy defects is the experimental fact that in TCR crystals where the initial F-center concentration is only one order of magnitude smaller, the colloid band is not produced when the crystal is subsequently reduced, even though the F annihilation kinetics remain similar [57]. This means that there are *two* space scales in the F center kinetics: the aggregation scale l_0 and the sink scale l. If $l_0 < l$, colloid formation is controlled by l_0, and l determines the kinetics at longer times (colloid destruction). In the opposite case, $l_0 > l$, it is the sink scale which controls the F annihilation kinetics since defects become trapped *before* they have a chance to meet each other during random walks. The fact that the colloid band is not already formed at $n_0 \approx 10^{17}$ cm^{-3} indicates that l is not very much smaller than l_0. Dislocations, grain boundaries and impurities are likely to be the main internal sinks for the F centers.

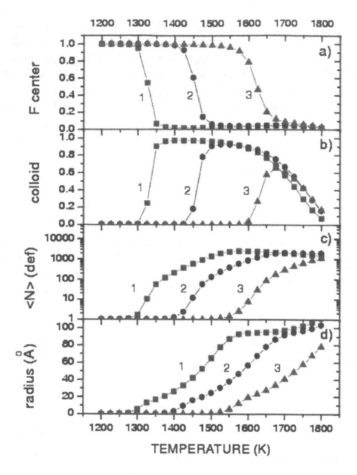

Figure 11. Calculated annealing kinetics for a defect interaction energy of 0.04 eV, $n_0 = 10^{18}$ cm^{-3}, and $n_T = 10^{17}$ cm^{-3}. Curves 1, 2 and 3 correspond to the diffusion pre-exponential factors $D_0 = 10^{-3}$, 10^{-4}, and 10^{-5} cm^2 s^{-1}, respectively. b) gives a fraction of all *F* centers aggregated into colloids.

In this study, we use the simplest model of point unsaturable defect sinks. Its concentration, n_T, determines the sink scale *l*. The *F*-center kinetics is formulated as follows. (i) Single *F* centers perform random walks between the nearest lattice sites characterized by the diffusion coefficient D(T). We fixed the activation energy $E_a = 3.5$ eV and varied the pre-exponential parameter D_0, in order to reproduce semi-qualitatively the experimental kinetics. It should be noted that $D_0 = 10^{-3}$ cm^2 s^{-1} is close to the maximum observed value. Thus, it is not realistic to significantly decrease E_a and to compensate with a corresponding increase of D_0. (ii) NN *F* centers attract each other, which is characterized by the interaction energy ε associated with the elastic attraction of the two close defects due to the overlap of the relevant lattice deformation fields [63]. Due to its short-range nature, the defect interaction is modeled on the lines of the Ising model (NN interaction). (iii) Clusters of *F* centers are *dynamic* formations. Any *F* center on the periphery can detach from the cluster. The delicate balance between aggregation of defects into colloids and colloid destruction at higher temperatures is controlled by the dimensionless factor ε/kT. The probabilities of jumps between two sites on the defective lattice are determined not only by the diffusion activation energy, E_a, characteristic of the perfect lattice, but also by the difference of the defect *interaction* energies with nearest neighbors in these two sites. (iv) Those *F* centers, which turn out to be NN of sinks (traps) instantly disappear. Traps with concentration n_T are randomly distributed over the sample.

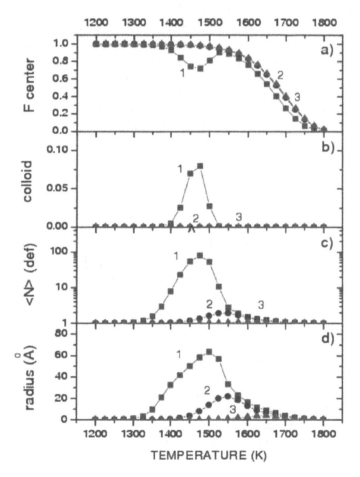

Figure 12. Calculated annealing kinetics for $\varepsilon = 0.03$ eV, $n_0 = 10^{18}$ cm^{-3}, and $n_T = 10^{17}$ cm^{-3}. Curves 1, 2 and 3 correspond to the diffusion pre-exponential factors $D_0 = 10^{-3}$, 10^{-4}, and 10^{-5} cm^2 s^{-1}, respectively.

Summing up, we have only three key parameters -D_0 , and n_T- which makes the solution of the problem quite straightforward. Our previous experience with similar problems suggests that the kinetics depend very much on the value of the defect interaction energy , which is typically quite small, of the order of 0.03 eV [82]. The small value of this parameter is also supported by the experimental data in Fig. 10: the F center concentration drops to zero near T_2, whereas the intensity of the colloid band reaches its maximum around 1600 K when the F center concentration is reduced by only $\approx$ 20%. This means that not every encounter of two F centers results in their aggregation, and defect pairs (dimers) easily separate. This is why only a fraction of the F centers aggregate into very loose clusters. The strong sensitivity of the kinetics with the interaction parameter makes it possible to determine its value quite precisely.

Calculations were performed using the computer code KINETICA described in detail in ref. [71]. This code was earlier successfully applied to a number of problems, including accumulation of Frenkel defects under irradiation of ionic solids [71] and catalytic surface reactions [82]. The most important calculated properties include: (i) concentrations of single and dimer F centers, (ii) concentration of colloids (defined as clusters containing more than three defects), (iii) mean colloid size and number of defects therein, all as a function of temperature. Not going into details, it should be noted here that we use microscopic formalism treating all elementary processes at atomic scale. A qualitatively new feature of this approach (described in detail in ref. [64]) – unlike the usual macroscopic rate-equations – is a direct incorporation of the effects of relative spatial distribution of the F centers, which is not assumed to be random. As a result, the reaction rates become dependent on both time and reaction-induced spatial distribution of the F centers which we calculate.

Another new feature of our theory is the incorporation of the defect elastic interaction, which considerably reduces the average mobility of the F centers due to their dynamical aggregation. As a result, our kinetic equations contain functionals of the joint correlation functions, unknown in the usual rate-equation approach. This makes our kinetic equations strongly non-linear and complicated. It should be mentioned that the F center annealing under study cannot be characterized by any certain order of the kinetics since the reaction rates now are time-dependent, and, unlike usual simple mono- or bimolecular reactions, here we have a competition of the reversible F center aggregation and their irreversible annihilation at traps, which are treated on equal ground in our model.

3.3.2 Results

In our calculations we fixed the initial concentrations of F centers, $n_0 = 10^{18}$ cm^{-3}, and traps, $n_T = 10^{17}$ cm^{-3}, as well as the diffusion energy $E_a = 3.5$ eV. The variable parameter is the defect interaction energy. Figs. 11 and 12 show the results of the calculations for $= 0.04$ eV and 0.03 eV, respectively. The three curves in each window correspond to three different values of the diffusion pre-exponential factor D_0. As indicated in the figure captions, from one curve to the next, the diffusion coefficient changes by one order of magnitude. The results in Fig. 11 were obtained for a strong F-center attraction: F centers aggregate very abruptly as soon as they are

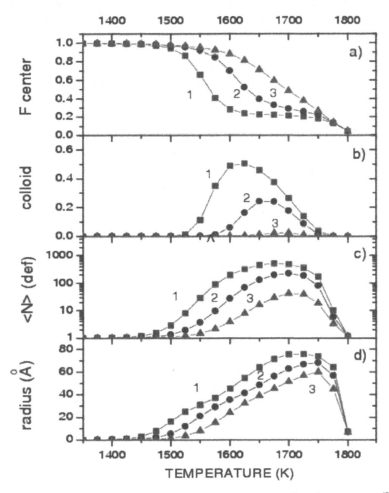

Figure 13. Calculated annealing kinetics for $= 0.035$ eV, $n_0 = 10^{18}$ cm^{-3}, and $n_T = 10^{17}$ cm^{-3}. Curves 1, 2 and 3 correspond to the diffusion pre-exponential factors $D_0 = 10^{-4.25}$, $10^{-4.5}$, and $10^{-4.75}$ cm^2 s^{-1}, respectively.

mobile and the colloid radius is not a bell-shaped function of the temperature, which contradicts the experimental findings. Curves 1 to 3 show that diminishing the F-center mobility by one order of magnitude increases the colloid formation temperature by 150-200 K. Even for such a large mutual attraction energy, the F centers are not completely bound in colloids: Figs. 11c, d and 2d show that there is some re-distribution of F centers among colloids of different sizes due to a dynamic detachment-attachment process. As a result, the mean colloid radius and the number of defects therein grow.

A relatively small variation of the attraction energy leads to qualitative changes. For $\varepsilon = 0.03$ eV, the F-center annihilation takes place in a wide temperature interval of 300-400 K (Fig. 12a), in better agreement with the experimental results. However, the colloid formation turns out to be a very inefficient process and the colloids contain a very small number of defects. Large colloids no longer grow at the expense of small colloids. They dissolve and the released F centers prefer to go to traps. Obviously, the true situation lies in-between the two cases shown in Figs. 11 and 12. In Fig. 13 we present results for the intermediate value of $= 0.035$ eV, and a smaller variation of the diffusion pre-exponential factor. Here, curves 2 resemble the experimental data: the temperature interval for the F-center decay is 300 K, and the peak temperature of the colloid formation and colloid radius curves also resemble those of the experiment. In addition, the colloid formation curve reveals a bell-shaped temperature dependence, with 25% of the F-centers immersed in the aggregates. This translates to a typical colloid radius of about 70 Å containing a few hundred F centers.

This result is consistent with experiments using microdiffraction, x-ray microanalysis, and high resolution electron microscopy, which demonstrated that the observed defects are indeed nanocavities with their walls plated with magnesium. For the same D_0 pre-exponential factor, decrease of the initial F concentration by only one order of magnitude results in a reduction by a factor of 2-3 of the magnitude of the colloid formation. This result indicates that most of the mobile F centers are effectively trapped before they have a chance to aggregate.

Summing up, our model [83] provided a solid basis for analysis of the diffusion-controlled aggregation of the F centers in thermochemically reduced MgO single crystals. Coupled with experimental data, the model leads to three basic conclusions: (i) the pairwise interaction energy between two nearest F centers is one of the key factors of the colloid formation kinetics (a similar conclusion was drawn for the F-center accumulation under irradiation of strongly ionic solids; (ii) F centers do not reach the external sample surface but annihilate at internal defects (dislocations, subgrain boundaries and impurities), (iii) due to an effective trapping-aggregation competition Mg colloids can only be observed in crystals with very high initial F-center concentrations ($> 10^{18}$ cm^{-3}).

Despite the fact that in our microscopic theory we use the only interaction parameter - the *pairwise F* center attraction energy ε- the interaction energy between any single *F* center attached to the cluster of *F* centers is proportional to the number of its nearest neighbours and thus depends not only on the radius of the cluster (colloid) but also on its shape, which is properly incorporated into our theory. The efficiency of this approach was demonstrated in Sections 3.1 and 3.2 for metal colloid growth and Ostwald ripening of small colloids into large colloids in electron irradiated LiF and CaF$_2$. It should be stressed that the kinetic theory presented here is the first successful attempt of semi-quantitative modeling of nanovoid formation in MgO. Further thermodynamical analysis is important for the understanding in detail (of?) an interaction of the mobile *F* centers with dislocations.

3.4 Defect Aggregation in Tracks of Swift Heavy Ions

When passing through dielectric materials, swift (energetic) heavy ions induce a trail of electronic excitations and ionizations of the target atoms. In subsequent processes, the excitations can decay non-radiatively with creation of various types of damage structures in the lattice such as single point defects, defect clusters, local phase transitions, or even decomposition of the solid by radiolysis [84-88]. The peculiarities of the damage process due to energetic heavy ion are mainly determined by two facts: (1) the excitation density is very high, reaching up to several keV/nm^3 close to the ion path, and decreasing with the radial distance approximately as $1/r^2$, and (2) the energy deposition to the target electrons is extremely fast occurring on the time scale 10^{-16} - 10^{-14} s. Compared to this time scale, defects in the lattice are created at a much later stage, around $10^{-12} - 10^{-11}$s. In addition to these specific projectile characteristics, damage creation in a given material depends on many other aspects such as material properties (e.g., the nature of chemical bonding) and irradiation conditions (e.g., temperature).

Although several attempts have been made to give a general description for track formation [92,93], at present the interpretation of damage processes under heavy ion irradiation is difficult and still leaves many open questions. This is also true for track formation in ionic crystals, for which - in contrast to many other insulators - amorphization is not expected. Detailed knowledge exists concerning defect creation upon excitation of the electron subsystem by low ionizing radiation (e.g., gamma rays, electrons, or neutrons), in particular for alkali and alkali-earth halides. Their damage mechanism is based on self-trapping of excitons (either by relaxation of free excitons or by electron-hole recombination) and their subsequent non-radiative decay into Frenkel defects. In this Section 3.4 we focus on lithium fluoride (LiF) in which the most significant stable defects at room temperature are the F-centers (a halogen-ion vacancy with one trapped electron) and complex hole centers. The creation of complex color centers and aggregation of neighboring defects can lead to F_2-, F_3- or F_n- centers, eventually even to metallic alkali colloids, and to the aggregates of the complementary hole centers (V_n- centers and molecular fluorine aggregates). This scenario probably also holds for the irradiation with energetic heavy ions, but the high dose and thus the high density of Frenkel defects close to the ion path certainly plays a crucial role in defect diffusion and aggregation processes and therefore in the final damage structure in the ion tracks.

We discuss below modelling of the aggregation of primary defects created along the trajectory of heavy ions of several hundred MeV. As material, we used LiF crystals mainly due to the fact that a large body of experimental track data exists [87-92] thus providing various parameters relevant for our calculations [85,94].

3.4.1 Experimental results and a model of ion-induced damage in LiF crystals

Ion tracks in LiF have been studied by various techniques such as optical absorption spectroscopy, small-angle x-ray scattering (SAXS), chemical etching, and scanning force microscopy [26,85,87,88,95]. Combining the different results, the following track description can be given:

(1) In a large halo region around the ion trajectory, the dominant type of defects are single F- and dimer F_2-centers identified by their absorption bands in the ultraviolet and visible spectral range between 200 and 900 nm. The complementary hole centers are absorbing at room temperature in the vacuum ultraviolet region. They were not studied here but are known to coexist. The efficiency of the creation of single defects is approximately the same as under conventional irradiation. At higher fluences, the spectra become more complex due to track overlapping. Although the position of some F_n bands is known (Table 2), the analysis of such aggregates is not straightforward. Finally, it should be mentioned that the absorption bands of Li colloids are expected to be close to the F_n bands, but they are not so well known and it is difficult to identify them in complex optical spectra. From the evolution of the F-center concentration as a function of the ion fluence, the radius of the halo surrounding the track core can be deduced. It varies from 5 nm for light ions (e.g., S, 1.6 MeV/u, 4.3 keV/nm) up to 40 nm for heavier ions (e.g. Au, 11.4 MeV/u, 24 keV/nm) [85,87].

(2) If the ions surpass a critical energy loss of about 10 keV/nm, a new effect occurs, namely the creation of complex defect clusters in a narrow cylindrical core with a radius between 1 - 1.5 nm [27,85]. The size of this track core has been determined from analyzing the highly anisotropic SAXS pattern, which is due to a modified electronic density $\Delta\varrho$ in the tracks (the scattering is proportional to $\Delta\varrho^2$). Note that the single F-centers do not contribute to the SAXS contrast. Above this threshold, tracks can be attacked by a suitable etchant. Both phenomena are stable up to much higher temperatures than single F- and hole centers exist. Although the nature of the specific damage in the core is not studied so far, several observations indicate that the track consists of a quasi-cylindrical, discontinuous array of defect aggregates.

Track formation in LiF crystals was tested in a wide temperature range, from 15 to 750 K [26]. The most remarkable fact is that the track core is created even when the irradiation is performed at 15 K. This is in clear contrast to conventional irradiations (x-rays or fast electrons), where aggregation processes at such a low temperature are strongly suppressed because primary Frenkel defects are not mobile. Based on the weak temperature dependence of the SAXS radius, we estimated that during irradiation, the local temperature around the ion path increases by T = 1200±100 K [87].

From a general point of view, aggregation of F-centers into complex clusters is only possible under the following specific conditions [4,87]: 1) F-centers and hole centers must be separated in space, 2) the concentration of 'surviving' F-centers must be sufficiently high ($C_F \geq 10^{21}$ cm^{-3}), and 3) diffusion of F-centers must be possible.

The concentration of single F defects in the track core, before the aggregation process starts, can be deduced from optical measurements where we find a typical value of about $C_F \approx 2\times10^6$ F-centers in the halo region of a single track [87]. Monte Carlo calculations of the lateral energy distribution [87,96] demonstrate that about 30 % of the total ion energy is deposited in the track core and about 70 % in the halo. Assuming similar defect creation efficiencies in both regions, the number of single point defects in the track core should also be around 10^6. The volume of the track core is given by the cylinder radius (as deduced from the SAXS experiments) and the length, corresponding to the projected ion range. Taking a projectile of 10 MeV per nucleon with a typical range of 80 μm and a radius of 1.5 nm, the track volume is 5.7×10^{-16} cm^3 (corresponding to about 3.4×10^7 (Li$^+$ - F) ion pairs). It follows that the concentration of primary Frenkel pairs in the track core is about 2×10^{21} cm^{-3} which is about 3 % of all lattice sites. As mentioned above, this number is certainly high enough for the F center efficient aggregation [4]. From conventional irradiations it is known that defect aggregation takes place via the *radiation enhanced diffusion* (RED), i.e., F-centers migrate in the electronically excited state (2*p* in terms of the H-atom model) and thus, with a much lower activation energy than under conventional thermal activation. In alkali halides, a typical activation energy for *F*-center diffusion in the excited state is about $Q_F \approx 0.2$ eV [97,98]. If we assume diffusion of excited *F*-centers, the lifetime of the excited state τ_F determines the diffusion or aggregation time. In alkali halides, τ_F is typically of the order of nanoseconds [91,92]. However, the local temperature rise (thermal spike) to T* is expected to decrease this lifetime to about $\tau_F = 10^{-11}$ s due to thermal quenching [91].

Now we can estimate the diffusion length $<x_H>$ of primary *H*-centers. In contrast to the excited *F*-centers, the activation energy for *H*-center diffusion in the ground state is well known and has a smaller value of $Q_H \approx 0.1$ eV [92]. At the local temperature T* = 1215 K, the diffusion coefficient $D_H \approx 5.10^{-4}$ cm²/s and thus the diffusion length for the *H*-centers becomes $<x_H> \approx 2$ nm, larger than $<x_F>$. Considering that RED is also effective for hole centers, $<x_H>$ would be even larger. Thus the conclusion could be drawn that under given track conditions, *F*-centers and *H*-centers are obviously well separated in space and thus an efficient aggregation of *F*-centers can take place.

3.4.2   Computer modeling of defect aggregation

Based on the above-estimated experimental parameters, we performed Monte-Carlo computer simulations of the *F*-center aggregation kinetics. The basic model has the following assumptions [94]:

(i) Defects are created in a simple cubic lattice of size L in the *z* direction and of infinite size in the *x* and *y* directions. (The *z*-axis is parallel to the trajectory of the projectile.)

(ii) The initial distribution of defects is described by the exponential function c(r)= c_0 exp(-r/a), where *r* is the distance from the ion path, and the initial density c_0 at the track center and the radius *a* can be varied. The total number of *F*-centers in a whole track is given by

$$N(\infty) = \int_0^\infty c(r)\, L\, 2\pi\, r\, dr = 2\pi a^2 L c_0 \qquad (1)$$

and the fraction of defects inside a cylinder of some arbitrary radius *r* is

$$N(r)/N(\infty) = 1 - (1 + \frac{r}{a})\, e^{-\frac{r}{a}}. \qquad (2)$$

According to Eq. (2), the fraction of defects created inside a cylinder with radius *r* = *a* equals N(a)/N(∞) = 26%, in a qualitative agreement with the estimations of the lateral energy distribution [87].

(iii) Hole H-centers are more mobile than F-centers and are separated from the F-centers before these latter start to migrate. Note that in LiF, H-centers are known to aggregate to fluorine (F_2) molecules as point defects or as molecular gas bubbles. In this case, their recombination with F-centers is effectively prevented [99]. Based on this, our model calculations on the F-center aggregation do not take into account H-centers.

(iv) Single F-defects start to perform random hops at $t=0$ and become immobile if they meet another defect at one of the nearest lattice sites. As estimated above, the typical diffusion time is $\tau_F = 10^{-11}$ s, which corresponds to a number of F center hops M ≈ 25.

A series of simulations has been performed using the following dimensionless parameters. The length of the ion track was fixed as $L=100$, which is sufficiently large to exclude finite-size effects along the z-axis and restrict the calculations to a reasonable computing time (proportional to L^2). The distance r is given in F-F separation units of eq = 0.28 nm. The initial defect density c_0 was varied from 0.025 to 0.6 in steps of 0.025, including the above-estimated experimental value (c_0=0.03).

The distribution of the F- clusters, the fraction of surviving single defects and their radial distribution was recorded during random walks where the number of hops was varied between M=0 and 100. We define as a *cluster* F_n aggregate center with $n \geq 2$, i.e., containing two or more single F-centers. With this assumption, we do not include structural properties of smaller (F_2-) centers, or larger aggregate centers (Li colloids) in the lattice. Our simulations show that the results depend only weakly on the initial core radius a, therefore only data for $a=5.0$ are presented.

3.4.3 Main results

Let us discuss now the main results. Figure 14.a shows the fraction of single F-centers (C_F) as a function of the number of defect hops. Depending on the initial defect densities c_0, some of the F-centers are already statistically aggregated even *before* starting random walks: 13% at c_0=0.1, 35% at c_0=0.35, and 46% at c_0=0.55. In the course of diffusion, the number of single defects decreases significantly, reaching after 25 hops a fraction of $C_F = 29\%$, 12%, and 9%, respectively. The evolution of C_F versus the number of hops, proportional to the diffusion time t, follows a simple power-law $C_F \propto t^{-0.6}$. This asymptotic dependence is the same for all initial defect densities tested here.

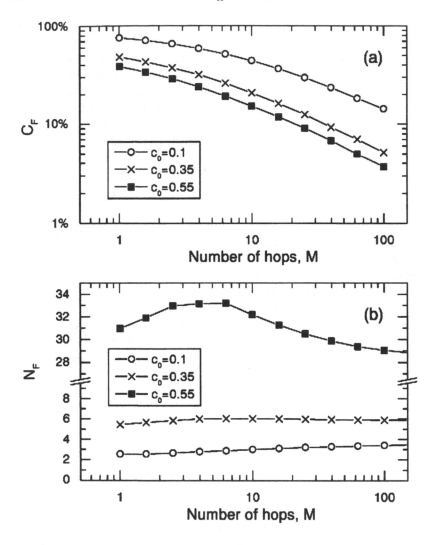

Figure 14. (a) The fraction of single F centers (C_F) with respect to all F centers, and (b) a mean number of F-centers in clusters (n) as a function of the number of hops for different initial defect densities c_0.

The average number of F-centers per cluster (N_F) as a function of the hop number is presented in Fig. 14.b. For a realistic initial defect density ($c_0 = 0.1$), n is extremely small (2-3) and remains nearly constant during the entire hopping process (up to M=100). In other words, the average cluster size varies very slowly with time. For a large defect density (c_0=0.55), N_F slightly increases in the initial stage and then, above M=6, returns back to the starting value. This means that formation of many small clusters in a halo start to dominate over larger clusters formed after first F jumps in the core.

The calculated fraction of single F-centers (C_F) as a function of the initial defect density c_0 can be fitted by a power law $C_F \propto c_0^{-\alpha}$ with the exponent $\alpha = -0.84 \pm 0.01$. The mean number of F centers in clusters versus c_0 is almost independent of the aggregation time but strongly increases for initial densities $c_0 \approx 0.4$.

Figure 15 shows the fraction of F-centers (C_F) in relatively big aggregates ($n \geq 10$) with respect to all F-centers (single and aggregated defects). For intermediate and high initial densities ($c_0 \geq 0.2$), defect hopping obviously results in a significant increase of big aggregates. Note however, that for the initial core density of $c_0 = 0.1$ (most realistic case), the fraction of F centers in big aggregates is only 0.005 and therefore extremely small.

The radial distributions of single F-centers for two different initial defect concentrations is presented in Figure 16.a ($c_0 = 0.10$) and Fig. 16.b ($c_0 = 0.35$). As mentioned above, the fraction of F centers which aggregated before the hopping process is 13% ($c_0 = 0.1$) and 35% ($c_0 = 0.35$) (cf. Fig. 14). With the increasing number of hops, the radial distribution of the single F-centers changes, moving out from the track axis to the larger distances. After $M = 100$, the distribution maximum has shifted to $(15-20)\lambda$ length units (corresponding to $\approx 4.2 - 5.6$ nm) for $c_0 = 0.1$ and even 25λ (≈ 7 nm) for $c_0 = 0.35$. For the small initial defects densities (Fig. 16.a), there is still a considerable amount of single F defects close to the ion path, while for large densities (Fig. 6.b), most single defects close to the ion path are aggregated even before the hopping process. In the legend of Fig. 16, we also indicated the fraction of single defects survived M hops. For example, after M=25 hops at the concentration $c_0 = 0.1$, C_F decreased from the initial 87% down to 29% (see also Fig. 14.a). This is in a qualitative agreement with the experimentally observed F-center concentration in the track's halo.

The dynamics of F-center aggregation, illustrated in Fig. 17, demonstrates in more detail quite different situations inside (core) and outside (halo) of a fixed track radius $a = 5\lambda$. In the initial stage, concentration of single F-centers within the core region (solid curve) exceeds that in the halo (broken curve). However, after 25 hops, C_F in a core decreases from initial 44% down to 5%. That is, single F-centers survive mainly in the halo (decrease from 32% down to 12%). This can be explained by the obvious fact that at a larger distance from the ion path, the density of F centers is much smaller and the probability for two F-centers to meet and thus to aggregate becomes less and less likely.

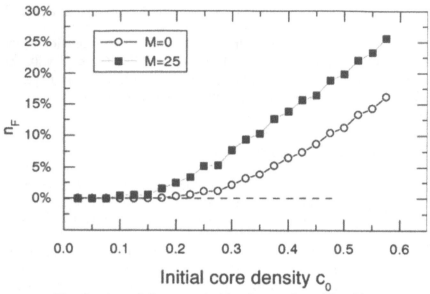

Figure 15. The fraction of *F*-centers in big clusters ($n > 10$) with respect to all F-centers (single and aggregated) versus the initial defect densities c .

Finally, in Fig. 18, we visualize the formation of large *F*-center aggregates ($n > 10$) in ion tracks for various initial defect densities. For $c_0 \cdot$ 0.25, the aggregates are well separated from each other and do not form a continuous trail. Therefore, an electrical conductivity based for example on percolating small metallic Li clusters should not be expected, in agreement with test experiments [100].

Summing this section up, Monte Carlo computer simulations were performed for the first time for aggregation kinetics of *F*-centers created along the trajectory of swift heavy ions. Main parameters, such as the migration activation energy, temperature in the track core, initial defect concentration, or diffusion time were estimated from the experimental data available for LiF. From the optical absorption measurements [87,88,94] one can conclude that in the core about 3 % of the lattice sites are occupied by the *F*-centers. In the initial state of damage creation, the number of defects is probably slightly larger. We therefore used $c_0 \approx 0.1$ as a realistic value. In order to explain the efficient aggregation of the *F*-centers, we have to assume that they diffuse in the electronically excited state. In addition, *F*-center aggregation requires a moderate local heating within the track. We expect that the temperature there does not exceed the melting point because exciton-based damage creation mechanism in alkali halides takes place in the solid phase. The short life time ($\approx 10^{-11}$ s) of the excited *F*-centers corresponds to about 25 random hops on the lattice. For a larger number of hops, the simulation results show only small modifications, i.e., within this time limit a quasi-stationary state of aggregation is already reached.

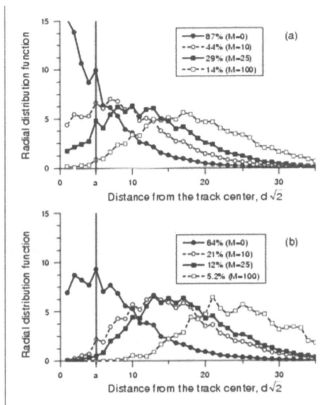

Figure 16. The radial distribution of isolated defects for defect density (a) $c_0=0.1$ and (b) $c_0=0.35$, for various hop numbers M. The percentage in the legend gives the fraction of single F-centers after the given number of hops (see also Fig. 14.a). The core radius a = 5 (~1.4 nm) is marked by the vertical line. All distribution curves are normalized to unity.

According to the calculations, most defect clusters are very small and typically consist of two-three F-centers. The fraction of single F-centers decreases with the aggregation time reaching a quasi-equilibrium of about 30 % after 25 hops. They occupy mainly lattice sites in the track's halo, i.e., a few nanometers away from the original ion path. Larger F-aggregates (e.g., $n > 10$) are in negligible concentration, even for moderate initial defect concentrations ($c_0<0.3$) and only could play a role for very large concentrations that are probably not realistic. The formation of extremely small defect aggregates along the track is in an agreement with experimental results from small angle x-ray scattering. It also explains the absence of any electron spin resonance as expected for metallic aggregates that are larger than 10 nm. Due to the small size of the aggregates and the absence of larger clusters, a discontinuous defect trail is formed. Such a track morphology also explains why tracks in LiF did not show any increase of electrical conductivity [100].

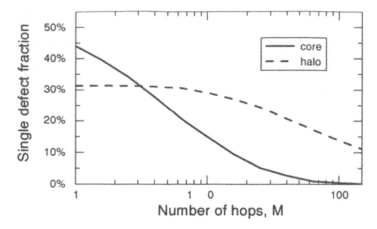

Figure 17. The fraction of single *F*-centers for the initial defect density $c_0=0.1$ as a function of number of hops M. The solid and dashed curves show the concentration kinetics of single *F* centers inside core and in a halo, respectively. Their sum is C_F in Fig. 14.a.

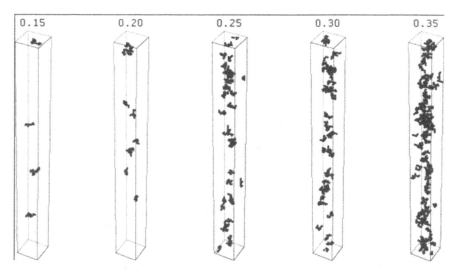

Figure 18. Visualization of the *F*-center aggregates formed along the ion trajectory for various initial defect densities c_0.

Although our model calculations gave good qualitative agreement with the experimental situation, several questions remain open. In particular, the diffusion of *F*-center in the electronically excited (*2p*) state and the efficient initial separation of the electron and hole centers have to be further investigated. It should be mentioned that at present, there is no direct experimental technique suitable for a study of the microstructure of such small *F*-center aggregates. Also, the local temperature increase on the time scale of 10^{-11} s is very difficult to access.

4   CONCLUSIONS

In this Chapter we presented results of theoretical analysis of metallic colloid formation created under quite different conditions, including low-energy electron irradiation, thermochemical reduction and irradiation with swift heavy ions. It is generally believed (see Kinoshita Chapters) that electron irradiation creates well-separated point defects. The same is in general true for   irradiation with a moderate density; neutrons produce more defect clusters, whereas swift heavy ions create tracks with high defect density in a core. To address these problems, we used analytical methods and Monte Carlo computer simulations. This approach could be applied also for a study of radiation damage in other materials, e.g., UO_2 nuclear fuel containing -emitters.

It should be stressed that metallic colloids in insulating matrices play an important role in many applications. For example, in silver halides they are responsible for latent imaging used in photography. In such optical materials as CaF_2, LiF, and MgF_2, which are used as deep-UV window materials, defect mobility and colloid study is important for the control of a radiation hardness. Colloid Li forms observed under irradiation at the elevated temperatures in Li_4SiO_4 are envisaged as Li-containing materials for the blanket zone of fusion reactors [101]. Colloid Li serves as an effective tritium scavenger forming LiT having a high decomposition temperature (about 1000 K).

Our microscopic computer modeling demonstrates that the initial stages of the *F* center aggregation kinetics and the relevant metal colloid formation depend on *both* the *F* center mobility (controlled by the activation energy for diffusion) and relatively weak defect attraction energies inside aggregates of *F* and *H* centers (E_{FF} and E_{HH}, respectively).

Similar defect aggregation and colloid formation strongly reduce the recombination rate of primary defects and thus permit to accumulate orders of magnitude larger total defect concentrations, as compared to the kinetics for non-interacting defects.

Acknowledgements. The authors are grateful to V. Kuzovkov, R. Gonzalez, M. Reichling, P. van Uffelen, V. Kashcheyevs, and K. Schwartz for collaboration and many stimulating discussions.

REFERENCES

[1]   A. E. Hughes, Radiation Effects, **74**, 57 (1983).

[2]   F. Seitz, Rev. Mod. Phys. **26**, 7 (1954).

[3]   J. H. Schulman and W. D. Compton, *Color Centers in Solids*, Pergamon Press, Oxford, 1962.

[4]   A.E. Hughes and S.C. Jain, Adv. Phys. **28**, 717 (1979).

[5]   C.F. Bohren and D.R. Huffman, Absorption and Scattering of Light by Small Particles, (Wiley, New York, 1983.)

[6]   K.K. Schwartz and Yu. Ekmanis, *Dielectric Materials: Radiation-induced Processes and Stability*, Zinatne, Riga, 1989 (in Russian).

[7]   L.H.Abu-Hassan, P.D.Townsend, R.A.Wood, Nucl Instr Meth. B **32**, 225 (1988); J. Nahum and D. A. Wiegand, Phys. Rev. **154**, 817(1967); J. Nahum, Phys. Rev. **174**, 1000–1003 (1968); R. M. Macfarlane, A. Z. Genack and R. G. Brewer, Phys.Rev.B **17**, 2821(1978).

[8]   V.V.Ter-Mikirtychev and T.Tsuboi, Progr. Quant. Electr. **20**, 219 (1996).

[9]   G.O Amolo, J.D. Comins, A.T. Davidson, Nucl Instr Meth B **218,** 244 (2004).

[10]   W.Hayes, *Crystals with Fluorite Structure*, Oxford, 1974.

[11]   W.B.Fowler, *Physics of Color Centers*, Academic Press, 1968.

[12]   M. Lambert, A. Guinier, C. R. Hebd, Seances Acad. Sci. **246**, 1678 (1958).

[13]   P. J. Ring, J. G. O'Keefe, and P. J. Bray, Phys. Rev. Lett. **1**, 453 (1958).

[14]   Y. W. Kim, R. Kaplan, and P. J. Bray, Phys. Rev. **117**, 740 (1960).

[15]   Ch. Ryter, Phys. Rev. Lett. **5**, 10 (1960).

[16]   M. Lambert, Ch. Mazieres, and A. Guinier, J. Phys. Chem. Solids **18**, 129 (1961)

[17]   A. Van den Bosch. J. Phys. Chem. Solids **25**, 1293 (1964)

[18]   A. Van den Bosch. Radiat. Eff. **19,** 129 (1973)

[19]   C. Taupin, J. Phys. Chem. Solids 28, 41 (1967).

[20]   S.G. L'vov, F.G. Cherkasov, A.Ya. Vitol and V.A. Silaev, Appl. Radiat. Isotopes **47**, 1615 (1996).

[21]   A. T. Davidson, J. D. Comins, T. E. Derry, and F. S. Khumalo, Rad. Eff. **98**, 305 (1986).

[22]   N. Seifert, S. Vijayalakshmi, Q. Yan, J. L. Allen, A. V. Barnes, R. G. Albridge, and N. H. Tolk, Phys. Rev. B **51**, 16 403 (1995).

[23]   Yu. I. Didchik, A.P. Shkandarevich, Yu.A. Ekmanis, Sov. Optics Spectr. **65**, 551 (1989).

[24]   A.T. Davidson, J.D. Comins, A.M. Raphuthi, A.G. Kozakiewicz, E.J. Sendezera, T.E. Derry, J. Phys.: Cond. Matter **7**, 3211 (1995).

[25]   N. Seifert, S. Vijayakshmi, Q. Yan, A. Barnes, R. Albridge, H. Ye, N. Tolk, W. Husinski, Rad. Eff.& Def. Solids, **128**, 15 (1994).

[26]   K. Schwartz, G. Wirth, C. Trautmann, T. Steckenreiter, Phys Rev B **56**, 10711 (1997).

[27]   C. Trautmann, K. Schwartz, O. Geiss, J. Appl. Phys. **83**, 3560 (1998); C. Trautmann, K. Schwartz, and T. Steckenreiter, Nucl. Instr. Meth., **B 156**, 162 (1999)

[28]   F. Sagastibelza, J.L. Alvarez Rivas, J. Phys.C: Sol. St. Phys. **14**, 1873 (1981).

[29]   T.A. Green, G.M. Loubriel, P.M. Richards, N.H. Tolk, R.F. Haglund, Phys. Rev. B **35**, 781 (1987).

[30]   N.Bouchaala, E.A.Kotomin, V.N.Kuzovkov, M.Reichling, Sol. St. Comm., **108,** 629 (1998)

[31]   G. Mie, Ann. der Physik **25**, 377 (1908).

[32]   M. Lambert, Ch. Mazieres, A. Guinier, J. Phys. Chem. Sol. **18**, 129 (1961).

[33]   S. Bronshteyn and A Protsenko, Rad. Eng. Electr. Phys. **15**, 677 (1970).

[34]   M. Huisinga, N. Bouchaala, R. Bennewitz, E.A. Kotomin, M. Reichling, V.N. Kuzovkov,W. von Niessen, Nucl. Inst. Meth., B **141**, 79 (1998).

[35]   M. Savostianova, Z.Phys. **64**, 262 (1930).

[36]   U. Jain and A.B. Lidiard, Phil. Mag., **35**, 245 (1977).

[37]   A.E.Hughes and B. Henderson, in: *Point Defects in Solids* (ed. J.H. Crawford and L.M.Slifkin), Plenum Press, London, New York, 1072.

[38]   A.E. Hughes and S.C. Jain, Adv. Phys. **28**, 717 (1979).

[39]   V.I. Dubinko, A.A. Turkin, D.I. Vainstein, H.W. den Hartog, J. Nucl. Mater. **304**, 117 (2002).

[40]   F.G. Cherkasov, S.G. L'vov, D.A. Tikhonov, H.W. den Hartog and D.I. Vainshtein. *J. Phys.: Condens. Matter* **14**, 7311 (2002).

[41]   H.W. den Hartog, D.I. Vainshtein, V.I.Dubinko, A.A.Turkin, V.V.Gann, J.Jacobs, *Radiation Damage in NaCl:retrievability, smart backfill materials, monitoring CORA Research Projectfor Dutch Ministry of Economic Affairs Final Report,* 1999.

[42]   H.W. den Hartog, J.C. Groote, and J.R. Weerkamp, Rad. Eff and Def. Solids, **139** (1996) 1; H.W. den Hartog, Rad. Eff and Def. in Solids, **150,** 167 (1999).

[43]   S. D. McLaughlan and H W Evans, Phys. Stat. Sol. **27**, 695 (1968).

[44]   R.Alcala and V.Orera, J.de Physique **C7** 520 (1976).

[45]   E. Johnson and L.T.Chadderton, Rad. Effects **79**, 183 (1983).

[46]   K. Noda, K. Uchida, T. Tanifuji and S. Nasu, Phys.Rev.B **24**, 3736 (1981). K. Noda, Y. Ishii, H. Matsui and H. Watanabe, Radiat. Eff. **97**, 297 (1986).

[47]   P. Vajda and F. Beuneu, Phys. Rev. B **53**, 5335 (1996).

[48]   F. Beuneu and P. Vajda, Phys. Rev. Lett **76**, 4544 (1996).

[49]   F. Beuneu, P. Vajda, G. Jaskierowicz and M. Lafleurielle, Phys. Rev. B **55**, 11263 (1997).

[50]   G. Krexner, M. Prem, F. Beuneu and P. Vajda, Phys. Rev. Lett. **91**, 135502 (2003)

[51]   P. Vajda, F. Beuneu, G. Krexner, M. Prem, O. Blaschko and C. Maier. Nucl. Instr. Meth., B **166/167,** 275 (2000)

[52]   T. Shikama and G. P. Pells, Philos. Mag. A **47**, 369 (1983).

[53]   G.P. Pells, Rad. Eff. **64**, 71 (1982).

[54]   M. H. Auvray-Gely, A. Dunlop, and L. W. Hobbs, J. Nucl. Mater., **133/134**, 230 (1985).

[55]   M. H. Auvray-Gely, A. Perez, and A. Dunlop, Philos. Mag. B **57** 137 (1988).

[56]   M.A. Monge, A.I. Popov, C. Ballesteros, R. González, Y. Chen, and E.A. Kotomin, Phys. Rev. B **62**, 9299 (2000).

[57]   A.I. Popov, M.A. Monge, R. González, Y. Chen, and E.A. Kotomin, Solid State Comm.**118**, 163 (2001).

[58]   F.E. Pretzel, D.T. Vier, E.G. Szklarz, and W.B. Lewis, Los Alamos Scientific Laboratory Report No. LA-2463, 1961.

[59]   A. Berthault, S. Bedere, and J. Matricon, J. Phys. Chem. Solids **38**, 913 (1977).

[60]   C. Vigreux, P. Loiseau, L. Binet, and D. Gourier, Phys. Rev B **61**, 8759 (2000).

[61]   O. Zogal, P. Vajda, F. Beuneu and A. Pietraszko, Eur. Phys. Journal **B2**, 451 (1998).

[62]   V.N. Kuzovkov and E.A. Kotomin, Rept. Progr. Phys., **51,** 1479 (1988).

[63]   E.A. Kotomin and V.N. Kuzovkov, Rept. Progr. Phys., **55**, 2079 (1992).

[64]   E.A. Kotomin and V.N. Kuzovkov, *Modern Aspects of Diffusion-Controlled Reactions*, vol. **34** in a series *Comprehensive Chemical Kinetics*, (Elsevier, Amsterdam, 1996).

[65]   S.J. Zinkle, Rad. Eff and Def. in Solids, **148,** 447 (1999).

[66]   J.R.W. Weerkamp, J.C. Groote, J. Seinen and H.W. den Hartog, Phys.Rev. **B50,** 9781 (1994)

[67] V.I. Dubinko, A.A Turkin, D.I. Vainshtein, H.W. den Hartog, Rad. Eff and Def. in Solids, **150**, 145; 173 (1999).

[68] G. Martin, Phil. Mag., **32**, 615 (1990).

[69] E.A. Kotomin, M. Zaiser, W.J. Soppe, Phil. Mag., A **70**, 313 (1994).

[70] J.A. D. Wattis, and P. V. Coveney, Phys Chem Chem Phys., **1**, 2163 (1999).

[71] V.N. Kuzovkov, E.A. Kotomin, W. von Niessen, Phys. Rev. B **58**, 8454 (1998)

[72] E.A. Kotomin, V.N. Kuzovkov, M. Zaiser, and W.J. Soppe, Rad. Eff and Def. in Solids, **136**, 209 (1995).

[73] E.A. Kotomin, V.N. Kuzovkov, Phys. Scripta, **50**, 720 (1994).

[74] N. Itoh and K. Tanimura, J. Phys. Chem. Sol., **51**, 717 (1990).

[75] K. Atobe, J. Chem. Phys. **71**, 2588 (1979).

[76] K. Bachmann and H. Peisl, J. Phys. Chem. Sol. **31**, 1525 (1970).

[77] V.M. Lisitsyn and V.Yu. Yakovlev, Rus. Phys. Journal, **23**, No 3, 110 (1980).

[78] K.A. Kalder and A.F. Malysheva, Sov. Optics Spectr., **31**, 252 (1971).

[79] R.T. Williams, M.N. Kabler, W. Hayes, and J.P.Stoll, Phys. Rev. **B14**, 725 (1976).

[80] A.I. Popov, E.A. Kotomin, M.M. Kukla, *Phys. Stat. Sol.*, **195**, 61 (1996).

[81] E.A. Kotomin and E.A. Popov, *Nucl. Inst. Meth.*, B **141**, 1 (1998).

[82] J. Mai, V.N. Kuzovkov, W. von Niessen, J. Phys A **29**, 6205, 6219 (1996).

[83] V.N. Kuzovkov, A.I. Popov, E.A. Kotomin, M.A. Monge, R. Gonzalez, and Y. Chen, Phys. Rev. **B 64**, 064102 (2001).

[84] R.L. Fleischer, Nuclear Tracks in Science and Technology, in: *Tracks to Innovation,* Springer,Berlin, 1998.

[85] C. Trautmann, M. Toulemonde, K. Schwartz, J. M. Costantini, and A. Müller, Nucl. Instr. Meth. **B 164**, 365 (2000).

[86] A.I.Popov and E.Balanzat, Nucl.Instr.Meth.**B 166-167**,545(2000). G.Szenes, F.Pászti, Á.Péter and A.I.Popov, Nucl.Instr.Meth.**B 166-167**, 949 (2000). K.Kimura, S. Sharma and A.I.Popov, Nucl.Instr.Meth.**B 191**,48(2002).

[87] K. Schwartz, C. Trautmann, T. Steckenreiter, O. Geiss, and M. Krämer, Phys. Rev. **B 58**, 11232 (1998).

[88] C. Trautmann, K. Schwartz, J. M. Costantini, T. Steckenreiter, and M. Toulemonde, Nucl. Instr. Meth. **B 146**, 367 (1998).

[89]   N. Itoh and K. Tanimura, J. Phys. Chem. Solids **51**, 717 (1990).

[90]   Ch. B. Lushchik, Creation of Frenkel pairs by exciton decay in alkali halides, in: *Physics of Radiation Damage*, Elsevier, Amsterdam, 1986, pp. 473–525.

[91]   F. Agullo-Lopez, C. R. A. Catlow, and P. Townsend, "Point defects in materials," Academic Press, London, 1988.

[92]   N. Itoh and A. M. Stoneham, *Materials Modification by Electronic Excitation*, Cambridge University Press, Cambridge, 2000.

[93]   M. Toulemonde, Ch. Dufor, A. Meftah, and E. Paumier, Nucl. Instr. Meth. **B 166–167**, 903 (2000).

[94]   E.A. Kotomin, V.Kashcheyevs, V.N. Kuzovkov, K. Schwartz, and C. Trautmann, Phys. Rev. **B 64**, 144108 (2001).

[95]   D. A. Young, Nature **182**, 375 (1958).

[96]   R. Katz, K. S. Loh, L. Daling, and G. R. Huang, Rad. Eff. Def. Sol. **114**, 15 (1990).

[97]   A. Rascón, J. L. Alvarez Rivas, J. Phys. C: Solid State Phys. **16** (1983) 241.

[98]   O. Salminen, P. Riihola, A. Ozols, and T. Viitala, Phys. Rev. **B 53**, 6129 (1996).

[99]   A. T. Davidson, J. D. Comins, and T. E. Derry, Rad. Eff. Def. Sol. **90**, 213 (1985).

[100] G. Wirth, K. Schwartz, and C. Trautmann, private communication.

[101] G. Kizane, J. Tiliks, A. Vitin, and J.Rudzitis, J. Nucl. Mater. **329-333**, 1287 (2004).

Chapter 8

MICROSTRUCTURAL EVOLUTION
OF IRRADIATED CERAMICS

Chiken Kinoshita

Kyushu University, Fukuoka, JAPAN

1. INTRODUCTION

The interaction of charged particles such as electrons and ions, with matter is based on both Coulomb interactions and elastic collisions. At high initial particle velocities, a large amount of energy is spent in the excitation of electrons bound to lattice atoms in non-metals. As the particle slows down, the energy communicated to such electrons becomes less important and before it comes to rest, elastic collisions with lattice atoms are dominant.

The distinguishing feature about a neutron is the fact that it carries no charge, and therefore suffers no energy loss by Coulomb interaction with the electrons in non-metals. In interactions with the nuclei of the solid, on the other hand, it must be anticipated that there will be energy transfer through elastic and inelastic collisions. The elastic collision results in producing an energetic ion denoted as a "primary knock-on atom (PKA)" with the maximum energy of about $4E/A$ (E: the neutron energy, A: the atomic weight of host lattice atoms). The inelastic collision produces transmutants, or impurity atoms. Sometimes the neutron is absorbed by the struck atom in the collision, and all of its energy is given up to the atom in the ensuing nuclear reaction. In violent nuclear reactions such as fission, a great amount of energy is carried away by two large atomic "fragments", and these are highly charged.

When non-metals such as ceramics are subjected to radiation, numerous changes occur in their structure and properties. The end products of structure changes can be classified in terms of three categories of defects: (a) electronic defects, which involve changes in valence states; (b) ionic defects, which consist of displaced lattice ions; and (c) gross imperfections, such as dislocations and voids.

K.E. Sickafus et al. (eds.), Radiation Effects in Solids, 193–232.

Inorganic non-metals are mostly ionic or covalent crystals and have their own particular sets of rules, which reflect both the displacement and kinetic processes. Several rules are the effect of electronic excitation, the polyatomicity, the charge state of point defects and the radiation-induced bias. In the case of ionic crystals, the necessity to preserve electrical neutrality and the reciprocity between defect structure and stoichiometry are added to those rules. Those particular sets of rules make radiation-induced phenomena in non-metals much more complicated than in metals.

The fundamental aspects of radiation damage and their characteristics in non-metals such as ceramics are described in Sections 2 and 3 in terms of the displacement of the secondary processes of radiation defects. In Section 4, characteristics of defect clusters formation are described in terms of mass and mobility of anions and cations, structural vacancies and/or stable nuclei of defect clusters. Effects of impurities, electronic excitation and electric field on microstructural evolution are also described in Section 5.

2.   THE DISPLACEMENT PROCESS OF RADIATION DEFECTS

2.1   The Primary Displacement Event and Its Threshold Energy

Radiation produces defects in solids in the form of permanent atomic displacements. Atomic displacements in non-metals may be produced either by direct momentum transfer to an atom nucleus from the irradiation particle or quantum, which is denoted as a "knock-on displacement", or as a response to alteration of atomic electronic states by ionizing radiation, which process is called "radiolysis". A synergistic response may also exist in the form of ionization-assisted knock-on displacement.

When large amounts of momentum are transferred, the PKA will initiate a "cascade" of subsequent atomic displacements. The number of displacements per PKA (N_d) is given by the modified Kinchin-Pease relation[1];

$$N_d = 0.8(E_p - E_e)/2E_d, \qquad (1)$$

where E_p is the PKA energy, E_e the energy dissipated in electronic excitations within the cascade, and E_d the displacement energy. The displacement energy E_d is defined as the threshold energy required to displace the lattice atom, and depends on the bond strength, the space available for accommodating an interstitial atom in the structure, and the form of the interstitial. Two features distinguish the displacement responses of non-metals from those of metals. First, non-metals are, with a few exceptions, polyatomic solids with different displacement probabilities for each atom type; this feature results in more complex dynamics and resultant primary displacement spectrum in displacement cascades. Second, the specificity of bonding in ionic or covalent solids permits, in some cases, exceptionally efficient utilization of electronic excitation energy in radiolytic displacement mechanisms, and in most cases, strong aversion to anti-site disorder.

Since non-metals generally consist of multiple sublattices, E_d must be separately measured for each sublattice. Almost all of E_d measurements have monitored the behavior of a particular point defect under electron irradiation, which produces rather isolated defects. A typical example is shown in Figure 1, where the displacement cross-section converted from the increment of electrical resistivity in TaC is shown as a function of energy of fast-electrons. The values of E_d for Ta and C are evaluated to be 44 eV and 28 eV, respectively [2], based on Equation (1). Table 1 lists recommended displacement energies for some non-metals [3].

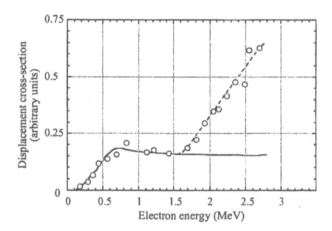

Figure 1. Comparison of experimental data with theoretical curves of displacement cross-section versus irradiating electron energy. The experimental cross-section is based on electrical resistivity-increment under electron irradiation and the theoretical ones are calculated under the assumption that displacement energies of Ta and C are 44 eV and 28 eV, respectively [2].

Table 1.   Recommended displacement energies in non-metals [3]

Material	Threshold displacement energy (eV)		
α-Al$_2$O$_3$	$E_d^{Al} \sim 20$		$E_d^{O} = 50$
MgO	$E_d^{Mg} = 55$		$E_d^{O} = 55$
MgAl$_2$O$_4$			$E_d^{O} = 60$
ZnO	$E_d^{Zn} \sim 50$		$E_d^{O} = 55$
BeO	$E_d^{Be} \sim 25$		$E_d^{O} \sim 70$?
UO$_2$	$E_d^{U} = 40$		$E_d^{O} = 20$
SiC	$E_d^{Si} \sim 40$?		$E_d^{O} = 20$
Graphite	$E_d^{C} = 30$		
Diamond	$E_d^{C} = 40$		

2.2   The Displacement Cascade and the Thermal Spike

When the energy from a PKA becomes partitioned among its neighboring atoms, the collective motion of the atoms must be considered to determine the atomic arrangements. This highly excited phase of the cascade corresponds to Seitz-Koehler's concept of "a thermal spike" [4]. In the absence of thermal spike effects, the number of point defects produced in a cascade is reasonably well approximated by Equation (1). The thermal spike, however, stimulates point defect motion and enhances recombination of Frenkel pairs. The remaining number of point defects in the matrix(N_d') is therefore expressed by

$$N_d' = \eta(T,E_p)N_d, \tag{2}$$

where $\eta(T, E_p)$ is the surviving defect fraction (also known as "defect production efficiency"), and depends on irradiation temperature (T) and E_p.

The values of $\eta(T, E_p)$ in α-Al$_2$O$_3$ and MgO are 0.1-0.5 at all PKA energies between 0.1 and 90 keV[3], and decrease with increasing irradiation temperature, as shown in Figure 2 [5].

In some crystals, such as silicon and germanium, which have covalent bonds, much higher density of point defects than those predicted by Equation (1) have been observed and are

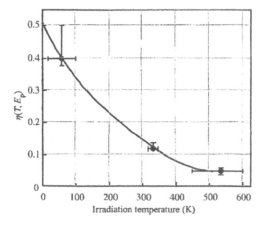

Figure 2. Irradiation temperature dependence of the surviving defect fraction $\eta(T, E_p)$ for

believed to be due to the thermal spike effect. It is well known that energetic heavy ions frequently induce an amorphous phase in cascade regions of those crystals. The size of damage cascades depends strongly on the combination of irradiation projectiles and target crystals, affecting their structure. Howe and Rainville [6] have systematically shown that the ratio of the average diameter of damage cascades ($<D>$) to the transverse transport straggling of projectiles ($2<y^2>^{1/2}$) is scaled in terms of the energy deposited per atom in a damage cascade (θ_v) irrespective of the combination of projectiles and targets.

Figure 3 is their results including ours, showing the normalized diameter of damage cascades as a function of θ_v for the silicon crystal. This figure is based on TEM images, which show the structure factor contrast corresponding to the amorphous phase. The value of $<D>/2<y^2>^{1/2}$ becomes greater than unity at a value of higher than about 1 eV/atom for the silicon crystal. The results strongly suggest the importance of the parameter θ_v which is related to the structure and the stability of damage cascades. When $\theta_v > 1$ eV/atom, very large fractions of transport cascade regions are rendered amorphous. In other words, thermal spikes in high energy cascades extend amorphous regions over the transport cascade volume and even beyond it.

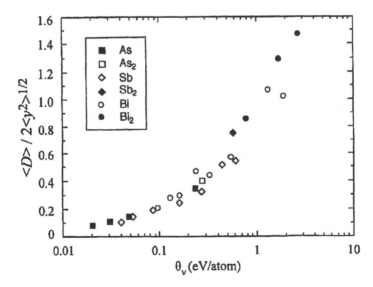

Figure 3. The ratio of average diameter of damage cascades ($<D>$) to the transverse transport straggling of projectiles ($2<y^2>^{1/2}$) as a function of the energy deposited per atom in a damage cascade (θ_v) for the silicon crystal irradiated with various kinds of ions [6].

2.3 Displacement Process through Electronic Excitation (Radiolysis)

In certain ionic crystals, defect production is most likely due to the conversion of electronic excitation energy into a form capable of manufacturing lattice defects rather than to elastic collisions. This phenomenon is called "radiolysis". This is important in some non-metals, such as alkali halides, alkaline earth fluorides and silica, but is generally concluded to be insignificant in many other non-metals. However, "swift heavy ions" radiation damage has been observed in some non-metals that are considered to be resistant to radiolysis (*e.g.* $MgAl_2O_4$ and a-Al_2O_3) [7].

In radiolysis defect production, there must be at least three steps [8]: (a) an electronic excitation resulting at least momentarily in the creation of a polarized or charged electronic defect in the lattice, (b) the conversion of this energy into kinetic energy of a lattice ion in such a way that the ion moves, and (c) the motion and stabilization of the ion. The mechanism of each step is discussed in detail in books written by Sonder and Sibley [8], and by Clinard and Hobbs [9].

3 THE SECONDARY PROCESS OF RADIATION DEFECTS

3.1 Defect Configuration and Mobility

In non-metals, there may be a number of possible interstitial and vacancy configurations, influenced by the poly-atomicity, the complexity of crystal structure and the charge state of defects. The stability and mobility of such point defects are indispensable for describing the defects kinetic process. The mobility of point defects is characterized in terms of their migration energies. Table 2 lists the migration energies for interstitials and vacancies identified with reasonable certainty in several respective non-metals, based on the tables summarized by Clinard and Hobbs [9] and by Zinkle and Kinoshita [3].

Various charge states for defects may have a large influence on defect mobilities. Although no definitive experimental work has been done, a theoretical calculation for oxygen interstitials in MgO for instance has reported that the migration energy decreases from ~1.45 eV for O_i^- to 0.5-1.2 eV for double-ionized O_i^{2-} ions [10]. A further complication is that the long range migration of point defects can be strongly affected by cation and anion impurities [11]. A typical example of the impurity effect will be described in Section 5.

Table 2. Probable defect migration energies in some non-metals [3,9]. Notation follows that of Kröger and Vink.

Solid	Defect species	Migration energy (eV)
C (graphite)	V_C (*a*-axis)	3.1
	V_C (*c*-axis)	5.5
	C_i (*a*-axis)	0.4
	C_i (*c*-axis)	2.9
Si	$V_{Si}^{\bullet\bullet}$	0.18
	$V_{Si}^{\bullet}$	0.45
	V_{Si}^{x}	0.33
	$[2Si]_{Si}^{\bullet}$	0.3
	$[2Si]_{Si}^{\prime}$	1.5
	$[2Si]_{Si}$	0.85
	Si_i	~0
Ge	V_{Ge}	0.3-1.0
	Ge_i	0.2
MgO	V_0	2.0-2.3
	V_{Mg}	2.0-2.5
	O_i or Mg_i	0.5-1.5
a-Al$_2$O$_3$	V_0	1.8-2.0
	V_{Al}	1.8-2.1
	O_i or Al_i	0.2-0.8
MgAl$_2$O$_4$	V_0	2.7
NiO	$V_{Ni}^{\bullet\bullet}$	1.43
	$V_{Ni}^{\bullet}$	1.58
	$V_0^{\bullet\bullet}$	1.7
UO$_2$	O_i	0.6
SiO$_2$	$(O_2)_i$	1.2
ZrO$_2$	$V_0^{\bullet\bullet}$	1.3

3.2   Defect Aggregates and Their Configuration

In many cases, it is possible to produce aggregates of interstitials and vacancies directly through damage cascades, or by prolonged irradiation or subsequent heat treatment. The term "aggregate" can include groups of interstitials or vacancies from two to many. Aggregation of interstitials can proceed further only by the nucleation and growth of dislocation loops commonly with an edge component which involves atoms from all sublattices in stoichiometric ratio [9]. For the case of biased displacements, non-stoichiometric interstitial aggregates lead to colloids or gas bubbles. Examples will be shown in Section 4.

Large three-dimensional aggregates of interstitials are unlikely because the strain associated with such aggregates becomes rapidly prohibitive. Volume inclusions therefore arise only from condensation of vacancies or substitutional defects. For the case of stoichiometric vacancy aggregates, condensation in three dimensions forms voids which are unstrained inclusions, while planar condensation results in collapse to form dislocation loops. Fission gas or radiation-produced insoluble gas often serves to convert voids to gas bubbles. For the case of non-stoichiometric vacancy aggregates, condensation results in colloids or gas bubbles in accordance with predominant anion or cation. Condensation of vacancies of only one of the atomic species present leads to precipitation of the remaining species in a new phase.

The classic studies are in alkali halides (M^+X^-) in which condensation of F-center halogen vacancies precipitates alkali metal colloids (Figure 4). Condensation of V-center alkaline vacancies, on the other hand, precipitates gas bubbles, when halogen atoms have elemental gaseous phase at the temperature of irradiation.

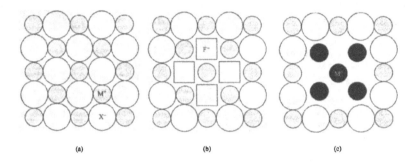

(a)   (b)   (c)

Figure 4. A model crystal of alkali halides (M^+X^-) (a) in which condensation of F-center halogen vacancies (b) precipitates alkali metal colloid (c).

3.3 Irradiation-Induced Phase Transformation

In the atomic displacement process, replacement collision sequences play an important role as a medium for transporting energy and mass through crystalline materials. Most displacements occur via the replacement collision sequences, leaving a vacancy at the beginning of the sequence and depositing an excess atom as an interstitial at the end of the sequence. When super-lattice alloys including non-metals are irradiated with high-energy radiation, the replacement and/or displacement collisions are accompanied by "anti-site disorder" (which is analogous to "chemical disordering" in metals). Since the anti-site disorder is extremely unstable especially in ionic crystals, it has scarcely been observed in non-metals. However, a cation sublattice disorder has been observed in $MgAl_2O_4$ irradiated with high fluence neutrons [12,13].

It is well known that many non-metals undergo amorphization under heavy-ion irradiation. Historically Naguib and Kelly [14] suggested that substances having the ionicity of less than 0.47 amorphized on ion impact, while substances with values larger than 0.59 were stable in crystalline form, although lots of exceptions to this "rule" have been found. Although the exact nature of the amorphization process is not well defined for most of these phases, amorphization appears to occur heterogeneously in these phases by several possible mechanisms: (a) directly in the displacement cascade, (b) from the local accumulation of high defect concentrations due to the overlap of cascades, or (c) by composite or collective phenomena involving more than one process.

The current states of knowledge on amorphization of non-metallic phases, on structure of the amorphous state, on amorphization processes, and on amorphization kinetics have been reviewed by Weber et al. [15]. Thermal spikes play a role for recrystallizing rather than keeping an amorphous phase in some cases where the cooling rate is sufficiently low to nucleate crystal embryo in the amorphous phase.

Radiation-induced precipitation is another example that should be included in this section. Several examples of precipitation of anion and cation species as separate phase have been discussed in Section 4. Irradiation produces Frenkel pairs of one or more species and selectively enhances their diffusion, leading to redistribution of constituent elements. Enrichment or depletion of the elements is predominant in the vicinity of point defect sinks and sometimes induces precipitation [16].

3.4 Volume Changes due to Defect Accumulation

Radiation-induced volume changes in crystalline phases can result from the accumulation of point defects in the crystalline structure, solid-state phase transformations (e.g. amorphization), and the evolution of microstructural defects (e.g. dislocation loops, voids, colloids, gas bubbles, and microcracks). The macroscopic swelling is the overall sum of these effects. In many crystals, barriers to recombination and aggregation can lead to the accumulation of significant point defect concentrations. The accumulation of these point defects in the crystalline structure results in an increase in unit cell volume. In metals, swelling due to point defects is typically limited to maximum values of ~0.1 vol.%; however, ionic crystals exhibit relatively large dilatational swelling (up to ~3 vol.%). This may appear to be a consequence of positive dilation due to vacancies and interstitials, although the former defects induce negative dilatation in metals [17].

Accumulation of a high concentration of point defects induces an amorphous phase in some non-metallic crystals. Volume change via radiation-induced amorphization has been extensively studied in most of the crystalline non-metals of interest as actinide-host phases. The macroscopic swelling in these materials increases under irradiation up to saturation ranges from about 1% to nearly 20% [15]. The temperature dependence of volume changes accompanying amorphization results in less swelling at higher temperatures, due to instability of the amorphous phase. At intermediate temperatures, where no amorphization occurs but only interstitials are mobile, crystalline materials may contain vacancies and high densities of interstitial dislocation loops or tangles. Since conversion of interstitials to lattice atoms represents addition of new lattice sites, the material swells accordingly. At higher temperatures where both interstitials and vacancies are mobile, excess vacancies aggregate to form voids. Swelling in the temperature range of void formation results from creation of new, occupied atom sites. Figure 5, which was summarized by Furuya et al. [18], shows the temperature dependence of swelling of several non-metallic crystals irradiated with various neutron fluences.

The fundamental responses to irradiation result in significant changes to physical properties, such as strength, toughness, electrical and thermal conductivity, dielectric response, and optical behavior. A sophisticated description of those properties appears in a review paper [19].

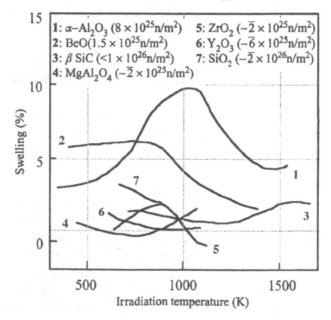

Figure 5. Temperature dependence of the swelling of some non-metals irradiated with fast neutrons (fluences are given in parentheses) [18].

4   CHARACTERISTICS OF DEFECT CLUSTERS FORMATION IN CERAMICS

In order to describe the kinetic process of microstructural evolution under irradiation, the stable configurations of defects and defect aggregates, in addition to the displacement energy, recombination volume for a Frenkel pair, and migration energy, should be known. Although little is known about general non-metals, there is a large body of literature on oxides, such as MgO, α-Al_2O_3, ZrO_2 and $MgAl_2O_4$, that suggests that both stoichiometric and non-stoichiometric interstitial aggregates can be formed, depending on irradiation conditions.

In addition, it is well known that structural vacancies can suppress the formation of defects and defect aggregates in materials such as TiC_{1-x} and MgO[16]. Research on the radiation resistance of magnesia spinel suggests that crystalline non-metals will be more radiation resistant when the concentration of equilibrium structural vacancies is high, barriers to interstitial-structural vacancy recombination are low, the defect energies of compositional disorder are small, and the critical nucleus for the formation of stable interstitial loops is large [17].

4.1   Formation Kinetics of Interstitial Clusters in MgO-Type Crystals

First, we describe characteristics of the cluster formation process in MgO wherein a dimer of Mg and O is more preferable due to similar masses and mobilities of anions and cations. Micrographs of pure and doped MgO, taken during irradiation at various temperatures with various 1000 keV electron fluxes, show the rapid nucleation and the subsequent growth of 1/2<10-1>{10-1} interstitial dislocation loops [20,21]. Figure 6 is a typical example of sequences of kinematical bright-field electron micrographs showing the behavior of loops in pure MgO [22].

The area density of dislocation loops in wedge-shaped specimens was measured and converted to the volume density. The volume density is shown as a function of irradiation time in Figure 7, which is an example for pure MgO, and it keeps constant for increasing irradiation time after the rapid nucleation. As the irradiation temperature is increased, the density of loops decreases and they grow more rapidly into well-defined loops.

Radiation Effects in Solids

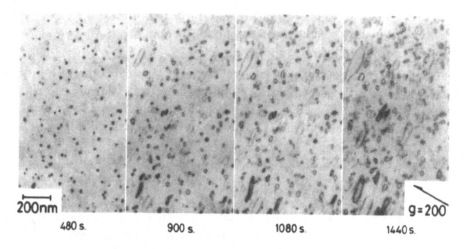

200nm   g=200

480 s.   900 s.   1080 s.   1440 s.

Figure 6. Formation and growth of interstitial dislocation loops in the pure Tateho MgO irradiated approximately along <001> at 1020 K with a 1000 keV electron flux of 2.0×10^{23} e/m²s. Loops are perfect dislocation loops with 1/2<10-1> Burgers vectors lying on {10-1} planes [22].

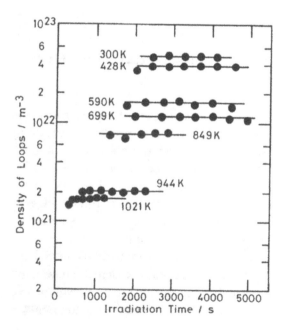

Figure 7. Time dependence of the volume density of dislocation loops during irradiation of the pure Tateho MgO with a 1000 keV electron flux of 2.0×10^{23} e/m²s. Irradiation temperatures are shown in the figure [22].

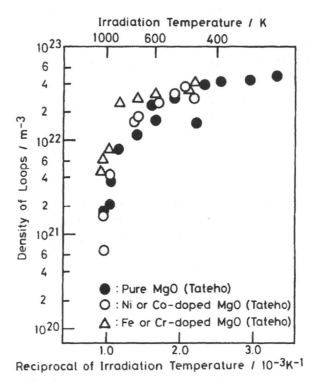

Figure 8. Temperature dependence of the saturated volume density of dislocation loops in the pure and the doped Tateho MgO under irradiation with a 1000 keV electron flux of 2.0 x 10^{23} e/m²s [22].

The saturated volume density for MgO is plotted against the reciprocal irradiation temperature in Figure 8, together with that for Ni-, Co-, Fe- or Cr-doped MgO, and it decreases with ever increasing rate as the reciprocal temperature is decreased. The empirical relations on the density C_L and the diameter D of individual loops for irradiation at temperatures below and above about 900 K (respectively) can be described by the following theoretical equations based on an assumption that a pair of Mg- and O-interstitials forms a stable nucleus of the loop:

$$C_L \propto t^0 (\sigma\phi)^{1/2} M_I^{-1/2} , \quad D \propto t^{1/3} M_I^{1/6} , \tag{3}$$

and

$$C_L \propto t^0 , \quad D \propto t(\sigma\phi)^{1/2} M_v^{-1/2} \text{ (thick foil)} \tag{4}$$

or $$D \propto t\sigma\phi \text{ (thin foil)}, \tag{5}$$

where t is the irradiation time, ϕ the electron flux and σ the displacement cross section. M_I and M_v are the mobilities of interstitials and vacancies, corresponding to the slower species, and expressed in terms of the irradiation temperature T and the migration energies of interstitials and vacancies, respectively.

The theory predicts that Equation (3) is for irradiation at temperatures where only interstitials are mobile, and that Equations (4) and (5) are for temperatures where vacancies as well as interstitials are mobile [20-22]. According to this model the temperature dependence of the volume density below 900 K and that of the growth rate of loops above 900 K, respectively, give the migration energies of interstitials and vacancies. The growth rate of each loop was measured as a function of T, and gives 2.03 ± 0.17 eV as the migration energy of slower vacancies in pure MgO [21,22]. The growth rate of loops around the thin area where surfaces are dominant sinks, on the other hand, is proportional to the displacement cross section as shown in Equation (5), and energy dependence of growth rate of each loop gives the displacement threshold energy. A value for the displacement threshold energy at 1173 K along <001> directions in MgO is given to be 30 ± 8 eV [22].

Equations (3), (4) and (5) were originally proposed for describing the kinetic behavior of dislocation loops in pure metals [23] and have been confirmed to be effective for the kinetics in concentrated alloys, such as Nb-3.1at%Zr, Fe-7.5 and 14.0at%Mo and Cu-1.3 and 4.9at%Ti [24,25].

4.2 Formation Kinetics of Interstitial Clusters in TiC-Type Crystals

Second, we describe results for TiC-type crystal which consist of crystals containing a high concentration of structural vacancies. Examples are TiC_{1-x}, $Ni_{1-x}O$, $FeAl_{1-x}$ and others, containing $x/2$ fraction of structural vacancies due to non-stoichiometry. In these crystals, most of the displaced atoms recombine with structural vacancies, and the nucleation and growth of interstitial loops can be extremely suppressed. If di-interstitials act as stable nuclei of loops, the density of loops is expressed by

$$C_L \propto (\sigma\phi)^2 t/(M_I C_{Vo}^2),\qquad(6)$$

where C_{Vo} is the concentration of structural vacancies [22,26]. The volume density, C_L, depends on irradiation temperature through the mobility of interstitials M_I. The diameter D, on the other hand, does not depend on the irradiation temperature and it is given by

$$D \propto \sigma \phi t / C_{Vo}^2 . \tag{7}$$

Irradiation with low flux 1000 keV electrons induces no defect clusters, as expected, but high flux irradiation nucleates interstitial loops after some incubation time. Figure 9 shows the volume density and the average diameter of dislocation loops in $TiC_{0.85}$ as a function of irradiation time after the incubation time t_{inc}. The incubation of detectable loops continue until the concentration of interstitials reaches a saturation level. Both C_L and D increase linearly with increasing time, and the growth rate of loops is independent of irradiation temperature, but the volume density is higher at lower irradiation temperatures, as theoretically expected. Experimental results on the flux dependence of C_L and D have also confirmed the model [22,26]. Based on the experimental results of $TiC_{0.85}$ and the theoretical prediction, it is concluded that structural vacancies control the nucleation and growth process of loops and are extremely effective to suppress the nucleation and growth of loops.

TiC_{1-x} irradiated with fission neutrons in Joyo (Fast Breeder Testing Reactor in Japan) shows no voids and a relatively low density of loops or dislocations up to 1-9 x 10^{25}n/m² at 673, 773 and 873 K. In TiC type crystals, the size of nuclei of loops is the same as in the MgO type, but the kinetic process of loops is much different through the structural vacancy.

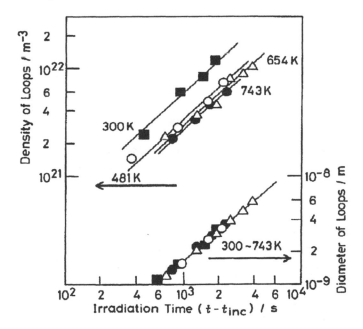

Figure 9. Time dependence of the volume density and the diameter of dislocation loops in $TiC_{0.85}$ irradiated with a 1000 keV electron flux of 1.9×10^{23} e/m²s, where t_{inc} is the incubation time depending on irradiation temperatures which are shown in the figure [22,26].

4.3   Formation Process of Vacancy and Interstitial Clusters in α-Al₂O₃ and MgAl₂O₄

The nuclei for the formation of the stable interstitial clusters are supposed to be anti-Shottkey quintettes for α-Al₂O₃ and anti-Shottkey septets for MgAl₂O₄[16].   In order to get insight into the characteristics, the formation process of interstitial and vacancy clusters in α-Al₂O₃ and MgAl₂O₄ irradiated with fast neutrons will be shown.

It is now well established that MgAl₂O₄ exhibits a strong resistance to void swelling during neutron irradiation, as described in Section 3.4. Figure 10 compares the swelling of single crystalline MgAl₂O₄ and α-Al₂O₃ as a function of fission neutron fluence [27-30], and shows the resistance of MgAl₂O₄ to void swelling up to more than 200 dpa in contrast to the high swelling of α-Al₂O₃.

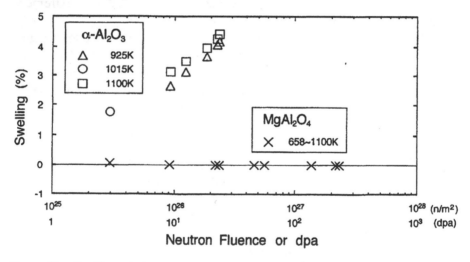

Figure 10.   Swelling of single-crystalline α-Al₂O₃ and MgAl₂O₄ as a function of neutron fluence or dpa based on data from Clinard et al. [27,28] and Garner et al. [29,30]. The displacement level is estimated based on an equivalence of 1 dpa per 10^{25}n/m²(E>0.1 MeV).

Side-by-side irradiations of nearly stoichiometric MgAl₂O₄ and α-Al₂O₃ in fission reactors have shown that the radiation-induced microstructural evolution proceeds by very different paths in these two materials. The large difference in dislocation loop evolution appears to account for the ease of void or cavity swelling in α-Al₂O₃[27,28,31,32] and the strong resistance to void or cavity formation in MgAl₂O₄[27,28,31-35]. Irradiation of MgAl₂O₄ to very high doses (up to 230 dpa) confirms the details of the dislocation evolution, which involves a progressive change in

Burgers vectors and habit planes as interstitial loops increase in size. Specifically, the nucleation and growth of loops in neutron irradiated $MgAl_2O_4$ proceeds via the following steps:

$$1/6[111](111) \rightarrow 1/4[110](111) \rightarrow 1/4[110](101)$$
$$\rightarrow 1/4[110](110) \rightarrow 1/2[110](110),$$

changing their character as they grow larger [32,35,36].

At lower neutron fluence levels and lower temperatures, unstable nuclei of $1/6<111>\{111\}$ loops appear and a few of them grow into well-defined loops following the sequence outlined above. The elimination of $1/6<111>$ loops has been observed in other studies. For instance, such loops induced by 6 keV Ar^+ ions are eliminated during irradiation with 0.1 to 1 MeV electrons [22]. It has been noted that the $1/6<111>$ loop has both anion and cation faults and cannot preserve stoichiometry and charge balance in either normal or inverse $MgAl_2O_4$ [27,33]. Because of the non-stoichiometric component involved in the construction of $1/6<111>$ loops, they are unstable. However, partial inversion of inserted cation layers makes $1/6<111>$ loops stable against any deviation from either stoichiometry or charge neutrality.

The nucleation of $1/6<111>$ loops may be easier than that of $1/4<110>$ loops, because the $1/6<111>$ loop has a smaller Burgers vector than does the $1/4<110>$ loop, and the nucleus of $1/6<111>$ loops allows varying compositions [33,37]. On the other hand, it is possible that the stacking sequence of (111) or (101) planes preserves their stoichiometry by partly changing cation distributions, a process referred to as "cation disordering". The stacking sequence of $1/6[111](111)$ has both anion and cation faults and those of $1/4[110](111)$ and $1/4[110](101)$ have only cation faults in $MgAl_2O_4$ crystals [37,39]. The change in Burgers vectors and habit planes of loops during growth is strongly controlled by the stacking fault energy.

No voids or cavities have been observed in single crystal $MgAl_2O_4$ irradiated up to 56 dpa at 1023 K, but tiny voids whose size is 2 to 3 nm in diameter have been found along the $1/4<110>\{110\}$ stacking faults at 138 dpa [32,36]. The void size increased to 6-8 nm after a dose of 217 dpa. From the size and the density of cavities, the void swelling is estimated to be only 0.07% at 217 dpa, which is in good agreement with macroscopic swelling measurements [29].

As for the neutron irradiation behavior of α-Al_2O_3, many studies [27,28,31,39-41] have shown that irradiation results in the formation of interstitial dislocation loops with a Burgers vector of b = 1/3 <10-11>, that lie on both the {10-10} and (0001) planes. Along with the formation dislocation loops and networks, numerous studies [27,28,31,32,39,40] have found the presence of voids or cavities. Three types of interstitial loops, 1/3[0001](0001), 1/3<10-10>{10-10} and 1/3<10-11>(0001), are formed in α-Al$_2$O$_3$. The size of loops increases with increasing fluence, changing the relative population of each type of loop. Upon growing, the population of the 1/3<10-11>(0001) type of loops increases, with a corresponding decrease in the 1/3[0001](0001) and 1/3<10-10>{10-10} types of loops. The dislocation loop evolution in α-Al$_2$O$_3$ involves the unfaulting of 1/3[0001] and 1/3<10-10> loops by shearing in the loop plane via the following reactions [31,42]:

$$1/3[0001](0001) \quad + \quad 1/3<10\text{-}10> \quad \rightarrow \quad 1/3<10\text{-}11>(0001)$$
$$\text{(faulted loop)} \qquad \text{(partial shear)} \qquad \text{(unfaulted loop)}$$
$$1/3<10\text{-}10>\{10\text{-}10\} + \quad 1/3[0001] \quad \rightarrow \quad 1/3<10\text{-}11>\{10\text{-}10\}$$

Figure 11 summarizes the character of defect clusters in MgAl$_2$O$_4$ and α-Al$_2$O$_3$ as a function of irradiation temperature and fission neutron fluence [27,28,31,32,35,36]. The curved lines in the figure denote the threshold for observable void or cavity formation in MgAl$_2$O$_4$ and α-Al$_2$O$_3$. It should be emphasized that the threshold fluence for void or cavity formation in MgAl$_2$O$_4$ is about two orders of magnitude higher than α-Al$_2$O$_3$. A high density of interstitial loops, including unfaulted perfect loops, is formed in α-Al$_2$O$_3$ irradiated to relatively low neutron fluences. The early nucleation of interstitial loops produces a supersaturation of vacancies, which promotes the nucleation of cavities. The bias factor for preferential absorption of interstitials at loops is proportional to the magnitude of the Burgers vector of the dislocation loop [43]. Therefore, the perfect 1/3<10-11> loops that appear in α-Al$_2$O$_3$ after a dose of ~1 dpa act as more efficient interstitial sinks than do faulted loops. The threshold fluence for void or cavity formation in α-Al$_2$O$_3$ therefore corresponds to the fluence required for formation of perfect loops.

In MgAl$_2$O$_4$ irradiated to much higher exposures, however, a low density of faulted loops remain to high neutron fluences, and the appearance of perfect loops is not always correlated with the formation of voids or cavities as seen in Figure 11. It appears that the formation of stable interstitial loops in MgAl$_2$O$_4$ occurs infrequently under neutron irradiation conditions, mainly because of the more effective direct recombination of interstitials and structural vacancies due to non-stoichiometry [32]. This reduced formation rate of interstitial loops is also due to the large critical nucleus of stable interstitial loops. Decreased formation of stable interstitial loops enhances the recombination of interstitials and vacancies and thereby suppresses the formation of vacancy clusters. The general absence of

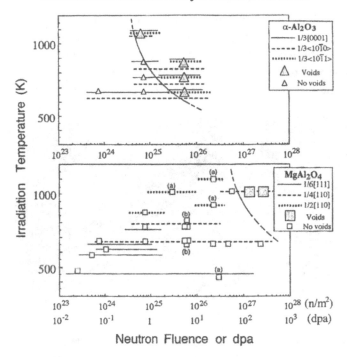

Figure 11. A summary of the character of dislocation loops and the critical fluence for the formation of voids or cavities in α-Al$_2$O$_3$ and MgAl$_2$O$_4$ as functions of irradiation temperature and neutron fluence or dpa, based on the results by Clinard et al. [27,28], Youngman et al. [31] and Kinoshita et al. [32,35,36]. The displacement level is estimated based on an equivalence of 1 dpa per 10^{25}n/m^2(E>0.1 MeV).

perfect dislocation loops and dislocation networks is one of reasons why MgAl$_2$O$_4$ is resistant to void or cavity swelling, although it is not the critical one.

4.4   Formation Kinetics of Interstitial Clusters in ZrO$_2$- Type Crystals

Due to the large mass difference between Zr and O atoms in ZrO$_2$, the elastic displacement cross-section of O atoms is much larger than Zr atoms under irradiation with 100-1000 keV electrons. Yasuda and Kinoshita [44] have found an anomalous formation of defect clusters under irradiation with 100-1000 keV electron beams subsequent to ion irradiation, such as 100 keV He$^+$ and 300 keV O$^+$ions, through transmission electron microscopy. Figure 12 shows typical examples of bright-field images of defect clusters in yttria stabilized zirconia (YSZ) irradiated with 200 keV electrons at 370, 420 and 520 K subsequent to 300 keV O$^+$ ion irradiation at 470 K. The defect clusters induce strong black/white lobes contrast with size up to about 300 nm in diameter, preferentially around focused electron beams. It is important to emphasize that these defect clusters are not observed in YSZ irradiated solely with 300 keV ions and that they are observed only after the subsequent irradiation with a focused electron beam.

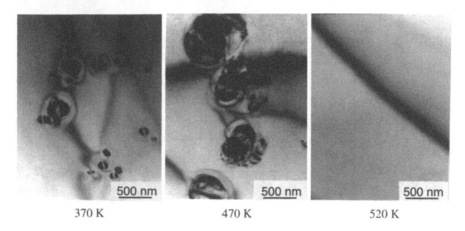

370 K   470 K   520 K

Figure 12. Bright-field image of YSZ irradiated with focused 200 keV- electron-beams subsequent to 5.1 x 10^{17} 300 keV O$^+$-ions irradiation at 470 K. Electron irradiation temperatures are shown [44].

A sequential change of defect clusters is shown in Figure 13, where defect clusters grow with increasing irradiation time. The growth rate of defect clusters is extremely fast and estimated to be about 1 nm/s. One can realize the contrast changing from black/black lobes to dislocation networks within a short time (cf. (d) and (e)). Several segments of dislocations are generated after the transformation (cf. (e)), and a prolonged electron irradiation induces new black/black lobes preferentially at or near dislocation lines. The processes of nucleation, growth and transformation are repeated under electron irradiation [44].

The nucleation and growth process of the defect clusters are completely different from that of neutral-type dislocation loops. The defect clusters are believed to be oxygen platelets having an accumulated electric charge. Their characteristic features are summarized as follows [45]: (1) very large black/black lobes contrast, which indicates an existence of strong strain-field around the defect clusters, (2) very large size up to 1.0-1.5 µm in diameter produced by very rapid growth rates of nearly 1.0 nm/s at an electron flux of 1.5x10^{23}, (3) preferential formation at the periphery of the focused electron beam, and (4) the transformation from the black/black lobes contrast to dislocation network a critical diameter of ~1.2 µm.

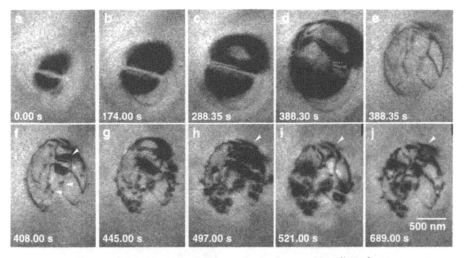

Figure 13. Growth process of a defect cluster in YSZ under 5×10^{21} e/m^{2}s 200 keV electron irradiation at 470 K. The specimen was originally irradiated with 5.1×10^{17} 300 keV O$^+$-ions at 470 K. Real time from the time when the picture (a) was taken is shown on each picture [45].

For the explanation of this phenomenon, a new theoretical model is suggested for the explanation of the growth kinetics and instability of charged dislocation loops in ceramic materials such as YSZ, taking into account an effective charge on dislocation loops through the trapping of electrons in dislocation cores. The strain and stress fields induced by internal electrical field near the charged dislocation loops (oxygen platelets) are also presented. The theoretical calculations show that the induced stress around the charged dislocation loops can be comparable with theoretical yield stress of YSZ. A critical condition for the beginning of plastic deformation near the charged dislocation loop (multiplication of dislocation network) is determined, which can be used for the estimation of critical radius of charged dislocation loop to be unstable. Details appear in the original papers [44,46].

5 EFFECTS OF IMPURITIES, ELECTRONIC EXCITATION AND ELECTRIC-FIELD [11]

5.1 Effect of OH⁻ Ions on the Microstructural Evolution in MgO under Electron Irradiation

Irradiation with high energy electrons induces 1/2<101>{101}-type dislocation loops in pure and cation-doped MgO, as discussed earlier in Section 4.1.

In thin foils of MgO subjected to immersion in boiled water, on the other hand, no defect clusters or bubbles are formed under irradiation with high energy electrons, depending on immersion time [47] Figure 14 shows typical examples of dislocation loops and cavities in MgO first rinsed in cold water and boiling water, respectively, for a few minutes and subsequently irradiated with high energy electrons [47].

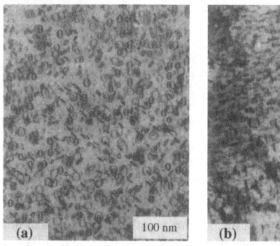

Figure 14.   Bright-field micrographs showing dislocation loops or cavities in MgO irradiated at 1200 K with 1000 keV electrons. (a) and (b) are for MgO specimens rinsed in cold water and boiling water for a few minutes, respectively [47].

It is well known that MgO reacts with H_2O to form $Mg(OH)_2$ even under atmospheric conditions. Furthermore, H_2O dissolution in MgO leads to the creation of structural vacancies in the lattice. Many experiments [48,49] have been performed to determine the mechanism of the dissolution process of H_2O in MgO. According to Freund and Wengeler [49], MgO reacts with H_2O to form vacancies on the Mg sublattice and OH^- ions in the crystal. Figure 15 illustrates schematically the lattice configuration in MgO for a fully-charge-compensated point defect complex consisting of a cation vacancy and two OH^- ions based on the mechanism that when an H_2O molecule is dissolved in a given volume of MgO, one oxygen ion and one extrinsic cation vacancy, V_{Mg}'', are added to the lattice. The missing divalent cation is charge-compensated by the two protons from the H_2O molecule, to form hydroxyl ions (OH^-), on O^{2-} lattice sites. The preferred configuration will be the fully-OH^--compensated cation vacancy, neutral to the surrounding lattice, $[OH^{\cdot}V_{Mg}''HO^{\cdot}]^x$. It is expected that such cation vacancy complexes have a great influence on the formation of radiation-induced defects.

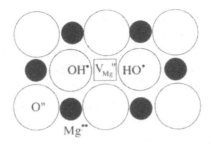

Figure 15. Schematic diagram illustrating a stabilized structure of a vacancy on the Mg sub-lattice associated with two OH⁻ ions, to form the fully OH⁻-compensated vacancy [OH˙V$_{Mg}$"HO˙]ˣ in MgO crystal [49].

Bulk Tateho MgO stored in vacuo (Crystal A) and that kept in air at room temperature for more than 10 years (Crystal B) and a vacuum-annealed Crystal B were subjected to FTIR measurements to detect the concentration of cation vacancies associated with OH ions [11]. Figure 16 shows examples of FTIR spectra in the range of 2800~4000cm⁻¹ for Crystals A and B, and Crystal B annealed at 1223 K in vacuum. The absorbance in the range of 3600~3800cm⁻¹, is proportional to the concentration of fully-OH⁻ compensated cation vacancies, [OH˙V$_{Mg}$"HO˙]ˣ.

This concentration is greatest in Crystal B, followed by annealed Crystal B, and Crystal A. The FTIR absorbance near 3300cm⁻¹, on the other hand, corresponds to the formation of the half-OH⁻-compensated cation vacancies, [OH˙V$_{Mg}$"]ˣ, created through the dissolution of [OH˙V$_{Mg}$"HO˙]ˣ. This absorption feature implies that the reaction [OH˙V$_{Mg}$"HO˙]ˣ to [OH˙V$_{Mg}$"]' occurs in Crystal B during annealing at 1223 K in vacuum.

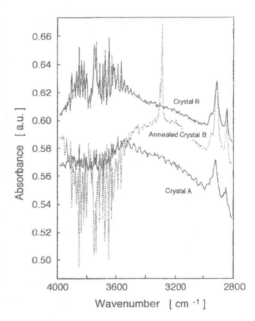

Figure 16. Typical examples of FTIR spectra in the range of 2800~4000 cm⁻¹, obtained from crystals A and B and from crystal B vacuum annealed at 1223 K for 3600 s. The absorbance in the range of 3600~3800 cm⁻¹ is in proportion to the concentration of [OH˙V$_{Mg}$"HO˙]ˣ, and the absorbance around 3300 cm⁻¹ corresponds to [OH˙V "]' [11].

The integral FTIR absorption in the range of $3600 \sim 3800 \text{cm}^{-1}$ and around 3300cm^{-1} was measured for Crystals A and B and for OH⁻-doped Crystal A (kept in an autoclave for 1-3 weeks at 673 K and 0.1 Mpa) before and after annealing in a vacuum at temperatures from 300 K to 1473 K; these results are plotted in Figures 17 and 18 as a function of annealing temperature [11]. The concentration of fully compensated cation vacancies is higher in Crystal B than in Crystal A and it decreases with increasing annealing temperature; meanwhile the concentration of half-OH⁻-compensated cation vacancies increases. The annealing behavior of the OH⁻-doped Crystal A clearly indicates that a high concentration of fully-OH⁻-compensated structural vacancies is converted to half-compensated cation vacancies, although there might be some sub-reactions around 450 K and 800 K or 1000 K.

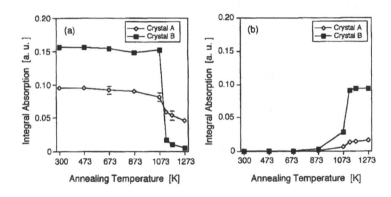

Figure 17. Vacuum-annealing-temperature dependence of the integral FTIR absorption in the range $3600 \sim 3800 \text{ cm}^{-1}$ (a) and around 3300 cm^{-1} (b), which correspond to the concentration of $[\text{OH}^{\cdot}\text{V}_{Mg}\text{"HO}^{\cdot}]^x$ and that of $[\text{OH}^{\cdot}\text{V}_{Mg}\text{"}]'$ for the Tateho MgO Crystals A and B, respectively [11].

Freund and Wengeler [49] performed isochronal annealing experiments in vacuum for MgO crystals containing large amounts of OH⁻ ions, and proposed a mechanism for the annealing of fully-OH⁻-compensated vacancies, $[\text{OH}^{\cdot}\text{V}_{Mg}\text{"HO}^{\cdot}]^x$. According to their results, hydrogen molecules and oxygen atoms begin to be released from the crystal around 573K and 823K, respectively. In this process, an $\text{OH}^{\cdot}\text{V}_{Mg}\text{"HO}^{\cdot}$ complex dissociates to produce a hydrogen molecule and an $\text{O}^{\cdot}\text{V}_{Mg}\text{"O}^{\cdot}$ vacancy complex; then, the $\text{O}^{\cdot}\text{V}_{Mg}\text{"O}^{\cdot}$ complex is converted into $\text{O}^{\cdot}\text{V}_{Mg}\text{"}$ and $\text{O}^{\cdot}$. From these results, together with the results of this study, it is concluded that the fully-compensated cation vacancies are converted to half-OH⁻-compensated vacancies, concurrently sweeping out OH⁻ ions [11].

Annealed specimens of Tateho MgO (Crystals A and B) were observed in-situ under irradiation with 1000 keV electrons in HVEM [11]. Figure 19 shows typical examples of bright-field electron micrograph sequences revealing the formation of 1/2<101>{101}-type interstitial dislocation loops in Crystal B, which was subjected to annealing for 3600 s at 1173 K or 1473 K, under irradiation with a 1000 keV electron flux of 2.8×10^{23} e/m²s at 1073 K. In Figure 19, the formation of dislocation loops is shown for specimens having almost identical thickness. The nucleation and the growth rates of dislocation loops are different in these two specimens.

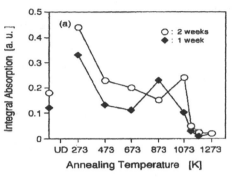

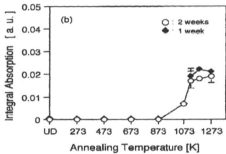

Figure 18. Same as in Fig. 17, except for the OH⁻-doped Crystal A, which was kept in an autoclave for 1 or 2 weeks. The identical specimen was used through the FTIR measurements for the OH⁻-doped MgO kept in the autoclave for 1 or 2 weeks and the annealed MgO at various temperatures. The result for each un-doped (UD) MgO before OH⁻-doping is also shown for comparison.

The volume density and the average diameter of loops were measured and the results are shown in Figure 20, together with results for an un-annealed Crystal B. Analogous results on the OH⁻-doped Crystal A are shown in Figure 21. Both the nucleation rate and the growth rate increase with increasing annealing temperature in vacuum. Furthermore, the nucleation rate is nearly proportional to the irradiation time; the nucleation rate for OH⁻-doped crystals is the lowest, as illustrated by Figure 21. The nucleation and growth process follows Equations (6) and (7) to within experimental error. These equations are derived for crystals containing a high concentration of structural vacancies. For the experiments presented here, it is postulated that the high concentration of vacancies on the Mg sub-lattices in non-annealed Crystal B provides recombination sites for displaced atoms, and so suppresses the formation of interstitial dislocation loops. In some cases where MgO contains a much higher concentration of structural vacancies, they might aggregate as voids or enhance the formation of bubbles associated with oxygen molecules [11].

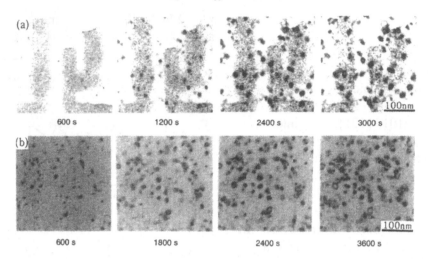

Figure 19. Bright-field electron micrographs illustrating the nucleation and growth process of interstitial dislocation loops in Crystal B, subjected to annealing in a vacuum for 3600 s at 1173 K (a) or 1473 K (b), and irradiated along the <001> direction with a 1000 keV electron flux of $2.8 \times 10^{23} e/m^2 s$ at 1073 K. The thicknesses of both specimens were nearly identical [11].

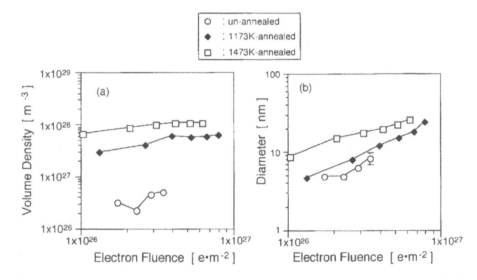

Figure 20. Electron fluence dependence of the volume density (a) and the diameter (b) of dislocation loops in Crystal B and that annealed in vacuum at 1173 K and 1473 K for 3600 s, for electron irradiations using a 1000 keV electron flux of $2.8 \times 10^{23} e/m^2 s$ [11].

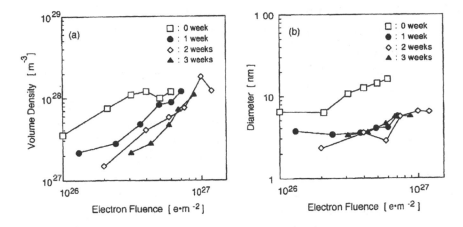

Figure 21. Same as in Figure 20, except that Crystal A was kept in an autoclave for 1, 2 or 3 weeks at 673 K and 0.1 Mpa [11].

The 1/2<101>{101} dislocation loops in the ORNL MgO and the Tateho MgO (Crystal A) grow in an anisotropic, elongated mode along <010> directions, in contrast to relatively circular loops in the OH⁻-doped Tateho and the purified Norton MgO, as shown in Figure 6. The elongated growth mode was observed in NaCl [50] and explained in terms of a difference between mobilities of jogs along the <101> and <010> directions [51]. This elongated growth mode may also be explained as due to a difference in the point defect/dislocation interaction between jogs formed on the <110> and <001> sides of the edge dislocations. Although the details of the atomic configuration that leads to this difference are not understood, the morphological difference between dislocation loops in MgO with and without OH⁻-ions suggests that OH⁻-ions make all the sites on the perimeter of the loop equivalent, and therefore no preferential growth occurs in the <010> directions.

As for the effects of cation impurities, the dislocation nucleation and growth process has also been investigated for Fe-, Cr-, Co- and Ni-doped Tateho MgO crystals [20]. However, no significant difference in the formation processes between doped and un-doped MgO has been detected, as was found in MgO containing OH⁻ ions. The doped Tateho MgO crystals are expected to contain a large amount of OH⁻ ions, which eliminate the cation impurity effects against the formation process of defect clusters. MgO crystals containing less cation vacancies associated with OH⁻ ions are required to reach conclusive remarks on the cation impurity effects. Research into the effects of cation and anion impurities may be useful for controlling the formation of defect clusters and influencing the radiation resistance of ceramics.

5.2   Effects of Simultaneous Displacive and Ionizing Radiation on the Formation Process of Defect Clusters [52]

Isolated Frenkel defects and displacement cascades are produced in materials depending on the energy of the PKA. Kinetic energy transfer less than E_d is known to cause athermal diffusion of point defects, or radiation-induced diffusion (RID). The inelastic interaction excites orbital electrons and ionizes lattice atoms. The nucleation and the growth of radiation-induced defects in irradiated materials are more or less induced by the simultaneous interaction of elementary processes, such as generation of isolated point defects, displacement cascades and ionization.

The simultaneous effects of those elementary processes on the irradiation effects are called irradiation spectrum effects, which can be categorized into two main components: (1) the effect of the recoil energy spectrum of PKAs and (2) the effect of non-displacive collisions. Numerous studies have revealed that defect production and migration in ionic and covalent crystals are very much influenced by the second effect described above.

A pioneering work of the effects of ionizing and displacive radiation in $MgO \cdot Al_2O_3$ crystals was done by Zinkle [53,54]. The density and the size of dislocation loops have been systematically investigated by cross sectional transmission electron microscopy in MgO, α-Al_2O_3 and $MgO \cdot Al_2O_3$ crystals after irradiation with a variety of MeV ions. The formation of interstitial dislocation loops is suppressed under irradiation with ions whose electronic stopping power is high. The ratio of electronic to nuclear stopping power (ENSP ratio), S_e/S_n (S_e: electronic stopping power, S_n: nuclear stopping power), or the electron-hole pairs produced per dpa has been proposed as a convenient parameter to describe the irradiation conditions that induce dislocation loops. Figure 22 shows the dislocation loop density in $MgO \cdot Al_2O_3$, MgO and α-Al_2O_3 at 923 K as a function of electron-hole pairs/dpa ratio at damage rates between 10^{-6} and 10^{-3} dpa/s [54]. It is clearly seen in Figure 17 that $MgO \cdot Al_2O_3$ is most sensitive to the suppression of dislocation loop formation. Critical values of electron-hole pairs/dpa ratios for suppressing dislocation loops have been proposed as ~50 for $MgO \cdot Al_2O_3$, around ~3000 for MgO and around ~10^4 for α-Al_2O_3. The physical mechanisms responsible for the suppression of dislocation loop formation have been discussed in terms of the increase in the mobility of point defects due to ionization-induced diffusion (IID), and/or the ionization-enhanced recombination of point defects. However, the reasons for the difference in the critical values for MgO, α-Al_2O_3 and $MgO \cdot Al_2O_3$ have not been clarified in those papers [53,54].

Another approach to investigate the simultaneous irradiation effects has been the use of transmission electron microscopes interfaced with ion accelerators [55-57]. Concurrent irradiation with a focused electron-beam and a homogeneously distributed ion-beam enables one to vary ENSP ratio (or S_e/S_n) with a variety of irradiation conditions (species, energy, flux and fluence of ions and electrons, irradiation temperature, crystal orientation, etc.) under simultaneous observations of the formation of defect clusters.

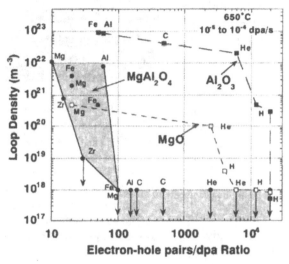

Figure 22. Dislocation loop density in MgO•Al$_2$O$_3$, MgO and α-Al$_2$O$_3$ as a function of electron-hole pairs/dpa ratio irradiated with various kinds of MeV-order ions at 930 K [53,54].

An example of bright field images in MgO•Al$_2$O$_3$ irradiated concurrently with 300 keV O$^+$ and 200 keV electrons at 870 K is shown in Figure 23 [56]. Few dislocation loops of interstitial-type are formed in regions irradiated with ions and electrons (the central area of the micrograph), while a higher density of dislocation loops is formed in regions irradiated solely with ions (the outer area of the micrograph). This clearly indicates that the dislocation loop nucleation is suppressed in MgO•Al$_2$O$_3$ by the simultaneous irradiation with 200 keV electrons.

If we assume that the displacement energy of constituent ions in MgO•Al$_2$O$_3$ is 25 eV for cation sublattice (Al- and Mg-ions) and 60 eV for anion sublattice (O-ions) [3], very few displacements are expected under 200 keV electron irradiation. Retardation of loop formation has, therefore, been attributed to the non-displacement process, that is, to ionization induced diffusion and/or radiation-induced diffusion.

As originally pointed out by the present author, in neutron- and ion-irradiated $MgO \cdot Al_2O_3$, tiny dislocation loops of interstitial-type (typically less than around 10 nm) are unstable and disappear after low energy electron irradiation, which induces no damage displacements [57]. Figure 24 illustrates a sequence of the elimination process for dislocation loops, initially induced by 6 keV Ar^+-ion irradiation at 300 K in α-Al_2O_3, under irradiation with 200 keV electrons at 300 K [58]. The elimination processes for dislocation loops are shown in Figure 25 for $MgO \cdot Al_2O_3$, $MgO \cdot 2.4Al_2O_3$ and α-Al_2O_3 as a function of electron fluence at 300 K, with electron energies from 100 to 1000 keV

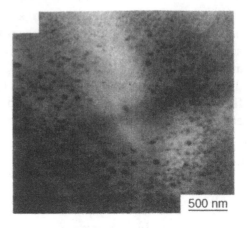

Figure 23. Bright-field image of $MgO \cdot Al_2O_3$ irraddiated concurrently with 30 keV O^+ ions and 200 keV electrons at 870 K. The fluences of 300 keV O^+ ions and 200 keV electrons are 2.4 x 10^{19} ions/m^2 and 1.1 x 10^{25} e^-/m^2, respectively. The central area of the micrograph is irradiated concurrently with O^+ ions and electrons, and the outer region is irradiated solely with O^+ ions [56].

and fluxes from 0.5 to 4.0 x 10^{23} e/m^2s [57]. Dislocation loops in all crystals of $MgO \cdot Al_2O_3$, $MgO \cdot 2.4Al_2O_3$, and α-Al_2O_3 disappear under electron irradiation with 100-200 keV electrons. Electron irradiation with 1000 keV

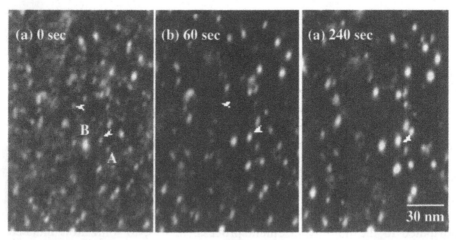

Figure 24. A sequence of weak-beam dark-field images of α-Al_2O_3 irradiated originally with 6 keV Ar^+ at 300 K followed by 200 keV electrons at 300 K for 0 sec (a), 60 sec (b) and 180 sec (c) with a flux of 4.0 x 10^{23} e/m^2s. The diffraction condition is (g, 6g) with a g-vector of g = 11-20 [58].

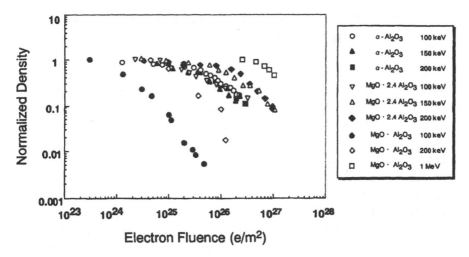

Figure 25. The normalized density of dislocation loops in MgO•Al$_2$O$_3$, MgO•2.4Al$_2$O$_3$ and α-Al$_2$O$_3$ under irradiation at 300 K with 100 to 1000 keV electrons. Electron flux ranges from 0.5 to 4.0 x 10^{23} e⁻/m^{2}s [57].

of MgO•Al$_2$O$_3$ also eliminates dislocation loops, whereas it does not eliminate loops but nucleates new ones in α-Al$_2$O$_3$ (data of α-Al$_2$O$_3$ are not shown in Figure 25). It is worthwhile to note in Figure 25 that lower energy electrons eliminate dislocation loops faster than higher energy electrons, and dislocation loops in MgO•Al$_2$O$_3$ are remarkably more unstable than in MgO•2.4Al$_2$O$_3$ and α-Al$_2$O$_3$.

As previously reported in numerous papers, MgO•Al$_2$O$_3$ exhibits excellent radiation resistance to void swelling, amorphization and dislocation loops formation [59-62]. Dislocation loop densities in MgO•Al$_2$O$_3$ irradiated with 300 keV O$^+$ ions at 870 K are more than one order lower than in α-Al$_2$O$_3$ irradiated under the same irradiation conditions [57]. Accordingly, on the basis of the results for the stability and production rate of dislocation loops, the difference in the suppression of loop formation between MgO•Al$_2$O$_3$ and α-Al$_2$O$_3$ is explained qualitatively by a competitive process of production of dislocation loops under ion-irradiation and their elimination by electron-irradiation. In particular, the suppression of loop formation in MgO•Al$_2$O$_3$ is related to the lower production rate and/or the lower stability of dislocation loops. The physical mechanism responsible for the disappearance of dislocation loops during electron irradiation is considered to be either by (1) absorption of radiation-induced vacancies or (2) dissociation into isolated interstitial atoms, since the nature of the defect clusters is of interstitial-type.

5.3   Effect of Electric Field on the Formation Process of Defect Clusters [63,64]

There have been numerous investigations of radiation effects in ceramics for applications such as fission reactors, nuclear waste disposal, fusion devices and space hardware. Among these applications, some ceramics are used as electrical components and are required to maintain their integrity under irradiation with respect to displacement damage and electronic excitation under applied electric fields [60]. In this section, we report the formation process of radiation-induced defects in α-Al$_2$O$_3$, irradiated at 760 K with 100 keV He$^+$ ions under an electric field of 100 kV/m. Microstructural observations were performed on the ion-irradiated specimens with a transmission electron microscope at an accelerating voltage of 200 kV. Hereafter, we describe the value of the electric field with nominal values, such as $E = 100$ and 300 kV/m. However, the real values of E at the localized regions observed by transmission electron microscopy might be different from those, since we used wedge-shaped thin-foil specimens. A finite element analysis by using ANSYS [65] has revealed that the value of E changes depending on the relative position to the electrodes for about ±50% from the nominal value.

Figure 26 shows typical examples of WBDF images from a wedge-shaped α-Al$_2$O$_3$ specimen irradiated with 100 keV He$^+$ ions at 760 K with and without an electric field of 100 kV/m at thicknesses of 40, 100 and 170 nm. A high density of tiny dot contrasts from 2 to 8 nm in size can be seen for both irradiation conditions, as have been reported in α-Al$_2$O$_3$ irradiated with electrons higher than 300 keV and in one irradiated with high energy ions, where the nature of those defect clusters have been demonstrated to be 1/3[0001](0001) and 1/3<10-10>{10-10} interstitial-type dislocation loops [66,67]. Judging from the analogous contrast of the defect clusters observed here and the similar irradiation conditions, the defect clusters in Figure 26 are probably dislocation loops of interstitial type. Interesting features in Figure 26 are that the density of dislocation loops is lower and the size of loops is larger in specimens irradiated with the electric field than those without the field.

In order to get further information on the interstitial kinetics, we have evaluated the number of interstitials included in dislocation loops. We define a parameter f as the number of interstitials in dislocation loops at a thickness (d) in a specimen irradiated with the electric field, normalized by the number of interstitials in loops without the field. This is expressed by the following equation:

$$f(d) = \{C^E(d){<}D^E(d){>}^2\} \, / \, \{C^U(d){<}D^U(d){>}^2\}, \qquad (8)$$

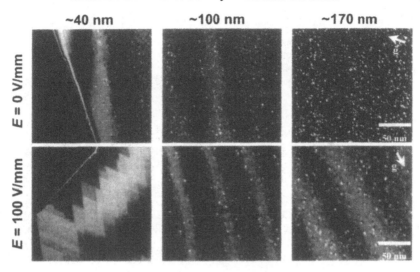

Figure 26. Weak-beam dark-field images of α-Al₂O₃ irradiated at 760 K with 100 keV He⁺ ions to a fluence of 1 x 10²⁰ He⁺ /m² with and without an electric field E = 100 kV/m at specimen thicknesses of 40, 100 and 170 nm. The diffraction condition used for these micrographs is $g - 3g$ (g = (0002)), and the specimen thickness was determined from the corresponding thickness fringes at the diffraction condition [63].

where $C(d)$ is the density of loops at d and $D(d)$ the average diameter of loops at d. The suffixes of E and U represent the irradiation conditions with and without the electric field, respectively. Figure 27 shows that the value of f is lower in the thin region, specifically $f \sim 0.1$ at d = 40 nm, and increases linearly with thickness saturating at around 1 at larger values of d ($\sim$150 nm). This indicates that a smaller fraction of displaced ions (interstitials) are included in dislocation loops under irradiation with the electric field especially in thin regions. Therefore, a large fraction of interstitials is considered to have disappeared at surface sinks under irradiation conditions with the electric field.

A clear difference appeared at an irradiation temperature of 870 K between microstructure with and without an electric field of $|E|$=100 kV/m, as shown in Figure 28. A high density of tiny dot contrasts (indicated as S in Figure 28(a)), whose average size is 2.0 nm, is formed under irradiation without the electric field. The relatively large contrast features with an average size of 4.2 nm (indicated as L in Figure 28(b)), which are considered to be interstitial-type dislocation loops, are also seen. On the other hand, only large contrast features similar to defect L are observed with the presence of an electric field of $|E|$=100 kV/m. The areal density of both defect clusters is shown in Figure 29 as a function of specimen thickness. Defects **S** are seen to possess a distinct peak at a depth of around 100 nm, whereas the density of defects L, which are formed in both

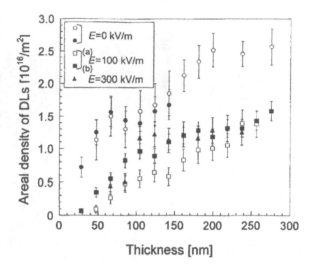

Figure 27.   Ratio of the number of interstitials contained in dislocation loops (*f*) in α-Al₂O₃ irradiated with and without the electric field *E* = 100 kV/m as a function of specimen thickness. *f* is defined by Equation (8) in the text.

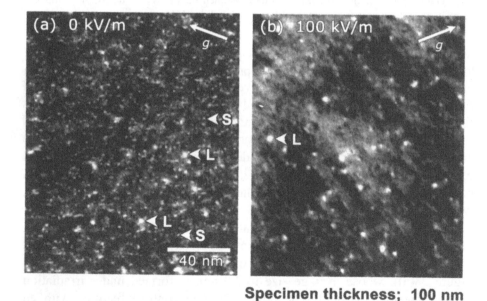

Specimen thickness; 100 nm

Figure 28.   WBDF images of α-Al₂O₃ irradiated at 870 K to a dose of 0.5 dpa, without (a) and with (b) an electric field of 100 kV/m. The diffraction condition is g-3g with g= 0002. Two kinds of defect contrasts are seen in (a) indicated with **S** and **L** in the micrograph. The specimen thickness is around 100 nm.

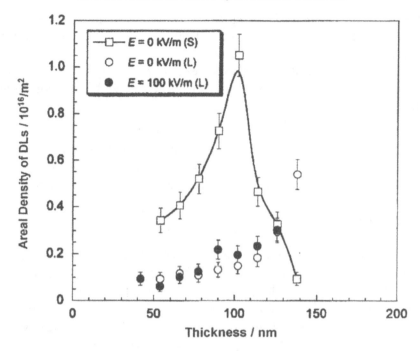

Figure 29. Areal density of dislocation loops in α-Al₂O₃ against specimen thickness irradiated at 870 K to a dose of 0.5 dpa. Defect S and L (see the definition in Figure 28(a)) are separately plotted for specimen irradiated without the electric field.

irradiation conditions, is seen to increase gradually with thickness. Although the nature of defect **S** is uncertain at this moment, there have been reported similar thickness-dependent microstructures in wedge-shaped Cu thin-foils irradiated with 20–100 keV He⁺ ions [68], in which vacancy-type clusters were formed near the specimen surface at a limited thickness of the specimen. This has been discussed in terms of a bias of vacancies near surfaces, which is caused by a difference in the mobility between vacancies and interstitials. A possible explanation for the nature of defects **S** is, therefore, vacancy-type clusters formed by a bias of vacancies, though further work is necessary to prove the speculation. It is, however, worthwhile to note that the formation of defect S is retarded with the presence of an electric field.

In a classical model [69], an electric field increases or decreases the migration energy of interstitials (E_m^i) by an amount $Z_i eEa$, depending on the sign of the electric charge of the ions and on the direction of the electric field. Here, Z_i is the valence of i-ions, e the electronic charge, and a the interatomic spacing. An estimation of $Z_i eEa$ in α-Al₂O₃ reveals ~10⁻⁴ eV with $|E| = 100$ kV/m and $a = 0.5$ nm, which is much lower than the reported values of E_m^i in α-Al₂O₃, namely 0.2-0.8 eV[3]. Therefore, the change in the interstitial kinetics observed in Figure 22, or the apparent increase in the mobility of interstitials, is not explained by the decrease in the absolute value of E_m^i.

The total flow of interstitials species i (J_i) driven by both the electric field and the concentration gradient can be described by the following equation:

$$J_i = Z_i e \mu_i C_i E - D \nabla C_i,$$ (9)

where μ_i is the mobility, C_i the concentration and D_i the diffusion coefficient of interstitials. The mobility is related to the diffusion coefficient via the Einstein relation, $\mu_i = D_i/kT$. Equation (9) can therefore be rewritten as

$$J_i = D_i(Z_i e C_i E /kT - D \nabla C_i).$$ (10)

We have made a rough estimate of values of the first term (related to the driving force by the electric field) and the second term (the concentration gradient) in the parentheses of Equation (10). In the first approximation, the value of C_i is assumed to be 0.1 (at%), which is a typical saturation value of interstitials in irradiated materials, and ∇C_i is assumed to be constant in the specimen with a thickness of 200 nm, though ∇C_i is considered to be dependent on the position in the specimen. The first term is estimated to be 4.6 x 10^3 (m^{-4}) with $E = 10^5$ (V/m), $T = 760$ K and $Z_i = 3$ (for example as Al^{3+} interstitials), whereas the second term is 10^4 (m^{-4}). The flow of interstitials driven by the electric field is, therefore, about one half of that driven by the concentration gradient at 760 K with a value of $E = 10^5$ V/m. Even when one assumes a lower value of C_i, the ratio of J_i driven by the electric field and the concentration gradient will be identical with the assumptions mentioned above.

This leads to the conclusion that the total flow of interstitials caused by the electric field is not negligible. Since the electric field causes a directed migration of interstitials, most of those will reach the surface sinks at the thin regions of wedge-shaped specimens (a typical wedge angle is 8 degrees for the specimens used in the present study). The lower values of f in the thin regions of the wedge-shaped specimens irradiated in the presence of an electric field are, therefore, explained by the directed migration of interstitials driven by the electric field to the surface sinks. On the other hand, an electric field of 100 kV/m has been shown to retard the formation of defect S (it is speculated to be vacancy-type cluster) at 870 K. This is presumably due to the change in the balance between vacancies and interstitials in α-Al$_2$O$_3$ with the electric field. A directed migration of charged vacancies and interstitials is considered to change the I-V recombination rate and/or surface sink strength, leading to the suppression of defect S formation.

6. SUMMARY

When a material is irradiated with energetic radiation, defects and imperfections are created. The spatial arrangement of defects is generally described as "radiation damage", and any physical property changes, which may result from the introduction of the damage are called "radiation effects". The fundamental aspects of radiation damage and their characteristics in materials are described in terms of their own particular sets of rules, such as the effect of electronic excitation, the polyatomicity, the charge state of point defects and the radiation-induced bias.

Special emphasis is placed on the characteristics of defect clusters formation in MgO, TiC, α-Al$_2$O$_3$, MgAl$_2$O$_4$ and ZrO$_2$, which are characterized in terms of mass and mobility of anions and cations, structural vacancies and/or stable nuclei of defect clusters. In MgO, a dimer of Mg and O is the most stable nucleus of interstitial loop based on similar masses and mobilities of anions and cations. In TiC type crystals, which contain a high concentration of structural vacancies due to non-stoichiometry, most of the displaced atoms recombine with structural vacancies, and the nucleation and growth process of interstitial loops is extremely suppressed. Furthermore, the formation of non-stoichiometric defect clusters is associated with cation disordering or biased cation- or anion-displacements in α-Al$_2$O$_3$, MgAl$_2$O$_4$ and/or ZrO$_2$.

The effects of impurities, electronic excitation and electric field on microstructural evolution are also described. The nucleation and growth process of dislocation loops in MgO is much influenced by hydrogen and carbon impurities. The stability of non-stoichiometric defect clusters is very much influenced not only by ionizing radiation but also by electric field.

REFERENCES

[1] G. H. Kinchin and R. S. Pease, *Reports on Progress in Physics*, 18(1955), 1.

[2] C. H. de Novion and J. Morillo, *J. Atomic Energy Soc. of Japan*, 30(1988), 585.

[3] S. J. Zinkle and C. Kinoshita, *J. Nucl. Mater.*, 251(1997), 200.

[4] F. Seits and J. S. Koehler, *Solid State in Physics*, 2(1956), 305.

[5] G. P. Pells, *J. Nucl. Mater.*, 155-157(1988), 67.

[6] L. M. Howe and M. H. Rainville, *Nucl. Instr. Meth. Phys. Res.*, B19/20(1987), 61.

[7] S. J. Zinkle and V. A. Skuratov, *Nucl. Instr. Meth. In Phys. Res.*, B141(1998), 737.

[8]   E. Sonder and W. A. Sibley, In: J. H. Crawford Jr and L. M. Slifkin (eds.) Point Defects in Solids (1972), (Plenum Press, New York-London.)

[9]   F. W. Clinard Jr. and L. W. Hobbs, In: R. A. Johnson and A. N. Orlov (eds.) *Physics of Radiation Effects in Crystals (1986)*, Elsevier Science Publishers BV.

[10] T. Brudevoll, E. A. Kotomin and N. E. Chritensen, *Phys. Rev.*, **B53**(1996), 7731.

[11] C. Kinoshita, T. Sonoda and A. Manabe, *Phil. Mag.*, **A78**(1998), 657.

[12] K. E. Sickafus, A. C. Larson, N. Yu, M. Nastasi, G. W. Hollenberg, F. A. Garner and R. C. Bradt, *J. Nucl. Mater.*, **219**(1995), 128.

[13] T. Soeda, S. Matsumura, J. Hayata and C. Kinoshita, *J. Electron Microsc.*, **48**(1999), 531.

[14] H. M. Naguib and R. Kelly, *Rad. Effects*, **25**(1975), 1.

[15] W. J. Weber, R. C. Ewing, C. R. A. Catlow , T. D. de la Rubia, L. W. Hobbs, C. Kinoshita, Hj Matzke, A. T. Motta, M. Nastasi, E. K. H. Salje, E. R. Evance and S. J. Zinkle, *J. Mater. Res.*,**13**(1998), 1434.

[16] C. Kinoshita, *J. Electron Microsc.*, **40**(1991), 301.

[17] C. Kinoshita, *J. Nucl. Mater.*, **191-194**(1992), 67.

[18] K. Fukuya, M. Terasawa and K. Ozawa, *J. Atomic Energy Soc. Jpn.,* 30(1988), 657.

[19] L.W. Hobbs, F. W. Clinard Jr.,S. J. Zinkle and R. C. Ewing, *J. Nucl. Mater.*,**216**(1994), 291.

[20] C. Kinoshita, K. Hayashi and S. Kitajima, *Nucl. Instr. Meth. in Phys. Res.*, **B1**(1984), 209.

[21] C. Kinoshita, K. Hayashi and T. E. Mitchell, *Adv. Ceram.*,**12**(1984),490.

[22] C. Kinoshita and K. Nakai, *Jpn. J. Appl. Phys.*, Series **2**(1989), 105.

[23] M. Kiritani and H. Takata, *J. Nucl. Mat.,* **69&70**(1978), 277.

[24] K. Nakai, C. Kinoshita, Y. Muroo and S. Kitajima, *Phil. Mag.*, **A48**(1983), 215.

[25] K. Nakai, C. Kinoshita and S. Kitajima, *Phil. Mag.*, **A52**(1985), 115.

[26] C. Kinoshita, T. Mukai and S. Kitajima, In: J. Takamura, M. Doyama and M. Kiritani (eds.) *Point Defects and Defects Interactions in Metals(1982)*, University of Tokyo Press, Tokyo , p. 887.

[27] F. W. Clinard, Jr., G. F. Hurley and L. W. Hobbs, *J. Nucl.Mater.*,**108-109**(1982), 655.

[28] C. A. Parker, L.W. Hobbs, K. C. Russell and F. W. Clinard, Jr., *J. Nucl. Mater.*, **133-134** (1985), 741.

[29] F. A. Garner, G. W. Hollenberg, F. D. Hobbs, J. L. Ryan, Z. Li, C. A. Black and R.C.Bradt, *J. Nucl. Mater.*, **212-215** (1994), 1087.

[30] C. A. Black, F. A. Garner and R. C. Bradt, *J. Nucl. Mater.*, **212-215** (1994) 1096.

[31] R. A. Youngman, T. E. Mitchell, F. W. Clinard, Jr. and G. F. Hurley, *J. Mater. Res.*, **6** (1991), 2178.

[32] C. Kinoshita, K. Fukumoto, K. Fukuda, F. A. Garner and G. W. Hollenberg, *J. Nucl. Mater.*,**219** (1995) ,143.

[33] L. W. Hobbs and F. W. Clinard, *Jr.*, *J. Phys.*, **41** (1980), C6-232.

[34] S. J. Zinkle, *Nucl. Instr. and Meth.*, **B 91** (1994),234.
[35] K. Nakai, K. Fukumoto and C. Kinoshita, *J. Nucl. Mater.*, **191-194** (1992), 63.

[36] K. Fukumoto, C. Kinoshita and F. A. Garner, *J. Nucl. Sci. and Technol.*, **32** (1995), 773.

[37] P. Veyssiere, J. Rabier and J. Grilhe, *Phys. Stat. Sol. (a)*, **31** (1975), 605.

[38] P. Veyssiere, J. Rabier, H. Garem and J. Grilhe, *Philos. Mag.*, **38** (1978), 61.

[39] R. S. Wilks, J. A. Desport and R. Bradley, *Proc. Brit. Ceram. Soc.*, **7** (1967), 403.

[40] R. S. Wilks, *J. Nucl. Mater.*, **26** (1968), 137.

[41] D. J. Barbat and N. J. Tighe, *J. Am. Ceram. Soc.*, **51** (1968), 611.

[42] W. E. Lee, M. L. Jenkins and G. P. Pells, *Philos. Mag.*, **A 51**(1985), 639.

[43]A. H. Cottrel and B. A. Bilby, *Proc. Phys. Soc.*, **62** (1949), 49.

[44] A. I. Ryazanov, K. Yasuda, C. Kinoshita and A. V. Klaptsov,*J. Nucl. Mater.*, **307** (2002), 918.

[45] K. Yasuda, C. Kinoshita, S. Matsumura and A. I. Ryazanov, *J. Nucl. Mater.* **319**(2003) ,74.

[46] A. I. Ryazanov, K. Yasuda, C. Kinoshita and A. V. Klaptsov,*J. Nucl. Mater.*, **323** (2003), 372.

[47] C. Kinoshita, *J. Nucl. Mater.,* **179-181**(1991), 53.

[48] A. M. Glass and T. M. Searle, *J. Chem. Phys.,***46**(1967),2092.

[49] F. Freund and H. Wengeler, *J. Chem. Solids*, **43**(1982),129.

[50] L. W. Hobbs, In: J. A. Venable (eds.) *Development in Electron Microscopy and Analysis(1976)*, Academic Press, New York, p. 287.

[51] R. A. Youngman, L.W. Hobbs and T. E. Mitchell, *J. Phys.*,Paris, 41(1980), C6-227.

[52] C. Kinoshita, K. Yasuda, S. Matsumua and M. Shimada, *Metall. and Mater. Trans.*,**35A**(2004)2257.

[53] S. J. Zinkle, *Nucl. Instr. Meth. Phys. Res.*, **B91**(1992), 67.

[54] S. J. Zinkle, *Rad. Effects Defects Solids,* **148**(1999), 447.

[55] C. Kinoshita H. Abe, S. Maeda and K. Fukumoto, *J. Nucl. Mater.*, **219**(1995), 152.
[56] K. Yasuda, C. Kinoshita, R. Morisaki and H. Abe, *Phil. Mag.* ,**A78**(1998), 583.

[57] K. Yasuda, C. Kinoshita, M. Ohmura and H. Abe, *Nucl. Instr. and Meth. Phys. Res.,* **B166-167**(2000), 107.

[58] K. Yasuda and C. Kinoshita, *Nucl. Instr. and Meth. Phys. Res.,* **B191**(2002), 559.

[59] C. Kinoshita, K. Fukumoto, K. Fukuda, F.A. Garner and G.W. Hollenberg, *J. Nucl. Mater.*, **219**(1995), 143.

[60] C. Kinoshita and S.J. Zinkle , *J. Nucl. Mater.,* **233-237**(1996), 129.

[61] K.E. Sickafus, N. Yu and M. Nastasi, *Nucl. Instr. and Meth. in Phys. Res.,***B116**(1996), 85.

[62] K. Yasuda, C. Kinoshita, K. Fukuda and F.A. Garner, *J. Nucl. Mater.,* **283-287**(2000), 937.

[63] K. Yasuda, T. Higuchi, K. Shimada, C. Kinoshita, K. Tanaka and M. Kutsuwada, *Philos. Mag. Lett.,* **83**(2003), 21.

[64] T. Higuchi, K. Yasuda, K. Tanaka, K. Shiiyama and C. Kinoshita, *Nucl. Instr. and Meth. in Phys. Res.,* **B206**(2003), 103.

[65] http://www.ansys.com/

[66] G.P. Pells and Stahopoulos, *Rad. Effects,* **74**(1983), 181.

[67] W.E. Lee, M.L. Jenkins and G.P. Pells, *Philos. Mag.,* **A51**(1985), 639.

[68] K. Yasuda, C. Kinoshita, M. Kutsuwada and T. Hirai, *J. Nucl. Mater.,* **233-237**(1996), 1051.

[69]W.D. Kingery, H.K. Bowen and D.R. Uhlmann, *Introduction to Ceramics 2nd Edition*, 1975.

Chapter 9

OPTICAL & SCINTILLATION PROPERTIES OF NONMETALS: INORGANIC SCINTILLATORS FOR RADIATION DETECTORS

Vladimir N. Makhov
Lebedev Physical Institute, Moscow, Russia

1   INTRODUCTION

At present, scintillation detectors are widely used for the detection of nuclear radiation, especially in high-energy physics and nuclear medicine. A scintillation detector is a device that consists of a scintillator and a light detector, usually a photomultiplier tube (PMT). The role of the scintillator is to convert the energy that the ionizing radiation has lost into pulses of light. For effective absorption of penetration radiation such as γ–rays, a scintillator requires high density and a high-effective atomic number Z. For good energy resolution, a high light yield, i.e., a high number of photons emitted per unit of absorbed energy, is necessary. For good time resolution, a scintillator needs a short light pulse. The scintillator should also be transparent for emitted light so that the light could reach the light detector.

Many inorganic crystals meet these requirements, in principle. A good inorganic scintillator should first effectively stop and convert the incoming high-energy photon or particle into a shower consisting of the largest possible number of elementary electronic excitations (electron-hole pairs), then provide an efficient mechanism by which the energy of the electron-hole pairs is transferred to the luminescent centers, and finally create an environment in which these centers luminescence efficiently and quickly. In this paper, two kinds of inorganic crystals will be discussed that are considered as promising fast scintillators and that are widely studied during past years: the cross-luminescence (CL) crystals and crystals doped with triply ionized, or trivalent, rare earth (RE) ions. However, first it is necessary to discuss the physical processes that occur in any kind of inorganic scintillator after the absorption of a high-energy photon or particle

K.E. Sickafus et al. (eds.), Radiation Effects in Solids, 233–257.
© 2007 *Springer.*

and to consider the factors that determine the main characteristics of the scintillation detector.

2   COMMON PROPERTIES OF INORGANIC SCINTILLATORS

There are three basic processes by which a high-energy photon interacts with matter: photoelectric effect, Compton interaction, and electron-positron pair production. The intensity of monoenergetic beam of γ-rays is reduced after entering the scintillator of thickness d according to

$$I = I_0 \, exp(-\mu d) , \qquad (1)$$

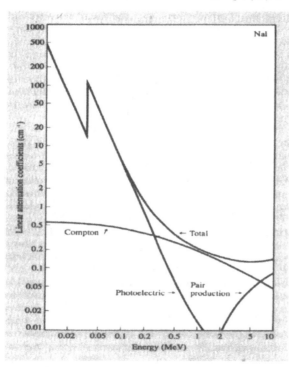

where the total linear attenuation coefficient μ is the sum of all interaction cross sections associated with photoelectric effect, Compton interaction, and pair production. The photoelectric effect (absorption of the quantum by the atom with the ejection of photoelectrons from some inner shells) is most important for low energies. For this kind of interaction, the complete energy of the incident photon is absorbed in the scintillator, resulting in the appearance at the detector of

Figure1. Linear attenuation coefficients μ in NaI:Tl for the different interaction mechanisms as a function of the γ–ray energy.

photopeak. For intermediate energies, the Compton scattering (i.e., scattering of γ-quantum on atomic electrons) is most important. For Compton interaction, the resulting energy spectrum extends between zero and some maximum energy, which is called Compton edge. If the energy of γ-quantum exceeds 1.02 MeV, the electron-positron pair production becomes possible. The positron annihilates with another electron with the emission of two photons whose energy is 0.511 MeV. At even higher energies, this channel of interaction becomes the dominant one. Figure 1 shows the dependence of partial cross sections and total absorption on photon energy for the well-known inorganic scintillator NaI:Tl. The probability of photoelectric interaction per atom is proportional to Z^p where p ranges between 4 and 5. For the Compton effect, the probability of interaction is proportional to Z. Thus the photofraction in absorption strongly increases with Z of the absorber.

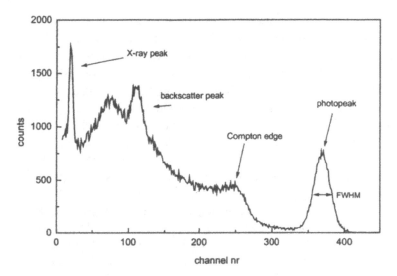

Figure 2. Typical pulse height spectrum for γ-rays coming from a radioactive source (662 keV γ-rays from a ^{137}Cs source) recorded with a scintillator crystal.

The basic principle of a scintillation detector is that the light yield of a scintillator is proportional to the energy deposited in the crystal. The standard scintillation detector consists of a scintillator coupled to some light detector, usually PMT, in which the photons produce photoelectrons that are then multiplied by the dynodes of the PMT. The pulse amplitude from PMT will be proportional to the number of photons emitted by the scintillator. The spectrometer consists of preamplifier, amplifier, and multichannel analyzer. Obviously it is necessary that all the components have a linear response. An example of the pulse height spectrum from a scintillation detector irradiated by 662 keV γ-rays is shown in Fig. 2. For γ-ray spectrometry, the photopeak-to-total ratio should be as large as possible.

The next stage in a scintillator, which occurs after the absorption of a high-energy photon or particle and the creation of elementary electronic excitations, is the energy transfer to emission centers. The mechanism and efficiency of the energy transfer are the main factors that control the light yield of the scintillator. In an ideal case, assuming that all electrons and holes created after the absorption of a high-energy photon with energy E_γ will be transferred to the recombination centers (which serve simultaneously as emission centers) and assuming that the intrinsic luminescence quantum efficiency of the emission centers is unity, the light yield will be determined by the number of created electron-hole pairs n_{e-h}. This can be expressed as [1–4]:

$$Light\ yield \sim n_{e-h} = E_\gamma / \beta \cdot E_g \qquad (2)$$

Thus, the fundamental limit on the light yield is determined by the band gap of the scintillator material E_g and the value for β. The latter indicates the factor by which the energy necessary to create one electron-hole pair exceeds the band gap. As was found from experiments and also theoretically, β is close to 2.5 for most of inorganic scintillators.

The fundamental limit of the light yield is smallest for fluorides with the largest band gaps and much larger for sulfides and other inorganic materials with small band gaps. In particular, the well-known scintillator ZnS:Ag has light yield near 100,000 photons/MeV, which is close to the fundamental limit for this scintillator material. Within sulfides, iodides, bromides, and chlorides, namely, within small band-gap materials, many scintillators with light yield approaching the fundamental limit have been found. On the other hand, oxides and fluorides almost always show much less light yield than the fundamental limit. In particular, this holds for the well-known and widely studied fast scintillator CeF_3. The reason for this effect is that the standard recombination mechanism of emission centers excitation (i.e., Ce^{3+} ions) as a result of consequent capture on the center of the hole and electron (or electron and hole) is not efficient in such crystals because the high-lying energy position of the Ce^{3+} ground state above the valence band in fluorides prevents efficient hole trapping by the Ce^{3+} ions [5]. The dominant mechanism of Ce^{3+} excitation in such systems is the impact mechanism when fast photoelectrons excite the Ce^{3+} ions directly (by impact) [6,7]. For such a mechanism, the above formula does not work correctly, and the light yield is much smaller than calculated by this simple method. This formula also cannot be applied for CL scintillators, which will be discussed below.

Energy resolution is the ability of the detector to discriminate between γ-rays (or particles) with slightly different energy. The maximal obtainable energy resolution (for the fixed energy of ionizing radiation) is mainly determined by the statistics in the number of photons detected by the photon detector. However, some contribution may also arise because of the nonproportional response of the scintillator and to other more technical factors such as inhomogenities in the crystal, nonuniformity in the light collection efficiency, nonuniformity in the photocathode performance, etc. For an ideal scintillation detector, i.e., if the influence of the above-mentioned factors is negligible, the fundamental limit of the energy resolution (FWHM) can be written as [4,8,9]

$$\Delta E/E = 2.35 \sqrt{(1+v(M))/N_{phe}} \tag{3}$$

where N_{phe} is the number of photoelectrons arriving at the first dynode of the PMT, and $v(M)$ is the variance in the PMT gain that is usually of the order of 0.1–0.2. Thus, the energy resolution is inversely proportional to the square root of the number of photoelectrons created at the PMT after the absorption of scintillation photons. In fact, the main factor limiting energy resolution is the light yield of the scintillator. The best energy resolutions were obtained for scintillators based on small band-gap materials with high light yield, as expected, in particular for the well-known scintillators $NaI:Tl$, $CsI:Tl$, and $CaI_2:Eu^{2+}$ (~5%) as well as for recently found scintillators $LaCl_3:Ce$, $LaBr_3:Ce$, and $RbGd_2Br_7:Ce$ [4]. The energy resolution of the $YAlO_3:Ce$ (YAP) scintillator was remarkably improved during several years of research and has almost reached the fundamental limit. On the other hand, the energy resolution of another well-known scintillator, $Lu_2SiO_5:Ce^{3+}$ (LSO), has not been improved at all despite considerable research and is still quite far from its theoretical maximum.

The time resolution of a scintillation detector is the ability to define accurately the moment of absorption of the ionizing particle or quantum. Good time resolution is very important in high counting rates (high intensity of radiation) or if the time-of-flight technique is applied. Clearly, for better time resolution, the scintillation pulse should be short; decay time of emission from the emission center should be small.

Discussed below are two kinds of luminescence mechanisms that are now widely used in different scintillation detectors, namely, CL and luminescence due to interconfigurational 5d-4f transitions in triply ionized RE ions.

3   CROSSLUMINESCENCE

3.1   Simple Model and Main Properties

Figure 3 shows the scheme of energy bands and electronic transitions in BaF_2 crystal as an example that describes the mechanism of CL. CL is a specific kind of intrinsic luminescence in ionic crystals. The luminescence is caused by radiative recombination of electrons from the valence band with the holes in the uppermost core band created in the crystal by VUV radiation which energy exceeds the ionization edge of the uppermost core band [10]. CL is observed in ionic crystals with low ionization energy of the uppermost core band (such as BaF_2, CsF, CsCl, CsBr, RbF, and KF) when the Auger-decay of the holes in the uppermost

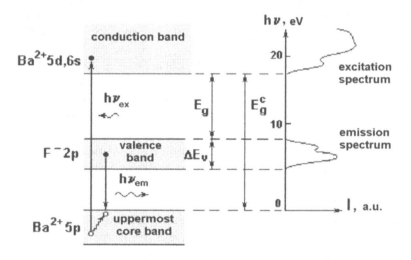

Figure 3. Simple model of crossluminescence in BaF_2.

core band is energetically forbidden, i.e., when the energy that can be released as a result of nonradiative recombination of the valence electron with the uppermost core hole is not sufficient for the escape of another electron from the valence band to the conduction band. This is the case if the energy separation between the tops of the valence and uppermost core bands is less than the band gap of the crystal. In such crystals the radiative decay of the core band becomes the dominant decay channel. The term crossluminescence introduced in [11] reflects the fact that at CL transition, the electron is transferred from one ion (anion) to another one (cation). In ionic crystals, the valence band is formed from the p-type states of the anion, and the uppermost core band is formed from the p-type states of the cation. As explained above for the following terms, this kind of luminescence is also called a core-valence luminescence [12] or an Auger-free luminescence [13]. The results of CL research have been reviewed before in [14–17].

One of the main properties of CL is the fact that the excitation edge of CL coincides with the edge of the holes creation in the uppermost core band [10,12,13,18–22]. Just the presence of such edge in luminescence excitation spectrum is the criterion of the CL existence in the particular crystal. The spectrum of CL usually consists of a few wide bands situated mainly in the UV and VUV region and corresponds in the energy scale to the spectrum of transitions from the whole valence band to the top of the uppermost core band. We infer that the core hole relaxes to the top of the uppermost core band before the radiative transition takes place. The characteristic decay time for CL is ~1 ns that corresponds to radiative transitions probabilities for allowed transitions with the energies typical for CL spectrum. The decay law is the same for all emission bands in the CL spectrum from the crystal: all emission bands correspond to transitions from the same initial state of the core hole but to different final states of the hole in the valence band. An important property of CL is its very high temperature stability, namely, for most of the CL-active crystals, the CL does not show thermal quenching up to temperatures around 600–700 K.

Actually, CL is observed only in six binary ionic crystals mentioned above, but it is also observed in many ternary and multicomponent crystals based on these six binary compounds. The number of CL crystals will be very large if one considers the impurity CL [23–26] when the CL-active cation (Ba^{2+}, Cs^+, Rb^+, and K^+) is introduced into some halide crystal that does not posses intrinsic but will possess extrinsic, not intrinsic, CL due to radiative recombination of valence electrons with the holes in the uppermost level of the impurity cation. The spectra of six main binary and one ternary ($KMgF_3$) CL crystals are shown in Fig. 4.

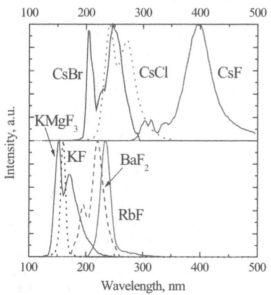

Figure 4. Crossluminescence spectra of six main binary crystals and $KMgF_3$.

Because the radiative lifetime of the core hole in CL-active crystals is relatively long ($\tau \sim 10^{-9}$ s) compared with characteristic times of lattice vibrations, not only the complete relaxation of the electronic system occur before radiative transition (thermalization of the core hole to the top of the uppermost core band), but also relaxation (deformation) of the lattice will occur near the core hole, resulting in the localization (self-trapping) of the core hole near some (cation) site in the crystal. Therefore, CL should be treated as a radiative recombination of the valence electrons with the localized core holes. However, in principle, some energy barrier can exist between free and localized states of the core hole, and accordingly, the core hole can have some mobility before its localization [27]. On the other hand, in the case of the impurity CL, the hole created on the impurity CL-active cation is always localized because of its nature, but impurity and intrinsic CL are characterized by similar properties.

3.2   Core Hole Relaxation and Related Phenomena

Thus, the core hole created in the uppermost core band of some CL-active crystal should be considered as some local center, but the final state of the hole after the CL transition can be both localized and delocalized. Because the width of the spectrum of CL corresponds well enough to the width of the valence band, the final states of the holes after the CL transitions should be free states in the valence band. In the space of the configurational coordinate, the CL transition can be presented as the transition of the hole localized state in the uppermost core band to the valence band. In Fig. 5 such a process is shown for CsF crystal as an example [28]. In such a scheme, the radiative CL transition can be represented as some charge-transfer transition in the CsF molecule, namely, as the transition of the hole from the Cs^+ ion to the F^- ion from the bond (i.e., localized) state to the unbond (i.e., delocalized) state.

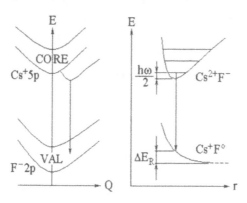

Figure 5. The adiabatic potential curves for the hole in CsF in single configurational coordinate scheme (left). The adiabatic potential curves for excimer molecule $(CsF)^+$ (right). 1/2 $\hbar\omega$ - zero-temperature vibrational energy. ΔE_R - relaxation energy.

The local nature of CL was used as the basis for theoretical calculations of CL spectra with different types of cluster approximations [29–31] as well as with the band model by considering the core hole relaxation [32,33]. Such calculations give usually rather good simulation of CL spectra for most of CL-active crystals. However, many features of CL, e.g., a low-energy tail in CL spectra or the existence of some (more narrow) bands in the CL spectra, were not reproduced theoretically by such calculations. The low-energy tail in CL spectra was assigned to the effect of lattice relaxation around the core hole [34]. The narrow bands in the spectra of some CL crystals were ascribed in [35] to radiative transitions of the free core holes to the valence band. Probably a more precise theoretical description of CL properties can be derived from using the cluster approach and the energy band model by considering lattice relaxation.

One more manifestation of the local nature of CL is the temperature broadening of CL emission bands. This broadening is well described in the framework of the model of phonon broadening for the local optical center in the crystalline environment in the limit of strong electron-lattice coupling. The FWHM for the emission band of such a local optical center can be represented as [28]:

$$W(T) = W_0 \times [coth(\hbar\omega/2k_BT)]^{1/2}, \qquad (4)$$

where $W(T)$ and W_0 are the spectral widths of the emission band at temperatures T and $T = 0$, respectively, ω is the phonon frequency, and k_B is Boltzmann constant.

Figure 6 shows this broadening for the CL spectrum of BaF_2 crystal as an example [36]. However, temperature broadening can differ for different emission bands in the CL spectrum of the same crystal and is almost independent of temperature in some situations. So a more complicated model of optical center should be applied for the description of temperature broadening of CL emission bands. However, from the comparison of the features of phonon broadening for CL and luminescence of self-trapped excitons, one can conclude that the lattice structure of the emitting center for CL probably has the on-center character: lattice deformation is symmetric around the core hole localized on the cation [37]. Further studies are necessary to clarify all the features of electronic and lattice relaxation in the process of CL.

3.3   Crossluminescent Crystals as Fast Scintillators

Among all inorganic scintillators studied so far, the best time resolution (few hundred picoseconds) was obtained for scintillation detectors based on the CL crystal BaF_2. However, CL crystals have many disadvantages as scintillators, namely, relatively low-light yield (~1000 photons/662 keV γ–quantum); the presence of a slow-emission component (usually emission of self-trapped excitons); and rather low density, moderate physicochemical properties. The low-light yield of CL scintillators is caused by the specific fundamental mechanism of CL excitation [38]; formula (2) cannot be used directly for the estimation of the light yield for CL scintillators.

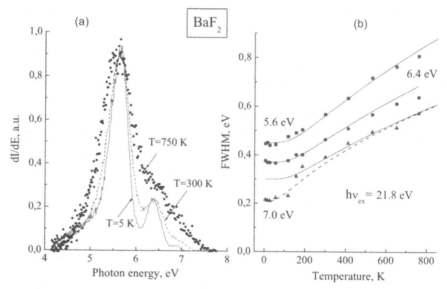

Figure 6. (a) Crossluminescence spectra of BaF_2 measured at different temperatures. Spectra are normalized with respect to their maxima. (b) Temperature dependencies of the width of the bands in the crossluminescence spectrum of BaF_2. Dots are experimental data, lines are calculated curves with ħω = 38 meV (solid lines) and 18 meV (dashed line). Excitation energy is 21.8 eV.

On the other hand, the specific properties of CL can be used in some special kinds of radiation detectors. In particular, CL is not excited (it is completely quenched) in the large local density of electronic excitations [39–41] in the ionization track, which occurs if the high-energy particles with high linear energy transfer (LET) are created in the CL scintillator. This principle was proposed for neutron-gamma discrimination in scintillators possessing both CL and some other kind of fast luminescence that is not quenched in the absorption of high LET radiation, e.g., in $LiBaF_3$ doped with Ce^{3+} [42].

CL scintillators can find (and have already found) application in some specific detectors, mainly if a very high-time resolution is required or for neutron-gamma discrimination.

4 INTERCONFIGURATIONAL 5d– 4f TRANSITIONS IN RARE EARTH IONS DOPED INTO WIDE BAND-GAP HOSTS

4.1 Energy-Level Structure of Rare Earth Ions in Crystalline Environment

It is well known that RE atoms, or lanthanides, are characterized by filling the inner 4f shell. For triply ionized RE ions, the 4f shell is the only incompletely filled shell, and so optical properties of trivalent RE ions (in the visible spectral range) are determined by radiative transitions between different states inside $4f^n$ electronic configuration, i.e., by intraconfigurational 4f-4f transitions. In the free ion, the electric-dipole electronic transitions inside the same electronic configuration are parity-forbidden. But in the crystalline environment, the spherical symmetry of the free ion is destroyed by the interaction of the 4f electrons with the electrostatic field of surrounding ions also called ligands. This results in the crystal field (Stark) splitting energy levels of RE^{3+} ions, and also some 4f-4f transitions become partially allowed. The 4f shell is shielded by completely filled outer 5s and 5p shells, and the influence of the crystal field on energy levels of RE^{3+} ions inside the $4f^n$ electronic configuration is small. This means that in first approximation the electronic states of $4f^n$ configuration for RE^{3+} ions in the crystalline environment can be treated as those in the free ions, and crystal field can be considered as a small perturbation. Accordingly, the energies of the corresponding levels of $4f^n$ configuration are only weakly sensitive to the type of the crystal host. However, the interaction with the crystal field mixes electronic states of different parity into 4f wave-functions. As a result, the electric-dipole transitions within the $4f^n$ electronic configuration become partially allowed, but radiative lifetimes for 4f-4f transitions lie in the microsecond and millisecond range.

Except for intraconfigurational, parity-forbidden $4f^n$-$4f^n$ transitions, in RE^{3+} ions the parity-allowed interconfigurational $4f^n$-$4f^{n-1}5d$ transitions are also possible when one 4f electron of the RE^{3+} ion is promoted to 5d orbital. Accordingly, one can observe luminescence due to radiative transitions from the lowest level of $4f^n5d$ electronic configuration to the ground state and to some excited levels of $4f^n$ electronic configuration. Radiative lifetimes for these parity-allowed transitions lie in the nanosecond range (only for spin-allowed [SA] transitions; see below). The crystals emitting this fast nanosecond luminescence due to interconfigurational 5d-4f transitions in RE^{3+} ions can be considered as promising fast scintillators.

Because the 5d electrons are not effectively shielded by other electrons, the crystal field influence on the energy levels of $4f^{n-1}5d$

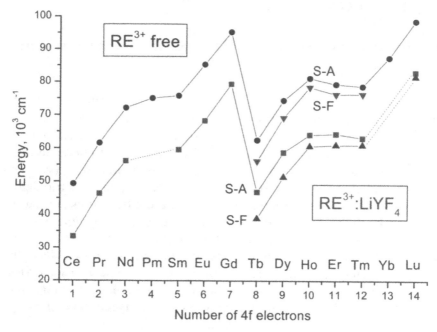

Figure 7. Energy of the lowest-energy spin-allowed (SA) and spin-forbidden (SF) $4f^n$-$4f^{n-1}$ 5d transitions in free lanthanides RE^{3+} and RE^{3+} doped into $LiYF_4$ crystal.

electronic configuration is strong. Accordingly, crystal field splitting of 5d levels is large, and the energies of levels within $4f^{n-1}5d$ electronic configuration can strongly differ for different crystal hosts. Figure 7 shows the energies of the lowest levels of $4f^n5d$ electronic configuration for free RE^{3+} ions and RE^{3+} ions doped into $LiYF_4$ crystal. This energy E (energy needed to excite an electron from the $4f^n$ ground state to the lowest $4f^{n-1}5d$ level) varies irregularly with n through the lanthanide series. The large variation of E through the lanthanide series is almost entirely determined by the ionization energy of the $4f^n$-shell. This ionization energy is large for the more stable half-filled and the fully filled 4f-shells of Gd^{3+} and Lu^{3+}, whereas the values of E for Ce^{3+} and Tb^{3+} are $\sim 40\ 000$ cm^{-1} lower. These ions have ground-state configurations $4f^1$ and $4f^8$: ionization of these ions results in stable completely filled $4f^0$ or half-filled $4f^7$ electronic configurations; accordingly, the ionization energy of the 4f-electrons in these ions is relatively small.

The first overview of $4f^n$ energy levels for all RE^{3+} ions was given in [43,44] where the $4f^n$ energy level diagram in the range 0–42000 cm^{-1} was constructed (the Dieke's diagram). This energy-level diagram was extended up to 50000 cm^{-1} in [45] and up to 70000 cm^{-1} in [46]. The first report on the lowest 4f^n-4f^{n-1}5d absorption bands for RE^{3+} ions in CaF$_2$ was given in [47]. The first experiments on luminescence spectroscopy of 4f^{n-1}5d excitations in RE^{3+} ions with synchrotron radiation as an excitation source were performed in [48,49]. VUV luminescence due to 4f^{n-1} 5d-4f^n transitions in Nd^{3+}, Er^{3+}, and Tm^{3+} was first observed and identified in [50] and was then extensively studied with high spectral and time resolution (see, e.g., [51–62]).

The energy of the lowest 5d level is small enough for only the Ce^{3+} ion. The respective 5d-4f emission spectrum lies in the UV or visible spectral range depending on the crystal host, i.e., among all RE^{3+} doped crystals only Ce^{3+}-doped scintillators are suitable for the effective detection of the light from such scintillators by a typical PMT. However, as will be discussed below, in some special schemes of radiation detectors, the VUV scintillators are needed. For the Ce^{3+} ion which has only one 4f electron, the emission spectrum consists of two wide bands corresponding to transitions from the lowest 5d level of Ce^{3+} to $^2F_{5/2}$ and $^2F_{7/2}$ states of Ce^{3+} 4f^1 ground configuration (Fig. 8). Decay times for 5d-4f radiative transitions in Ce^{3+}, which are parity and spin allowed, lie in the range from a few to several tens of nanoseconds for different crystal hosts. Decay time is shorter for wider band-gap materials because of shorter emission wavelength and accordingly faster decay (in rough approximation, decay time is proportional to the third power of wavelength).

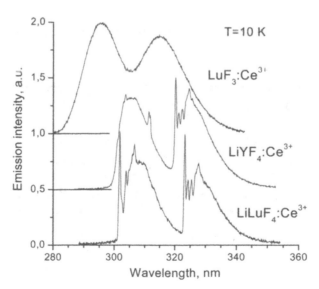

Figure 8. Emission spectra of LiYF$_4$:Ce^{3+}, LiLuF$_4$:Ce^{3+}, and LuF$_3$:Ce^{3+} crystals.

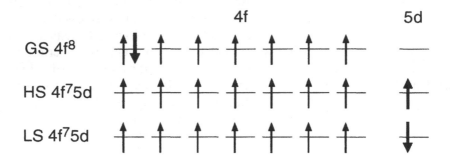

Figure 9. Schematic electron configurations for the ground state $4f^8$, the lowest energy high spin $4f^7 5d$ state and the lowest energy low spin $4f^7 5d$ state for Tb^{3+}.

The necessary condition for the observation of $4f^{n-1}5d-4f^n$ radiative transitions in the RE^{3+} ion is the existence of large enough energy gap between the lowest level of $4f^{n-1}5d$ configuration and the closest lower-lying excited level of $4f^n$ configuration. The width of this gap should exceed at least five times the maximum energy of phonons in the host crystal to prevent fast nonradiative multiphonon relaxation from the $4f^{n-1}5d$ levels to lower-lying $4f^n$ levels. Up to now, emission due to $4f^{n-1}5d-4f^n$ transitions has been observed only for seven RE^{3+} ions, namely, for Ce^{3+}, Pr^{3+}, Nd^{3+}, Gd^{3+}, Er^{3+}, Tm^{3+}, and Lu^{3+} (for Gd^{3+} and Lu^{3+} only at low temperature $T < 200$ K). In the case of Gd^{3+}, some specific features of energy level structure for this ion are responsible for the existence of $4f^6 5d-4f^7$ luminescence [63].

4.2   Spin-Allowed and Spin-Forbidden 5d-4f Transitions in Rare Earth Ions

Purely nanosecond decay for 5d-4f radiative transitions is observed only for RE^{3+} ions in the first half of the lanthanide series. For RE^{3+} ions in the second half of the lanthanide series, the 5d-4f transitions from the lowest 5d level are spin-forbidden (SF), and corresponding lifetimes lie in the microsecond range. As an example, in Fig. 9 (reproduced from [53]) the first RE ion in the second half of lanthanide series Tb^{3+} is considered, which has electronic configuration $4f^8$.

In the ground state of $4f^8$ configuration, the maximum number of unpaired parallel spins is 6, which results in a total spin quantum number $S = 3$ and therefore a spin multiplicity $2S + 1 = 7$. When one electron is promoted to the 5d shell, it can orient its spin in two ways: either parallel with the seven remaining 4f electrons, giving rise to a high-spin state (HS) with $S = 4$ and $2S + 1 = 9$, or antiparallel, yielding a low-spin state (LS) with $S = 3$ and $2S + 1 = 7$. According to the Hund's rule, the HS state will be lower in energy. Thus, the transition from the ground state to the lowest $4f^7 5d$ state will be SF and therefore relatively weak compared to the higher-energy $4f^8$-$4f^7 5d$ excitation to the LS state, which is spin-allowed (SA). Accordingly, the respective radiative $4f^7 5d$ -$4f^8$ transitions will be also SF or SA. Depending on the ion (Er^{3+}, Tm^{3+}, or Lu^{3+}) and the host, either slow SF emission from the lowest HS $4f^{n-1}5d$ state or fast SA emission from the lowest LS $4f^{n-1}5d$ state dominates or both emissions coexist.

As an example, time-resolved VUV emission spectra resulting from the interconfigurational $4f^{10}5d$-$4f^{11}$ transitions in Er^{3+} ions doped into $LiYF_4$ crystal are shown in Fig. 10. The spectrum consists of two series of bands: fast emission bands with nanosecond decay (decay constants lie in the range from a few to tens of nanoseconds for different hosts) and/or slow ones shifted to longer wavelengths (decay constants lie in the microsecond range). The fast and slow components arise from SA and SF $4f^{10}5d$-$4f^{11}$ transitions in Er^{3+}, respectively.

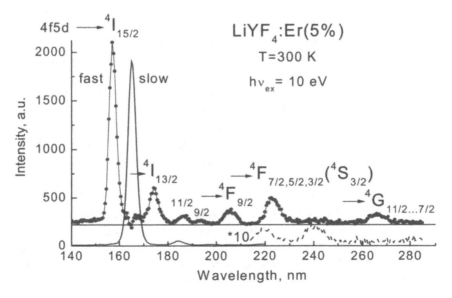

Figure 10. VUV/UV emission spectrum of $LiYF_4$:Er^{3+} crystal due to spin-allowed (fast component) and spin-forbidden (slow component) $4f^{10}5d$-$4f^{11}$ transitions in Er^{3+}.

4.3   Crystal Field Splitting and Electron-Lattice Interaction

The energy of the lowest $4f^{n-1}5d$ level of the RE^{3+} ion, and accordingly, the energy of radiative $4f^{n-1}5d$-$4f^n$ transitions in RE^{3+}, is determined by the centroid shift of $4f^{n-1}5d$ configuration of the RE^{3+} ion in the crystal compared with the free RE^{3+} ion and the value of crystal field splitting. The centroid shift and the value of the crystal field splitting are about the same for all RE^{3+} ions doped into the same host crystal [64]. Centroid shift depends on many parameters of the host crystal such as the type of crystal lattice, the sizes of the ions, interion distances, CN, but it is mainly determined by the type of the anion. In a first approximation, the crystal field splitting of RE^{3+} $4f^{n-1}5d$ levels within some family of compounds (e.g., in fluorides) decreases with the distance from an RE^{3+} ion to its first neighbor anions that forms its coordination polyhedron: the shorter the distance, the stronger the splitting, and short distances correspond to a low coordination number (CN), i.e., to low values of neighboring anions. For fluoride crystals, the CN of the RE^{3+} ion can change from 11 to 6, e.g., LaF_3 (CN = 11), $KGdF_4$ (CN = 9), CsY_2F_7 (CN = 8), K_2LuF_5 (CN = 7), and Cs_2NaYF_6 (CN = 6). So for fluorides, the strongest crystal field splitting, and accordingly, the lowest energy of the first $4f^n$-$4f^{n-1}5d$ transition, can be expected for RE^{3+} ions doped into elpasolite crystals of a type Cs_2NaYF_6 and the smallest crystal field splitting (the highest energy of the first $4f^n$- $4f^{n-1}5d$ transition) for RE^{3+} doped into LaF_3. As a result, the longest wavelength of RE^{3+} 5d-4f luminescence will be observed in elpasolite-type hosts and the shortest wavelength in trifluoride hosts. However, the real situation is more complicated because the centroid shift is different for different crystallographic systems, and RE^{3+} ions can occupy one or several crystallographic sies, for example, one site in $LiKYF_5$ or six sites in KYF_4.

As an example of different energies for the lowest RE^{3+} 5d level, emission spectra due to 5d-4f radiative transitions in the Ce^{3+} ion doped into $LiYF_4$, $LiLuF_4$, and LuF_3 hosts are shown in Fig. 8 [65]. The Ce^{3+} 5d crystal-field splitting is expected to be smaller in LuF_3 than in $LiYF_4$ or $LiLuF_4$ because of the higher coordination number for Ce^{3+} (9 instead of 8); accordingly, the Ce^{3+} 5d-4f luminescence is observed at higher energy in the LuF_3 host than in the $LiYF_4$ and $LiLuF_4$ hosts. Figure 11 shows another example of different energies for the lowest RE^{3+} 5d level, emission spectra due to $4f^65d$-$4f^7$ radiative transitions in Gd^{3+} ion in the $LiGdF_4$ and GdF_3 hosts [63]. The Gd^{3+} 5d crystal-field splitting in GdF_3 (CN = 9) is smaller than in $LiGdF_4$ (CN = 8); accordingly, the Gd^{3+} $4f^65d$-$4f^7$ luminescence is observed at higher energy in GdF_3 than in $LiGdF_4$.

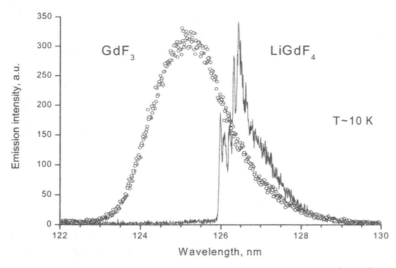

Figure 11. Emission spectra from LiGdF₄ and GdF₃ crystals due to $4f^65d$-$4f^7$ interconfigurational transitions in the Gd^{3+} ion.

Because of the more localized nature of inner 4f electrons and more extended 5d outer orbitals, the vibrations of the host lattice have a different influence on the 4f-4f and 5d-4f electronic transitions in the RE^{3+} ions. For the illustration of this electron-lattice interaction, the model of the single configurational coordinate [66] is usually used. This diagram (see Fig. 12 for Ce^{3+} as an example) shows the potential energy curves of 4f ground state, 4f excited states, and the 5d states of RE^{3+} ion as a function of configurational coordinate Q, which can be considered in first approximation as the mean distance between the central RE^{3+} ion and the surrounding ligands. This model is based on the Frank-Condon principle, which is the assumption that electronic transitions are fast compared with nuclei motion and occur at fixed positions of nuclei, i.e., "vertically" in the configurational coordinate diagram. Stronger electron-lattice interaction for 5d states compared with 4f states means larger offset for the equilibrium configurational coordinate for the potential curve of the 5d state with respect to the potential curve of 4f states. Electron-lattice interaction results in broadening the absorption and emission bands for 4f-5d transitions and in a shift between maxima of absorption and emission bands ΔE_S, which is called the Stokes shift.

For the quantitative characterization of the strength of electron-lattice interaction, the Huang-Rhys parameter S is used which is defined as the ratio of relaxation energy (see Fig. 12) to phonon energy $\hbar\omega$ of lattice vibrations:

$$S = \Delta E_R / \hbar\omega \qquad (5)$$

If the shape of potential curves and phonon energies is the same for the excited and ground states, the relaxation energy ΔE_R is equal to half of Stokes shift $\Delta E_S/2$, the value of which can be found from the experiment.

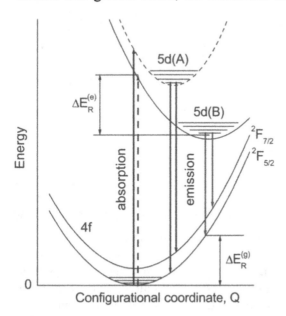

Figure 12. Single configurational coordinate diagram for 4f and 5d states of Ce^{3+} doped into two hosts A and B with different value of electron-lattice interaction. $\Delta E_R^{(g)}$ and $\Delta E_R^{(e)}$ are relaxation energies for the ground and excited states of Ce^{3+}, respectively, in host B.

In the limit of strong electron-lattice interaction ($S > 5$), the spectra become broad without any fine structure, and the absorption and emission bands becomes Gaussian-like. For intermediate strength of electron-lattice interaction ($S \sim 1\div5$) observed for 5d states in some matrices, the absorption and emission bands consist at low temperature of both narrow zero-phonon lines corresponding to electron origins of the respective transitions and wide vibronic side bands. For 4f-4f transitions, electron-lattice interaction is small ($S \ll 1$), and absorption and emission spectra show only narrow nonphonon lines.

As an example of intermediate strength of electron-lattice interaction, emission spectra of $LiYF_4:Ce^{3+}$ and $LiLuF_4:Ce^{3+}$ crystals can be considered (Fig. 8). These spectra, which are caused by Ce^{3+} 5d-4f radiative transitions, consist of sharp zero-phonon lines and broad vibronic sidebands. Three sharp lines in the higher-energy band and four sharp lines in the lower-energy band dominate over other features in the emission spectrum of $LiLuF_4:Ce^{3+}$ crystal (sharp lines in the higher-energy emission band of $LiYF_4:Ce^{3+}$ crystal are not so well pronounced) and are ascribed to crystal-field components in the S_4 symmetry of the $^2F_{5/2}$ and $^2F_{7/2}$ ground-state levels of Ce^{3+}, respectively. In the spectrum of $LuF_3:Ce^{3+}$, both emission bands have a smooth Gaussian-like shape, and no zero-phonon lines are observed in the spectrum because of a much stronger electron-lattice interaction between 5d states of Ce^{3+} ions and lattice vibrations in this crystal as compared with $LiYF_4{:}Ce^{3+}$ and $LiLuF_4{:}Ce^{3+}$.

VUV emission spectrum of the $LiGdF_4$ crystal (see Fig. 11) is another example of intermediate strength of electron-lattice interaction ($S \sim 1$) [63]. This spectrum corresponding to radiative transitions in the Gd^{3+} ion from the lowest level of $4f^65d$ configuration to the ground $4f^7\ {}^8S_{7/2}$ state clearly shows the fine structure due to zero-phonon and vibronic lines. The crystal-field splitting of the ground $4f^7\ {}^8S_{7/2}$ state of Gd^{3+} is extremely small, i.e., only a single zero-phonon line can be expected in the $4f^65d{-}4f^7$ emission spectrum of Gd^{3+}. Thus, the fine structure in the spectrum is due to one zero-phonon line corresponding to the electronic origin of the $4f^65d{-}4f^7$ ${}^8S_{7/2}$ transition plus vibronic replicas corresponding to different modes of host lattice vibrations. Contrary to this, the spectrum of GdF_3 has a smooth shape because of stronger electron-lattice interaction between the $4f^65d$ electronic configuration of the Gd^{3+} ions and the lattice vibrations in this crystal.

4.4 Development of Rare Earth Based Scintillators for Radiation Detectors in Nuclear Physics and Medicine

Up to now many RE^{3+} doped (mainly Ce^{3+} doped) inorganic crystals were proposed as fast and efficient scintillators (see, e.g., review paper [67]); however, research and development of new promising scintillators is still being continued. What further improvements can be expected? It is obvious that there is no best scintillator for all possible radiation detectors. It is necessary to use an optimal scintillator for a particular application. For applications where energy resolution is the most important, the light yield must be as large as possible. This can be achieved only for compounds with small enough band gaps, in particular, for iodides, sulfides, and some oxides. On the other hand, in narrow band-gap materials, Ce^{3+} luminescence cannot be observed at all. In any case, the light yield cannot exceed the fundamental limit, which is around 100,000 photons/MeV for practically all possible Ce^{3+} based inorganic scintillators [4]. For small band-gap materials, another restriction on energy resolution arises: in such compounds, Ce^{3+} luminescence is in the long wavelength part of the spectrum, where PMTs are usually less sensitive, giving a smaller number of photoelectrons and accordingly, wrose energy resolution. If high-time resolution is needed, the wider band-gap materials should be considered because of shorter emission wavelengths and accordingly faster decay of Ce^{3+} luminescence. Very important factors that influence the choice of the scintillator are also high enough density, high-effective atomic number Z (for large photofraction in absorption), good physicochemical properties, the possibility to grow large high-quality crystals, and a modest price per cubic centimeter.

Scintillation crystals emitting in the VUV spectral range due to 5d-4f radiative transitions in the RE^{3+} ions (Nd^{3+}, Gd^{3+}, Er^{3+}, Tm^{3+}, Lu^{3+}) can be used in some special schemes of gamma detectors. In such detectors, a solid scintillator is coupled with a multiwire proportional chamber filled with photosensitive vapor like TMAE, TEA, etc., or is coupled with a solid photocathode like CsI. Because these vapors (or photocathodes) are sensitive to VUV light only, it is necessary to use scintillators emitting in the VUV. The main problem of such detectors is their low efficiency caused by rather low-light yield of the used scintillators (usually BaF_2) and poor overlapping of the scintillation spectrum with quantum efficiency curves of photosensitive vapors or photocathods. So the search continues for new VUV scintillators with higher light yield, and the spectrum shifted to shorter wavelengths is necessary for improving detectors of this type. Much better spectral overlapping of Nd^{3+}, Er^{3+}, or Tm^{3+}, 5d-4f emission spectrum with quantum efficiency curves of photosensitive vapors and photocathodes compared with the spectrum of BaF_2 can give some advantage of using Nd^{3+}, Er^{3+} or Tm^{3+} containing crystals in such gamma detectors (Fig. 13) [68]. However, the light yield of VUV emission from such scintillators is usually rather low, except for only a few examples of Nd^{3+}-doped crystals [58,68]. One of the main directions of inorganic scintillator development for medicine is positron emission tomography (PET) [69]. In PET the imaging

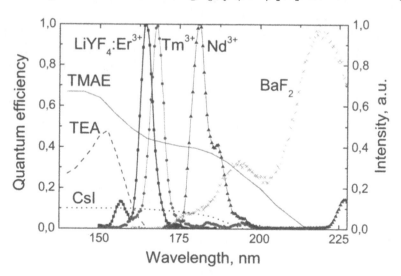

Figure 13. Quantum efficiency curves of TMAE and TEA vapors, CsI solid photocathod, and emission spectra of $LiYF_4$ crystals doped with Nd^{3+}, Er^{3+} and Tm^{3+} in comparison with emission spectrum of BaF_2.

is realized by means of two 511 keV gamma-quanta that are emitted collinearly when a positron, emitted by a radiopharmaceutical introduced into the patient, annihilates in the tissue. These two quanta are detected position sensitively and in coincidence. In such a scheme, the point of positron emission is determined as being situated on the line connecting the two detection positions. For position-sensitive detection, a PET system consists of thousands of scintillation detectors. The already available PET systems use $Bi_4Ge_3O_{12}$ (BGO) crystals as scintillators. BGO is rather good for PET, because it is dense, has high atomic number, and large probability of photoelectric effect for 511 keV annihilation quanta. In general, each scintillator has 4×4 mm^2 cross section and few cm length. This gives the position resolution of about 4 mm in the center and 5–6 mm at the edge of the system with the diameter 0.8 m. However there are some disadvantages of BGO. Because the response time of BGO is rather long (about 300 ns), this gives many random coincidences that strongly increases the background. Because of the same timing property of BGO, the dead time of the detector can be too long. To improve the PET system, a scintillator with a much faster response is needed, but the scintillator still must have large density and a high atomic number for good absorption and a high fraction of photoelectric effect; the emission spectrum must match the spectral sensitivity of the standard detector. Most of new scintillators proposed for PET are based on Ce^{3+} doped compounds with a decay time of the order of a few tens of nanoseconds. The most promising and widely studied scintillators for PET are LSO [70] and LuAlO$_3$:Ce (LuAP) [71]. The main problem is the difficulty of growing large enough and defect-free crystals. The problem is also the high price of Lu-containing materials. More research is needed before real application becomes possible.

A special interest is thermal neutron detection. Neutrons do not produce ionization directly in scintillation crystal, but they can be detected through their interaction with the nuclei of a suitable element. For an efficient neutron detection, the isotope with a high neutron capture cross section should be part of the chemical compound of the scintillator. The neutron capture cross section decreases strongly with neutron energy; therefore, the neutron capture reactions can be used only for low energy, mainly thermal neutron detection. Usually scintillators enriched with ^{6}Li or ^{10}B isotopes are proposed for neutron detection. However, it is well known that the highest cross section for neutron capture is for two isotopes of gadolinium. Thus, the Gd-based crystals can be very promising scintillators for thermal neutron detection. Luminescence of Gd^{3+} ion itself, which is caused by 4f-4f transitions, is very slow (few ms), and Gd-containing host material should be doped with Ce^{3+} for obtaining fast response. An example of Gd-based scintillator for neutron detection RbGd$_2$Br$_7$:Ce^{3+} was proposed in [72]. Because of the high thermal neutron cross section of ^{157}Gd isotope (255,000 b), the stopping efficiency for 500 μm $RbGd_2Br_7{:}Ce^{3+}$ layer is close to 100%. Such a small thickness is helpful in minimizing the γ-ray background effects. Obviously, more research is required to find an optimal gadolinium-based scintillator for thermal neutron detection.

REFERENCES

1. A. Lempichi, A.J. Wojtovicz, E. Berman, Nucl. Instrum. & Meth. A **333** (1993) 304

2. A. Lempichi, A.J. Wojtovicz, J. Lumin. **60–61** (1994) 942

3. P.A. Rodnyi, P. Dorenbos, C.W.E. van Eijk, Phys. Status Solidi B **187** (1995) 15

4. P. Dorenbos, Nucl. Instrum. & Meth. A **486** (2002) 208

5. C. Pedrini, D. Bouttet, C. Dujardin, A. Belsky, A. Vasil'ev, Proc. Int. Conf. on Inorganic Scintillators and their Applications, Delft University Press, The Netherlands (1996) 103

6. Yu.M. Aleksandrov, V.N. Makhov, M.N. Yakimenko, Sov.Phys. Solid State **29** (1987) 1092

7. V.N. Makhov, Proc. Int. Workshop on Heavy Scintillators for Scientific and Industrial Applications, Editions Frontieres, France (1993) 167

8. P. Dorenbos, J.T.M. de Haas, C.W.E. van Eijk, IEEE Trans. Nucl. Sci. **NS-42** (1995) 2190

9. C.W.E. van Eijk, Nucl. Instrum. & Meth. A **471** (2001) 244

10. Yu.M. Aleksandrov, V.N. Makhov, P.A. Rodnyi, T.I. Syrejshchikova, M.N. Yakimenko, Sov. Phys. Solid State **26** (1984) 1734

11. J.L. Jansons, V.J. Krumins, Z.A. Rachko, J.A. Valbis, Phys. Status Solidi B **144** (1987) 835

12. P.A. Rodnyi, M.A. Terekhin, Phys. Status Solidi B **166** (1991) 283

13. S. Kubota, M. Itoh, J. Ruan(Gen), S. Sakuragi, S. Hashimoto, Phys. Rev. Lett. **60** (1988) 2319

14. V.N. Makhov, Nucl. Instrum. & Meth. A **308** (1991) 187

15. P.A. Rodnyi, Sov. Phys. - Solid State **34** (1992) 1053

16. C.W.E. van Eijk, J. Lumin. **60–61** (1994) 936

17. P.A. Rodnyi, Radiation Measurements **38** (2004) 343

18. Yu.M. Aleksandrov, I.L. Kuusmann, P.Kh. Liblik, Ch.B. Lushchik, V.N. Makhov, T.I. Syrejshchikova, M.N. Yakimenko, Sov. Phys. - Solid State **29** (1987) 587

19. Yu.M. Aleksandrov, V.N. Makhov, T.I. Syrejshchikova, M.N. Yakimenko, Nucl. Instrum. & Meth. A, **261** (1987) 153

20. Yu.M. Aleksandrov, V.N. Makhov, N.M. Khaidukov, M.N. Yakimenko, Sov. Phys. Solid State **31** (1989) 1609

21. V.N. Makhov, N.M. Khaidukov, Sov. Phys. Solid State **32** (1990) 1978

22. M. Itoh, S. Kubota, J. Ruan (Gen), S. Hashimoto, Rev. Solid State Sci. **4** (1990) 467

23. I. Kuusmann, T. Kloiber, W. Laasch, G.Zimmerer, Rad. Effects and Defects in Solids **119** (1991) 21

24. Yu.M. Aleksandrov, I.L. Kuusmann, V.N. Makhov, S.B. Mirov, T.V. Uvarova, M.N. Yakimenko, Nucl. Instrum. & Meth. A **308** (1991) 208

25. A.S. Voloshinovskii, V.B. Mikhailik, P.A. Rodnyi, S.V. Syrotyuk, Phys. Status Solidi B **173** (1992) 739

26. A.S. Voloshinovskii, M.S. Mikhailik, V.B. Mikhailik, E.N. Mel'chakov, P.A. Rodnyi, C.W.E. van Eijk, G. Zimmerer, J. Lumin. **79** (1998) 107

27. N.Yu. Kirikova, V.N. Makhov, Sov. Phys. Solid State **34** (1992) 1557

28. V.N. Makhov, M.A. Terekhin, I.H. Munro, C. Mythen, D.A. Shaw, J.Lumin. **72–74** (1997) 114

29. I.F. Bikmetov, A.B. Sovolev, J.A. Valbis, Sov. Phys. Solid State **33** (1991) 1715

30. J. Andriessen, P. Dorenbos, C.W.E. van Eijk, Mol. Phys. **74** (1991) 535

31. T. Ikeda, H. Kobayashi, Y. Ohmura, H. Nakamatsu, T. Mukoyama, J. Phys. Soc. Jpn. **66** (1997) 1079

32. Y. Kayanuma, A. Kotani, J. Electron Spectrosc. Relat. Phenom. **79** (1996) 219

33. O.I. Baum, A.N. Vasil'ev, Proc. Fifth Int. Conf. on Inorganic Scintillators and Their Applications, Moscow State University Press, Russia (2000) 453

34. T. Matsumoto, K. Kan'no, M. Itoh, N. Ohno, J. Phys. Soc. Jpn. **65** (1996) 1195

35. A.N. Belsky, I.A. Kamenskikh, V.V. Mikhailin, A.N. Vasil'ev, J. Electron Spectrosc. Relat. Phenom. **79** (1996) 111

36. V.N. Makhov, I. Kuusmann, J. Becker, M. Runne, G. Zimmerer, J. Electron Spectrosc. Relat. Phenom. **101–103** (1999) 817

37. V.N. Makhov, V.N. Kolobanov, M. Kirm, S. Vielhauer, G. Zimmerer, International Journal of Modern Physics B **15** (2001) 4032

38. A.N. Vasil'ev, Proc. Fifth Int. Conf. on Inorganic Scintillators and Their Applications, Moscow State University Press, Russia (2000) 43

39. K. Kimura, J. Wada, Phys. Rev. B **48** (1993) 15535

40. R.A. Glukhov, C. Pedrini, A.N. Vasil'ev, A.M. Yakunin, Proc. Fifth Int. Conf. on Inorganic Scintillators and Their Applications, Moscow State University Press, Russia, (2000) 448

41. Ch.B. Lushchik, F.A. Savikhin, V.N. Makhov, O.V. Ryabukhin, V.Yu. Ivanov, A.V. Kruzhalov, F.G. Neshov: Physics of the Solid State **42** (2000) 1020

42. C.M. Combes, P. Dorenbos, R.W. Hollander, C.W.E. van Eijk, Nucl. Instrum. & Meth. A **416** (1998) 364

43. G.H. Dieke, H.M. Crosswhite, Appl. Opt. **2** (1963) 675

44. G.H. Dieke, *Spectra and Energy Levels of Rare Earth Ions in Crystals*, Interscience Publishers, New York (1968)

45. W.T. Carnall, G.L. Goodman, K. Rajnak, R.S. Rana, A Systematic Analysis of the Spectra of the Lanthanides Doped into Single Crystal LaF_3, Argonne National Laboratory, Argonne Illinois (1988)

46. R.T. Wegh, A. Meijerink, R.J. Lamminmäki, J. Hölsä, J. Lumin. **87–89** (2000) 1002

47. E. Loh, Phys. Rev. **147** (1966) 332

48. R.L. Elias, Wm.S. Heaps, W.M. Yen, Phys. Rev. B **8** (1973) 4989

49. Wm.S. Heaps, L.R. Elias, W.M. Yen, Phys. Rev. B **13** (1976) 94

50. K.H. Yang, J.A. DeLuca, Appl. Phys. Lett. **29** (1976) 499

51. R.T. Wegh, H. Donker, A. Meijerink, Phys. Rev. B **57** (1998) R2025

52. J. Becker, J.Y. Gesland, N.Yu. Kirikova, J.C. Krupa, V.N. Makhov, M. Runne, M. Queffelec, T.Y. Uvarova, G. Zimmerer, J. Lumin. **78** (1998) 91

53. R.T. Wegh, A. Meijerink, Phys. Rev. B **60** (1999) 10820

54. V.N. Makhov, N.M. Khaidukov, N.Yu. Kirikova, M. Kirm, J.C. Krupa, T.V. Ouvarova, G. Zimmerer, J. Lumin. **87–89** (2000) 1005

55. N.M. Khaidukov, M. Kirm, S.K. Lam, D. Lo, V.N. Makhov, G. Zimmerer, Optics Communications **184** (2000) 183

56. V.N. Makhov, N.M. Khaidukov, N.Yu. Kirikova, M. Kirm, J.C. Krupa, T.V. Ouvarova, G. Zimmerer, Nucl. Instrum. & Meth. A **470** (2001) 290

57. L. van Pieterson, M.F. Reid, G.W. Burdick, A. Meijerink, Phys. Rev. B **65** (2002) 045114

58. V.N. Makhov, N.Yu. Kirikova, M. Kirm, J.C. Krupa, P. Liblik, A. Lushchik, Ch. Lushchik, E. Negodin, G. Zimmerer, Nucl. Instrum. & Meth. A **486** (2002) 437

59. L. van Pieterson, M.F. Reid, A. Meijerink, Phys. Rev. Lett. **88** (2002) 067405

60. Y. Chen, M. Kirm, E. Negodin, M. True, S. Vielhauer, G. Zimmerer, Phys. Status Solidi B **240**, R1 (2003)

61. P.S. Peijzel, P. Vergeer, A. Meijerink, M.F. Reid, L.A. Boatner, G.W. Burdick, Phys. Rev. B **71** (2005) 045116

62. V.N. Makhov, N.M. Khaidukov, D. Lo, J.C. Krupa, M. Kirm, E. Negodin, Optical Materials **27** (2005) 1131

63. M. Kirm, J.C. Krupa, V.N. Makhov, M. True, S. Vielhauer, G. Zimmerer, Phys. Rev. B **70** (2004) 241101

64. P. Dorenbos, Phys. Rev. B **62** (2000) 15650

65. N.Yu. Kirikova, M. Kirm, J.C. Krupa, V.N. Makhov, E. Negodin, J.Y. Gesland, J. Lumin. **110** (2004) 135

66. B. Henderson, G.F. Imbusch, *Optical Spectroscopy of Inorganic Solids*, Clarendon Press, Oxford (1989)

67. C.W.E. van Eijk, Nucl. Instrum. & Meth. A **460** (2001) 1

68. V.N. Makhov, J.Y. Gesland, N.M. Khaidukov, N.Yu. Kirikova, M. Kirm, J.C. Krupa, M. Queffelec, T.V. Ouvarova, G. Zimmerer, Proc. Fifth Int. Conf. on Inorganic Scintillators and Their Applications, Moscow State University Press, Russia (2000) 369

69. C.W.E. van Eijk, Nucl. Instrum. & Meth. A **509** (2003) 17

70. C.L. Melcher, J.S. Schweitzer, IEEE Trans. Nucl. Sci. NS-**39** (1992) 502

71. B.I. Minkov, Functional Mat. **1** (1994) 103

72. K.S. Shah, L. Cirignano, R. Grazioso, M. Klugerman, P.R. Bennett, T.K.Gupta, W.W. Moses, M.J. Weber, S.E. Derenzo, IEEE Trans. Nucl. Sci. NS-**49** (2002) 1655

Chapter 10

RADIATION-INDUCED PHASE TRANSITIONS

Paolo M. Ossi
Politecnico di Milano, Milan, Italy

1   INTRODUCTION

Amorphous, or noncrystalline matter, is the condensed state representative of the absence of the long-range order typical of crystalline materials. It is the end point of a hierarchy of structures with progressively lower structural order, connecting ideal perfect crystals to entropy disordered crystals, incommensurate (aperiodic) structures, quasicrystals, glassy solids. Although in principle related to different situations, the terms amorphous, noncrystalline, glassy, and vitreous are used as equivalents to each other throughout this work.

It is a popular conception that amorphous materials are metastable; more precisely, if we cool down a melt at a sufficiently high rate as to kinetically avoid crystallization, the liquid falls into the supercooling region where it progressively loses internal mobility degrees, until at the glass transition temperature the (metastable) supercooled liquid deviates from the equilibrium line and enters into the (metastable) amorphous state. The equilibrium, stable phase underlying both the supercooled liquid and the glass is the crystalline phase.

Figure 1 provides a schematic map of free energy G versus the atomic configuration. It helps, by analogy with topographic maps of the Earth's surface, to understand those collective phenomena that determine the relative stability of liquids and glasses. In the figure, maxima, minima, and saddle points are evident, and they define a complex pattern in the configuration space. Minima have consistently different depths; with respect to each minimum we define a valley, namely, a set of configurations connected by a strictly downhill motion to the considered minimum. Such a region is the basin (b) of attraction. Contiguous basins have in common a boundary that contains at least a saddle (s) point, or transition state. Figure 1 allows comparing the free energies for a frozen (f) state, an unstable (u)

K.E. Sickafus et al. (eds.), Radiation Effects in Solids, 259–319.
© 2007 *Springer.*

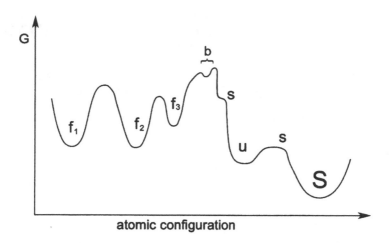

Figure 1. Schematic free energy G trend for representative configurations of a structurally disordered system; f: frozen state; u: unstable state; b: basin; s: saddle point; S: stable state.

state, and a stable (S) state. The difference between unstable and frozen states is essentially given by the height of the energy barrier in relation to the available thermal energy k_BT. Because *small* changes in temperature or in pressure induce reversible changes in glass properties, the frozen, amorphous state may in practice be considered stable, namely, its properties do not change over time. An important question concerns the number Λ of minima as a function of the number of particles of the system, N; in the simplest case of an elemental system

$$\Lambda(N) \propto N! \exp[cN] \qquad (1)$$

with c positive; predicting c value starting from known atomic structures and interactions is a present challenge. A macroscopic piece of glass shows an extremely complex G topography, and the system can explore a broad range of nearly equivalent configurations in spite of the progressively reduced dynamics in the supercooled liquid region.

The free energy scheme in Fig. 1 shows that it is necessary to drive a crystal to a metastable state, with free energy G higher than the free energy of the glassy state, before a phase transition sequence starts. This can include either direct precipitation to the stable crystalline phase or glass formation, possibly followed by transition to other intermediate states, such as a quasicrystal and a metastable crystal; ultimately, the crystalline stable state can be recovered.

To each transition step is associated a timescale *t*; the observability, or not, of a given intermediate state depends not only on thermodynamic constraints (a lowering of system free energy must be associated to the transition), but also on kinetic ones. The timescale over which a transition process occurs must be shorter than those of possible competing processes.

The study of amorphous materials can be reduced to two questions: *Why* does a solid become amorphous? *How* do the crystal-amorphous (C-A) transition and the reverse (A-C) one occur? The first question is a basic one, because it is equivalent to asking oneself why the "normal" condition of solid matter at low temperature is the crystalline state.

Alternatively, why at zero temperature does the equilibrium state of a system correspond to arranging the constituent particles in a periodic fashion? To this question we still do not have a complete, general answer. The second question, although more related to technological issues concerned with the production of amorphous materials and their thermal (meta)stability, sheds light on the mechanisms of amorphization and thus contributes to the solution of the former problem.

The principal aim of this work is to discuss the mechanisms underlying the C-A phase transition of a system that relaxes in short times toward equilibrium, starting from a highly energized, strongly nonequilibrium initial state, where it has been brought by a bombarding beam of fast-charged particles. The irradiation induced C-A phase transition was recognized to be a glass formation process already at the beginning of the twentieth century [1] when localized volumes lacking crystalline order were observed in minerals such as zircon and gadolinite. These minerals contain radioactive elements that gave rise in geological times to fission fragment damage. However, only since the end of 1960s it was possible to thoroughly explore bombardment-induced modifications of materials. Indeed right then charged particle irradiation techniques were developed up to the point that surface modification of a target could be induced in a controlled way. Across the past 40 years, a vast amount of experimental and theoretical work has accumulated in the field. Such an activity is justified, considering that energetic particle bombardment is probably the most controlled process to trigger amorphization. The sample holder is placed in a chamber under high vacuum to avoid ambient contamination and we choose the nature, energy, flux, and dose of the bombarding beam.

In this study, attention mainly will be focused on metallic materials. The usually simple crystalline reference structure and the associated isotropic long-range interatomic potential offer a simplified framework to introduce disordering processes, although the observed glass-forming ability is normally low, as compared with many nonmetallic elements and compounds. For such a reason, research on metal glasses is relatively young and has been constrained since its beginning within limits dictated by available cooling rates. In fact, if a high-enough cooling rate is available, every material can be brought into the amorphous state; all solid-state methods to prepare glassy alloys depend on the attainment of fast quenching, with the only notable exception of the near-equilibrium bulk interdiffusion reactions in certain metal-metal couples, where glass formation is attributed to the anomalous, fast diffusion of one alloy constituent in the other [2]. Otherwise, typical cooling rates range from about 10^9 Ks^{-1} for sputtering and vapor condensation onto a substrate kept at cryogenic temperature to 10^8 Ks^{-1} for laser melting and to 10^6 Ks^{-1} for splat quenching. Lower limits of about 10^3 Ks^{-1} are required to obtain metal glasses in the most favorable cases; thin samples are normally obtained, the attainable quenching rate being critically affected by sample thickness.

For energetic particle bombardment, the estimated quench rate of localized target regions—brought at temperatures by far higher than the compound melting point—is on the order of 10^{14} Ks^{-1}. Such kinetic conditions allow exploring the amorphization path opposite to interdiffusion reactions, whereby the nonequilibrium characteristics of the process are driven to extreme limits. For the above process features, the first consequence is the attainability of intimate atomic mixtures with controlled composition without the limitations imposed by conventional, equilibrium alloying techniques.

The crystal-amorphous transition essentially involves two different routes: in the first, which is typical, e.g., of liquid and vapor quenching processes, a randomized atomic structure is frozen. This path is associated to kinetics of atomic movements so slow that structural rearrangements are practically arrested before the long-range order (LRO) typical of a crystal structure is reconstructed.

The second route preserves the crystalline order while energy is continuously provided to the structure, e.g., through defect production and accumulation, up to a point where the energized crystal becomes unstable with respect to the structurally disordered state. Such a picture is believed to hold in all the above-mentioned processes grouped under the name solid-state amorphization, including irradiation-induced glass formation.

The distinction just outlined is somewhat idealized, and it is not always easy to ascribe processes to one or to the other category. An incoming highly energetic ion locally deposits in the target a large amount of energy; we are uncertain whether we are in the presence of a truly localized liquid- (or even vapor-) quenching event or of a direct crystal-glass transition in the solid state.

The common feature underlying the experiments of our interest is that irradiated systems are brought into a strongly off-equilibrium state by an external cause, the bombarding beam, through nonisothermal processes. Such externally driven processes can either produce an enthalpy excess in the initially homogeneous crystalline system, or they can lead to the release of chemical energy associated to the initial target configuration, as with ion beam mixing.

Every solid-state amorphization transition may be divided into two steps: the externally induced energy increase of the crystal and the subsequent disruption of the ordered structure. The attainment of the amorphous state is associated to a free energy decrease, whereas any specific process that is invoked to explain the first transition step should raise crystal energy.

Since 1962, there has been research in the field of amorphization of metallic targets induced by particle bombardment when the intermetallic compound U_6Fe was vitrified by self-bombardment induced by fast fission fragments [3]. Since then, vitrification of several alloys, after ion and electron irradiation, ion implantation, and ion mixing, was reported.

A general characteristic observed in irradiation-induced amorphization processes in metals is that only *alloys*, diluted or concentrated, can be vitrified. To this observation, four exceptions are presently known: the elements bismuth, gallium [4], silicon, and germanium [5]. The reason for the anomalous behavior is most likely associated to the bonding characteristics of silicon, germanium, and bismuth that display a significant degree of covalence, while the free energy difference among the various allotropic forms of gallium and amorphous gallium is small. For all other metallic systems, the co-presence of different constituents appears essential to produce and to maintain a highly energized state, still retaining a crystalline structure during the first step of the amorphization process.

Such a coincidence lends support to the role of *chemical* effects, including the chemical energy contribution associated to the different types of defect introduced in the system and to their agglomerates, to drive glass formation processes.

This paper is organized as follows: Section 2 is devoted to electron irradiation. A selection of experimental results is reviewed to highlight the mechanisms both of production of disorder and of amorphization. In Sections 3 and 4, ion implantation and ion-mixing experiments are introduced within the same conceptual frame adopted for the analysis of electron bombardment. The interplay and relative relevance to the C-A transition of process parameters, such as projectile nature and dose and irradiation temperature, are examined. We discuss also irradiation-induced formation of quasi-crystalline phases and its competition with amorphization. The peculiar structural features of such an intermediate state are correlated to the specific, quite stringent experimental conditions needed to obtain quasi-crystals by irradiation, as well as to the crystal-quasi-crystal-amorphous phase sequence and the experimental paths that give rise to it. The icosahedral short range order (SRO) often found in metallic amorphous structures can evolve to the long-range icosahedral order typical of quasi-crystals upon an activation process such as irradiation. Thus quasi-crystals represent unique intermediate order structures, lying in between the crystal and the amorphous state and allow following with more detail, the C-A transition itself. Section 5 is devoted to the interface migration-charge transfer (IMCT) atomistic model of phase stability in ion bombarded compounds, whose degree of interpretative success when compared with experimental results is high, both concerning metallic and nonmetallic systems.

2   ELECTRON IRRADIATION

At the beginning of 1980s it was discovered that electron irradiation by high-energy beams in the MeV range, at low (typically cryogenic) temperature induces amorphization of intermetallic compounds such as NiTi [6], Cu_4Ti_3 [7], and NiZr [8]. The intrinsic simplicity of the damage electrons produce allows understanding the specific role of extensive chemical disordering as the driving force for amorphization. Indeed, electron irradiation does not alter target chemical composition because impurity atoms are not introduced or created, in contrast to ion-beam processing. Second, electrons do not cause ionization damage, but rather they cause displacement damage [6]; the energy transferred to a target atom by electrons accelerated at 1 MeV is around the threshold displacement energy, i.e., 25 eV. Thus the energy transferred to the primary knock-on atoms (PKAs) can displace one, or at most, more rarely, two atoms from their equilibrium lattice sites.

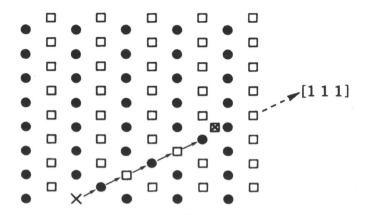

Figure 2. Replacement collision sequence (RCS) along [111] in a cubic B2 structure. Chemical disordering occurs via formation of a vacancy (X) and an interstitial (⊠). ●: A atoms; □: B atoms.

The induced damage consists essentially of a replacement collision sequence (RCS), produced along a lattice atomic chain, as shown in Fig. 2 and of the associated chemical disordering that is ascribed to the introduction of antisite defects. RCSs are found most likely in close-packed directions, such as [110] and [100] in fcc lattices and [111] in bcc lattices. If the RCS is long enough, the vacancy left at the position of the first displaced atom does not recombine with the sequence-ending interstitial, and a stable Frenkel pair is produced. This way, a homogeneous distribution of vacancies and interstitials can be achieved. Experimentally, specimens are irradiated with typically 1- to 2-MeV electrons in a high-voltage electron microscope (HVEM); modifications in the bright-field images and in the selected area electron diffraction patterns (SAED) are monitored in situ during irradiation.

The complete replacement of the crystalline spot patterns in the SAED by a diffuse halo pattern marks amorphization, whose evolution can be studied as a function of projectile dose and irradiation temperature.

It was observed that the critical temperature T_c, defined as the highest temperature at which the crystal transforms to amorphous, coincides with the maximum temperature for chemical disordering (Fig. 3). The most easily measured chemical-ordering parameter is the Bragg-Williams LRO parameter S that can be expressed, making reference to the intensity-ratio method, as

$$S = \left[\frac{<(I_s/I_f)>_{irr}}{<(I_s/I_f)>_{unirr}} \right]^{1/2} \tag{2}$$

where I_s and I_f are the intensities of superlattice and fundamental reflections, respectively, and their average ratio after a given irradiation time is normalized to their average ratio in the unirradiated material.

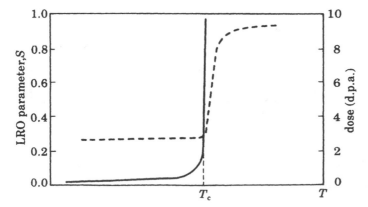

Figure 3. Schematic illustration of the coincidence between T_c for chemical disordering (----) and T_c for amorphization (——) under electron bombardment (adapted from [9]).

A substantial loss of chemical long-range order usually precedes the transition [10]: high-resolution electron microscopy shows that slightly before amorphization, a certain degree of residual chemical order is present, and it progressively disappears up to the transition point. The outlined chemical-disordering mechanism of transition must incorporate process kinetics. Thus, competing with the disordering mechanism, a temperature-dependent chemical reordering mechanism must exist. Indeed, the maximum degree of disorder attainable depends strongly on temperature, whereas irradiation-induced disorder is essentially temperature independent [11]. It is expected that thermally induced motion and recombination of point defects promotes chemical reordering; there are also indications that vacancies are by far more efficient than interstitials to promote ordering in a glassy metallic system [10]. The temperature dependence of the critical dose for amorphization of CuTi [12] displays the same features as the corresponding schematic curve in Fig. 3 (see also Fig. 11), namely, a weak dependence in the low-temperature region, followed by a steep increase up to the neighborhood of T_c, where the critical electron dose to induce amorphization tends to diverge.

Such a trend is common, and it was observed in Zr_2Fe and Zr_3Fe [13]; indeed, MeV electrons produce isolated vacancies and interstitials, and T_c is thought to be determined by migration of free vacancies [7]. The step observed at about 80 K in CuTi was ascribed to a recovery stage of an irradiation-induced defect, possibly interstitials [14]. The role of point defects to drive amorphization is evident, looking at Fig. 4; it shows the

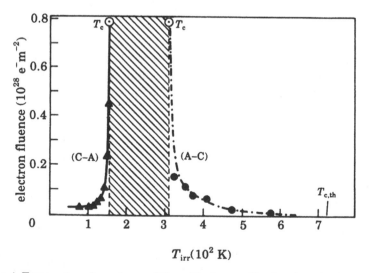

Figure 4. Temperature dependence of the critical amorphization dose for C-A (▲) and A-C (●) transitions in electron-irradiated Co_3B (adapted from [15]). ⊙: T_c. T_c(C-A): 160 K; T_c(A-C): 310 K. Dashed area: no transition in either sense.

temperature dependence of the fluence needed to the C-A and, respectively, the A-C transitions of Co_3B [15], the only system on which both transitions have been observed under electron irradiation. Here the appearance of small (around 10 nm in diameter) stoichiometric crystallites in an amorphous matrix marks the initiation of the A-C transition.

It is noteworthy that high-energy electron bombardment of an amorphous alloy usually accelerates the recrystallization process; such enhanced kinetics is ascribed to an increased rate of atomic migration in the amorphous structure. The atomic extramobility results from two mechanisms: an increment of "point defects" created by irradiation that act as diffusion carriers [16] and the nonthermal atomic migration induced by direct collisions with incident electrons, the so-called radiation-induced diffusion (see Chapter 6, Allen & Was).

Depending on their thermal mobility, the contribution of point defects becomes more or less important with respect to radiation-induced diffusion. The higher the rate of atomic diffusion in the vitreous matrix, the higher will be the temperature (well above T_c for C-A transition) at which crystalline nuclei can develop and grow into the amorphous material. In principle, T_c could coincide for the C-A and for the reverse A-C transition under electron irradiation. If, however, as in the case of Co_3B, atomic mobility is not sufficient to produce stable crystalline nuclei at temperatures above T_c for C-A transition, the fluence versus recrystallization temperature curve is displaced to higher temperatures, and an intermediate temperature region is observed that extends between the two T_cs for C-A and A-C transitions, respectively. Over such a region, the material remains in its phase (crystalline, or amorphous) independently of the irradiation process. The dependence of the amorphization dose on electron energy was investigated in Zr_3Fe at low temperature, between 25 K and 30 K [13], where the C-A transition is homogeneous [17]. Amorphization was observed even at energies as low as 250 keV at a dose of $6.5 \times 10^{26} e^{-}m^{-2}$; the dose quickly falls down to $8 \times 10^{25} e^{-}m^{-2}$ when projectile energy is doubled, then increasing electron energy up to 900 keV it further decreases slowly to about $5 \times 10^{25} e^{-}m^{-2}$.

The role of chemical disordering in an alloy before amorphization has been studied with particular care in Ni_4Mo irradiated with 1-MeV electrons at a dose rate of 5×10^{-3} displacements per atom (dpa) s^{-1}, between 50 K and 1050 K. This alloy is specifically indicated for such studies because both LRO (D1a, b.c.t. structure) and SRO (1 1/2 0) may be separately produced in it. It has been reported that at low temperature (200 K $\geq T_{irr}$) a disordered structure is produced that does not exhibit any SRO or LRO diffraction intensities after a dose of 0.5 dpa [11]. For samples initially ordered, the disruption of LRO is accompanied by formation of point defect clusters. Similarly to what is observed in Co_3B [15], over an intermediate temperature range between 450 K and 550 K, the irradiated material preserves its initial type of order, independently of the irradiation process. These results were confirmed by a study based on atom probe field ion microscopy, TEM, and HVEM [18] in which 1-MeV irradiation at 85 K of samples showing either LRO, or SRO reveals that at the same dose of 0.4 dpa Ni_4Mo completely disorders without intermediate short-range ordering of the initial LRO. The completely disordered alloy contains a random distribution of Mo atoms.

On the other hand, $NiZr_2$, which is also easily amorphized by electrons [8], was the object of a molecular dynamics (MD) simulation, specifically aimed at verifying the role of antisite defects to drive the C-A transition [19]. After selecting a configuration from the constant-temperature constant-pressure MD trajectories for the equilibrated alloy at 300 K, a given number of Ni and Zr atoms is exchanged, thus introducing antisite defects. The pair distribution functions $g(r)$ are then plotted for different values taken by the LRO parameter S, and the C-A transition is observed for $S < 0.6$ while the crystal is stable against chemical disorder for $S \geq 0.6$. MD thus proves that a moderate degree of chemical disorder is sufficient to induce the C-A transition in alloys undergoing amorphization at relatively low irradiation doses of less than 1 dpa.

Also simulated was quenching a melt from 1600 K down to 300 K, and the thermodynamic state of the subsequently annealed system was characterized again by calculating $g(r)$ functions. A comparison is thus

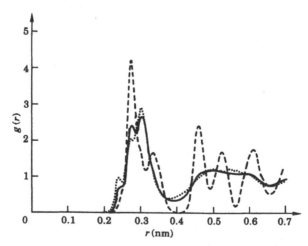

Figure 5. Global pair distribution functions g(r) for intermetallic $NiZr_2$ disordered up to different degrees in MD simulations. --- crystalline $NiZr_2$, — complete chemical disordering (S = 0), ⋯⋯ vitrification, simulating fast quench from the melt (adapted from [19]).

possible between amorphous states obtained through physically different paths. Chemical destabilization in the solid state and after liquid quenching produces disordered states whose $g(r)$ are nearly identical, as shown in Fig. 5. The only meaningful difference is in the partial Ni-Ni $g(r)$ from which average coordination numbers increase from $Z = 2$ for the crystal to $Z = 2.75$ for chemically disordered and to $Z = 3.72$ for liquid-quenched alloy. Nickel atoms are arranged more compactly, at shorter distances in the amorphous structures than in crystal, particularly when the metal glass is formed from the liquid state. Zr atoms are nearly unaffected by the disordering process of whatever kind.

Simulation results have to be considered with some caution, because they are influenced by the particular alloy considered, by the choice of interatomic potential and by the way of switching atom positions. For example, an MD study of amorphization in Cu-Ti alloys, including CuTi, $CuTi_2$, and Cu_4Ti_3 [20], shows that glass formation under irradiation consists of two steps: introduction of chemical disorder through point-defect recombination *and* build up of point defects. The conclusion contrasts with a previous model, based on the results of electron irradiation in the same compounds [9] where the role of chemical disordering to cause vitrification was stressed. On the other hand, the experimental difficulty is that it is practically impossible to produce chemical disorder, still avoiding the generation of point defects; thus it is hard to separate the importance of the two contributions.

In turn, MD simulations allow a separate introduction in the crystal Frenkel pairs, which result from ballistic disordering and random atom exchanges, producing truly chemical disorder. This was performed in NiZr [21]; the simulated $g(r)$ functions at the same doses of 0.16 dpa (ballistic) and 0.18 dpa (chemical) coincide with $g(r)$ for the quenched liquid, showing that both disordering mechanisms have the same efficiency to induce amorphization that occurs at the value $S = 0.56$ of the LRO parameter in reasonable agreement with the experimental datum [22].

Loss of chemical LRO not always precedes amorphization; the starting material structure certainly has an effect on the transition path. Initially fully martensitic, NiTi, e.g., transforms to austenite (b.c.c. B2 structure) [23], showing that martensite is highly unstable during irradiation, presumably because of its inability to chemically disorder and to resist radiation damage. Electron irradiation experiments indicate that direct amorphization of martensite is unfavored; NiTi amorphization is likely to occur through chemical disorder locally produced in the B2 phase, where amorphization subsequently takes place.

The correlation between chemical disordering and glass transition onset is evident in Fig. 6 [24]: the CLRO parameter S progressively lowers with irradiation dose. In this case, the dose sufficient for amorphization onset slightly changes from 0.4 dpa at 10 K to 0.6 dpa at 295 K. The effect of Kr^+ irradiation is also reported for comparison; notice that Kr^+ induced

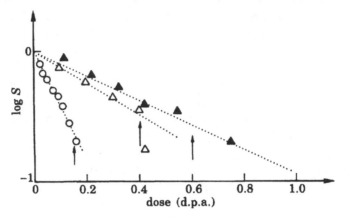

Figure 6. Correlation between chemical disordering of Zr_3Al irradiated with 1 MeV projectiles and damage introduced in the target. O·····O: Kr^+, T_{irr} = 295 K; △·····△: electrons, T_{irr} = 10 K; ▲·····▲: electrons, T_{irr} = 320 K. Arrows mark vitrification doses (adapted from [24]).

vitrification at 295 K requires only 0.15 dpa. Amorphization sets on only after S substantially decreases down to $S < 0.4$, as shown by the markers in Fig. 6.

An analysis of electron diffraction patterns from single grains reveals that the presence of diffuse streaks along given directions marks the softening of a particular shear elastic constant; the streaks are observed to grow, while at the same time chemical disordering takes place.

In Zr_3Al, considerable softening of the shear elastic constant $C' = \frac{1}{2}$ $(C_{11} - C_{12})$, with respect to C_{44} is observed. In a cubic structure, C' is the elastic constant for pure shear along [110] induced by a stress applied on the $(1\bar{1}0)$ plane, while C_{44} is the shear along [100], with the stress on the (010) plane. C' softening indicates that atoms lying in the $(1\bar{1}0)$ plane can be sheared along [110] more easily than atoms on the (010) plane can be along [100]. In principle, diffuse streaks could be due to thermal vibrations, but these are ruled out because streak intensity is temperature independent over the range between 10 K and 295 K.

The softening could be the result of point defect formation; however, such a contribution is expected to be irrelevant, analogous to what is observed in irradiated pure metals [25]. Thus, C' softening with respect to

C_{44} is due to *static* displacement of atoms on the $(1\bar{1}0)$ plane, along $[110]$ and other equivalent planes and directions.

After the onset of the C-A transition, at less than 1 dpa, amorphization of Zr_3Al develops very slowly in contrast to e.g., CuTi, NiTi, and Cu_4Ti_3, which all require doses of less than 1 dpa to completely vitrify. In such alloys, neither defect aggregation nor metastable intermediate crystalline phases are observed before C-A transition. By contrast, after the transition onset, S values continue to decrease in Zr_3Al, while the development of the halo intensity is only gradual at 10 K, being even slower at 57 K and at 295 K.

As a progressive development and a strong temperature dependence of halo intensity are *not* observed when chemical disordering alone is present, also point defect accumulation [26] besides chemical disordering must be considered to be a mechanism effective to drive complete amorphization of Zr_3Al. The observed aggregation of point defects associated to their migration proceeds at a rate increasing with temperature and retards vitrification; indeed, the volume fraction of amorphous phase, produced by the same electron dose, decreases. Such mechanisms account for the large, anomalous dose required to complete the C-A transition in Zr_3Al.

When investigating the leading amorphization mechanism, we are implicitly asked a question about the very primordial transition stage. We mentioned that at low temperature the C-A transition is homogeneous, and it occurs at rather low levels of radiation damage [18]. At higher temperatures, up to the critical temperature T_c for amorphization, the transition path is inhomogeneous. The initial assertion that vitrification starts at dislocation lines and proceeds through the formation of amorphous cylinders of matter centred around dislocations, as observed in NiTi [23], was somewhat relaxed. In addition, grain boundaries, defects related to martensitic transformations, antiphase boundaries, and free surfaces were included among nucleation sites [27, 28].

Recent careful in situ observations in a high resolution-high voltage electron microscope of NiTi undergoing vitrification by electron irradiation were combined with image analyses of MD-simulated atomic configurations [29]. Nanometer-sized clusters randomly form and annihilate throughout the irradiated region; the fluctuating process is likely to be responsible for structural fluctuations at the nanometer scale, related to transition among correlated structures with different lifetimes of the associated clusters.

Extensive experimental work on electron irradiation of about 70 binary compounds [30] allowed correlating the tendency of a given system toward amorphization with its position in the phase diagram; if it lies close to the bottom of a valley of the liquidus curve, enhanced amorphization tendency results.

The observation that is strongly reminiscent of a general glass formation criterion stresses again that a relevant factor controlling vitrification is of *chemical* nature. The manifold in coordination, namely, the extent to which the atomic coordination of the material can be changed, plays a crucial role in determining the relative ease of vitrification. In electron irradiation, such an ability must be intrinsic of the material because it depends only on the starting stoichiometry. The same kind of capability to modify atomic coordination was reported to deeply influence glass formation of metalloid implanted transition metal-metalloid (TM-m) alloys; in the latter case, however, ion bombardment alters sample composition and can bring it to optimal values to enhance the effect of local atomic coordination changes [31].

3   ION BOMBARDMENT

3.1   Fundamentals

From the historical point of view, amorphous metallic phases were first obtained by high-fluence and medium-high energy (several tens to a few hundreds keV) ion implantation in the middle 1970s [32, 33]; since then, two experimental configurations have been commonly adopted: ion irradiation and ion implantation. In the former, a chemically homogeneous target is bombarded with ions that *do not alter* the chemical characteristics of the sample. The scope is obtained either with inert ions (rare gases or ions of one of the target species) or by a selection of ion energies high enough that sample thickness is smaller than the projected range R_p. Average sample composition is fixed in irradiation experiments, so these are the most suited to explore the nature of the damage processes associated with the slowing down of a massive energetic particle in a solid and their importance to vitrification.

Suppose a species B is introduced in the form of accelerated ions into a crystalline target A up to a given depth, that depends on the implantation energy and on the mass of A and B species. The process is known as implantation—a system of composition $A_{1-x}B_x$, either a compound or a possibly saturated solid solution, amorphous, or crystalline may result. Quite frequently, alterations of target composition of the order of some tenths of percent are obtained at fluences in the 10^{21} i$^+$m^{-2} range and greater. In the implantation case, amorphization processes depend on radiation damage as well as on alteration of the chemical nature of the target. They are expected to be considerably more complicated than in irradiation experiments.

To understand such processes, it is necessary to follow the leading interactions experienced by an energetic ion entering a solid. Over the usual energy and mass range, nuclear energy loss dominates. Target atoms are put in motion, gaining energy and momentum in ion-nuclei elastic collisions governed by a repulsive Coulomb potential screened by a Thomas-Fermi function. A large fraction of the PKAs set in motion by the primary ion through elastic collisions gain sufficient energy to subsequently displace other target atoms through several secondary and higher order collisions. The branched sequence of energy sharing collisions constitutes a collision cascade.

If the number of moving atoms during the relevant stages of the ion stopping process is small, the cascade is dilute or linear. For medium-high projectile mass, however, all atoms of the target within the volume encompassed by the collision are put in motion and define a dense cascade. During the process of energy sharing within a cascade, at a certain point atoms are no more sufficiently energetic to produce further displacements, yet their average energy (1–2 eV) corresponds to very high temperatures. This final evolution stage of the cascade is called thermal spike.

In Table 1 are listed the relevant mechanisms contributing to ion irradiation and to ion implantation (first two rows); the third row shows also the mechanisms operating in the delayed regime, typical of ion mixing.

Table 1. Summary of the relevant mechanisms that contribute to ion irradiation and to ion implantation, arranged according to their typical timescales (first and second rows). The mechanisms active in the prompt and intermediate regimes are effective in ion mixing as well, besides the delayed ones (third row).

Regime	Timescale t	Characteristics	Mechanisms
Prompt	< 1 ps	Ballistic mixing High-energy Athermal	Collision cascades Recoil mixing Isotropic mixing Ballistic diffusion
Intermediate	1 ÷ 100 ps	Intermediate energy Thermally assisted	Collision cascades Thermal spikes Thermochemical effects
Delayed	100 ps ÷ 3600 s	Low energy Thermally assisted	Radiation-enhanced diffusion (RED) Radiation-induced segregation (RIS)

The diagnostics following an ion-bombardment experiment provide the opportunity to characterize the various ion paths by different parameters: these are the total range $R = \Sigma\, r_m r_{m+1}$, the projected range R_p, and the transverse range $R_\perp$, as schematized in Fig. 7.

Because of the statistical nature of the process, the distribution of the rest positions of ions in the target is experimentally found to be nearly

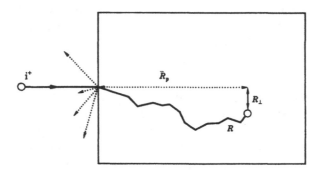

Figure 7. Schematic representation of the total range R, the projected range R_p and the transverse range $R_\perp$ as functions of sample depth for an ion $(o \to i^+)$ impinging at normal incidence on target surface. For clarity, higher-order recoil events have been omitted. Dotted arrows emerging from target surface: sputtered particles of different energy.

Gaussian; thus, besides the average R, R_p, and $R_\perp$ values, also the corresponding range stragglings ΔR, ΔR_p, and $\Delta R_\perp$ are of interest.

The displacement processes result in radiation damage of the material, basically through the production of vacancies and interstitial atoms, while in the near-surface region, atoms can gain sufficient energy to escape from the solid (sputtering). Frenkel pairs are produced by RCSs along close-packed directions [34].

RCSs produce a vacancy at the first knock-on atom site and a self-interstitial atom (SIA) at a certain distance. Although important for cascades at all energies, RCS length distribution (which, e.g., in ordered alloys quantitatively defines the disorder produced by Frenkel pairs) is not well known. As an order of magnitude, MD simulations show that in copper the average RCS length produced by a Cu+ of 5-keV energy is of the order of 23 nm [35]. By comparison, MD on silicon yields an RCS average length ranging from 1.1 to 1.7 nm [36].

An incident ion with energy in the keV range affects a region with typical dimensions of some nanometers, involving a volume of about 10^{-20} cm³; thus, although typical timescales over which cascades evolve (1 ps or less for the fast regime events) are in principle accessible, the involved volumes are exceedingly small to be observable in real time. Yet the features of collision cascades make them an ideal playground for MD simulations.

Computer simulations play a relevant role to provide a coherent picture of several important processes that contribute to the establishment of radiation damage when dense cascades form. The most significant advances have been obtained in recent years through MD, involving significant atom numbers (a few hundred thousands) and after recording the time evolution of cascades characterized by progressively higher energies, up to tens of keV. The main results have been critically reviewed recently [37].

Collision evolution is followed for about 10 ps [38], looking at the locations of SIA defects; most of the replacements are located in a dense cluster near the cascade centre. The vacancies form a compact *depleted* zone, in agreement with field ion microscopy studies [39]. Such features of the defective structure are present already during the first stage of cascade evolution, in linear regime, both for Cu and for Ni.

The high atomic replacement density in the core region of the cascade, which is thus strongly disordered, suggests that atomic mixing is caused by *local melting*. Indeed, looking at Fig. 8 we see that the simulated pair distribution functions $g(r)$ for the core region have the same features for both Cu and Ni, and in the case of Cu, $g(r)$ exactly mimics that for liquid Cu. The local temperature at the cascade centre ($r < 1.3$ nm) after 0.25 ps is near 4000 K for both metals, with an outward-directed

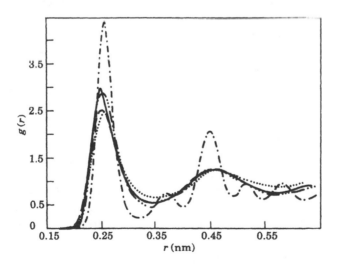

Figure 8. Pair distribution functions $g(r)$ for MD simulated cascade quenching in Cu and Ni, at different times (1.1 ps and 3.84 ps for Cu; 1.1 ps and 2.8 ps for Ni) after ion impact (adapted from [38]) ⋯⋯ Cu, 1.1 ps; —— Cu, 3.84 ps; – ⋯ – Ni, 1.1 ps; – · – · – Ni, 2.8 ps.

temperature gradient of about 1400 K nm^{-1}, the starting cooling rate being in the 10^{15} Ks^{-1} range. It is noteworthy that while in Cu *g(r)* is globally unaltered after 3.84 ps, in Ni sharp peaks rapidly develop after 2.8 ps. This qualitatively different behaviour is attributed to a lower defect production and higher atomic disordering in Cu cascades than in Ni ones.

Notice that both simulations were performed under the standard assumption of negligible coupling between the electronic subsystem and the lattice during cascade lifetime. Rapid coupling could however cause faster quenching of the thermal spike through energy transfer from the ionic to the electronic system and through energy transport out of the cascade region by electronic thermal conduction. The weak coupling assumption is valid if the electron mean free path λ is greater than the average cascade size; in liquid Cu at the melting point, $\lambda \approx 4.5$ nm [40], while its value is reduced in transition metals with a high electronic DOS at the Fermi level, such as Ni.

MD of cascade evolution in semiconductors (Si) [45], in heavy elemental targets (Au), and in alloys (β-NiAl) [37] allowed drawing some conclusions with general validity. First, thermal spike melting is ubiquitous and, second, its effect is more prominent with increasing atomic number Z and with decreasing cohesive energy of the target. Such trends are correlated to a stronger energy localization in the cascade core and a lower melting temperature in a crystalline softer matrix.

MD simulations of 25-keV displacement cascades in Cu at 10 K [41] indicate that distinct subcascades form and interact with each other, leading to overlap and formation of a single molten zone. Besides the already observed vacancy-rich core of the cascade, surrounded by a cloud of SIAs, also a distribution of SIAs, correlating with the spatial distribution of the two original subcascades exists. SIA clusters move away from the cascade surface, ahead of the interface between the molten zone and the solid, with a speed just below that of a C_{44} wave. As RCSs travel at a speed higher than the sound longitudinal speed, the observed SIAs cannot result from RCSs.

On the basis of such observations, *two* distinct defect production mechanisms appear to operate in dense cascades, namely, RCSs and, as the energy density (i.e., the pressure in the compressed region at cascade periphery) increases, an increasing number of SIAs produced in clusters, which sum up to the isolated SIAs produced through RCSs. Indeed, in Cu [37] after 1.41 ps the cascade core experiences about twice the melting temperature and a density of defects around 15–20% is found; this is nearly twice the defect density at the equilibrium melting point. Again, MD simulations on Cu show a density wave travelling at a speed as high as 4 nm ps^{-1} from cascade core to the periphery [42]. Direct observation of defects produced within cascades, in Au and Al implanted with Al$^+$, Ar$^+$, and Xe^+ up to energies of 400 keV, in the temperature range from 120 K to 800 K, give results in essential agreement with MD simulations [43].

Atomic mixing is largely explained by enhanced diffusion in the locally melted region associated to the thermal spike; in bombarded alloys, atomic mixing in turn is strictly related to disordering of the initially ordered alloy structure. Even in the case of a structure with low-chemical ordering energy, such as Cu_3Au, which is known to resist amorphization, simulation showed [35] that the LRO parameter S lowers to a value $S \approx 0.17$ in the centre of the cascade, in agreement with observed values of $S \approx 0.2$ for highly disordered alloy regions implanted with 10 keV Cu^+ at fluences as low as 10^{15} i^+m^{-2} [44].

3.2   Amorphization Mechanisms

The basic amorphization mechanisms, in the case of elastic ion-target collisions, are determined by the complex dynamical evolution of the defective cascade structure just outlined. A tempting model [46] involves fast quenching of a dense cascade (or of subcascades where sufficient energy density has been achieved, as in light-ion-induced cascades), occurring with estimated cooling rates of the order of 10^{14} Ks^{-1}. The extension to the dilute cascade case implies that only a fraction of the collisions is able to produce a threshold energy concentration. Accordingly, amorphization in implanted metals should be a fluctuating phenomenon that occurs only in subcascades generated by high-energy head-on recoils; an explanation to the experimentally observed high amorphization doses is thus provided [47].

Looking more closely at cascade evolution, the rapid relaxation driven by internal stresses of the nonequilibrium defect structure formed in a cascade is thought to produce a glassy zone. By a shock wave mechanism, displacement cascades are formed [48] thereby a compression wave, travelling at high speed ahead of the incoming ion is associated to mass transport from cascade center to the periphery. The mechanism is similar to the results of MD simulations [42]—temperature and pressure attain very high values in the shock wave front. Such conditions are effective to induce local melting also when light ions are implanted into metals.

The easy amorphization of metalloid (m) implanted transition metals (TM) led to the suggestion [31] that cascade-induced disorder is stabilized by metalloid impurities. Most of the ion-generated defects in a pure metal are unstable (pure metals are not amorphizable) in the nuclear stopping regime. Only around m^+, which are stopped at certain positions, small amorphous clusters with typical volumes of $10^{-21}cm^3$, survive, provided that covalent TM-m bonds have been formed. The accumulation of such single ion generated glassy clusters leads to a network of vitreous islands that ultimately turns to a uniform amorphous layer. By HRTEM spheroidal amorphous clusters were directly observed after Fe implantation with Al^+ at medium-high doses, between 10^{20} and 10^{21} i^+m^{-2}, with ion energy 100 keV [49]. Bombardment temperature was fixed at 90 K, 300 K, and 420 K, respectively; a lower cluster size was found, indicating that a critical Al concentration is needed to stabilize the amorphous structure. A dynamical evolution of clusters was observed under bombardment, with a cluster diameter increasing with increasing temperature from 1.2 to 1.55, to 2.4 nm, corresponding to cluster volumes between 1 and $7 \times 10^{-21} cm^3$ and to critical Al concentrations between 3.8 and 4.6 at.%.

Successive ion implantations allowed monitoring vitrification in a quasicontinuous fashion, as a function of alloy composition [50]. It was observed that if impurity concentrations are small or intermediate (around 10 at.%), the implanted impurity atoms are incorporated interstitially or substitutionally, depending on the alloy system, and generally they result in the formation of supersaturated solutions.

Although interstitial incorporation always produces an increase of lattice parameter, substitutional incorporation causes lattice dilation, or contraction, depending on the atomic size misfit of the impurity in the host lattice. X-ray analysis shows that considerable lattice strain accumulates with increasing impurity concentration until at a threshold that depends on the system, accumulated strains are suddenly released at the amorphization onset, as shown in Fig. 9.

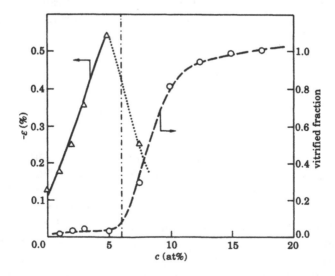

Figure 9. Development of strain ε (Δ,——,······) and of the fraction of vitrified material (O ---), vs. implanted species concentration, for B^+ implanted Mo (adapted from [50]). The dot-dashed line marks the impurity concentration value at which amorphization kinetics abruptly changes from an incubation stage to fast growth.

If impurity concentration is raised further, the glassy fraction also grows over an extended concentration range; this is because of the statistical distribution of implanted atoms over target volume, thereby regions with different local impurity densities are formed. Once the critical local impurity concentration, corresponding to a critical local strain level, is exceeded over a certain volume, the latter is vitrified.

Thus, impurities act both to generate strain when incorporated in the host lattice and to stabilize the amorphous phase when the C-A transition occurs.

In a statistical model proposed for the kinetics of implantation-induced amorphization in the prototypical B^+ implanted Ni [51], the minimum cluster volume V_c and the critical solute concentration c_c to trigger glassy cluster formation are extracted from the dependence on solute concentration of the amorphous volume fraction f_a during the C-A transition. In the model, the region at the implantation depth is divided into small identical cells with volume V_c; amorphization is taken to occur when the number of B atoms in a cluster is at least N_c, with $<N>$ average number of B atoms in the cluster volume. At any B concentration c_B, f_a is

$$f_a = \sum_{N=N_c}^{\infty} (<N>)^N \exp[-<N>]/N! \qquad (3)$$

The equation for f_a is obtained considering that the actual number of B atoms within V_c, i.e., the probability distribution for N around $<N>$ obeys Poisson's statistics. $<N>$ is proportional to V_c and to the average B concentration in the sample $<c_B>$. Calling f_c the critical B fraction corresponding to N_c, the experimental curve $f_a = (1-f_c)$ is fitted to (3), varying the parameters V_c and c_c. In Ni single-crystal [51] V_c is 15×10^{-22} cm^3 and c_c 9.3 at.%; it is noteworthy that in experiments with polycrystalline Ni [52] V_c is 4.52×10^{-22} cm^3 and c_c 8.7 at.%. The latter slightly lower values can be attributed to the abundance of grain boundaries in the polycrystal that supply more nucleation sites than the single-crystal target.

Again along the philosophical approach of defect accumulation in the host lattice is an amorphization mechanism [53] by which a vancacy-interstitial complex is assumed to form either by a diffusional process or directly in the displacement cascade. The experimental rationale behind this model is that low critical defect concentrations c_c are required to induce amorphization, thus implying very small recombination volumes. Notice however that the estimate of $c_c = 2\%$ is too low with respect to many experimental observations [50].

If point defects are the accumulating species, vacancies and interstitials must couple together in order to achieve the needed buildup. Such ordered units tend to locally restore chemical order in contrast to the significant LRO disruption by ion bombardment. However, as soon as the local concentration of defect complexes exceeds a threshold, a small volume of defected crystal is destabilized, turning amorphous. Amorphization kinetics is calculated and the effect of varying T_{irr} is reproduced. The real difficulty of the mechanism is that until now no experimental evidence for the existence of such complexes has been found, nor has their importance to drive C-A phase transition been definitely assessed.

Although the kinetics of C-A transition highlights the transition mechanisms, comparatively few experimental data are available. We focus on a detailed study [54] concerning U_6Fe, vitrified by self-fission damage and NiTi, amorphized by 390-keV Ni^+ [55]. Looking at the progress of amorphization by TEM, the glassy areas appear as spots in bright field images; it is thus possible to plot the fraction of crystalline material that has been vitrified as a function of the irradiation dose expressed in dpa, as shown in Fig. 10. In NiTi, the slope of the curve before the saturation stage differs from unity, implying that the transition kinetics is not first order. The greater than unity slope indicates that cascade effectiveness to amorphize the material increases when cascade overlapping becomes significant.

In U_6Fe, the C-A transition involves the bulk without the limitations given by geometry and volume constraints in charged particle irradiations.

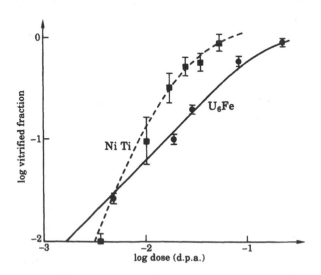

Figure 10. Amorphization kinetics in U_6Fe, damaged by self-fission ($\bullet$—— [54]) and in Ni^+ bombarded NiTi ($\blacksquare$ --- [55]) vs. irradiation dose. Fitting curves [56] were obtained taking into account both direct vitrification and cascade overlapping.

As with NiTi, the dependence of the fraction of amorphized material on irradiation dose is best fitted when both direct amorphization in the fission fragment cascades and cascade overlap are considered [56].

Until now we did not even mention the existence of inelastic electronic collisions. Indeed, in metals and semiconductors, these are usually low energy processes that do not cause atomic displacements and can be considered a supplementary damage energy source.

However, it was recognized that structural changes due to electronic energy loss $(dE/dx)_e$ of highly energetic heavy ions do occur in amorphous metallic alloys [57, 58]. Above a threshold $(dE/dx)_e$ the cross sections for defect production are up to two orders of magnitud e higher than for elastic coll-isions.Here we discuss only some experiments that allow comparing the effects on target structure of electronic energy loss with those of nuclear energy loss.

The crystalline alloy prototype Ni_3B [59] was irradiated at 10 and at 80 K with GeV heavy ions, changing the average $(dE/dx)_e$ value between 12 keVnm^{-1} and 72 keVnm^{-1}. Electrical resistance was measured in situ, whereas electron diffraction was performed at room temperature after the irradiations were completed. An abrupt increase of electrical resistivity above a threshold $(dE/dx)_e$ value was attributed to partial sample amorphization. The number of elastic dpa induced by the irradiations was about 2×10^{-4}, with ion fluences as low as less than 4×10^{16} i$^+$m^{-2}, in front of the experimentally determined level of 0.1 dpa necessary for the C-A transition by nuclear energy loss, which is thus ruled out. The emerging picture is that of an amorphized damage track, surrounded by a disordered, but still crystalline region.

The C-A transition associated to electronic energy loss is difficult to be induced, because free electrons quickly dissipate the perturbation provoked by the projectile.

Besides thermal spike formation by electron-phonon interaction, the observed amorphization may be accounted for by a Coulomb explosion mechanism: the space charge produced by the high ionization density along the ion path lasts for a time sufficient to cause collective repulsion of the ionized atoms, with energy sufficient to cause atomic displacements [60]. An atomistic model of structural stability of a target irradiated in the electronic stopping power regime was also provided, both for metallic [61] and for nonmetallic targets [62].

Again using heavy ions with energies near 1 GeV (with typical ratios of about 2×10^3 between electronic and nuclear energy losses at fluences between 10^{15} i$^+$m^{-2} and 10^{17} i$^+$m^{-2}), several metallic alloys were bombarded choosing both systems with low chemical-ordering energy, such as Cu_3Au, and alloys with high-ordering energies, thus being amorphizable by low-energy ion or high-energy electron irradiations; these included NiTi, Zr_3Al, and $NiZr_2$ [63].

In Cu_3Au and Zr_3Al, only displacement cascades, induced by elastic collisions, are revealed by electron microscopy and x-ray diffraction. In $NiZr_2$ at low fluences, separated amorphous tracks are formed that overlap at higher fluences, causing complete amorphization and the same kind of anisotropic growth observed in metallic glasses [57].

In NiTi, equivalently to what is observed with high-energy electron irradiation and with low-energy ion implantation, the initial low temperature martensitic structure changes to cubic B2 structure before amorphization; the latter occurs at a dose similar to that required to vitrify $NiZr_2$, corresponding to a threshold close to 40 keV nm^{-1}.

3.3   Selected Experiments

Ion treated, both irradiated and implanted targets are rather thin, with typical thickness of the order of a few hundreds of nanometers or less. Usually, structural information on such films is obtained from electron diffraction and EXAFS, mainly from the analysis of the radial atomic distribution function as well as from TEM observations of local atomic environments. The most relevant information is extracted from the static structure factor $S(k)$, as determined by the normalized coherently scattered electron intensity; by Fourier transforming $S(k)$, the pair-distribution function $g(r)$ is obtained.

To test the structural characteristics of amorphous layers produced by ion beams, TM-m binary alloys are specifically suited. Such alloys are easily vitrified by various techniques at compositions near TM_4-m; this allows comparing with each other differently prepared amorphous samples. Moreover, the structure of TM-m alloys was simulated both by physical structural models consisting of dense random packings of hard spheres and by computer models. In particular, Fe and Ni were amorphized by high fluence (10^{21} i^+m^{-2} and 5×10^{21} i^+m^{-2}) B^+ and P^+ implantation [64]; both $S(k)$ and $g(r)$ of ion-treated films were determined. Comparing the obtained curves with those for the corresponding liquid alloys and for glassy samples obtained via fast quenching from the melt, we see that different TM-m alloys amorphized by ion implantation are structurally similar to each other. In particular, the $g(r)$ general features of differently prepared glassy alloys and of liquid alloys are in reciprocal agreement except for the presence of a shoulder in the second peak of alloys vitrified by implantation. The shoulder is the signature of irradiation-induced structural inhomogeneity; indeed the topological SRO correlation length, as deduced from $g(r)$ curves, is shorter than that for the corresponding rapidly quenched alloys but longer than in the same compounds in the liquid state. The normalized interatomic distances are larger than predicted by structural models based on packing tetrahedral units, which shows that the degree of structural imperfection frozen in the material during implantation is higher than during rapid quenching from a melt.

A careful x-ray analysis of strain development versus film vitrification in Mn^+ implanted Al [65] shows that the forced introduction of this kind of impurity in the metallic matrix produces a lattice contraction, whereas strain and static atomic displacements increase up to a *critical* implantation level. Beyond it, strains are released while atomic displacements tend to saturate the lattice parameter turning back to its original value. In turn, grain size shrinks continuously with growing Mn content. Amorphization starts at about 8 at.% Mn and at room temperature it is completed at about 20 at.% Mn. Such a relatively broad composition range over which the transition occurs is likely to be caused by significant defect mobility at the relatively high-radiation temperature adopted that destabilizes the developing glassy phase.

Also Mo, W, and Nb single crystals, both (111) and (110) oriented, upon implantation at room temperature with 100 keV O^+ at fluences between 10^{21} i^+m^{-2} and 10^{22} i^+m^{-2} turn vitrified, concurrent lattice dilation being observed [66].

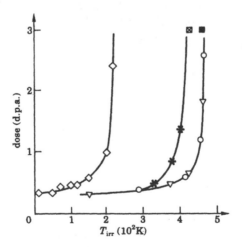

Figure 11. Dependence on T_{irr} of the dose needed to vitrify CuTi, bombarded with several projectiles accelerated at 1 MeV. ◊ electrons, ∗ Ne^+, ∇ Kr^+, O Xe^+, ⊠ partial amorphization (Ne^+), ■ no amorphization (Ne^+), (adapted from [12]).

It is expected that amorphization kinetics strongly depends on projectile mass, as the dominant mechanism (basically whether dense collision cascades are present) depends on the locally deposited energy density, thus on the ion species. A demonstration of this dependence comes from the comparison among the effects of various bombarding particles accelerated at the same energy of 1 MeV on amorphization of nearly equiatomic CuTi [12]. The projectiles include electrons and progressively more massive ions, Ne^+, Kr^+, and Xe^+, respectively. The critical temperature for amorphization increases in a monotonic way with projectile mass from 185 K for electrons, to 540 K for Kr^+ and Xe^+, as shown in Fig. 11; also the shape of the curves is the same, with an abrupt rise to infinity of the critical amorphization dose on approaching the critical vitrification temperature.

Electron irradiation produces isolated vacancies and interstitials, while, when cascades form, vacancies are supposed to cluster within cascade core, thus achieving mobility at higher temperatures (see MD results). Both the stability of a cluster, and consequently the critical amorphization temperature, increase with cluster size, i.e. with projectile mass.

The above picture is confirmed by the results in Fig. 12; here, again in CuTi, the fractions of glassy phase formed by room temperature Ne^+ and Kr^+ irradiations are reported as functions of the dose. Both curves present a linear interval in the low-dose region, with slopes of three (Ne^+) and one (Kr^+), while the slope for electron irradiation is considerably higher than that for Ne^+. As the slope gives the average number of cascades that have to overlap to induce vitrification [67], a single Kr^+ produces a cascade large enough to induce glass formation, while three cascades generated by Ne^+ must overlap to vitrify the material.

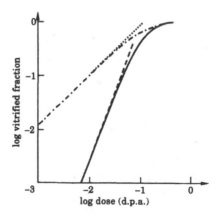

Figure 12. Dose dependence of the fraction of vitrified CuTi, irradiated with Ne^+ (——) and with Kr^+ (–·–·–). $T_{irr} = 295K$. The slopes of the dotted and of the dashed lines are one and three, respectively: they indicate the number of overlapping cascades required to amorphize the target (adapted from [12]).

Strongly ordered alloys when subjected to ion irradiation may undergo substantial disordering, without turning amorphous, as exemplified by Ni_3Al, which remains disordered and crystalline even when T_{irr} is as low as 80 K [68]. On the other hand, equiatomic NiAl, irradiated at 10 K with 360 keV Xe^+ experiences two distinct phase transitions [69], namely, chemical disordering with the concurrent formation of a premartensitic phase at low fluences (up to 3×10^{17} Xe^+m^{-2}) and progressive amorphization above such threshold until the transition is accomplished at about 10^{18} Xe^+m^{-2}.

The intermediate premartensitic phase that is found also irradiating FeAl, isomorphous to NiAl [70] is strain-induced. According to the direct impact amorphization kinetics observed in NiAl, when ion bombardment produces high-energy deposition in the cascades, strain results from both chemical disordering and a high density of displaced atoms within cascade volume.

However, increasing Xe^+ fluence, such as when cascade overlap becomes more and more relevant, growth of the glassy phase coincides with progressive weakening of the basic reflections of the ordered B2 structure, while martensite superlattice reflections suddenly disappear. Such a correspondence confirms that martensite is unable to resist irradiation (see Section 2), presumably because of a limited possibility to accommodate strain beyond relatively low limits.

The path followed by a system toward amorphization is not uniquely determined: cases are known where no loss of LRO is observed before ion-induced vitrification. Ni^+ irradiation of NiTi over the temperature range from 300 K to 825 K does not result in any intermediate crystalline disordered phase below the critical amorphization temperature that lies between 525 K and 625 K [71]. At quite low fluences the starting two-phase austenitic-martensitic structure transforms to a simple B2 structure, progressively converted (directly) to the glassy phase. A possible reason why no observable disordering of the B2 phase occurs is the low amorphization dose required in NiTi, lower than 0.1 dpa at room temperature. Indeed the cascade region formed by a heavy ion can be escaped only by a small fraction of the total number of produced defects. As a consequence of rapid development of vitrification, defects do not have sufficient time to extend chemical disorder to the B2 phase.

Point defects are the main product of radiation damage in intermetallic compounds; the microstructure of systems prone to amorphization evolves in a way absolutely different from that of irradiated compounds that retain a crystalline structure. Usually no defect clustering is observed in glass-forming alloys, perhaps with the exception NiTi [71] and of Zr_3Al [72]. Normally the C-A transition proceeds *inhomogeneously* throughout the crystalline matrix; grain boundaries and dislocations are preferential sites where glassy clusters form. Indeed localized compositional changes are likely to occur at such structurally defected sites.

When we consider *direct* amorphization after projectile impact and when the *nucleation* of glassy clusters is considered, the C-A transition evolves continuously. Experiments focused on the mechanisms that possibly trigger the transition in several intermetallics, such as Zr_3Al, FeTi, NiAl [72], and Nb_3Ir [73].

In Zr_3Al irradiated with 1 MeV Kr^+ by surface Brillouin scattering (SBS), a dramatic decrease by about 50% of shear modulus as well as chemical disordering and lattice dilation are found to precede amorphization as shown in Fig. 13. A linear relation connects the measured shear modulus and lattice dilation, closely analogous to the Grüneisen relation for the high-temperature contribution of anharmonic lattice vibrations.

Based on the similarities to melting [75], a shear elastic instability of the disordered crystal lattice was proposed to be the leading amorphization mechanism.

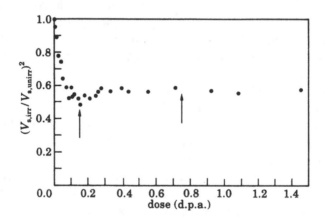

Figure 13. Change of shear elastic constant vs. increasing irradiation dose in Zr_3Al bombarded with 1 MeV Kr^+ at room temperature. The arrows mark the dose range over which vitrification occurs (adapted from [74]).

There are indications that the elastic modulus softening mechanism accompanying vitrification is ubiquitous. A C_{44} decrease by 35% was observed both in irradiated single crystal $MgAl_2O_4$ spinel [76] and in multilayered ion mixed Au-Ta films before amorphization. In the latter case, two distinct stages of alloying and subsequent vitrification as imaged by TEM were associated to a critical softening of the phase velocity of surface waves, both Rayleigh and Sezawa [77].

A comparison between the behavior of FeTi and NiAl is illuminating. The structure of both alloys is cubic B2, with a strong tendency to chemical order, as revealed by the stability of the B2 phase up to the melting point. However, while FeTi is amorphized by heavy ion irradiation at room temperature, NiAl resists vitrification and maintains chemical order under the same conditions [78]. TEM samples of both alloys were irradiated at room temperature with 1.7 MeV Kr^+, whereas shear modulus was measured by SBS in thicker samples irradiated to the same fluences as those observed with TEM.

SAED patterns show amorphization of FeTi [72]; before transition to the glassy state, the relative intensity of superlattice spots with respect to that of the fundamental spots decreases significantly, indicating that chemical disordering precedes amorphization. On the contrary, neither disordering nor amorphization is observed in NiAl.

SBS shows that in both compounds lattice dilation rapidly increases with irradiation, and it saturates at about 0.1 dpa, but it is by far larger (~ 6%) in FeTi than in NiAl (~ 2%). The shear modulus decreases accordingly, the reduction entity being about 40% in FeTi and less than 10% in NiAl. Plotting the shear modulus vs. lattice dilation, similarly to the Zr_3Al case, a linear decrease of modulus with increasing dilation is observed both in FeTi and in NiAl; the trend coincides with that observed in pure metals heated to the melting point. Thus a strong correlation is apparent between chemical disordering, lattice dilation, shear constant softening, and glass transition, which suggests that an elastic shear instability driven by lattice dilation and chemical disordering is a good candidate to trigger amorphization.

The results of MD simulation of chemical disorder induced amorphization [20] display an increase of the mean square atomic displacement (i.e., lattice dilation) that accompanies chemical disorder before vitrification. Because an increase of mean square displacement results in a shear modulus decrease, the agreement between model and experiment appears good. Yet it is noticeable that Nb_3Ir, irradiated with 1.8 MeV α-particles [73], shows a strong sound velocity change, corresponding to a 40% decrease in shear elastic constant C, although it remains crystalline even at the highest irradiation doses.

The softening trend of C in a crystalline material with antisite (chemical) disorder is represented as

$$C = C_1 + C_2 S^2 \tag{4}$$

with S LRO parameter; the equation gives an excellent fit to experimental data for both Nb_3Ir and Zr_3Al.

However, in the former case, the observed softening is consistent with irradiation-produced changes in the electronic DOS at the Fermi level, as experimentally deduced by changes in the superconducting transition temperature.

MD simulations [79] support the similarities between amorphization and melting; every crystal can in principle undergo two kinds of melting transition, with the same peculiar features observed in irradiation-induced vitrification of ordered alloys. The first kind involves heterogeneous nucleation of the liquid-glassy phase at extended lattice defects, such as grain boundaries, surfaces, voids, and dislocations. The interface between the new phase and the solid matrix propagates through the crystal by a thermally activated mechanism. On the opposite side, homogeneous melting, in the absence of thermally activated atomic mobility, arises from the Born mechanical instability [52]. Chemical and structural disordering upon irradiation can drive the crystal to a critical point of the volume-temperature map, where transformation to the glassy phase either by heterogeneous or by homogeneous nucleation is favored.

4   ION MIXING

4.1   Basic Concepts

The main, unavoidable drawbacks of ion implantation are the prolonged irradiation times required to introduce a significant impurity amount into the matrix and the often relevant sputtering of the constituents whose importance increases with beam energy and ultimately poses an unsurmountable limit to the maximum thickness of the treated surface layer.

Ion mixing (IM) allows the ability to overcome such problems. The idea is to irradiate a compositionally inhomogeneous target with inert noble gas ions with typical energies in the hundreds of keV range. A typical configuration is a multilayer $ABABA$.... No chemical effects are expected by such a doping, whereas spatial homogenization of the sample results in $A_{1-x}B_x$ compound, the relative constituent concentrations being adjustable by calibrating the thickness of the single A and B layers, usually taken in the 10 nanometer range.

As an order of magnitude, mixing a multilayer a few tens of nanometers thick requires fluences of a few 10^{19} Xe^+m^{-2}, considerably lower than typical implantation doses and corresponds to irradiation times of few minutes.

Although a detailed treatment of IM is offered in Chapter 13 (Nastasi & Mayer), the relevant processes that contribute to IM are recalled in Table 1, according to their typical timescales. Collision-induced, short-lived mechanisms are the same that operate in implantation. The intermediate and delayed time regimes are dominated by chemical-thermodynamic effects, ultimately determining the result of the vast majority of IM experiments performed at medium and at high temperatures. Interfaces play a major role in this kind of bombardment. Given an initially sharp *A-B* interface, its evolution can be characterized by an experimental mixing parameter Ω that gives the number of mixed atoms from an independent compositional profile obtained by Rutherford backscattering spectrometry (RBS)

$$\Omega = \frac{1}{\varepsilon'}\frac{\Delta\sigma^2}{\Delta\Phi} \tag{5}$$

where ε' is the total nuclear energy density deposited along the projectile track, which is estimated as the average nuclear stopping cross section at the *A-B* interface, σ^2 is the variance of the atomic distribution function at the same interface, and Φ is the ion fluence. The mixing rate is obtained by (5) through the slope of the σ^2 vs. Φ curve. In the literature, due to formal similarities to diffusion experiments, also the quantity

$$\frac{d(4Dt)}{d\Phi} = 2\frac{\Delta\sigma^2}{\Delta\Phi} \tag{6}$$

is used. In (6) Dt is the diffusion length, with t irradiation time; its proportionality to the fluence indicates the random nature of the process, while its amplitude depends on the amount of energy deposited through elastic collisions, i.e., the nuclear energy density. To compare mixing efficiencies under different experimental conditions, the most suitable parameter is F_D, the energy deposited per unit length in elastic collisions.

Temperature-independent mixing that exceeds ballistic mixing estimates was observed in several heavy metal bilayers (with average atomic number $\langle Z \rangle$ around 50, or more) irradiated at 77 K with 600-keV Xe[+][80]. Under the hypothesis that diffusion is strongly biased by chemical effects and assuming cylindrical shape for the thermal spike [81], the mixing rate is

$$\frac{\Delta\sigma^2}{\Delta\Phi} = \frac{K_1}{2}\frac{\varepsilon'^2}{\rho^{5/3}(\Delta H_c)^2}\left(1 + K_2\frac{\Delta H_m}{\Delta H_c}\right) \tag{7}$$

where σ, Φ, ε' have the usual meaning, ρ is the macroscopic material density, and ΔH_c is the system cohesive energy, given by the average over the constituent cohesive energies plus the heat of mixing ΔH_m.

Equation (7) provides a scaling function for the mixing rate vs. chemical driving forces; it yields agreement with experiment within about 25%, choosing as fit parameters $K_1 = 0.0034$ nm and $K_2 = 27$.

Under the approximate hypothesis of thermal equilibration within spike volume that allows using thermodynamic concepts such as temperature and activated diffusion, the relevant mixing appears to occur at average atom kinetic energies of the order of 1 eV. Such a mixing depends on almost the total nuclear energy deposited in the spike volume rather than on the contribution from elastic collisions, F_D.

The interplay of defect kinetics, thermodynamic factors, and interfaces was explored in a classical MD simulation of IM induced by 5-keV cascades across bilayer (111) interfaces in Co-Cu and Ni-Cu [82]. Cobalt and nickel have similar mass, and the melting point of Ni (1726 K) is nearly equal to that of Co (1768 K); however, the heat of mixing Ni with Cu, around +4 kJmol^{-1}, is consistently lower than that of Co with Cu, about +12 kJmol^{-1}.

During the intermediate regime, on the average, simulations show that more Cu impurities are found in Co than vice versa in all inspected cascades, regardless of where the recoil started. The effect can be understood considering that the cascade core is enriched in vacancies and interstitials are found toward the periphery; an analysis of defect locations showed that most Cu impurities in Co lie at the interface and close to the center of the cascade and to the vacancies. The impurity flux is balanced by a vacancy flux in the opposite direction; we are in the presence of an inverse Kirkendall effect.

A companion asymmetry is observed in Ni-Cu, again with more Cu impurities in Ni than vice versa. An explanation of the effect is provided by the difference between melting temperatures (T_m of Cu is 1357 K); the different T_ms originate different recrystallization rates of the metals because during cascade quenching, vacancies are pushed into the metal with lower T_m, while more atoms of the constituent with lower T_m are introduced in the high T_m metal.

The combined effect of the defect creation rate of the two constituents and of the attainment of the compound-critical solid solubility that in turn determines an asymmetric growth at the *A-B* interface was explored in IM experiments on Ni-Sc and Ni-W multilayers of different compositions [83].

IM allowed obtaining crystalline phases, solid solutions with solubility limits much extended with respect to equilibrium alloy formation techniques, and glassy phases. The first ion mixed amorphous alloys date back to the infancy of the technique [84]; since then, several tens of alloys, mostly binary, have been reported to amorphise under IM conditions.

4.2 Amorphization Mechanisms

Phase formation under IM depends at the same time on the irradiation induced excitation-relaxation dynamics of the target and on its chemical-thermodynamic properties. The former involves various mixing mechanisms (see Chapter 13) and their relative weight in different experimental conditions, while the latter requires to take into account the enthalpies of formation and migration of point defects, the features of the phase diagram of the system, as well as the formation and mixing enthalpies, ΔH_f and ΔH_m.

The equilibrium phase diagram of a two-constituent *A-B* system usually contains at specific compositions different intermetallic crystalline phases, each characterized by a value of the free energy *G* and by a more or less complicated spatial atomic arrangement with definite occupation probabilities for the atoms of each kind.

Schematically, the early irradiation stages of ballistic mixing evolve along lines similar to those described when discussing ion implantation; it is atomic interdiffusion during the thermal spike life in the intermediate regime to dominate phase formation during IM.

In general, given the small cascade size (≈ 10 nm) and its short life before quenching to ambient temperature (less than 10 ps), a kinetic frustration to the nucleation of new phases or to atomic rearrangements occurs. This leads to several effects of IM, namely, transition from complex crystalline structures to simpler ones (usually cubic, fcc, or bcc, with very few atoms per unit cell) [80]; transformation of a crystalline phase to an extended, disordered solid solution; and vitrification.

The reason why structures simpler than the initial target are attained is that the time required to form a nucleus of such a size that persists and grows scales with the *square* of the number of atoms contained in the nucleus, rather than with the number of contained atoms, as observed in polymorphic (i.e., without long-range diffusion) liquid-solid transitions.

The solution formed at high temperature within the cascade polymorphously freezes to ambient temperature: as time is lacking to allow for long-range diffusion before cascade freezing, stoichiometry changes are not possible with respect to the compositional profile evolved *inside* the cascade volume. Such a picture explains both dramatically extended solid solutions often reported and glass formation.

According to Gibbs criteria for thermodynamic stability, in the temperature *(T)*-composition *(c)* plane, the $T_o(c)$ curve is the locus of the points where the free energy curve for each crystalline phase, at a given *T* value, crosses the free energy curve for the liquid.

When a solution whose composition lies outside the corresponding $T_o(c)$ curve is formed in a thermal spike, thermodynamics indicates that a glassy phase is likely to occur. The evidence for local melting within dense cascades, with the liquid state persisting for several picoseconds (see Section 3.1), provides a scenario to interpret glass formation due to the fast quench rates (10^{14} Ks^{-1}) that prevent subsequent decomposition into crystalline phases. Besides, if irradiation temperature lies below the glass temperature $T_g(c)$ curve, nucleation and growth of intermetallics are thermally inhibited.

If the irradiation temperature is so low that interstitials and vacancies produced in the cascade cannot thermally migrate, every other kind of IM induced-phase transition other than vitrification is frustrated; indeed, at low irradiation temperatures amorphous phases easily form.

In principle, the only exception to the above phase formation rule is represented by systems with large and positive ΔH_m values whose solutions are expected to be immiscible even by the powerful mechanisms acting in a collision cascade.

Yet experiments specifically investigating IM in binary systems with large and positive ΔH_ms, which sometimes are immiscible even in the liquid state, produced either glassy phases or single-phase solid solutions or extensions of the terminal mutual solubilities, depending on irradiation temperature, starting composition of the multilayers, and projectile mass and energy [85]. A recent compilation of experimental results is available [86].

The major importance of temperature to define the attainment or not of glassy phases during IM was correctly considered by a theory of radiation-induced chemical disordering [87]; if the irradiation temperature T_{irr} is lower than the critical point at which RED (see Chapter 6) sets on, i.e., system mixing does not depend on temperature, attainment of a structurally disordered phase is easier.

Concerning the influence of T_{irr} on attainment and metastability of a glassy phase, some exemplary studies are available. Ge_4Au and Ge_3Au_2 glassy multilayers, obtained through Xe^+ irradiation, were found to recrystallize within a few days at room temperature [88]. Amorphous Ni_3Al recrystallizes after about 10 s at 473 K because of the probable persistence of mobile defects that promote fast recovery of the crystalline state [89]. Several systems were amorphized only at low temperature (at least 77 K, in several instances 4 K); among them, e.g., Al_3Fe [90], $Co_{5\div35}Zr_{95\div65}$ [91], Nb_4Zr [92], $Au_{65}Mo_{35}$ [93], Mo_4Ag [94], and Al_2Au [95]. It is noteworthy that Al_2Au was amorphized both by light projectiles (He^+) and by massive ones (Ar^+) at very low T_{irr}: on such grounds it was proposed that amorphous phase formation in ion-mixed binary metallic systems does not necessarily imply formation of dense cascades if the temperature is low enough. Thus role of the temperature should be to activate either the thermal spike mechanism or the progressive defect accumulation one to drive the transition to the glassy state.

4.3 Selected Experiments

The thermodynamic model of dense-cascade cooling [80], whose validity was discussed with reference to mixing mechanisms and their effectiveness (see Section 4.2), also plays a key role when phase formation and stability are analyzed as functions of such diverse IM conditions as projectile mass and energy, irradiation temperature, target constituents, and designed average sample stoichiometry; all of these parameters can be bselected over road ranges.

One of the main difficulties found when the validity of the Vineyard-Johnson model is investigated stems from the uncertainties in the numerical values of basic thermodynamic quantities entering the model, such as heats of formation and heats of mixing at various alloy compositions.

Nonequilibrium phase fields, calculated starting from reliable thermodynamic data, were compared with experimental phase formation to study the influence of thermodynamics on IM-induced phase formation in Al-Nb alloys [96]. Multilayered films of total thickness of 80 nm, composed of layers each not thicker than 15 nm, were irradiated with 500 keV Xe^+ in the temperature range from 40 K to 620 K at a fluence of 2×10^{20} i^+m^{-2} to achieve complete mixing. Examination by TEM of the samples indicates that the main process products is an extended bcc solid solution of Al in Nb and an amorphous phase, which is found at low T_{irr} values on the Al-rich side of alloy compositions.

The stoichiometry giving the most stable glassy phase is $Al_{64}Nb_{36}$, that is single phase and can be produced at T_{irr} as high as 570 K. The regions of amorphous and bcc phase are separated by a rather wide (near 15 at.%) two-phase region, which contains a relative amount of glassy phase continuously increasing with Al content, at any given T_{irr} value.

With the exception of $NbAl_3$, that forms at T_{irr} = 620 K, only supersaturated bcc and amorphous phases are found, depending on the kinetic constraints during irradiation. These are determined by the T_{irr} value and they change ongoing from low temperatures (40 K and 78 K) to high temperatures (above 300 K).

At low T_{irr} values, phase formation is basically determined by the processes occurring after the impact of every *single* bombarding particle, while at high temperatures, enhanced thermal diffusion caused by migrating, irradiation-induced defects is increasingly important. The wide two-phase region reported only for T_{irr} higher than room temperature was already observed during IM of Fe-Zr multilayers at Zr concentrations between 6 at.% and 18 at.% [96].

Intermetallic compound formation by irradiation is inhibited by the fact that Nb_2Al and Nb_3Al have complicated crystal structures, which is difficult to establish during IM due to kinetic constraints. $NbAl_3$ is an ordered, line compound. It is often observed that compounds with narrow phase fields are easily amorphized under irradiation [77] because local off-stoichiometry compositions produce large free energy increases. The latter, together with the high frozen defect concentration induced by mixing, destabilize the ordered structure of $NbAl_3$ unless T_{irr} is sufficiently high.

In the frame of the thermal spike model, vitrification is expected to be favored at low T_{irr} where it is attributed to a polymorphic phase transition during cascade quenching, thus in the absence of long-range diffusion.

Amorphization was also studied in the temperature-dependent mixing regime where significant interdiffusion occurs, and the characteristics of such a process were compared with those of amorphization induced by diffusion reactions in the solid state (SSAR) performed in thermal equilibrium [1].

In the latter case, there is a kinetic condition holding nearly always, namely, one of the binary alloy constituents is an anomalously fast diffuser in the other. In IM, it was often observed that only *one* moving species is responsible for long-range intermixing and also for amorphization that is likely to occur if the critical temperature for compound formation is higher than the critical temperature for the onset of temperature-dependent mixing.

For example, in the case of Mo-Al bilayered thin films, irradiated with Xe^+ at 473 K, where mixing efficiency is 20 times higher than in the temperature independent regime and Al is the only migrating species [97], RBS indicates compound formation. However, x-ray diffraction does not reveal any peak from Al-Mo equilibrium compounds, whereas poor diffraction contrast is seen under TEM. An SAED pattern shows only a diffuse band, corresponding to the formation of an amorphous alloy. On such an experimental basis, a glassy phase grown by temperature-dependent mixing results because only one species undergoes thermally activated, ion-enhanced motion, thus providing a kinetic condition similar to that observed in an SSAR. For this alloy, temperature-dependent mixing starts at room temperature, and the critical temperature for equilibrium compound formation under irradiation is near 450 K. Thus, this experiment suggests that the growth of a vitreous phase can be assisted by enhancing, through temperature-dependent IM, the long-range diffusion of the dominant moving constituent in an SSAR. In this case, the kinetic path toward amorphization is the opposite with respect to the well-established fast quenching during cascade cooling.

To compare and to sequentially use different amorphization techniques, an amorphous NiB alloy was formed by a thermally assisted reaction at the interfaces between the layers of polycrystalline Ni and glassy B during electron beam deposition of Ni-B-Ni trilayers. In addition, the vitrified alloy regions were considerably extended by postdeposition IM, due to the highly nonequilibrium process conditions. An upper limit of nearly 40 nm was found for the thickness of the formed amorphous NiB layers, indicating that the glassy layer operates as a reaction, or respectively, as a diffusion barrier [98].

Specific investigations were performed to compare phase formation in systems easily ($\Delta H_m < 0$) or hardly ($\Delta H_m \geq 0$) miscible in the solid state. An illustrative example [99] is provided by mixing multilayers of Hf-Co (ΔH_m :-32 kJ mol^{-1}), Cu-Nb (ΔH_m :+25 kJ mol^{-1}), Ag-Ta (ΔH_m :+15 kJ mol^{-1}), Ag-V (ΔH_m :+17 kJ mol^{-1}), and Ag-Fe (ΔH_m :+27 kJ mol^{-1}).

Heavy ions (Xe$^+$) at an energy of 500 keV and fluences of 2×10^{20} i$^+$m^{-2} were used over the T_{irr} range between 77 K and 600 K. Phase formation was studied as a function of T_{irr} and sample composition.

Irrespective of the heat of mixing sign, single-phase amorphous structures were formed, whose compositional range is limited by enhanced terminal solid solutions, while the compounds present in the equilibrium phase diagram do not form. On the extreme side of positive ΔH_m values, coexisting vitreous and crystalline structures were observed in Ag-V, while only a mixture of the terminal crystalline structures was found in Ag-Fe.

The thermal spike model [80] assumes that in low-temperature irradiations long-range diffusion is inhibited, and only polymorphic transitions are present because of the very fast quenching rates. On this basis the phase sequence is predicted, starting from the calculation of the T_0 curve. In particular, in systems hard to mix, only phase separation should occur. Clearly, the model appears to be too simplified to explain experimental results, since *two-phase* structures are observed, which inevitably require some kind of chemical diffusion, in its turn inconsistent with theoretical predictions. Incidentally, one of the most commonly proposed pictures of amorphous phase formation cannot be carried out in systems with positive heat of formation. Progressive energy accumulation leading to C-A transition when the free energy of the defected crystal is higher than that of the amorphous phase fails. This is caused by the absence of compounds, whereas the simple crystalline phases observed cannot store enough energy to enhance the free energy of the disordered crystal above the value of the corresponding vitreous system.

IM of Fe-Cu [94] shows, besides quite easy mixing, amorphization at Fe$_{70}$Cu$_{30}$ composition, at comparably low irradiation fluences. Again, the critical fluence to achieve glass formation significantly depends on sample composition: for a film of average Cu_4Fe stoichiometry, low fluence irradiation produces homogeneous mixing; at high fluences, several phases coexist, indicating that high-fluence irradiation leads to phase separation. In this case, glassy $Fe_{70}Cu_{30}$ localized amorphous islands are interspersed in the matrix and are stable up to room temperature.

The interplay of ion fluence and alloy initial average stoichiometry to determine the result of ion irradiations is evident in Ag-Mo. IM at 77 K of multilayered films with composition ranging from $Ag_{24}Mo_{76}$ to $Ag_{84}Mo_{16}$ was performed [94]. With a Xe^+ dose of 7×10^{19} i^+m^{-2}, the films resulted in either amorphous, single phase, or mixtures of amorphous and metastable fcc phases. It is noteworthy that one such mixed phase film with composition $Ag_{24}Mo_{76}$ stored for 3 months at room temperature, spontaneously turned into single-phase, amorphous.

On the contrary, a single-phase amorphous $Ag_{20}Mo_{80}$ film was unable to resist further irradiation at the comparatively low dose of 5×10^{19} i^+m^{-2}, transforming into a mixture of crystalline Ag and metastable fcc phase. A companion amorphous sample, after aging for 3 months at room temperature, dissociated into crystalline Ag and Mo, respectively.

Also in Ni-Mo, ion mixing at fluences between 7×10^{18} Xe^+m^{-2} and 1×10^{19} Xe^+m^{-2} produces uniform glassy films for compositions $Ni_{65}Mo_{35}$, $Ni_{55}Mo_{45}$, and $Ni_{45}Mo_{55}$. Increasing ion fluences leads to phase separation and self-organization of the precipitated crystalline phases in the latter two alloys [100]; in $Ni_{65}Mo_{35}$, loops of odd lines appear in the still-surviving vitrified matrix.

Phase formation in the Au-Rh system with positive ΔH_m (+10.7 kJ mol^{-1}) depends both on ion mass and energy [85]; upon irradiation with 400 keV Kr^+, a single-phase metastable solid solution forms in contrast to no mixing produced by 400 keV He^+ bombardment. On the contrary, 30-keV He^+ irradiation yields Au- and Rh-rich solid solutions. Such data are explained assuming that thermal spike temperature lowers with decreasing projectile mass. This implies reduced intermixing and diffusion-induced demixing by 400 keV He^+ because of a negative value of the diffusion coefficient in (7). Accordingly, the different mixing efficiency between low- and high-energy He^+ bombardments is attributed to ballistic mixing, usually overwhelmed by the thermal spike contribution whose magnitude is, however, comparable to that of thermal spike diffusion at 30 keV.

4.4 Bombardment-Induced Formation of Quasicrystalline Phases

In quasicrystalline systems [101] over certain compositional ranges, the atoms of an alloy can arrange themselves in a quasiperiodic lattice with no translational symmetry yet in the presence of orientational LRO with icosahedral symmetry. Icosahedral local atomic coordination is central to structural models of metal glasses. Thus the main interest to study the amorphous-quasicrystal (A-QC), as well as the reverse transition under irradiation, lies in the unique possibility to follow, through an intermediate, well-defined state, the crystal to glass path and to shed light onto important features of the amorphous state itself.

The most detailed investigations on the A-QC transition were performed by high-energy electron irradiation; for an immediate comparison, we refer to the discussion on amorphization by electron bombardment and to the associated mechanisms (see Section 2).

Briefly, destruction of chemical order with increasing facility, the more complex is the crystal structure of the irradiated compound is of prime importance to achieve amorphization [15]. When a threshold degree of chemical disorder is attained on a local scale, structurally ordered clusters are destabilized and disappear as structural units, ultimately leaving a dense random packing (DRP) of alloy atoms that coincides with the glassy state.

Unless the temperature is raised up to a point where substantial thermal diffusion sets in, the glassy state persists whenever large clusters, or large unit cells, must re-form to obtain a crystal or a quasicrystal. This is true even when a specific kind of order requiring consistent atomic rearrangements is required.

The central point is the loss of chemical order, preceding the disruption of ordered atomic clusters. The amorphous structure preserves, essentially for steric constraints, a degree of tetrahedral as well as of icosahedral order, both short (0.2–0.5 nm) and medium (0.5–2.0 nm) range; the same kind of local order is reproduced also by fast cooling simulations of model liquids [102].

The QC-A transition of irradiated samples is marked by the development of intense halos located at the angular positions of the most intense QC peaks, i.e., {100000}, {110000}, and {101000}. Such a coincidence confirms that the same kind of tetrahedral and icosahedral SRO is present both in the QC and in the A phase. The pair correlation functions $g(r)$ of both phases are remarkably similar up to the second nearest neighbors [103]. The glassy structure of the quasicrystal forming prototype system $Al_{84}V_{16}$ does not differ from that of conventional amorphous materials. The positions of $g(r)$ peaks agree within few percents with those calculated for a structural model of glass, based on interpenetrating chains of tetrahedral units biased toward icosahedral atomic arrangements.

It is remarkable that pronounced icosahedral correlations are found in the relaxed structure obtained from a DRP of hard spheres with Lennard-Jones potential. The weight of such correlations increases with increasing undercooling [102], thus suggesting that if the quench is fast enough to preserve icosahedral SRO, the latter determines the structure of the amorphous state.

The impressive correlation between the positions of halos in A samples and reflection polygons in QC ones does by no means indicate that the A stucture is to any extent quasicrystalline. Rather, such a coincidence reflects the basic geometrical principles underlying the most effective atom packing, as well as the spherical symmetry of pair potentials in metals.

Besides the above structural features of the QC state, some general conclusions drawn from the bulk of experimental results can help with understanding the interrelation between QC and A phases.

First, low temperature electron irradiation invariably triggers the QC-A transition. A threshold fluence Φ_t, depending both on T_{irr} and on electron energy, is found; below Φ_t, the QC structure is stable, while above Φ_t it transforms to the A state. In $Al_{86}Mn_{14}$ [104] the Φ_t trend vs. particle energy is schematized in Fig. 14; saturation is evident at energies above about 300 keV. The temperature dependence of amorphization is clearly similar to that observed usually for glass formation under electron irradiation, where a critical temperature is found. In $Al_{86}Mn_{14}$ irradiated with 300-keV electrons above 75 K, the QC structure is stable, irrespective of the irradiation dose.

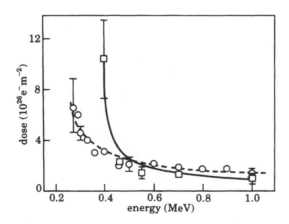

Figure 14. Stability of quasicrystalline vs. amorphous structure of $Al_{86}Mn_{14}$ (O; ---) and $Mg_{32}Al_{19}Zn_{30}$ (□;——), irradiated with electrons of different energy. T_{irr} = 20 K (adapted from [104]).

Second, if the glassy phase is properly annealed, the QC state results; the fact that during the transition the system avoids the crystalline state confirms that a preferential structural relation between QC and A phase exists.

Third, in several instances competition is observed in quasicrystal-forming systems between A and QC phase. High-quenching rates from the liquid phase favor A phase; lower-quenching rates result in a progressive development of QC structure through a nanometer-sized QC state. In vapor deposition, three temperature regions are defined: at high temperatures the samples turn out to be crystalline, in an intermediate temperature range they are nano-QC, and at low temperatures they have A structure.

A question immediately arises concerning the transition mechanism; this could involve grain growth, or alternatively, growth following nucleation with very high frequency. A careful characterization of the A-QC transition kinetics in $Al_{83}Mn_{17}$ and $Al_{72}Mn_{22}Si_6$ through isothermal DSC, TEM, and x-ray diffractometry [105] raised doubts about the easy qualification of the starting sputtered films as "amorphous." The monotonically decreasing calorimetric signal, typical of grain growth processes, indicates that the amorphous state is nano-QC with average grain size of about 2 nm. Thus the nucleation and growth process seems to be ruled out.

A further question concerns the way to describe atomic arrangements in an ensemble of QC grains when the correlation length is of the order of 1–2 Mackay icosahedra. Grain boundaries can be identified as planar discontinuities in the orientational order, but with increasingly finer grain size both the nano-QC phase and its A counterpart share the same *local* icosahedral order; it is more and more difficult to distinguish from each other.

QC phases were obtained through ion bombardment in several IM experiments on different Al-TM alloys. The most extensive investigations were performed on QC phase formation in ion-mixed AlMn alloys [106]; irradiating Al-Mn multilayers at a fluence of 1×10^{20} Xe^+m^{-2} results in complete, uniform mixing. At T_{irr} between 373 K and 473 K for Mn concentrations between 12 at.% and 21 at.%, a QC phase forms without any intermediate amorphous phase. At $T_{irr} = 333$ K an A phase results. Such a phase can be converted to QC by modest annealing in analogy to the transition path observed under electron irradiation. It is noteworthy that bombardment at 423 K of Al-Mn multilayers deposited on Si substrates induces moderate Si interdiffusion in the AlMn intermixed film, leading to formation of an A phase. Silicon is known to stabilize the vitreous AlMnSi phase; thus it also seems reasonable that when the QC phase forms directly, it is preceded by an A phase, which is, however, not stable enough to be observable.

The above experiments indicate that irradiation temperature and ion fluence play a concurrent role to QC phase formation. Indeed, low fluences, of about 2×10^{19} i^+m^{-2} produce phase mixtures due to inhomogeneous mixing of the samples, but if T_{irr} is high enough, even low doses may result in overall QC phase formation. Room temperature irradiation with Xe^+ of Al-Fe multilayers with carefully controlled average composition, chosen within a narrow interval between Al_4Fe and $Al_{86}Fe_{15}$, leads to an A phase. Upon modest annealing at 470 K, this is converted to QC $Al_{84}Fe_{16}$, while excess Al precipitates as pure metal [107].

A similar result was obtained on multilayers whose composition was centered around $Al_{65}Fe_{35}$ [108]. After Ar^+ irradiation at room temperature, film separation into a dominant equiatomic phase that contains many local areas of Al_4Fe stoichiometry occurs. SAED of such regions showed that they are globally amorphous with interspersed QC islands. No annealing of the amorphous structure was required, because QC $Al_{84}Fe_{16}$ directly precipitates from the matrix upon further irradiation. Such a feature seems to indicate that nano-QC instead of A phase was present in the films.

The peculiar relation between QC and A structure was explored in depth in $Fe_{70}Cu_{30}$ ion-mixed multilayers [109]. The starting Fe and Cu immiscible ($\Delta H_m = +13$ kJmol^{-1}) crystalline phases transform into a single $Fe_{70}Cu_{30}$ amorphous phase at a comparatively low dose of 8×10^{18} Xe^+m^{-2} and upon further irradiation to 5×10^{19} Xe^+m^{-2} quasicrystals locally grow within the amorphous regions where they are randomly distributed. HREM shows that such a QC phase is single phase, with composition $Fe_{50}Cu_{50}$; the observed QC grain size ranges between 4 and 10 nm.

At variance with the above-discussed A-QC phase sequence with increasing irradiation dose, in $Fe_{40}Nb_{60}$ and $Fe_{20}Ta_{80}$ multilayers that are ion mixed at room temperature with Xe^+, a phase sequence reversal is found [110]. At low doses, a single QC phase forms and further bombardment induces the transition from QC to A single phase.

5   MODELING STRUCTURAL STABILITY UNDER ION BOMBARDMENT

The empirical observation that some systems can be brought into the glassy state more easily than others stimulated research into criteria and models on the formation of noncrystalline phases that can both interpret and predict glass-formation ability. Behind these criteria there should be a firmly established glass transition theory that, however, is still lacking. This is why models have validity usually limited to certain classes of materials with no more than two constituents and to some specific amorphization processes. Also, in the case of particle bombardment, these models are more useful to interpret experimental data than to predict material behavior.

Before examining the interface migration-charge transfer (IMCT) atomistic model, we recall that kinetic and thermodynamic factors dominate when a system is amorphized by fast quenching, starting from the liquid phase. The connection between solid-state amorphization reactions and quenching processes from the liquid state is based on the fact that in the former heterogeneous nucleation of the noncrystalline phase is crucial. This is associated with the breakdown of the chemical LRO, similar to what is observed when a crystal melts.

The difference between the above two families of glass-forming processes is actually much less marked than it might appear, because first in many systems several traces of the short- and medium-range structural organization in the crystalline phase survive in the liquid phase. Second, in many amorphization reactions in the solid state, it is quite plausible for the system that is highly energized not to persist always in the solid state but for very short time intervals and at a local scale, to liquefy or even vaporize. In this sense MD simulations of collision cascades are exemplary.

Phenomenological models of structural stability can be divided into two broad categories: physicochemical models and thermodynamic-kinetic models. Attention was focused essentially on metallic binary systems for which various material parameters were considered. These include difference in atomic size [111], electronegativity [112], group number of compound constituents, degree of complexity of the compound crystal structure competing with the amorphous state, solute concentration [113], constituent crystal structure [114], position of the compound in the equilibrium phase diagram [115], width of the compositional range [78], estimated broadness of the glass formation range [116], and limiting values of the formation enthalpy of the compound ΔH_f [117]. Besides the minimum solute concentration criterion [113], amorphization in covalently bonded materials was rationalized using the degree of bond ionicity and the melting temperature [118].

All the above models are based on some physicochemical properties of compounds or of their constituents and do not specifically consider those mechanisms such as chemical disordering and defect formation, migration and accumulation, that operate during irradiation. Besides chemical disordering, another relevant mechanism of destabilization of the crystalline structure is *excess* defect accumulation, particularly vacancies and interstitials. Here the main difficulty is associated with the observation that the huge defect concentration required to induce amorphization (0.01–0.02) is near the athermal recombination limit. In homogeneous distributions of vacancies and interstitials throughout the target volume, each newly created defect should immediately recombine with a preexisting point defect. Point defect accumulation is favored whenever vacancies and interstitials are produced in separated regions during irradiation; this occurs when ion bombardment results in dense cascade formation. All these experimental facts were embodied in a model proposed for point-defect accumulation [119] whereby fast (normally, interstitials) defects annihilate at a surface, while slow ones (normally, vacancies) accumulate in the lattice, both mechanisms being temperature dependent.

A different representation of amorphization was provided to analyze bombardment-induced structural evolution in network forming nonmetallic compounds [120] where easy glass formation is related to material ability to organize itself in an extended, disordered, three-dimensional network with essentially the same SRO as its corresponding crystal. It is noteworthy that the small polyhedra that build up the network can be taken as clusters of atomic size, with a well-defined shape and stoichiometry. These clusters reciprocally arrange themselves to give extended, static configurations. The balance between the number of degrees of freedom of each polyhedron vertex and the number of constraints that result once the polyhedron is linked to its neighbors gives the degree of topological freedom. When this is positive, the structure is under-constrained and some arbitrariness of structural rearrangement is allowed, so that amorphization is favored over crystallisation.

We now turn our attention to the atomistic IMCT model that studies the structural stability of a target ion bombarded in the nuclear stopping regime, specifically considering the presence and evolution of dense collision cascades. We discuss the main features of the model and test its predictions with a representative set of metallic and nonmetallic binary compounds whose structural behavior under ion bombardment is known.

In the IMCT model, it is assumed that bombardment of an A_xB_y compound, called initial (*i*) compound, results in the development of displacement cascades along the projectile path. For about 10^{-13}s after ion impact (ballistic regime), each cascade is a region of concentrated displacement damage (see Table 1). Toward the end of this regime, the energy of higher-order recoils lies between the constituent displacement energies $E_d(A)$ and $E_d(B)$ (e.g., $E_d(A) < E_d(B)$) so that only atoms of the constituent with smaller E_d (in our example A atoms) are displaced. Thus the fraction of displaced A atoms, with respect to all displaced atoms, is greater than the atomic fraction of A constituent in the initial compound, namely, x. With increasing energy degradation, the target evolves to the thermal spike regime (10^{-11}s), characterized by a short-range random atomic motion and freezing of local, mutually competing, liquid-like configurations. The cooling rate, in the order of 10^{14} Ks^{-1}, provides suitable conditions to nucleate metastable, even amorphous phases. Each cascade, containing a few thousand atoms, is embedded in the crystalline solid

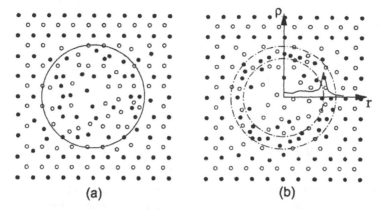

Figure 15. (a) Schematic in-plane view of a collision cascade in a binary equiatomic compound (ballistic regime). The region encircled by the line is the most strongly disordered. (b) Preferential short-range migration of one constituent (●) at cascade-matrix interface, as represented by the dashed ring. The transient density change across the cascade is shown (adapted from [42]).

matrix, adiabatically unperturbed during the cascade lifetime. Given the high number of incident ions, the initial cascade stoichiometry is assumed to coincide with the average compound stoichiometry because the exiguous cascade volume is considered to be compositionally homogeneous. Before studying cascade evolution, we observe that normally, equilibrium segregation makes the high temperature-surface composition of a binary compound different from the interior composition [121]. Also, when data for bombardment-induced Gibbsian segregation and for equilibrium segregation are available for a given compound, the segregating constituent is the same in both conditions and it is the one with the smaller surface tension, γ [122].

On such grounds, as schematized in Fig. 15, we assume that the matrix-cascade interface is equivalent to a free surface and that ion bombardment causes short-range migration to this interface of the constituent with smaller γ. The interface is a discontinuity between the crystal and the liquefied matter within the cascade volume [42], just as a free surface is a discontinuity between crystal and vacuum. Experiments [121, 122] indicate which is the migrating constituent. Interface enrichment causes a chemically different surface to form; this is a stabilizing process because the largest possible amount of energy is absorbed from the cascade region. Indeed the migrating constituent has the lower surface energy. As γ's scale with E_d's along the periodic table [123], a coherent explanation of interface enrichment is provided by both the above pictures. In the example,

$E_d(A)$ is lower than $E_d(B)$, and γ_A is lower than γ_B: thus the spike-matrix interface is enriched in A atoms. Interface enrichment evolves as

$$\frac{c_s^\infty(B)}{c_s^\infty(A)} = \frac{c_b^\infty(B)}{c_b^\infty(A)}\exp(-\Delta G^{seg}/k_B T) \qquad (8)$$

where the c_s^∞ s are the equilibrium surface concentrations of A and B and the c_b^∞ s are the bulk concentrations in the initial compound.

Thus, local atomic migration to the interface induces system energy lowering, formation of a composition gradient (the cascade core is depleted in the interface migrating constituent), and a nonequilibrium electronic density profile, caused by the composition gradient.

Reequilibration of the electronic density profile is simulated by elementary charge transfer reactions (CTRs) within dimers of atoms AB of the initial compound. The CTR scheme simulates the complicated mechanisms of charge reequilibration during spike lifetime, i.e., 10^{-12}-10^{-11}s. With the hypotheses that in each CTR *one* neutral A atom (atomic number Z_A) and *one* neutral B atom (atomic number Z_B) are involved and that *one* electron is transferred from the atom of the nonmigrating constituent to the atom of the interface migrating species, if γ_A is *less* than γ_B the CTR is

$$AB \xrightarrow{\gamma_A < \gamma_B} \begin{array}{l} A + 1e^- \to A^- \\ B - 1e^- \to B^+ \end{array} \qquad (9a)$$

Conversely, if γ_A is higher than γ_B

$$AB \xrightarrow{\gamma_A > \gamma_B} \begin{array}{l} A - 1e^- \to A^+ \\ B + 1e^- \to B^- \end{array} \qquad (9b)$$

This crude atomistic approach is justified by the high energy of the spike. A^-, B^+, A^+, B^- are singly charged ions. They are called effective elements and are introduced in the matrix in pairs, as dimers: these are the most elementary units of the equiatomic $[(Z_A + 1)(Z_B - 1)]$, or $[(Z_A - 1)(Z_B + 1)]$ compound, called the effective (*e*) compound. All CTRs tend to shell-closing in both atoms. Indeed γ values progressively diminish in elements with progressively filled electronic shells. Notice that a CTR occurs also between elements with the same degree of outer shell filling due to their different electronegativities [125].

The energy cost $\Delta E_{e^-}(i \to e)$ to create an effective compound dimer is calculated using elemental electron energies [125]. For example, for the CTR in (9a)

$$\Delta E_{e^-} = \Delta E(A \to A^-) + \Delta E(B \to B^+), \qquad (10a)$$

where the two terms on the r.h.s. have the same structure, and, for example,

$$\Delta E(A \to A^-) = \Delta E[Z_A \to (Z_A + 1)] = \sum_j \frac{E_j(Z_A + 1) - E_j(Z_A)}{n_j(Z_A + 1) + n_j(Z_A)}. \qquad (10b)$$

Each energy term in (10b) is the difference between the energies, E_j, of the valence electrons of type j ($j = s,p,d,f$) in $(Z_A + 1)$ and in Z_A, normalized to the total number of j electrons in the two elements. One considers s and d, and s and p electrons when initial compound constituents are part of groups IA and IIA, IIIA to IIB and IIIB to VIIB, respectively, in the periodic table. Positive ΔE_{e^-}s $(i \to e)$ values mean that the nucleation of effective compound dimers rises system energy with a destabilizing effect. Each dimer is an amorphization nucleus. Negative ΔE_{e^-}s mean that each effective compound dimer is a crystallization nucleus, contributing to system stability. A dimer is indeed small as compared with the size of stable nuclei able to survive thermal fluctuations; however, the IMCT model points at the beginning of the phase nucleation process over the very short timescale of cascade life. At a later stage it is likely that dimers coalesce to form bigger, more stable entities with a favorable surface to volume ratio. Also, we observe that the use of atomic energy levels, as well as the consideration of *free* atoms in (9), results from the fact that atoms in a high-energy cascade, including the thermal spike stage, have no memory of any solid structure and behave as free atoms.

Notice that the compositional changes assumed by the IMCT model require definitely short-range atomic migrations over distances at most in the order of cascade radius; thus, the atomic migration timescale is consistent with cascade lifetime [42]. Indeed, MD simulations of thermal spike evolution in copper show a matter wave propagating toward cascade periphery with a typical speed of 4 nm ps^{-1}, as shown in Fig. 15b. This mechanism is completely different from those that govern ion mixing, whose maximum efficiency is associated to *long-range* atomic migration, driven by defect migration, with timescales of many seconds after cascade formation. It is not immediate that the MD results in metals can be extended to borides, carbides, nitrides, and oxides; such an extension is supported by experiments and simulations on selected compounds. Au^{2+} bombardment of SiC, a material where thermal spikes are very brief, is exemplary;

amorphous clusters are formed *in* the cascade and are not recovered during annealing at temperatures as high as 1300 K [126].

The second step of the IMCT model considers the trend of variation of a global thermochemical parameter. Given the enthalpies of formation, $\Delta H_{f,i}$ of the initial compound and $\Delta H_{f,e}$ of the equiatomic effective compound, the enthalpy change $\Delta(\Delta H_f)$ associated with effective compound nucleation is calculated

$$\Delta(\Delta H_f) = \Delta H_{f,e} - \Delta H_{f,i}. \tag{11}$$

To use standard enthalpy values is in principle inadequate to the high temperature and pressure conditions in a cascade-thermal spike; their use is forced by the lack of thermodynamic data for the cascade regime. Positive $\Delta(\Delta H_f)$s indicate system destabilization (amorphization), whereas negative parameters correspond to system stabilization (crystallization). Indeed when $\Delta(\Delta H_f) < 0$, the *two* formation reactions are globally exothermic, thus corresponding to $\Delta G < 0$ and to a stabilizing entropy increase of the surroundings of the reacting system. When $\Delta(\Delta H_f) > 0$, the reaction is associated to $\Delta G > 0$, corresponding to the observed metastability of the amorphous state.

The third step of the IMCT model is to estimate the local volume change associated to ion formation by a CTR. Given the prototypical unit of the $A_x B_y$ crystal (e.g., $x < y$), with volume V_u [128], a CTR (9a) yields x couples of ions $(A^- + B^+)$, with volumes, V_{A^-} [129] and V_{B^+} [130] and $(y - x)$ B neutrals, with volume V_B [130]. The volume of the products of each CTR, V_{CTR} is

$$V_{CTR} = x\,(V_{A^-} + V_{B^+}) + (y - x)\,V_B. \tag{12}$$

In metals, neutrals are not considered because the system can resist ionization effects due to bond nondirectionality. The absolute value of the volume change $|\Delta V|$ in metallic alloys, with small volume changes, and of the relative volume change $|\Delta V|_{rel}$ in nonmetallic compounds, is calculated by comparing V_{CTR} and V_u

$$|\Delta V| = |\Delta V|. \tag{13a}$$

$$|\Delta V|_{rel} = |(V_{CTR} - V_u)/\,V_u|. \tag{13b}$$

Large local volume changes are associated to material destabilization.

Tables 2 and 3 are separately listed for metallic and nonmetallic binary compounds, initial compound, amorphized or crystalline upon ion bombardment, interface-migrating constituent, effective compound resulting from CTR, $|\Delta V|$, or respectively $|\Delta V|_{rel}$, $\Delta E_{e\text{-}}(i \rightarrow e)$, $\Delta(\Delta H_f)$.

Table 2 considers a representative set of 20 metallic compounds, 10 of which are reported in the literature to turn amorphous under some specific bombardment conditions. Simple, transition, and polyvalent metals are considered as well as compounds with formation enthalpies ranging from strongly negative to largely positive. In this last case in equilibrium conditions, the compound constituents are immiscibile even in the liquid state. Table 3 lists a set of 20 nonmetallic compounds, including borides, carbides, nitrides and oxides, 10 of which are amorphized by ion bombardment.

Looking back at our discussion, we recall that amorphization susceptibility is deeply affected by the chemical nature of the bombarding ions, their energy, the dose, and temperature of irradiation. In particular, some compounds, both metallic and nonmetallic, can be amorphized only at cryogenic temperatures, while they resist amorphization at higher temperature. Also, a peculiar dose dependence is sometimes found; a material can be completely, or partially amorphized when irradiated up to an upper threshold dose beyond which recrystallization is observed. For contrasting experimental data, in the IMCT model we assume that a compound is amorphized when at least under a specific set of experimental conditions partial, or complete amorphization was observed. A thorough exploration of irradiation conditions is not presently available for all compounds, so that a compound considered crystalline on the basis of present experimental data could be amorphized in a future experiment.

Radiation-amorphized compounds, both metallic and nonmetallic, have positive values of $\Delta E_{e\text{-}}(i \rightarrow e)$ and $\Delta(\Delta H_f)$, but for crystalline compounds, values of both quantities are negative, in agreement with model predictions. $\Delta E_{e\text{-}}(i \rightarrow e)$ is represented versus $\Delta(\Delta H_f)$ in Figs. 16a and 16b, respectively, for metallic and nonmetallic materials. In both figures filled symbols refer to compounds amorphized by irradiation, and all fall in the region of *positive* parameter values, but all data for crystalline compounds (open symbols) lie in the region of *negative* parameter values.

The trend of the electronic energy cost necessary to form an effective dimer agrees with the trend of a global thermochemical parameter such as $\Delta(\Delta H_f)$; this lends support to the idea that thermodynamic forces, here embodied in compound formation enthalpy, are effective in the late stages of collision cascade evolution.

Table 2. Collection of experimental data and calculated IMCT model parameters for ion-bombarded binary metallic compounds (a: amorphized; c: crystalline). (1): two-electron transfer in CTR.

Initial (i) Compound	Ref.	Migrating Constituent		Effective (e) Compound	$|\Delta V|$ $(\times 10^{-3} nm^3)$	ΔE_e- (i→e) (eV)	$\Delta(\Delta H_f)$ (eV at^{-1})
a – $Cr_{67}Zr_{33}$	[132]	Zr	[123]	V Nb	22.1	+0.43	+0.15
a – $Fe_{67}Cr_{33}$	[133]	Fe	[134]	Co V	34.3	+0.39	+0.20
a – $Fe_{75}Mo_{25}$	[135]	Fe	[136]	Co Nb	17.5	+0.08	+0.14
a – $Ta_{71;87}Ag_{29;13}$	[99]	Ag	[123]	Hf Cd	23.7	+0.58	+0.14
a – $Gd_{90-55}Co_{10-45}$	[137]	Gd	[131]	Tb Fe	15.6	+0.32	+0.26
a – $In_{75}Pd_{25}$	[138]	In	[122]	Sn Rh	10.2	+1.17	+0.08
a – $Nb_{50}Cr_{50}$	[139]	Cr	[123]	Zr Mn	40.3	+0.80	+0.15
a – $Pd_{70}Bi_{30}$	[140]	Bi	[122]	Ag Pb	26.0	+0.97	+0.24
a – $Ti_{52}Re_{48}$	[141]	Ti	[123]	V W	17.8	+0.15	+0.37
a – $Y_{60;56}Nb_{40;44}$	[142]	Nb	[123]	Sr Mo	23.8	+0.48	+0.78
c – $Ag_{70}Ir_{30}$	[143]	Ag	[131]	Cd Os	8.6	-0.17	-0.08
c – $Al_{87}Cr_{13}$	[144]	Al	[145]	Si V	10.7	-0.25	-0.44
c – $Co_{98-90}Ta_{2-20}$	[146]	Co	[123]	Ni Hf	6.2	-0.04	-0.36
c – $Cu_{75}Ir_{25}$	[143]	Cu	[123]	Zn Os	6.7	-0.17	-0.14
c – $Fe_{60}Co_{40}$	[147]	Fe	[148]	Ni Mn (1)	11.1	-0.85	-0.06
c – $Fe_{90}In_{10}$	[149]	In	[131]	Mn Sn	9.5	-0.16	-0.33
c – $Pd_{75}Ta_{25}$	[150]	Pd	[123]	Ag Hf	5.1	-0.23	-0.07
c – $Ti_{75}Ru_{25}$	[151]	Ti	[123]	V Tc	4.2	-0.08	-0.13
c - $V_{85}Ag_{15}$	[99]	Ag	[131]	Ti Cd	7.1	-0.69	-0.31
c - $Zr_{75}Y_{25}$	[152]	Y	[123]	Sr Nb	13.7	-0.36	-0.38

If we now consider volume changes, in Table 2 the threshold volume change value $|\Delta V|_t$ of $15 \times 10^{-3} nm^3$ [165] is confirmed; all amorphized systems show $|\Delta V|$ values greater than $|\Delta V|_t$, with the exception of $In_{75}Pd_{25}$, whereas $|\Delta V|$ is lower than the threshold for all crystalline alloys. In Table 3, a $|\Delta V|_{rel}$ threshold of 0.30 divides relative volume changes in amorphized nonmetallic compounds from $|\Delta V|_{rel}$ values in crystalline compounds. In the former, $|\Delta V|_{rel}$ values are greater than the threshold; in the latter they are lower than the threshold. Although the threshold parameter value does not bear any particular meaning, the indication that amorphization is associated with a local dilation exceeding an upper limit is in agreement with experimentally observed strains as high as 15% in vitrified alloys [120]. In Figs. 17a and 17b $\Delta(\Delta H_f)$ is represented against $|\Delta V|$ for all compounds considered here.

Table 3. Collection of experimental data and calculated IMCT model parameters for ion-bombarded binary nonmetallic compounds (a: amorphized; c: crystalline. (1): two-electron transfer in CTR. (2): ΔH_f not available for AuB.

Initial (i) Compound	Ref.	Migrating Constituent		Effective (e) Compound	$|\Delta V|_{rel}$	ΔE_{e^-} (i→e) (eV)	$\Delta(\Delta H_f)$ (eV at⁻¹)
a – FeB	[153]	B	[122]	Mn C	0.70	+0.61	+0.23
a – MoB	[154]	B	[122]	Nb C	0.75	+0.49	+0.10
a – TiB₂	[155]	Ti	[122]	V Be	0.65	+6.25	+2.58
a – B₄C	[156]	B	[122]	N Be (1)	0.38	+0.38	+0.16
a – Cr₃C₂	[157]	Cr	[158]	Mn B	0.65	+0.29	+0.09
a – GaN	[159]	N	[122]	Zn O	1.90	+0.31	+0.44
a – Si₃N₄	[160]	N	[122]	Al O	1.65	+1.52	+3.13
a – Bi₂O₃	[118]	O	[122]	Pb F	0.33	+2.42	+0.80
a – Fe₂O₃	[161]	O	[122]	Mn F	0.41	+1.79	+0.20
a – SiO₂	[118]	O	[122]	Al F	0.32	+0.56	+0.51
c – AuB	[154]	Au	[122]	Hg Be	0.28	-0.48	– (2)
c – ZrB	[154]	B	[122]	Y C	0.26	-0.55	-0.11
c – MoC	[154]	Mo	[122]	Tc B	0.15	-0.16	-0.27
c – TiC	[155]	C	[158]	Sc N	0.17	-0.93	-1.34
c – WC	[154]	W	[122]	Re B	0.13	-0.56	-0.13
c – Cr₂N	[162]	N	[122]	V O	0.13	-0.91	-3.24
c – Zr₃N₄	[163]	N	[122]	Y O	0.12	-0.95	-8.15
c – Cu₂O	[118]	O	[122]	Ni F	0.09	-0.22	-2.87
c – TiO	[118]	O	[122]	Sc F	0.23	-0.24	-1.87
c – Y₂O₃	[164]	O	[122]	Sr F	0.30	-0.39	-2.36

We recall that an old glass formation criterion developed for binary metallic compounds [166] relies on the idea that the crystal is destabilized by local lattice deformations. The relative atomic volume difference between initial compound constituents is considered. A large atomic volume mismatch is supposed to give rise to steric hindrances. The latter inhibit the formation of specific crystal structures during structural relaxation on fast cooling from the melt. On the contrary, in the IMCT model the volume occupied by a prototypical unit of unirradiated crystalline compound is compared with the volume of compound constituents after dissociation and ionisation processes occurred at cascade-matrix interface.

The physical picture underlying IMCT model deserves a few comments. First, because the displacement cascade and its structural evolution play a central role to the model we are interested in the very atomic structure of a cascade and its time evolution, including both the

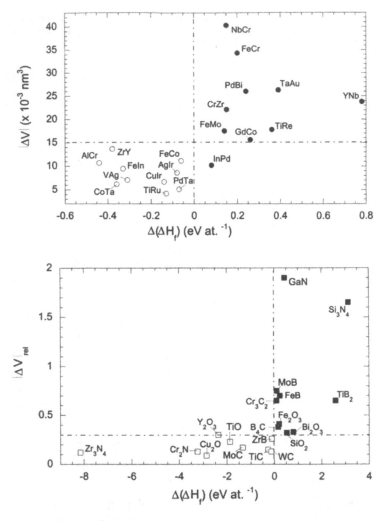

Figure 17. (a) Relationship between $\Delta(\Delta H_f)$ and $|\Delta V|$ for vitrified (filled circles) and crystalline (open circles) ion-bombarded binary metallic compounds. In all compounds the second element is the solute (see Table 2). (b) Relationship between $\Delta(\Delta H_f)$ and $|\Delta V|_{rel}$ for vitrified (filled squares) and crystalline (open squares) ion-bombarded binary nonmetallic compounds (see Table 3).

ballistic and the thermal spike stages. MD simulations [37, 42] show that the cascade is a liquefied matter droplet at high temperature and very likely at high pressure. The structural evolution of this system depends on the competition between the rapidly increasing (viscosity increases) relaxation time τ and the recrystallization rate. The first is related to the leading atom rearrangement mechanism toward low-energy disordered configurations. Regarding recrystallization, it is reasonable that it is initiated at the cascade-crystalline matrix interface. No essential differences are to be expected between a fast- quenching liquid and cascade quenching as far as we are concerned with the mechanisms that govern the bifurcation of the liquid toward either a crystal, or a supercooled liquid, ultimately resulting in an amorphous structure. Although only a few studies are available, a convincing proof that the mechanisms involved in liquid quenching and in cascade quenching are the same is provided by the essentially indistinguishable structure of amorphous samples of the same material prepared by the two techniques [167]

Second, in the IMCT model each cascade is associated to a crystal-matrix interface where most physical phenomena occur. Interface importance is expected because in a collision cascade the surface-to-volume ratio is large. Epitaxial crystal growth presumably starts at this interface, because the barrier to crystal nucleation is nearly zero. However, in our model localized compositional changes driven by preferential short-range interface migration of one compound constituent occur. These compositional changes are at the origin of the structural (liquid to amorphous, or liquid to crystal respectively) evolution of matter in the cascade volume. That a surface chemical transition, such as an interface enrichment, can give rise to a structural transition was observed experimentally [168] and simulated theoretically [169].

Finally, in the IMCT model both the crystalline and the amorphous nuclei that are formed, given their very small size, are supposed to undergo strong space-time fluctuations when they arrange themselves in the extended crystalline or amorphous structure, as observed [29]. This is the result of the statistical nature of configuration evolution in the fast-cooling liquid. Thus many local atomic configurations with similar energy and different structure compete with each other, and the amorphous state is a frozen in statistical distribution of such microscopic configurations. This description finds support in recent Raman spectroscopy investigations of damage onset in implanted diamond [170]. Experiments show that amorphous clusters of atomic size nucleate in the damaged, but still mainly single-crystalline bombarded volume, whereas point defects are interspersed between the amorphous clusters and the largely undamaged diamond lattice. The number and size of amorphous zones increase with advancing target irradiation.

In conclusion, energy, enthalpy, and volume changes induced by a CTR all show thresholds that separate crystalline from amorphized binary ion-bombarded compounds. The theory that points at dimers is a simplified approach to the entangled problem of phase stability under irradiation. Yet general trends in compound properties are envisaged by calculating quantities with a clear physical meaning. Because the basic mechanisms that govern cascade evolution in nonmetallic compounds are the same as in metals over the timescale of IMCT model evolution, the model provides a unified framework to study structural evolution under ion bombardment in both classes of materials.

REFERENCES

1. A. Hamberg, Geol. Förh. **36** (1914) 31

2. R.B. Schwarz and W.L. Johnson, Phys. Rev. Lett. **51** (1983) 415

3. J. Bloch, J. Nucl. Mater. **6** (1962) 203

4. U. Görlach, P. Ziemann, and W. Bockel, Nucl. Instrum. Meth. **209–210** (1983) 235

5. J.W. Mayer, in *Radiation Effects in Semiconductors*, J.R. Holland and G.D. Walkins, editors (Gordon & Breach, New York, 1971) 367

6. G. Thomas, H. Mori, H. Fujita, and R. Sinclair, Scr. Metall. **16** (1982) 589

7. D.E. Luzzi, H. Mori, H. Fujita, and M. Meshii, Acta Metall. **34** (1986) 629

8. H. Mori, H. Fujita, M. Tendo, and M. Fujita, Scr. Metall. **18** (1984) 783

9. D.E. Luzzi, H. Mori, H. Fujita, and M. Meshii, Scr. Metall. **19** (1985) 798

10. D.E. Luzzi and M. Meshii, Res. Mech. **21** (1987) 207

11. S. Banerjee, K. Urban, and M. Wilkens, Acta Metall. **32** 299 (1984)

12. J. Koike, P.R. Okamoto, M. Meshii, J. Non-Cryst. Sol. **106** (1988) 230

13. A.T. Motta, L.M. Howe, and P.R. Okamoto, Mater. Res. Soc. Symp. Proc. **373** (1995) 183

14. J. Koike, D.E. Luzzi, P.R. Okamoto, and M. Meshii, Mater. Res. Soc. Symp. Proc. **74** (1987) 425

15. T. Sakata, H. Mori, and H. Fujita, Acta Metall. Mater. **39** 425 (1991) 817

16. J.L. Brimhall, J. Mater. Sci. **19** (1984) 1818

17. D.E. Luzzi, H. Mori, H. Fujita, and M. Meshii, Scr. Metall. **18** (1984) 957

18. M.K. Miller, E.A. Kenik, and T.A. Zagula, J. Phys. (Paris) **48** C-6, (1987) 385

19. C. Massobrio, V. Pontikis, and G. Martin, Phys. Rev. B **41**, (1990) 10486

20. M.J. Sabochick and N.Q. Lam, Phys. Rev. B **43** (1991) 5243

21. R. Devanathan, N.Q. Lam, M.J. Sabochick, P.R. Okamoto, and M. Meshii, Mater. Res. Soc. Symp. Proc. **235** (1992) 539

22. G. Xu, P.R. Okamoto, L.E. Rehn, and M. Meshii, Mater. Res. Soc. Symp. Proc. **235** (1992) 317

23. H. Mori, H. Fujita, and M. Fujita, Jpn. J. Appl. Phys. **22** (1983) L94

24. J. Koike, P.R. Okamoto, L.E. Rehn, and M. Meshii, Metall. Trans. A **21** (1990) 1799

25. K.H. Robrock and W. Schilling, J. Phys. F **6** (1976) 303

26. Y. Limoge and A. Barbu, Phys. Rev. B **30** (1984) 2212

27. H. Mori and H. Fujita, in *Proc. Yamada Conf. on Dislocations in Solids,* H. Suzuki, editor (University of Tokyo, Tokyo 1985) 563

28. D.E. Luzzi, J. Mater. Res. **6** (1991) 2059

29. S. Watanabe, H. Takahashi, and N.Q. Lam, Mater. Res. Soc. Symp. Proc. **792** (2004) R.8.6.1

30. H. Mori and H. Fujita, in *Ordering and Disordering in Alloys,* A.R. Yavari, editor (Elsevier, London 1992) 277

31. B. Rauschenbach and K. Hohmuth, Phys. Stat. Sol. A **75** (1983) 159

32. P.V. Pavlov, E.I. Zorin, D.I. Tetelbaum, V.P. Lesnikov, V.P. Ryzhkov, and A. V. Pavlov, Phys. Stat. Sol. A **19** (1973) 373

33. R. Andrew, W.A. Grant, P.J. Grundy, J.S. Williams, and L.T. Chadderton, Nature **262** (1976) 380

34. J.B. Gibson, A.N. Goland, M. Milgram, and G.H. Vineyard, Phys. Rev. **120** (1960) 1229

35. T. Diaz de la Rubia, R.S. Averback, R. Benedek, and W.E. King, Phys. Rev. Lett. **59** (1987) 1930

36. M.-J. Caturla, T. Diaz de la Rubia, and G.H. Gilmer, Mater. Res. Soc. Symp. Proc. **316** (1994) 141

37. R.S. Averback and T. Diaz de la Rubia, in *Solid State Physics* **51** H. Ehrenreich and F. Spaepen, editors (Academic Press, San Diego, 1998) 282

38. T. Diaz de la Rubia, R.S. Averback, R. Benedek, and I.M. Robertson, Rad. Eff. Def. Sol. **113** (1990) 39

39. M.I. Current, C.-Y. Wei, and D.N. Seidman, Phil. Mag. A **47** (1983) 407

40. T.E. Faber, in *Physics of Metals,* vol. 1, J.M. Ziman, editor (Cambridge University Press, Cambridge 1969) 284

41. T. Diaz de la Rubia and M.W. Guinan, Phys. Rev. Lett. **66** (1991) 2766

42. H. van Swigenhoven and A. Caro: Phys. Rev. Lett. **70** (1993) 2098

43. S. Ishino, Rad. Eff. Def. Sol. **113** (1990) 29

44. M.L. Jenkins and M. Wilkens: Phil. Mag. A **34** (1976) 1155

45. T. Diaz de la Rubia and G.H. Gilmer, Phys. Rev. Lett. **74** (1995) 2507

46. D.A. Thompson, Radiat. Eff. **56** (1981) 105

47.   F.F. Komarov and N.V. Moroshkin, Rad. Eff. Lett. **86** (1984) 133

48.   G. Carter, Nucl. Instr. Meth. **209–210** (1983) 1

49.   B. Rauschenbach, Nucl. Instr. Meth. B **58** (1991) 283

50.   A. Seidel, G. Linker, and O. Meyer, J. Less. Comm. Met. **145** (1988) 89

51.   A.V. Drigo, M. Berti, A. Benyagoub, H. Bernas, J.C. Pivin, F. Pons, L. Thomè, and C. Cohen, Nucl. Instr. Meth. B **19–20** (1987) 533

52.   P.C. Liu, P.R. Okamoto, N.J. Zaluzec, and M. Meshii, Phys. Rev. B **60** (1999) 800

53.   D.F. Pedraza, Mater. Sci. Eng. **90** (1987) 69

54.   D.M. Parkin and R.O. Elliott, Nucl. Instr. Meth. B **16** (1986) 193

55.   P. Moine, J.P. Riviere, M.O. Ruault, J. Chaumont, A. Pelton, and R. Sinclair, Nucl. Instr. Meth. B **7–8** (1985) 20

56.   R.O. Elliott, D.A. Koss, and B.C. Giessen, Scr. Metall. **14** (1980) 1061

57.   S. Klaumünzer and G. Schumacher, Phys. Rev. Lett. **51** (1983) 1987

58.   S. Klaumünzer, Ming-Dong Hou, and G. Schumacher, Phys. Rev. Lett. **57** (1986) 580

59.   A. Audouard, S. Balanzat, S. Bouffard, J.C. Jousset, A. Chamberod, A. Dunlop, D. Lesueur, G. Fuchs, R. Spohr, J. Vetter, and L. Thomè, Phys. Rev. Lett. **65** (1990) 875

60.   C. Trautmann, S. Klaumünzer, and H. Trinkhaus, Phys. Rev. Lett. **85** (2000) 3648

61.   P.M. Ossi, Phil. Mag. B **74** (1997) 541

62.   P.M. Ossi, Nucl. Instr. Meth. B **209** (2003) 55

63.   A. Barbu, A. Dunlop, D. Lesueur, and G. Jaskierowicz in *Ordering and Disordering in Alloys*, A.R. Yavari, editor (Elsevier, London 1992) 295

64.   B. Rauschenbach, G. Otto, K. Hohmuth, and V. Heera, J. Phys. F **17** (1987) 2207

65.   C.A. Majid, Phil. Mag. A **61** (1990) 769

66.   B.G. Eristavi, E.M. Diasamidze, R.N. Dekanosidze, N.I. Maisuradze, E.R. Kutelia, A.V. Sichinava, ans N.E. Menabde, Acta Metall. Mater. **39** (1991) 1703

67.   J.F. Gibbons, Proc. IEEE **60** (1972) 1962

68.   J. Eridon, G.S. Was, and L.E. Rehn, J. Mater. Res. **3** (1988) 626

69.   C. Jaouen, M.O. Ruault, H. Bernas, J.-P. Riviere, and J. Delafond, Europhys. Lett. **4** (1987) 1031

70.   J.P. Riviere, M.O. Ruault, M. Schack, and J. Chaumont, Radiat. Eff. **79** (1983) 275

71. P.J. Masziasz, D.F. Pedraza, J.P. Simmons, and N.H. Packan, J. Mater. Res. **5** (1990) 932

72. J. Koike, P.R. Okamoto, L.E. Rehn, R. Bhadra, M.H. Grimsditch, and M. Meshii, Mater. Res. Soc. Symp. Proc. **157** (1990) 777

73. M. Grimsditch, K.E. Gray, R. Bhadra, R.T. Kampwirth, and L.E. Rehn, Phys. Rev. B **35** (1987) 883

74. P.R. Okamoto, L.E. Rehn, J. Pearson, R. Bhadra, and M.H. Grimsditch, J. Less Comm. Met. **140** (1988) 231

75. R.W. Cahn and W.L. Johnson, J. Mater. Res. **1** (1986) 724

76. R. Devanathan, N. Yu, K.E. Sickafus, M. Nastasi, M. Grimsditch, and P.R. Okamoto, Phil. Mag. B **75** (1997) 793

77. M.G. Beghi, C.E. Bottani, P.M. Ossi, R. Pastorelli, M. Poli, B.K. Tanner, and B.X. Liu, Surf. Coat. Technol. **100–101** (1998) 324

78. J.L. Brimhall, H.E. Kissinger, and L.A. Charlot, Radiat. Eff. **77** (1983) 237

79. D. Wolf, P.R. Okamoto, S. Yip, J.F. Lutsko, and M. Kluge, J. Mater. Res. **5** (1990) 286

80. W.L. Johnson, Y.T. Cheng, M. Van Rossum, and M.-A. Nicolet, Nucl. Instr. Meth. B **7–8** (1985) 657

81. G.H. Vineyard, Radiat. Eff. **29** (1976) 245

82. K. Nordlund and R.S. Averback, Phys. Rev. B **59** (1999) 20

83. L. Hu, Z.F. Li, W.S. Lai, and B.X. Liu, Nucl. Instr. Meth. B **206** (2003) 127

84. B.Y. Tsaur, S.S. Lau, L.S. Hung, and J.W. Mayer, Nucl. Instr. Meth. **182–183** (1981) 67

85. W. Hiller, M. Buchgeister, P. Eitner, K. Kopitzki, V. Lilienthal, and G. Mertler, Nucl. Instr. Meth. B **48** (1990) 508

86. B.X. Liu, W.S. Lai, and Q. Zhang, Mater. Sci. Eng. Rep. (2000) 1

87. R.H. Zee and P. Wilkes, Phil. Mag. A **42** (1980) 463

88. B.X. Liu and M.-A. Nicolet, Phys. Stat. Sol. A **70** (1982) 671

89. G.S. Was and J.M. Eridon, Nucl. Instr. Meth. B **24–25** 557 (1987)

90. C. Jaouen, J.P. Riviere, J. Delafond:,Nucl. Instr. Meth. B **19–20** (1987) 549

91. J. Bøttiger, K. Pampus, and B. Torp, Nucl. Instr. Meth. B **19–20** (1987) 696

92. A. Cavalleri, F. Giacomozzi, L. Guzman, and P.M. Ossi, J. Phys.: Condens. Matt. **1** (1989) 6685

93.   F. Pan and B.X. Liu, J. Phys.: Condens. Matt. **8** (1996) 383

94.   O. Jin, Z.J. Zhang, and B.X. Liu, Appl. Phys. Lett. **67** (1995) 1524

95.   A. Schmid and P. Ziemann, Nucl. Instr. Meth. B **7–8** (1985) 581

96.   K. Pampus, K. Dyrbye, B. Torp, and R. Bormann, J. Mater. Res. **4** (1989) 1385

97.   E. Ma, C.Q. Nieh, M.-A. Nicolet, and W.L. Johnson, J. Mater. Res. **4** (1989) 1299

98.   A.N. Campbell, J.C. Barbour, C.R. Hils, and M. Nastasi, J. Mater. Res. **4** (1989) 1303

99.   L.U. Aaen Andersen, J. Bøttiger, and K. Dyrbye, Nucl. Instr. Meth. B **51** (1990) 125

100. B.X. Liu, Nucl. Instr. Meth. B **40–41** (1989) 603

101. D. Schechtman, I. Blech, D. Gratias, and J.W. Cahn, Phys. Rev. Lett. **53** (1984) 1951

102. P.J. Steinhardt, D.R. Nelson, and M. Ronchetti, Phys. Rev. B **28** (1983) 784

103. J. Mayer, K.Urban, J. Härle, and S. Steeb, Zeits. Naturforsch. A **42** (1987) 113

104. M. Bauer, J. Mayer, M. Hohenstein, and M. Wilkens, Phys. Stat. Sol. A **122** (1990) 79

105. L.C. Chen, F. Spaepen, J.L. Robertson, S.C. Moss, and K. Hiraga, J. Mater. Res. **5** (1990) 1871

106. D.M. Follstaedt and J.A. Knapp, J. Less Comm. Met. **140** (1988) 375

107. B. Rauschenbach and K. Hohmuth, Nucl. Instr. Meth. B **23** (1987) 323

108. B.X. Liu, G.A. Cheng, and C.H. Shang, Phil. Mag. Lett. **55** (1987) 265

109. G.W. Yang, W.S. Lai, C. Lin, and B.X. Liu, Appl. Phys. Lett. **74** (1999) 3305

110. C. Lin, G.W. Yang, and B.X. Liu, J. Appl. Phys. **87** (2000) 2821

111. K. Affolter, M. von Allmen, H.P. Weber, and M. Wittmer, J. Non-Cryst. Sol. **55** (1983) 387

112. S.S. Takayama, J. Mater. Sci. **11** (1976) 164

113. D.E. Luzzi and M. Meshii, Scr. Metall. **20** (1986) 943

114. B.X. Liu, W.L. Johnson, M.-A. Nicolet, and S.S. Lau, Nucl. Instrum. Meth. **209–210** (1983) 229

115. H. Mori, H. Fujita, *Non-Equilibrium Solid Phases of Metals and Alloys* (Japan Institute of Metals, Sendai, 1989) 93

116. B.X. Liu, Mater. Lett. **5** (1987) 322

117. J.A. Alonso and S. Simozar, Sol. St. Comm. **48** (1983) 765

118. H.M. Naguib and R. Kelly, Radiat. Eff. **25** (1975) 1

119. A.T. Motta and D. Olander, Acta Metall. Mater. **38** (1990) 2175

120. L.W. Hobbs, J. Non-Cryst. Sol. **182** (1995) 27

121. P.M. Ossi, Surf. Sci. **201** (1988) 519

122. R. Zangwill, *Physics at Surfaces* (Cambridge University Press, Cambridge 1988) 11

123. B.J. Garrison, H. Winograd, D. Lo, T.A. Tombrello, M.H. Shapiro, and D.E. Harrison, Jr., Surf. Sci. **180** (1987) L129

124. U. Mizutani, M. Sasaura, Y. Yamada, and T. Matsuda, J. Phys. F: Met. Phys. **17** (1987) 667

125. A.A. Radzig and B.M. Smirnov, *Reference Data on Atoms, Molecules and Ions* (Springer, Berlin 1985) 463

126. W. Jiang, W.J. Weber, S. Thevuthasan, and V. Shutthanandan, J. Nucl. Mater. **289** (2001) 96

127. F.R. de Boer, R. Boom, W.C.M. Mattens, A.R. Miedema, and A.K. Niessen, *Cohesion in metals: Transition Metal Alloys* (North Holland, Amsterdam, 1989) 758

128. M. Binnewies and E. Milke, *Thermochemical Data of Elements and Compounds* (Wiley-VCH, Weinheim, 1999) 928

129. *Metals and Alloys Indexes* (Intl. Centre for Diffraction Data, Swarthmore, Pennsylvania, 1992)

130. K.D. Sen and P. Politzer, J. Chem. Phys. **91** (1989) 5123

131. *Periodic Table of the Elements*: (IUPAC 1998)

132. L.M. Howe, D. Phillips, H. Zou, J. Forster, R. Siegele, J.A. Davies, A.T. Motta, J.A. Faldowski, and P.R. Okamoto, Nucl. Instr. Meth. B **118** (1996) 663

133. F. Lefebvre and C. Lemaignan, J. Nucl. Mater. **165** (1989) 122

134. P.A. Dowben, M. Grunze, and D. Wright, Surf. Sci. **134** (1983) L524

135. Z.J. Zhang, J. Appl. Phys. **81** (1997) 2027

136. P. Dumoulin and M. Guttmann, Mater. Sci. Eng. **42** (1980) 249

137. Z.H. Yan, B.X. Liu, and H.D. Li, Nucl. Instr. Meth. B **19–20** (1987) 700

138. A. Plewnia, B. Heinz, and P. Ziemann, Nucl. Instr. Meth. B **148** (1999) 901

139. J.H. Hsieh, W. Wu, R.A. Erck, G.R. Fenske, Y.Y. Su, and M. Marek, Surf. Coat. Technol. **51** (1992) 212

140. R. Durner, B. Heinz, and P. Ziemann, Nucl. Instr. Meth. B **148** (1999) 896

141. J.L. Brimhall, H.E. Kissinger, and L.A. Charlot, Mater. Res. Soc. Symp. Proc. **7** (1982) 235

142. Y.G. Chan, B.X. Liu, and W. Zhang, J. Phys. Condens. Matt. **9** (1997) 389

143. E. Peiner, K. Kopitzki, Nucl. Instr. Meth. B **34** (1988) 173

144. R.C. Da Silva, R.C. Sousa, O. Conde, M.F. Da Silva, and J.C. Soares, Surf. Coat. Technol. **83** (1996) 60

145. M.B. Kasen, Acta Metall. **31** (1983) 489

146. Z.J. Zhang and B.X. Liu, J. Phys. Condens. Matt. **6** (1994) 9065

147. J.F. Dinhut, F. Benhameur, and J.P. Riviere, Nucl. Instr. Meth. B **71** (1992) 191

148. G. Allié, C. Lauroz, and P. Villemain, Surf. Sci. **104** (1981) 583

149. M. Neubauer, K.P. Lieb, P. Schaaf, and M. Uhrmacher, Thin Sol. Films **275** (1996) 69

150. N. Bibic, Z.H. Jafri, M. Milosavljevic, and D. Perusko, Vacuum **46** (1995) 899

151. Y.T. Cheng, W.L. Johnson, and M.-A. Nicolet, Mater. Res. Soc. Symp. Proc. **37** (1985) 565

152. B.X. Liu, Z.J. Zhang, O. Jin, and F. Pan, Nucl. Instr. Meth. B **106** (1996) 17

153. V.V. Uglov, J.A. Fedotova, J. Jagielski, G. Gawlik, and J. Stane,: Nucl. Instr. Meth. B **159** (1999) 218

154. K. Hohmuth, B. Rauschenbach, A. Kolitsch, and E. Richter: Nucl. Instr. Meth. **209–210** (1983) 249

155. M. Soltani-Farshi, H. Baumann, D. Rück, E. Richter, U. Kreissig, and K. Bethge, Surf. Coat. Technol. **103–104** (1998) 299

156. H.-Y. Chen, J. Wang, H. Yang, W-Z. Li, and H-D. Li, Surf. Coat. Technol. **128–129** (2000) 329

157. T. Fujihana, Y. Okabe, and M. Iwaki, Nucl. Instr. Meth. B **127-128**, (1997) 660

158. T. Fujihana, Y. Okabe, and M. Iwaki, Surf. Coat. Technol. **66** (1994) 419

159. C. Liu, B. Mensching, M. Zeitler, K. Voltz, and B. Rauschenbach, Phys. Rev. B **57** (1998) 2530

160. S.J. Zinkle and L.L. Snead, Nucl. Instr. Meth. B **116** (1996) 92

161. A.R. González-Elipe, F. Yubero, J.P. Esponós, A. Caballero, M. Ocaña, J.P. Holgado, and J. Morales, Surf. Coat. Technol. **125** (2000) 116

162. W. Ensinger and M. Kiuchi:,Surf. Coat. Technol. **84** (1996) 425

163. L. Pichon, T. Girardeau, A. Straboni, F. Lignou, J. Perriere, and J.M. Frigerio, Nucl. Instr. Meth. B **147** (1999) 378

164. A. Traverse, P. Parent, J. Mimault, N. Thromat, M. Gautier, J.P. Durand, A.M. Flank, A. Quivy, and A. Fontaine, Nucl. Instr. Meth. B **86** (1994) 270

165. P.M. Ossi, and R. Pastorelli, Nanostr. Mater. **11** (1999) 739

166. T. Egami and W. Waseda, J. Non-Cryst. Sol. **64** (1984) 113

167. D. Lee, J. Cheng, M. Yuan, C.N.J. Wagner, and A.J. Ardell, J. Appl. Phys. **64** (1988) 4772

168. Ch. Konvicka, Y. Jeanvoine, E. Lundgren, G. Kresse, M. Schmid, J. Hafner, and P. Varga, Surf. Sci. **463** (2000) 199

169. F. Berthier, J. Creuze, R. Tetot, and B. Legrand, Appl. Surf. Sci. **177** (2001) 243

170. J.O. Orwa, K.W. Nugent, D.N. Jamieson, and S. Prawer, Phys. Rev. B **62** (2000) 5461

Chapter 11

INTRODUCTION TO MATHEMATICAL MODELS FOR IRRADIATION-INDUCED PHASE TRANSFORMATIONS

Kurt E. Sickafus

Los Alamos National Laboratory, Los Alamos, NM 87545 USA

1   INTRODUCTION

The purpose of this chapter is to introduce students to a few simple rate theory models for describing phase transformations due to radiation damage. More comprehensive reviews have been provided elsewhere (see, e.g., a recent review by Weber [1]). This chapter introduces the reader to the concepts underlying the most oft-cited models. Most of the models to be presented here were initially developed to describe amorphization phase transformations. However, the models are more general to the extent that they are also viable to describe irradiation induced crystal-to-crystal solid state phase transformations.

Figure 1 shows a schematic diagram of a highly idealized damage region or damage cascade surrounding the path of an energetic ion (perhaps a primary knock-on atom or PKA) as it traverses a crystalline solid. The primary damage region is represented by a cylinder (shaded). In rate theory models, it is assumed that within this cylindrical volume the material experiences a phase transformation (such as amorphization) or that this volume suffers an increment of damage that advances it towards an inevitable phase transformation. We also show a larger dotted cylinder in Fig. 1 to represent schematically the following concept. In some models, it is postulated that a portion of the damaged cylindrical volume (namely at the periphery of the cylinder) is able to recover from the radiation-induced disordered state after the cascade thermalizes. This recovery is sometimes assumed to be complete, so that the material between the inner and outer cylinders is like that of the unirradiated crystalline solid; or sometimes this material is assumed to only partially recover, so that it is left in some incremental damage state.

321

K.E. Sickafus et al. (eds.), Radiation Effects in Solids, 321–352.
© 2007 *Springer.*

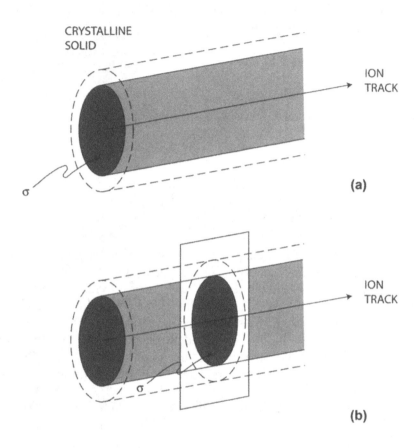

CRYSTALLINE
SOLID

ION
TRACK

σ

(a)

ION
TRACK

σ

(b)

Figure 1. (a) Idealized cylindrical damage volume associated with an energetic ion traversing a crystalline solid. (b) Same as (a) but showing a section through the cylindrical damage volume at some arbitrary position along the ion track. The area σ is the damage cross-section for the incident ion.

In an ion irradiation experiment, the process described above is repeated over and over as ions impinge on randomly varying positions about the sample surface. Eventually, at sufficient ion fluence, the entire sample surface will have experienced one or more damage events. Similarly, in a neutron irradiation experiment, PKAs are born continuously within the sample with increasing neutron exposure, so that damage tracks accumulate randomly throughout the bulk of the crystalline solid.

Eventually every atom in the sample will be contained within at least one cylindrical damage volume. The purpose of rate theory models is simply to predict the time dependence of any specific structural alteration that may be found to accompany irradiation.

Most simple rate theory models are based on two-dimensional (2-D) descriptions for the transformations of interest. Thus, in Fig. 1b, we show a planar section through the cylindrical damage volume at some arbitrary distance along the path of the energetic ion. This section through the shaded cylinder in Fig. 1b inscribes a circle of area σ. This area is known as the damage cross-section if the material within is incrementally damaged, or as the transformation cross-section if the material within actually experiences a phase transformation after the cascade thermalizes. We then assume that similar circular damage regions accumulate during irradiation on an imaginary plane normal to the ion tracks. For this reason, simple rate theory models are only strictly applicable to ion irradiation experiments, wherein all ions impinge on the crystalline solid along the same (beam) direction. So, we will assume that the rate theory models we develop next are intended to describe radiation damage accumulation in mono-energetic ion irradiation experiments. Furthermore, we assume that at some depth d below the sample surface, each incident ion induces a circular damage region of area σ.

The first models we will consider are athermal in the sense that the damage within each cross-sectional area σ is assumed to be trapped during the quench of the cascade, with no opportunity for kinetic recovery. Later, we will briefly consider models that allow for thermal recovery mechanisms within each ion cascade. For simplicity, we will assume that our models describe some specific crystalline-to-amorphous transformation. However, keep in mind that these models are more generally applicable to any discrete solid-state transformation.

Finally, all of the models considered in this chapter have a common mathematical feature: each model involves the solution to a time-dependent differential equation or a simultaneous set of differential equations. Though numerous techniques are available to solve sets of differential equations, such as Laplace transforms and eigenvalue/eigenvector techniques, we will use the simple and elegant technique of the *matrix exponent* for most of our solutions (see, e.g., Williamson and Trotter [2]). Furthermore, many computer programs are now available to manipulate and solve equations in symbolic form and some of these programs offer a matrix exponent operator designed to obtain analytic solutions to simultaneous differential equations. In our following discussion of rate theory models, we will use the program *Mathematica* [3] to illustrate the simplicity of this computer-based solution technique.

2   ATHERMAL TRANSFORMATION MODELS

The first few athermal models that we present in this chapter were first introduced by Gibbons [4] to describe ion irradiation-induced damage accumulation and amorphization in semiconductor materials. These models involve either intra-cascade phase transformation or transformation following cascade overlap. We will later consider models that include both of these mechanisms and then a model that allows for concurrent damage recovery.

2.1   The Direct Impact (or *Black Spot*) Model

Figure 2 shows the important features of the direct impact model for radiation-induced amorphization. We assume that each incident ion produces a region of area σ_a in which the material experiences a crystalline-to-amorphous phase transformation after the ion induced damage cascade thermalizes.

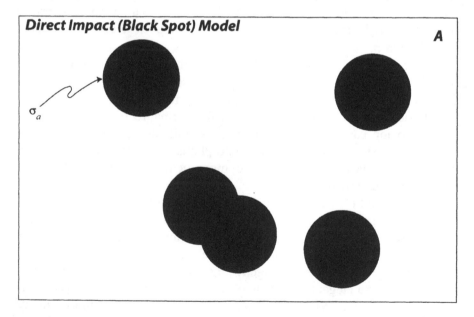

Figure 2. Black spots representing damage regions in a crystalline material that have undergone ion-induced phase transformations. The area σ_a is the transformation cross-section per incident ion. A is the projected area of the irradiated sample.

This transformation region is represented by a black circular spot in Fig. 2. With increasing ion fluence, more black spots form and overlap. The sample is rendered fully-amorphous when the entire planar area A in Fig. 2 is black (A represents the projected area of the irradiated sample).

The rate at which the crystalline-to-amorphous phase transformation proceeds can be described mathematically as follows. Let $f_u(t)$ represent the fraction of area A that is undamaged due to irradiation at time t (white). Similarly, let $f_a(t)$ represent the fraction of area A that is amorphized due to ion irradiation damage at time t (black). Let ϕ represent the ion flux in units of [ions/area/s] and let σ_a represent the amorphization cross-section per ion (units of [area]). The probability that a given ion impinging on the sample surface at time t hits an undamaged (pristine) region is $f_u(t)$ while the probability that it hits an amorphous region is $f_a(t)$, where $f_u(t) + f_a(t) = 1$. Thus, f_a grows with time (a positive derivative) in proportion to f_u, while f_u decreases with time (a negative derivative) in proportion to f_u. With these definitions, the time rates of change of f_u and f_a are given by:

$$\frac{df_u}{dt} = -\sigma_a\,\phi\,f_u \tag{1a}$$

$$\frac{df_a}{dt} = +\sigma_a\,\phi\,f_u \tag{1b}$$

We define the quantity $\sigma_a\,\phi$ as the production rate for amorphization, P_a:

$$P_a = \sigma_a\,\phi \tag{2}$$

Then we rewrite the pair of simultaneous differential equations in Eqn. (1) as a linear system of differential equations in constant coefficients:

$$\frac{df_u}{dt} = -P_a\,f_u + 0\,f_a \tag{3a}$$

$$\frac{df_a}{dt} = +P_a\,f_u + 0\,f_a \tag{3b}$$

This is the starting point for the *matrix exponent* solution technique to this pair of simultaneous differential equations. We desire to solve for $f_u(t)$ and $f_a(t)$, which taken together we can write in vector notation:

$$\mathbf{f} = \begin{pmatrix} f_u(t) \\ f_a(t) \end{pmatrix} \tag{4}$$

Likewise, the constant coefficients on the right-hand sides of Eqns. (3a) and (3b) can be combined in matrix notation:

$$A_{mat} = \begin{pmatrix} -P_a & 0 \\ +P_a & 0 \end{pmatrix} \tag{5}$$

With Eqns. (4) and (5), we can rewrite the set of simultaneous equations in Eqn. (3) as:

$$\frac{d\mathbf{f}}{dt} = A_{mat}\mathbf{f} \tag{6}$$

The only other requirement for the matrix exponent procedure is a boundary condition vector representing the starting conditions on $f_u(t)$ and $f_a(t)$. Our initial conditions are:

$$f_u(t=0) = 1 \tag{7a}$$
$$f_a(t=0) = 0 \tag{7b}$$

so that we can define a starting condition vector $\mathbf{f}_{init}$ by:

$$\mathbf{f}_{init} = \begin{pmatrix} 1 \\ 0 \end{pmatrix} \tag{8}$$

Using A_{mat} and $\mathbf{f}_{init}$ from Eqns. (5) and (8), respectively, we can solve for $f_u(t)$ and $f_a(t)$ using the matrix exponent algorithm provided in the program *Mathematica*. The *Mathematica* output is shown in Figure 3.
So, the solutions for $f_u(t)$ and $f_a(t)$ are given by:

$$f_u(t) = e^{-P_a t} \tag{9a}$$
$$f_a(t) = 1 - e^{-P_a t} \tag{9b}$$

The expressions in Eqn. (9) are plotted in Figure 4. For this plot, it was assumed that the amorphization (transformation) cross-section is given by $\sigma_a = 6.25 \times 10^{-14} \ cm^2$ and the irradiation flux is given by $\phi = 10^{12} \ ions/_{cm^2 \cdot s}$, so that the production rate for amorphization, P_a, is $0.0625 \ s^{-1}$ (Eqn. (2)).

The Black Spot Model

```
Amat = {{-Pa, 0},
        {+Pa, 0}};
```

```
MatrixForm[Amat]
```

$$\begin{pmatrix} -P_a & 0 \\ P_a & 0 \end{pmatrix}$$

```
finit = {1, 0}
```

$$\{1, 0\}$$

```
MatrixForm[finit]
```

$$\begin{pmatrix} 1 \\ 0 \end{pmatrix}$$

```
fractions = MatrixExp[t Amat].finit
```

$$\{e^{-t P_a},\ e^{-t P_a}(-1+e^{t P_a})\}$$

```
fu = fractions[[1]]
```

$$e^{-t P_a}$$

```
fa = fractions[[2]]
```

$$e^{-t P_a}(-1+e^{t P_a})$$

Figure 3. Mathematica input (bold) and output for the matrix exponent solution to the simultaneous pair of differential equations shown in Eqn. (3).

The direct impact or *black spot* model is most often used to describe transformation rates in materials that are highly susceptible to amorphization (such as covalently bonded crystalline solids) or for heavy ion irradiation conditions wherein the nuclear energy density within each cascade volume is large. Also note the inherent heterogeneous nature of this growth rate model.

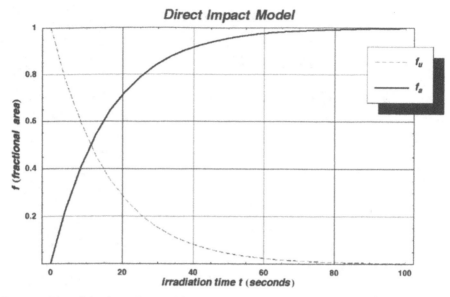

Figure 4. Plot of the dependence of fractional areas f_u and f_a on irradiation time, t, based on Eqn. (9).

2.2　The Single Cascade Overlap (or *Gray Spot*) Model

Figure 5 shows the important features of the single cascade overlap model for radiation-induced amorphization. In this case, we assume that each incident ion produces a region of area σ_d in which the material experiences structural damage but no phase transformation, after the ion induced damage cascade thermalizes. This damage region is represented by a gray circular spot in Fig. 5. With increasing ion fluence, more gray spots form and overlap. In regions where gray spots overlap, this model assumes that the critical defect density required to initiate a crystalline-to-amorphous transformation is exceeded. Such a region is shown in black in Fig. 5. This is intended to indicate that amorphization has occurred in this area following cascade overlap. The sample is rendered fully-amorphous when the entire planar area A in Fig. 5 is black.

There are three types of areas in this model: undamaged, damaged and amorphized (white, gray and black, respectively). We will represent these fractional areas by $f_u(t)$, $f_d(t)$ and $f_a(t)$. The probability that a given ion impinging on the sample surface at time t hits an undamaged (pristine) region is $f_u(t)$ while the probabilities that it hits a damaged or amorphous region are $f_d(t)$ and $f_a(t)$, respectively, where $f_u(t) + f_d(t) + f_a(t) = 1$. Thus, f_a grows with time (a positive derivative) in proportion to f_d; f_u decreases with time (a negative derivative) in proportion to f_u; lastly, f_d grows with time in proportion to f_u, but decreases with time in proportion to f_d. With these definitions, the time rates of change of f_u, f_d and f_a are given by:

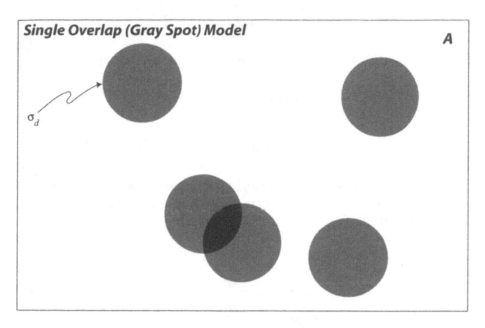

Figure 5. Gray spots representing regions in a crystalline material that have been damaged by incident energetic ions but have not undergone an ion-induced phase transformation. Where the gray spots overlap, the material has suffered an amorphization transformation. This is indicated by black shading. The area σ_d is the damage cross-section per incident ion. A is the projected area of the irradiated sample.

$$\frac{df_u}{dt} = -\sigma_d \, \phi \, f_u \qquad (10a)$$

$$\frac{df_d}{dt} = +\sigma_d \, \phi \left(f_u - f_d \right) \qquad (10b)$$

$$\frac{df_a}{dt} = +\sigma_d \, \phi \, f_d \qquad (10c)$$

We define the quantity $\sigma_d \, \phi$ as the damage production rate, P_d:

$$P_d = \sigma_d \, \phi \qquad (11)$$

As before, ϕ represents the ion flux. Based on the simultaneous set of differential equations in Eqn. (10), the matrix for use in the *matrix exponent* solution procedure is given by:

$$A_{mat} = \begin{pmatrix} -P_d & 0 & 0 \\ +P_d & -P_d & 0 \\ 0 & +P_d & 0 \end{pmatrix} \tag{12}$$

The initial condition vector for the *matrix exponent* method is given by:

$$\mathbf{f}_{init} = \begin{pmatrix} f_u(t=0) \\ f_d(t=0) \\ f_a(t=0) \end{pmatrix} = \begin{pmatrix} 1 \\ 0 \\ 0 \end{pmatrix} \tag{13}$$

Using A_{mat} and $\mathbf{f}_{init}$ from Eqns. (12) and (13), respectively, the *Mathematica matrix exponent* solutions to $f_u(t)$, $f_d(t)$ and $f_a(t)$ are shown in Figure 6. So, the solutions for $f_u(t)$, $f_d(t)$ and $f_a(t)$ are given by:

$$f_u(t) = e^{-P_d t} \tag{14a}$$

$$f_d(t) = P_d\, t\, e^{-P_d t} \tag{14b}$$

$$f_a(t) = 1 - e^{-P_d t} - P_d\, t\, e^{-P_d t} \tag{14c}$$

The expressions in Eqn. (14) are plotted in Figure 7. For this plot, it was assumed that the damage cross-section is given by $\sigma_d = 6.25 \times 10^{-14}$ cm^2 and the irradiation flux is given by $\phi = 10^{12}$ ions/cm$^2 \cdot$s, so that the damage production rate, P_d, is 0.0625 s^{-1} (Eqn. (11)).

The single cascade overlap or *gray spot* model is most often used to describe transformation rates in materials that are moderately susceptible to amorphization or for semi-heavy ion irradiation conditions wherein the nuclear energy density within each cascade volume is only moderately high. We should also note that the "sigmoidal" shape of the $f_a(t)$ curve in the linear ordinate – linear abscissa plot shown in Fig. 7 is a typical characteristic of cascade overlap rate theory models.

The Gray Spot Model

```
f_mat = {{-P_d, 0, 0},
          {+P_d, -P_d, 0},
          {0, +P_d, 0}};
```

MatrixForm[f_mat]

$$\begin{pmatrix} -P_d & 0 & 0 \\ P_d & -P_d & 0 \\ 0 & P_d & 0 \end{pmatrix}$$

f_init = {1, 0, 0};

fractions = MatrixExp[t f_mat].f_init

$$\{e^{-t\,P_d},\; e^{-t\,P_d}\,t\,P_d,\; e^{-t\,P_d}\,(-1 + e^{t\,P_d} - t\,P_d)\}$$

f_u = fractions[[1]]

$$e^{-t\,P_d}$$

f_d = fractions[[2]]

$$e^{-t\,P_d}\,t\,P_d$$

f_s = fractions[[3]]

$$e^{-t\,P_d}\,(-1 + e^{t\,P_d} - t\,P_d)$$

Figure 6. Mathematica input (bold) and output for the *matrix exponent* solution to the simultaneous set of differential equations shown in Eqn. (10).

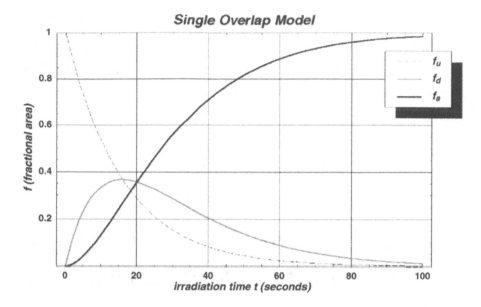

Figure 7. Plot of the dependence of fractional areas f_u, f_d and f_a on irradiation time, t, based on Eqn. (14).

2.3   Higher Order Cascade Overlap Models

Gibbons [4] showed that the general solution for the amorphization rate, $f_a(t)$, for cases in which n-tuple cascade overlap is required to produce an amorphous region is given by:

$$f_a(t) = 1 - e^{-P_{d(n)}t} \sum_0^n \frac{\left(P_{d(n)}\, t\right)^k}{k!} \qquad (15)$$

where n is the number of cascade overlaps required to render a region of the sample amorphous, and $P_{d(n)}$ is the incremental damage production rate. Eqn. (15) is easy to verify using the *matrix exponent* technique.

Consider, for instance, a double cascade overlap model ($n = 2$). Figure 8 shows schematically the double-overlap model for radiation-induced amorphization. In this case, we assume that each incident ion produces a region of area $\sigma_{d(2)}$ in which the material experiences structural damage but no phase transformation, after the ion induced damage cascade thermalizes. This damage region is represented by a light gray circular spot in Fig. 8. With increasing ion fluence, more gray spots form and overlap. In regions where gray spots overlap twice, this model assumes that the critical defect density required to initiate a crystalline-to-amorphous transformation is exceeded. Such a region is shown in black in Fig. 8. This is intended to indicate that amorphization has occurred in this area following cascade overlap.

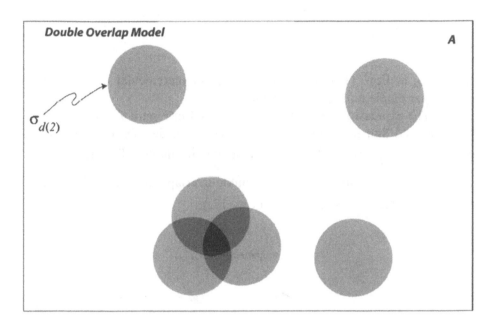

Figure 8. Light gray spots representing regions in a crystalline material that have been damaged by incident energetic ions but have not undergone an ion-induced phase transformation. Where the gray spots overlap twice, the material suffers an amorphization transformation. This is indicated by black shading. The area $\sigma_{d(2)}$ is the incremental damage cross-section per incident ion. A is the projected area of the irradiated sample.

As in earlier discussions, the sample is rendered fully-amorphous when the entire planar area A in Fig. 8 is black.

There are four types of areas in this model: undamaged, semi-damaged, highly-damaged and amorphized (white, light gray, gray and black, respectively). We will represent these fractional areas by $f_u(t)$, $f_{d1}(t)$, $f_{d2}(t)$ and $f_a(t)$. Based on the *matrix exponent* solutions for these fractional areas (not shown here), $f_u(t)$, $f_{d1}(t)$, $f_{d2}(t)$ and $f_a(t)$ are given by:

$$f_u(t) = e^{-P_{d(2)}t} \qquad (16a)$$

$$f_{d1}(t) = P_{d(2)}\, t\, e^{-P_{d(2)}t} \qquad (16b)$$

$$f_2(t) = \tfrac{1}{2} P_{d(2)}^{\,2}\, t^2\, e^{-P_{d(2)}t} \qquad (16c)$$

$$f_a(t) = 1 - e^{-P_{d(2)}t} - P_{d(2)} t\, e^{-P_{d(2)}t} - \frac{1}{2} P_{d(2)}^2\, t^2\, e^{-P_{d(2)}t} \quad (16d)$$

where the incremental damage production rate for double-overlap, $P_{d(2)}$, is given by:

$$P_{d(2)} = \sigma_{d(2)}\, \phi \qquad (17)$$

Area $\sigma_{d(2)}$ in Eqn. (17) is the double-overlap incremental damage cross-section per incident ion and ϕ is the ion flux.

The expressions in Eqn. (16) are plotted in Figure 9. For this plot, it was assumed that the double-overlap incremental damage cross-section is given by $\sigma_{d(2)} = 6.25 \times 10^{-14}\ \mathrm{cm}^2$ and the irradiation flux is given by $\phi = 10^{12}\ \mathrm{ions}/\mathrm{cm}^2 \cdot \mathrm{s}$, so that the double-overlap incremental damage production rate, $P_{d(2)}$, is 0.0625 s^{-1} (Eqn. (17)).

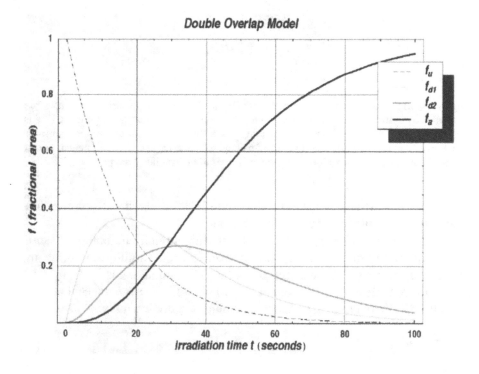

Figure 9. Plot of the dependence of fractional areas f_u, f_{d1}, f_{d2} and f_a on irradiation time, t, based on Eqn. (16).

The double cascade overlap model is most often used to describe transformation rates in materials with modest susceptibility to amorphization or for light-ion ion irradiation conditions wherein the nuclear energy density within each cascade volume is low. We also call attention to the "sigmoidal" shape of the $f_a(t)$ curve in the linear ordinate – linear abscissa plot shown in Fig. 9, similar to that of the gray spot model (Fig. 7).

Figure 10 shows log $f_a(t)$ versus log t for several orders of the cascade overlap model (Eqn. (15)): specifically, the direct impact or *black spot* model ($n = 0$); and four orders of the cascade overlap or *gray spot* model ($n = 1,2,3,4$). The amorphous fraction calculations in Fig. 10 are identical to the f_a calculations shown in Figs. 4, 7, and 9, for $n = 0$, 1, and 2, respectively, only the plot in Fig. 10 is log-log instead of linear-linear as in the previous figures.

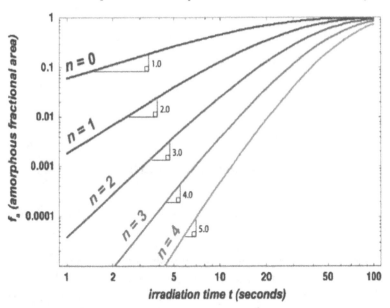

Figure 10. Plot of log (f_a) versus log (t) for several orders of the cascade overlap model. The general solution to the cascade overlap model is summarized in Eqn. (15). For these calculations, it was assumed that the incremental damage cross-section is given by $\sigma_{d(i)} = 6.25 \times 10^{-14}$ cm^2 and the irradiation flux is given by $\phi = 10^{12}$ $ions/_{cm^2 \cdot s}$, as in previous examples. Also shown are the slopes of the curves at small irradiation times (low dose).

Figure 11 shows the derivatives of the curves from Fig. 10. These curves represent the log-log slopes of the curves in Fig. 10 at any time, t. It should be noted that at low irradiation dose (small t) the derivatives asymptotically approach integer values 1, 2, 3, 4 and 5 for $n = 0$, 1, 2, 3 and 4 overlaps, respectively (as indicated previously in Fig. 10). These slopes are different than the slopes originally published by Gibbons [4].

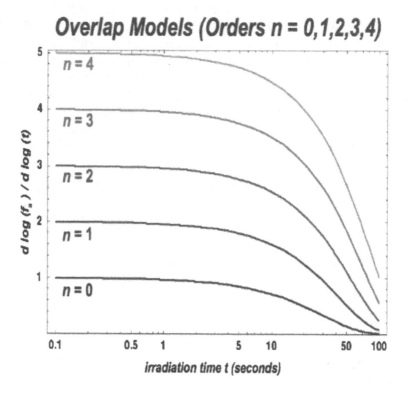

Figure 11. Plot of $d \log (f_a) / d \log (t)$ versus irradiation time (t) for the cascade overlap model curves shown in Fig. 10.

These 'low-dose' slopes are independent of cascade size. We illustrate this by considering the double cascade overlap model ($n = 2$) in Figures 12 and 13, for different sized damage cascades. Here, we assume that the cascade diameter, D, varies from $D = 12.5$ Å to 25 Å to 50 Å. Furthermore, for simplicity we assume that $\sigma_{d(2)}^2 = p(D/2)^2$, so that $\sigma_{d(2)}$ varies from $1.23 \cdot 10^{-14}$ cm^2 to $4.91 \cdot 10^{-14}$ cm^2 to $1.96 \cdot 10^{-13}$ cm^2 as D increases. It should be noted that the slopes of the curves shown in Fig. 12 all approach asymptotically a value of 3 at low dose or small t (see Fig. 13).

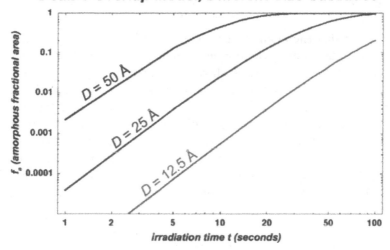

Figure 12. Plot of log (f_a) versus log (t) for the double cascade overlap model using three different damage cascade sizes: $\sigma_{d(2)} = 1.23 \cdot 10^{-14}$ cm^2, $4.91 \cdot 10^{-14}$ cm^2 and $1.96 \cdot 10^{-13}$ cm^2, from smallest to largest (corresponding to cascade diameters $D = 12.5$ Å, 25 Å and 50Å, respectively). The irradiation flux in these calculations was assumed to be $\phi = 10^{12}$ $\frac{ions}{cm^2 \cdot s}$, as in previous examples.

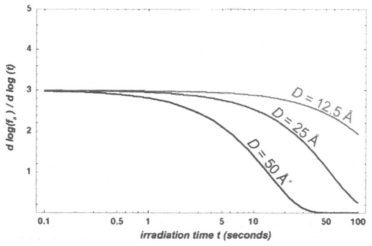

Figure 13. Plot of $d \log (f_a) / d \log (t)$ versus irradiation time (t) for the double cascade overlap model curves shown in Fig. 12. Note that at low dose (small t) these "slope" curves all approach asymptotically the integer value of 3.

2.4   A Composite (*Black and Gray Spot*) Phase Transformation Model

Imagine that each damage cascade leaves behind some material that *has* suffered a phase transformation and some material that is damaged but *not* transformed. This concept is shown schematically in Figure 14. We assume here that the material at the periphery of the cascade resists transformation because some of the defects produced during the ballistic phase of the cascade are able to escape into the bulk or annihilate as the damage cascade thermalizes (thus preventing an irradiation-induced transformation). We further assume in this model that if damage regions overlap singly, a phase transformation occurs in the overlapped region. This is indicated in Fig. 14 where overlapped gray regions combine to form a black area.

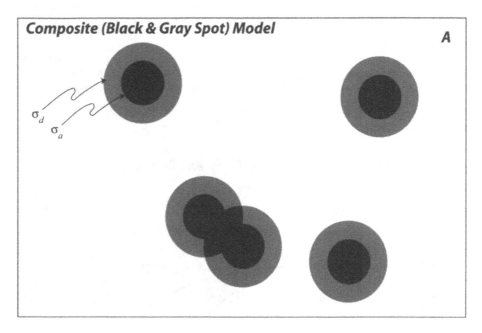

Figure 14. Composite black and gray spots representing regions in a crystalline material that have been damaged by incident energetic ions. The central portion of each composite spot (area σ_a) has been subjected to a phase transformation (either amorphization or a crystal-to-crystal transformation) while the gray periphery of each spot is damaged but has not undergone an ion-induced phase transformation. Where the gray regions overlap, the material has succumbed to an amorphization (or crystal-to-crystal) transformation. This is indicated by black shading. The areas σ_a and σ_d are the amorphization and damage cross-sections per incident ion, respectively. A is the projected area of the irradiated sample.

As in the gray spot model (Section 2.2), there are three types of areas in this model: undamaged, damaged and amorphized (white, gray and black, respectively). The time rate of change of fractional areas $f_u(t)$, $f_d(t)$ and $f_a(t)$ in this composite model are given by:

$$\frac{df_u}{dt} = -\left(P_d + P_a\right)f_u \tag{18a}$$

$$\frac{df_d}{dt} = +P_d\, f_u - \left(P_d + P_a\right)f_d \tag{18b}$$

$$\frac{df_a}{dt} = +P_a\, f_u + \left(P_d + P_a\right)f_d \tag{18c}$$

where $P_d = \sigma_d\, \phi$ (Eqn. (11)) and $P_a = \sigma_a\, \phi$ (Eqn. (2)). As before, ϕ represents the ion flux. The *matrix exponent* solution to the simultaneous set of differential equations in Eqn. (18) is given by:

$$f_u(t) = e^{-\left(P_d + P_a\right)t} \tag{19a}$$

$$f_d(t) = P_d\, t\, e^{-\left(P_d + P_a\right)t} \tag{19b}$$

$$f_a(t) = 1 - e^{-\left(P_d + P_a\right)t} - P_d\, t\, e^{-\left(P_d + P_a\right)t} \tag{19c}$$

If $P_d = 0$, the solution given in Eqn. (19) simplifies to the solution to the direct impact or black spot model (Section 2.1, Eqn. (9)). If $P_a = 0$, the solution given in Eqn. (19) simplifies to the solution to the single overlap or gray spot model (Section 2.2, Eqn. (14)). Figure 15 shows $f_u(t)$, $f_d(t)$ and $f_a(t)$ from Eqn. (19) plotted versus t, assuming that the damage cross-section, σ_d, is given by $\sigma_d = 2.25 \times 10^{-14}$ cm^2 and the amorphization (transformation) cross-section, σ_a, is given by $\sigma_a = 4.00 \times 10^{-14}$ cm^2. The corresponding damage production rate, P_d, is 0.0225 s^{-1} (Eqn. (11)), and the production rate for amorphization, P_a, is 0.0400 s^{-1} (Eqn. (2)). We assumed here that the irradiation flux is given by $\phi = 10^{12}$ ions$/_{\text{cm}^2 \cdot \text{s}}$. We can also define a composite cross-section, σ_c, given by:

$$\sigma_c = \sigma_d + \sigma_a \tag{20}$$

so that in Fig. 15, $\sigma_c = 6.25 \times 10^{-14}$ cm^2. Similarly, we can define a corresponding composite production rate, P_c, given by:

$$P_c = P_d + P_a \qquad (21)$$

so that in Fig. 15, $P_c = 0.0625$ s^{-1}.

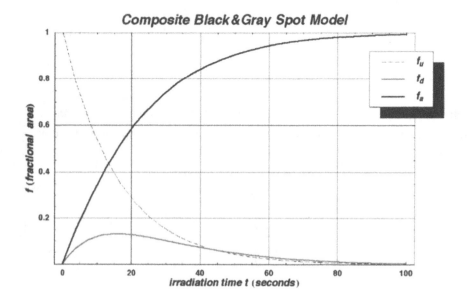

Figure 15. Plot of the dependence of fractional areas f_u, f_d and f_a on irradiation time, t, based on Eqn. (19).

One method to test the validity of a composite black and gray spot model, in relation to a set of experimental observations, is to plot the experimental data as log (f_a) versus log (t) and measure the log-log slope of the transformation growth curve at low dose. We introduced this log-log plot in Section 2.3, Fig. 10. According to Fig. 10, the log-log slope for a composite model should lie between 1 and 2. These limits correspond to direct impact ($n = 0$) or single overlap ($n = 1$) transformation mechanisms, respectively.

Interestingly, the log-log slope described above does not follow a rule-of-mixtures, as the portion of the composite production rate due to damage (i.e., P_d / P_c), varies from 0 to 1. To see this, let x P_c represent the fraction of the composite production rate due to damage, and ($1-x$) P_c represent the fraction of the composite production rate due to amorphization:

$$P_d = x \, P_c \tag{22a}$$

$$P_a = (1-x) P_c \tag{22b}$$

The relationships in Eqn. (22) can be substituted into Eqn. (19) to obtain solutions to fractional areas f_u, f_d and f_a as a function of the damage fraction, x, per incident ion. The resultant slopes of log (f_a) versus log (t) plots at small irradiation time (low dose) as a function of damage fraction, x, is shown in Figure 16.

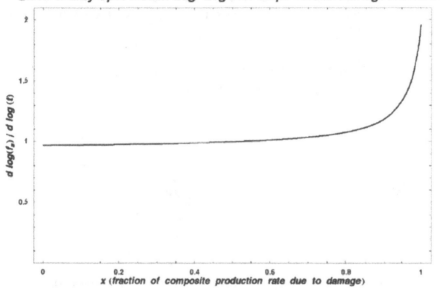

Black & Gray Spot Model: Log–Log Plot Slope versus Damage Fraction x

Figure 16. Plot of d log(f_a) / dt ln(t) versus damage fraction, x, based on the composite (black and gray spot) model. Each point on this curve represents the slope of a plot of log(f_a) versus ln(t) for a given value of x and specifically for irradiation time $t = 1$ s.

Notice in Fig. 16 that while the log-log slope varies from 1 to 2 as expected, the slope does not begin to deviate from 1 until $x > 0.8$ (i.e., until more than 80% of the composite damage production rate is due to damage production, P_d, rather than amorphization production, P_a). This suggests that the slope of a plot of log(f_a) versus ln(t) will be a poor indicator to determine whether or not a composite rate model effectively describes the transformation behavior observed in a particular experiment.

2.5   A Composite Transformation Model with an Athermal Recovery Component (*Black & Gray Spots and an Eraser*)

Consider a black and gray spot composite model as in the previous section, but consider also the addition of a damage *eraser*. This concept has been used previously to account for possible athermal damage recovery mechanisms. Such recovery is often attributed to the electronic stopping component of the ion stopping power, or to other sources of ionization energy. For instance, Abe et al. [5] performed ion irradiation experiments on semiconductors with concurrent electron irradiation, and used a damage eraser concept to describe the role of the electrons in effecting some damage recovery during ion exposure. The Abe model is shown schematically in Figure 17.

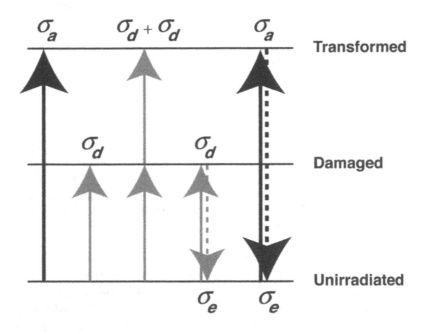

Figure 17. Schematic diagram showing incremental damage states for a composite model consisting of black spots (amorphization or transformation cross-section σ_a), gray spots (damage cross-section σ_d) and an eraser (recovery cross-section σ_e). The eraser cross-section, σ_e, is assumed to be the same for both amorphous (transformed) regions and damaged regions. In other words, the eraser induces recovery in both damaged and transformed regions with the same efficiency and it renders both regions "fully-recovered" (i.e., indistinguishable from unirradiated material).

As in the previous section, we will assume that the composite damage rate, P_c, is the sum of the transformation production rate, P_a, and the damage production rate, P_d (Eqn. (21)). We will also assume, as in Eqn. (22), that xP_c is the fraction of P_c that produces damage and $(1-x)$ P_c is the fraction of P_c that produces amorphization (transformation). Now, if we let P_e represent the recovery production rate (the eraser), then the time rate of change of fractional areas $f_u(t)$, $f_d(t)$ and $f_a(t)$ in this model are given by:

$$\frac{df_u}{dt} = -P_c\, f_u + P_e\, f_d + P_e\, f_a \tag{23a}$$

$$\frac{df_d}{dt} = +x\, P_c\, f_u - (P_c + P_e)f_d \tag{23b}$$

$$\frac{df_a}{dt} = +(1-x)P_c\, f_u + P_c\, f_d - P_e\, f_d \tag{23c}$$

The *matrix exponent* solution to the simultaneous set of differential equations in Eqn. (23) is given by:

$$f_u(t) = \frac{P_e}{P_c + P_e}\left(1 + \frac{P_c}{P_e}e^{-(P_c + P_e)t}\right) \tag{24a}$$

$$f_d(t) = x\,\frac{P_c\, P_e}{\left(P_c + P_e\right)^2}\left[1 - e^{-(P_c + P_e)t}\left(1 - \left(P_c + \frac{P_c^2}{P_e}\right)t\right)\right] \tag{24b}$$

$$f_a(t) = \frac{P_c\left(P_c + (1-x)\,P_e\right)}{\left(P_c + P_e\right)^2}\left[1 - e^{-(P_c + P_e)t}\left(1 + \frac{x\,P_c\left(P_c + P_e\right)}{P_c + (1-x)\,P_e}t\right)\right] \tag{24c}$$

In the limit $P_e = 0$ (no recovery mechanism), if we set $x = 0$, Eqn. (24c) for f_a reduces to the solution for the direct impact (black spot) model (Eqn. (9b)), while if we set $x = 1$, Eqn. (24c) reduces to the solution for the single overlap (gray spot) model (Eqn. (14c)).

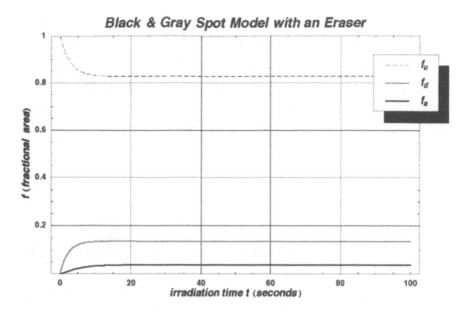

Figure 18. Plot of the dependence of fractional areas f_u, f_d and f_a on irradiation time, t, based on Eqn. (24). For this calculation, we assumed $P_c = 0.0625$ s^{-1}, $P_e = 0.3$ s^{-1}, and $x = 0.95$. These are close to the values used by Abe et al. [5] to fit this model to amorphization rate data for germanium (Ge) irradiated concurrently with 30 keV Xe$^+$ ions at an ion flux of 3.0×10^{15} ions/m^{2}s and 1 MeV electrons at a flux of 2.2×10^{23} e/m^{2}s.

Some aspects of the curves in Fig. 18 are noteworthy. For instance, when a recovery mechanism is operative (specifically P_e in this model), the amorphization fraction f_a does not saturate at a value of 1. Only partial amorphization (or phase transformation) is possible in this model. Also, the fractional areas f_u, f_d and f_a, all reach a steady state after some time (this time depends on the production rates P_c and P_e). Once the steady state is reached, no further change in microstructure occurs with increasing irradiation dose.

3   THERMAL TRANSFORMATION MODELS

In this section, we present two simple kinetic models to describe the temperature dependence of ion irradiation-induced damage accumulation and phase transformations. These models allow for concurrent damage recovery during irradiation. Also, these models assume that the damage recovery mechanisms are temperature dependent.

3.1 A Direct Impact Model Including Thermal Recovery (The Morehead-Crowder Model)

Figure 19 shows schematically damage accumulation and amorphization (transformation) progress for a rate model originally proposed by Morehead and Crowder [6]. This model consists of direct impact amorphization along with kinetic recovery within each cascade region.

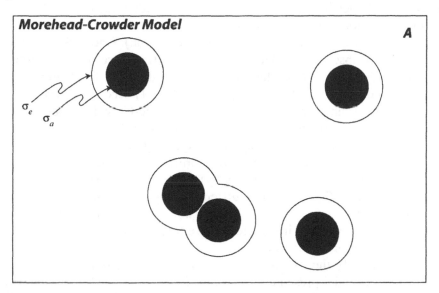

Figure 19. Composite black and white spots representing regions in a crystalline material that have been damaged by incident energetic ions. The central portion of each composite spot (area σ_a) has been subjected to a phase transformation (either amorphization or a crystal-to-crystal transformation) while the white periphery of each spot (area σ_e) is assumed to be damaged during the ballistic phase of the cascade, but then to recover as the cascade thermalizes. The areas σ_a and σ_e are the amorphization and recovery (or erasure) cross-sections per incident ion, respectively. A is the projected area of the irradiated sample.

The Morehead-Crowder model assumes that each cascade is represented by a constant cascade cross-section, σ_c, which is sub-divided into two areas, an area of recovery (or erasure) which is temperature dependent and given by $\sigma_e(T)$, and a residual amorphous (or transformed) area, given by $\sigma_a(T)$, such that:

$$\sigma_a(T) = \sigma_c - \sigma_e(T) \qquad (25)$$

This model assumes that the entire cascade cross-sectional area, σ_c, would be rendered amorphous, were it not for temperature-dependent recovery.

The recovery cross-section, $\sigma_e(T)$, is intended to represent an annular region at the periphery of each cascade, inside of which some sort of temperature-dependent recovery mechanism operates as the cascade thermalizes. This recovery process prevents amorphization or phase transformation. After the quench of the cascade, this area is assumed to be either lightly damaged or undamaged (in fact, the mathematics of the Morehead-Crowder model assumes that this region is indistinguishable from unirradiated material). As temperature is increased, σ_e becomes an increasingly larger fraction of σ_c. When $\sigma_e = \sigma_c$, recovery is complete and no amorphization can occur. This temperature is known as the *critical* temperature, T_c, for amorphization (or transformation). One must be below T_c in order to observe complete amorphization (or phase transformation). Another way to say this is that at T_c, the ion dose (or irradiation time) necessary to observe complete amorphization ($f_a = 1$) becomes infinite.

The most important feature of the Morehead-Crowder model is that the recovery mechanism is assumed to be an *intra-cascade* process. Recovery is confined to previously undamaged regions. Recovery does not act on already amorphous (or transformed) regions. In this way, amorphization can proceed (eventually) to completion ($f_a = 1$) at any temperature below T_c. This is quite different compared to the black and gray spot model with an eraser presented in Section 2.5. In the latter model, any finite eraser, P_e, results in incomplete phase transformation ($f_a < 1$), even at infinite ion dose (or irradiation time).

To quantify the Morehead-Crowder model, we combine aspects of the black spot model presented in Section 2.1 and the black and gray spot with an eraser model presented in Section 2.5. Let P_{a0} represent the amorphization (transformation) production rate per ion (equivalent to $\sigma_c\,\phi$, where ϕ is the ion flux[§]) and $P_e(T)$ represent the recovery (erasure) production rate per ion (equivalent to $\sigma_e(T)\,\phi$). Let f_u represent the fraction of undamaged material at any time t and let f_a represent the fraction of amorphous material at this same time. With these definitions, the time rates of change of f_u and f_a for the Morehead-Crowder model are given by:

$$\frac{df_u}{dt} = -\left(P_{a0} - P_e(T)\right)f_u + 0\,f_a \tag{26a}$$

$$\frac{df_a}{dt} = +\left(P_{a0} - P_e(T)\right)f_u + 0\,f_a \tag{26b}$$

[§]	Note that we assume the amorphization rate, P_{a0}, is constant, i.e., independent of temperature. Thus, we have added a subscript 0 to indicate that this is the 'athermal' amorphization rate, equivalent to the actual amorphization rate at $T = 0$ K (i.e. in the absence of recovery).

The solution to this simultaneous pair of differential equations (boundary conditions $f_u = 1$ and $f_a = 0$ at $t = 0$) is given by:

$$f_u(t) = e^{-(P_{a0} - P_e(T))t} \tag{27a}$$

$$f_a(t) = 1 - e^{-(P_{a0} - P_e(T))t} \tag{27b}$$

Now, if we assume that the recovery process is related to the diffusion of some as yet unspecified species, we can replace $P_e(T)$ with a standard diffusion rate term that, as usual, depends exponentially on temperature:

$$P_e(T) = P_{e0}\, e^{\frac{-E_a}{kT}} \tag{28}$$

where P_{e0} is the recovery rate coefficient (a parameter to be fit to experimental data), E_a is the activation energy for the diffusion process, and k is Boltzmann's constant ($k = 8.617 \times 10^{-5}$ eV/K). Substituting Eqn. (28) into Eqn. (27), we find:

$$f_u(t) = e^{-\left(P_{a0} - P_{e0}\, e^{-\frac{E_a}{kT}}\right)t} \tag{29a}$$

$$f_a(t) = 1 - e^{-\left(P_{a0} - P_{e0}\, e^{-\frac{E_a}{kT}}\right)t} \tag{29b}$$

In the limit $T = 0$ K, Eqn. (29) reduces to the direct impact (black spot) model solution shown in Eqn. (9) (the athermal subscript 0 is suppressed in Eqn. (9)).

Now, we make the observation that as T increases, the recovery rate, $P_e(T)$ increases (Eqn. (28)). Eventually, $P_e(T)$ outpaces the amorphization rate, P_{a0}, and amorphization is no longer possible at any ion dose. The temperature at which this occurs is the so-called critical amorphization temperature, T_c. To calculate this temperature based on our Morehead-Crowder model solution (Eqn. (29)), we consider an arbitrary damage state, f_a^*, wherein ion-induced amorphization (or transformation) is only partially complete. We compare the irradiation time required to achieve this intermediate damage state at $T = 0$ K versus some higher temperature, T. We define the time to reach damage state f_a^* at $T = 0$ K as t_0 and the time to reach f_a^* at temperature T as t. Based on these assumptions, Eqn. (29b) yields:

$$\frac{f_a^*(t=t_0)}{f_a^*(t=t)} = \frac{1-e^{-P_{a0}\,t_0}}{1-e^{-\left(P_{a0}-P_{e0}\,e^{-\frac{E_a}{kT}}\right)t}} \tag{30}$$

Note that the term $P_{e0}\,e^{\frac{-E_a}{kT}}$ in Eqn. (29b) goes to zero in the limit $T = 0$ K.

Since the left-hand side of Eqn. (30) is equal to unity, we can readily solve for the irradiation time ratio, t_0/t:

$$\frac{t_0}{t} = 1 - \frac{P_{e0}}{P_{a0}}\,e^{\frac{-E_a}{kT}} \tag{31}$$

If we let the damage state f_a^* equal 1 (complete amorphization), then the corresponding time to reach this state at the critical temperature, T_c, approaches infinity ($t \to \infty$). Accordingly, the left-hand side of Eqn. (31) becomes equal to zero and so, we can solve for T_c:

$$T_C = \frac{E_a}{k \ln\left(\dfrac{P_{e0}}{P_{a0}}\right)} \tag{32}$$

This description of the Morehead-Crowder model is similar to that presented by Weber et al. [7].

Note that for Eqn. (32) to yield a physically permissible value for T_c (i.e., $T_c > 0$), it is a necessary condition that P_{e0} be greater than P_{a0}. Also, as $P_{e0}/P_{a0} \to 1$, $T_c \to \infty$. Usually when experimental data is fitted to the Morehead-Crowder model, the values for two parameters are determined: the activation energy, E_a, and the rate ratio P_{e0}/P_{a0}. In an Arrhenius-type plot of $\ln(1 - t_0/t)$ versus $1/kT$, based on Eqn. (31), the slope of the curve yields E_a, while the intercept along the ordinate gives the rate ratio, P_{e0}/P_{a0}. With this, the critical amorphization temperature, T_c, can be estimated from Eqn. (32).

Figures 20 and 21 show amorphization (phase transformation) accumulation as a function of irradiation time, t, for various irradiation temperatures, T. The parameters for these calculations are described in Fig. 20.

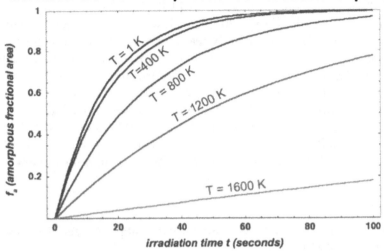

Figure 20. Plot of the dependence of the fractional amorphous area, f_a, on irradiation time, t, for selected temperatures, based on Eqn. (29b) for the Morehead-Crowder model. For this calculation, we assumed $P_{a0} = 0.0625$ s^{-1}, and $P_{e0} = 0.125$ s^{-1}, so that $P_{e0} / P_{a0} = 2$, and $E_a = 0.1$ eV.

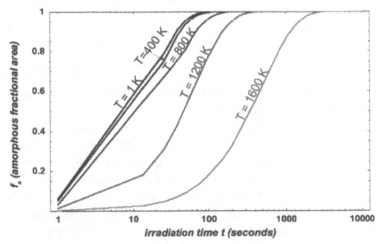

Figure 21. Fractional amorphous area, f_a, versus irradiation time, t, for the same conditions as plotted in Fig. 20. The only difference from Fig. 20 is that this plot shows the irradiation time on a log scale, so that eventual complete amorphization ($f_a = 1$) is apparent for all of the selected temperatures.

We note that the critical temperature for the conditions assumed in Fig. 20, and based on Eqn. (32), is T_c = 1674 K. So, all of the temperatures for which f_a is plotted in Fig. 20 are below T_c. For these irradiation temperatures, f_a will always eventually reach the value 1, which represents a fully-amorphized material (Figure 21 shows this). For any temperature above T_c, Eqn. (29b) produces negative, non-physical values for f_a. In other words, above T_c the Morehead-Crowder model yields *no* amorphous fraction at any ion dose (f_a = 0 for all irradiation times, t). This is quite different than the model presented in Section 2.5 which also included an eraser. In that model, the inclusion of an eraser prevented complete amorphization at all doses, but did allow for partial amorphization at any dose.

3.2   Direct Impact Model Including Thermally-Induced Interfacial Recrystallization

The last radiation-induced phase transformation model we will consider is one involving epitaxial recrystallization at interfaces. Consider a material that is susceptible to amorphization under irradiation, but that undergoes some recovery at crystal-amorphous (c/a) interfaces. We suppose that this recovery involves temperature dependent epitaxial recrystallization. For simplicity, we'll assume that the production rate for amorphization, P_a, is given by Eqn. (2), i.e. $P_a = \sigma_a \, \phi$. Also, we'll assume that the epitaxial recrystallization rate, $P_e(T)$, is given by the same expression that we used in Eqn. (28) to define the recovery rate in the Morehead-Crowder model, namely, $P_e(T) = P_{e0} \, e^{\frac{-E_a}{kT}}$. As usual, we define $f_u(t)$ as the fraction of undamaged material at any time, t, and $f_a(t)$ as the fraction of amorphous material at time t.

The trick to establishing the proper differential equations to solve for $f_u(t)$ and $f_a(t)$, is to note that for a recovery mechanism that operates at an interface (such as c/a), the rate of recovery must scale with the amount of interface, not with the fractional area of one of the phases on either side of the interface [1]. A term that scales appropriately with the amount of c/a interface is $f_a (1- f_a)$. This term increases with f_a initially, then peaks at f_a = 0.5 and diminishes back to zero (no c/a interface) as f_a approaches 1. This is in accordance with the behavior of interfacial area as a function of amorphous phase fraction.

By analogy to the development of Eqn. (3) in Section 2.1 (the direct impact model), we can write a modified pair of simultaneous differential equations, which describe the conditions appropriate to this recovery model:

$$\frac{df_u}{dt} = -P_a\, f_u + P_e(T)\, f_a\left(1 - f_a\right) \tag{33a}$$

$$\frac{df_a}{dt} = +P_a\, f_u - P_e(T)\, f_a\left(1 - f_a\right) \tag{33b}$$

These differential equations do not lend themselves to the matrix exponent solution technique. However, since there are only two fractional areas in this model, f_u and f_a, we note that $f_u(t) + f_a(t) = 1$, so that we need only solve for $f_u(t)$ or $f_a(t)$. If we substitute $f_u = 1 - f_a$ into Eqn. (33b), it is a simple matter to solve for $f_a(t)$ by the technique of separation of variables (with the boundary condition $f_a(t=0) = 0$). The solution is:

$$f_a(t) = \frac{1 - e^{-\left(P_a - P_e(T)\right)t}}{1 - \dfrac{P_e(T)}{P_a}\, e^{-\left(P_a - P_e(T)\right)t}} \tag{34}$$

This solution is equivalent to that obtained by Weber [1].

Figures 22 and 23 show plots of fractional transformation, f_a, versus irradiation time, t, for various irradiation temperatures. The parameters used for these calculations are given in Fig. 22 and are the same as those used for presenting the Morehead-Crowder model in Figs. 20 and 21.

This concludes our review of radiation-induced phase transformation rate models.

Epitaxial Recrystallization: Amorphization vs. Irradiation Temperature

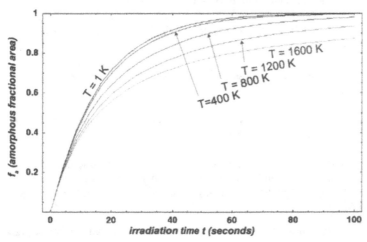

Figure 22. Plot of the dependence of the fractional amorphous area, f_a, on irradiation time, t, for selected temperatures, based on Eqn. (29b) for the Epitaxial Recrystallization model. For this calculation, we assumed the same parameter values as for the Morehead-Crowder model calculations presented in Fig.20, namely $P_a = 0.0625$ s^{-1}, and $P_{e0} = 0.125$ s^{-1}, so that $P_{e0}/P_a = 2$, and $E_a = 0.1$ eV.

Epitaxial Recrystallization: Amorphization vs. Irradiation Temperature

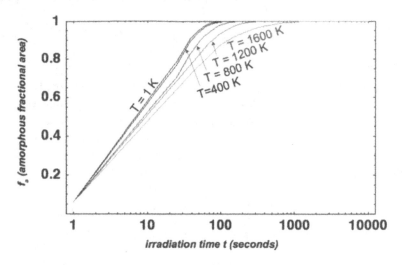

Figure 23. Fractional amorphous area, f_a, versus irradiation time, t, for the same conditions as plotted in Fig. 22. The only difference from Fig. 22 is that this plot shows the irradiation time on a log scale, so that eventual complete amorphization ($f_a = 1$) is apparent for all of the selected temperatures.

REFERENCES

[1] W. J. Weber, "Models and mechanisms of irradiation-induced amorphization in ceramics," Nucl. Instrum. and Meth. B **166-167** (2000) 98-106.

[2] R. E. Williamson and H. F. Trotter, Multivariable Mathematics (Prentice-Hall, Inc., Englewood Cliffs, NJ, 1974).

[3] S. Wolfram, The Mathematica Book, 4th ed. (Wolfram Media/Cambridge University Press, Cambridge, 1999).

[4] J. F. Gibbons, "Ion Implantation in Semiconductors - Part II: Damage Production and Annealing," Proc. IEEE **60 (9)** (1972) 1062-1096.

[5] H. Abe, C. Kinoshita, Y. Denda and T. Sonoda, "Effects of Concurrent Irradiation with Ions and Electrons on Accumulation Process of Damage Cascades in Germanium and Silicon," Proceedings of the Japan Academy Series B-Physical and Biological Sciences **69 (7)** (1993) 173.

[6] F. F. Morehead, Jr. and B. L. Crowder, "A Model for the Formation of Amorphous Si by Ion Bombardment," Rad. Effects **6** (1970) 27-32.

[7] W. J. Weber, R. C. Ewing and L.-M. Wang, "The radiation-induced crystalline-to-amorphous transition in zircon," J. Mater. Res. **9 (3)** (1994) 688-698.

Chapter 12

AMORPHOUS SYSTEMS AND AMORPHIZATION

Harry Bernas
CSNSM-CNRS, Université Paris XI, 91405-Orsay, France

1   INTRODUCTION

Since ion irradiation can lead to high densities of displaced atoms in solids, one is led to the conjecture that it may possibly amorphize an initially crystalline lattice. Under what conditions can this occur? What relation does ion-induced amorphization bear to the near-equilibrium thermodynamic conditions for obtaining those specific metastable solid solutions that qualify as "glasses"? This chapter approaches the second question first, deliberately putting ion irradiation effects in the perspective of materials science, rather than the reverse. In that sense, it is rather complementary to the presentation given in Chapter 10 (Ossi). I first recall some background information and references on the main features of the amorphous state. I then discuss amorphization dynamics – with and without irradiation – emphasizing those properties, which are specifically obtained under irradiation. The following aspects have guided my presentation.

It is supposedly "intuitive" that irradiation at a "sufficiently low" temperature should induce enough disorder in a material to drive it amorphous. In fact, this intuition requires qualification in most cases, and is often quite wrong. The reason is that although ion irradiation involves far-from-equilibrium dynamics, the requirements for a stable amorphous final state are not totally independent of quasi-equilibrium thermodynamics, just as the latter allows us (see Section 2) to make predictions regarding which metastable solid solutions may lead to an amorphous state by fast quenching. On the other hand, the particular nonequilibrium nature of irradiation – collision-induced atomic motion and/or heterogeneous atom implantation, which may be viewed as driving forces – does indeed have a major effect (see Section 3) on the stability and structures of the phases that may be accessed at a given composition, as well as on the phase transformation mechanism. In fact, there is not necessarily a stable driving force in the system. Particle beams provide a special type of control over

353

K.E. Sickafus et al. (eds.), Radiation Effects in Solids, 353–386.
© 2007 *Springer.*

statistical physics behavior, which has not always been appreciated. For example, by cutting off the ion beam an evolving system is left pending, at a point in phase (or configurational) space that depends on its prior history. By applying heat or upon resuming irradiation, the system then relaxes toward some new equilibrium configuration. To my knowledge, there has been no attempt to account for this type of behavior; this may be one underlying reason why ion beam effects are often viewed with some wariness by materials scientists.

In a word, we are required to navigate between the shores of equilibrium and nonequilibrium thermodynamics, particle irradiation providing us with a special type of statistical physics experiments which explore this *mare incognita*. In Section 4 of this Chapter, I discuss some specificities of ion irradiation in this regard, as well as a means of gathering new information on the nature and properties of the amorphous state. I include some remarks on how ion beam experiments might throw some light on the conjecture of a "glass phase transition", a very interesting, unsolved problem.

In this course I have attempted to emphasize basic problems, outline the present level of understanding, and stimulate discussion, rather than aim at completeness. My references are hence restricted to papers that emphasize selected points, or to useful textbooks or reviews from which the interested reader can find a way to the primary literature. Although I realize that this is something of a provocation in a school on radiation effects, I have deliberately underrated those features which depend on the details of irradiation-induced point defect evolution (I most often just refer to "collision-induced mobility"). This bias is counterbalanced by that of Chapter 10, which also provides many references to the ion beam literature.

2   SOME GENERAL PROPERTIES OF AMORPHOUS SYSTEMS

2.1   What is "Amorphous"?

How to define the amorphous (or "glassy") state? This basic question is still not completely resolved [1, 2]. Is it, or does it "resemble", a frozen liquid, and if so, how? If cooling is slow, the melt crystallizes below T_m via a well-defined first-order phase transition. Experience shows that the liquid must be supercooled to a temperature below T_g, far below the melting temperature T_m (Fig. 1a), at *sufficient speed* to avoid crystallization; i.e., in order to avoid nucleation of stable crystallites. Hence, an amorphous system

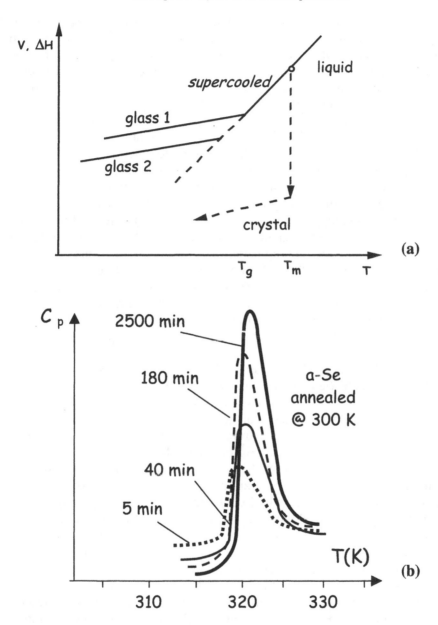

Figure 1. (a) Liquid-to-glass (vs. liquid-to-crystal) transition as detected from change in volume or enthalpy. Glass transition temperature T_g depends on supercooling speed. (b) The specific heat displays a peak near the glass transition, but this is not a standard second-order phase transition. All curves were obtained after quenching from the same melt temperature, at the same speed and to the same final temperature T_f, well below T_g. After relaxing at T_f for varying times (see figure), samples were annealed at constant speed: both height and position of specific heat peak depend on the system's relaxation towards its equilibrium configuration at T_f. Kinetics thus dominate the transition. Figure adapted from R. B. Stephens, J. Noncryst. Solids **20**, 75 (1976).

is basically thermodynamically unstable. It is produced by the slowing-down of relaxation processes that would otherwise return it to such thermodynamically stable states as the liquid (at high temperature) or some ordered crystalline structure(s) (at low temperature). On a microscopic scale, these processes involve the temperature dependence of atomic movements, lattice relaxation and lattice vibrations, which in turn differ for different materials, depending on the nature of the latter's bonds (electrostatic for insulating glasses, covalent for amorphous semiconductors, and involving hybridized conduction electrons for amorphous metals). Translated to the macroscopic scale, they involve the temperature evolution of the supercooled liquid's shear viscosity. As the liquid is cooled at a sufficiently fast rate, its viscosity increases; its volume and change of enthalpy decrease until - at a temperature termed the "glass temperature" T_g - both quantities show a marked deviation, differing increasingly from the extrapolated high-temperature curve as the temperature is reduced. T_g is the temperature at which the viscosity is sufficiently high (typically above 10^{13} P) that the material no longer flows: if the liquid is kept at some temperature below, but fairly close to, T_g, the relaxation processes will continue (albeit extremely slowly), and carry the system from the "glass" curve to the extrapolated "supercooled liquid" curve.

A crucial point is that the properties of a glass depend on its history, i.e., the initial state (temperature of the liquid), cooling rate and quench temperature. As a result, T_g for a given substance is *non-unique* (Fig. 1b), as opposed to such characteristic quantities as the melting or freezing temperatures. Thus, the glass transition first appears as a kinetic transformation leading to configurational freezing, rather than a thermodynamic phase transition. Since the relaxation processes involved are numerous and complex, we anticipate that when a given material becomes glassy, the final structural configuration on any scale may also be non-unique. This is indeed the case, as we shall see. As shown in Fig.1, the glass transformation's characteristics (notably the shift in T_g) largely depend on the ratio between the experimental time scale and the configurational relaxation kinetics. At a given temperature in the vicinity of (just below) T_g, the time evolution of the specific volume, for example, is determined by a distribution of structural relaxation times that is extremely temperature-dependent. Glass property analyses therefore depend strongly on the ratio of the experimental measuring time to the average structural relaxation time. This "time-window" effect is relevant to our topic, since flux-dependent irradiation-induced (or -enhanced, if thermal effects are present) atomic displacements are the source of structural relaxation in our case. Also, their influence will depend on the size of the structural unit involved in the relaxation process. On the other hand, it is an experimentally well-established fact that when a given glass can be prepared under such different conditions as melt-cooling, quench-condensation on a cold substrate, solid-state reactions, ion implantation or ion beam mixing, its ultimate, long-term properties show very little dependence (beyond the small shifts just mentioned) on the preparation technique. This is a strong indication that we may base our studies on the vast amount of work relating the existence of amorphous phases to free energy considerations and thermodynamic phase diagrams. We shall find that both of these apparently contradictory features of the glass transition are significant when discussing the effects of ion irradiation on amorphization.

2.2 Thermodynamics

As noted above, the structure of the final state for a quenched liquid is always the result of a competition between a (minimum energy) crystalline state and a glass. The overriding factor is the quenching speed, i.e., the speed at which atomic movements (including phonons) are slowed down. But it is not the sole factor. Although the glass-formation process is a nonequilibrium process, structural stability criteria and the statistics of atomic motion always play a role in determining whether the final state is amorphous or crystalline. It is therefore not surprising to find that thermodynamic effects influence the so-called glass formation ability (GFA). Here are a few important points, taken from examples of metallic alloys.

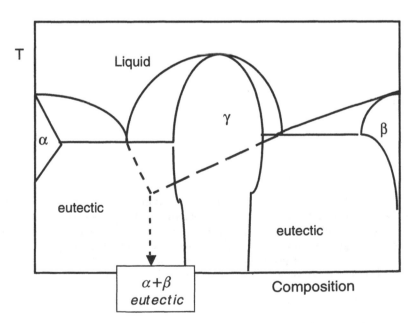

Figure 2. Example of a phase diagram in which fast kinetics leads to solidification of a two-phase component (eutectic), rather than to an amorphous alloy.

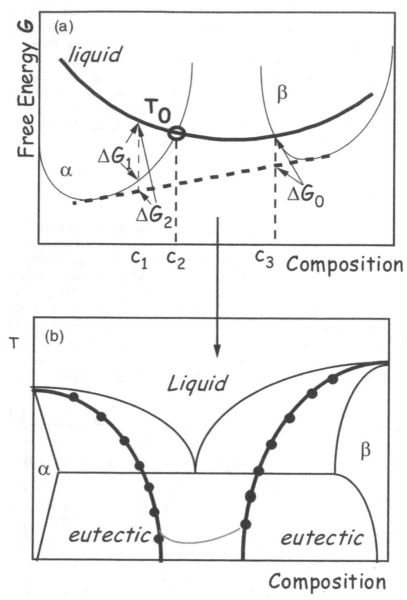

Figure 3. Role of free energy (FE) in favoring glass structure formation. (a) FE curves just above T_g for α, β and liquid phases near eutectic composition. The T_0's are the points where FE (α or β phase) = FE (liquid). The GFA is the outcome of competition between the driving force for glass formation ΔG_0 vs. that for α crystallization •G_1 and eutectic formation ΔG_2. (b) Resulting phase diagram: dotted lines are T_0 curves, between which GFA exists, if cooling is fast enough. Adapted from Ref. 4.

Standard thermodynamics of solutions provides us with free energy diagrams and the corresponding phase diagrams from which we may deduce the existence of regular solutions, unmixing (nucleation and growth of second phases), or the equilibrium between 2 phases [3]. From these, we can gain insight on the competition between the formation of a metastable solid solution versus that of forming a glass [4].

An example of a case in which a glass *cannot* form is shown in Fig. 2, where sufficiently rapid cooling from the melt avoids formation of the intermediate composition γ phase, but always favors the α+β eutectic over glass formation. On the other hand, Fig. 3 shows a phase diagram with two phases (α and β) and a so-called "deep eutectic". As seen in the free energy diagram (Fig. 3a), at some temperature near to (but above) T_g, there is a composition at which the free energy of the crystalline phase equals that of the liquid. The curve (Fig. 3b) representing all such points (termed T_0) in the (T;c) diagram draws the boundaries beyond which the solution's components separate out. If sufficiently rapid cooling has occurred and the system remains above the T_0 curve, atomic movements are frozen in, so that no crystallization may occur, possibly leading to the formation of an amorphous alloy in the region below the dotted line in Fig. 3b. The GFA in this composition range may be enhanced considerably by several additional factors: (1) the existence of different, competing incommensurate crystal structures for the α versus β phases frustrates crystallization (Fig. 4) by enhancing interfacial energies and inducing strain, which in turn oppose the chemical driving force for nucleation; (2) kinetic effects (the slowing-down of atomic movements by fast supercooling) enhance the influence of the nucleation rate and subsequent growth rate of solute phases, at the expense of the free energy difference between solute phases. The competition between the differing temperature dependences of these variables may also favor the formation of the metastable glass solution [5]. Since ion irradiation induces or enhances atomic mobility, it will modify all the properties mentioned here, but these thermodynamic properties provide a reasonable guide to the interpretation of essentially all ion irradiation experiments on alloys. The latter statement is confirmed by the previously mentioned quasi-identity of many glass-forming alloy composition ranges and properties when prepared via different techniques, including those involving ion beams.

On the other hand, at thermodynamic equilibrium a system is by definition closed, stationary and ergodic. Recall [6] that an ergodic system is such that its ensemble average is equal to the time average,

$$\langle y \rangle = \langle y(t) \rangle = \left(\frac{1}{N} \right) \sum_1^N \left[y^{(k)}(t) \right]$$

so that when evolving from state A of the system to state B, it goes through all paths AB in phase space. Now, as apparent from Fig. 1, this just does *not* happen in a glass. The relaxation of the glass via diffusion or inverse viscosity takes place by non-Arrhenian jumps [1] whose characteristics depend on its past history. The system does not progress all over phase space, but only travels through a part of it that differs according to the initial and quench conditions: this is a non-ergodic process, for which the existence and possible nature of a "glass phase transition" is an open question [7, 8]. How ion beam interactions relate to such questions will be touched upon later.

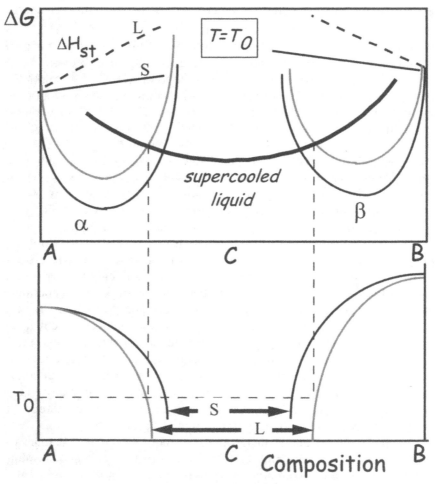

Figure 4. Same as Fig. 3, showing how strain eases the formation of a glass structure. The strain enthalpy ΔH_{st} is represented by the straight lines for small (S) and large (L) strain. The total free energy curves of the α and β phases are raised in the latter case, increasing the range of compositions where GFA is significant. Adapted from Ref. 4.

2.3   Structure: "Amorphous" Does Not Mean Random

On a near-neighbor scale, the atomic arrangement in any amorphous system (be it an oxide glass, an amorphous semiconductor or a metallic alloy "metglass") is determined, just as in a crystalline solid, by the atomic bonds. The features of the latter are by no means random and it is no surprise, from cohesion criteria, that the interatomic distances and bonding angles for a given system take rather well-defined values. As a result, short range order, or more precisely *chemical* short range order (CSRO), can be defined for an amorphous system, in the absence of long range order. From the schematic two dimensional example [1] given in Fig. 5, it is seen that rather narrow distributions in the number of atoms in a ring and among the bond angles on a short-range scale suffice to produce a solid with no long

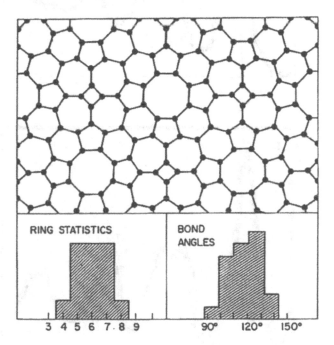

Figure 5. Two-dimensional continuous random network (upper part), and the corresponding ring size and bond angle distributions. From R. Zallen, *The physics of amorphous solids*, John Wiley, New York, 1983.

range order. This has been evidenced in X-ray and neutron diffraction experiments, as well as extended X-ray atomic fine structure (EXAFS) experiments performed on all types of glasses. Whatever the nature of their chemical bond, they confirm the existence of a well-defined CSRO [9, 10]: an example is shown in Fig. 6. On the other hand, the existence of *distributions* in the number of neighbors and interatomic distances suggests that the free energy of a glass structure is not a unique quantity, as opposed

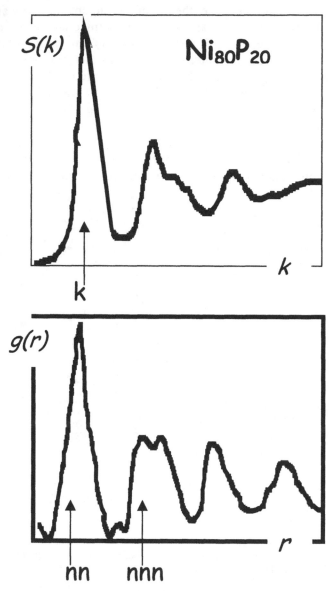

Figure 6. Typical pair distribution function (upper) and corresponding radial distribution function (lower) for an amorphous alloy (here, fast-quenched $Ni_{80}P_{20}$) as determined by X-ray scattering. For comparison, arrows show nearest neighbor and next-nearest neighbor positions in the crystalline lattice. The double peak in $g(r)$ at nnn position is a characteristic feature of the amorphous structure, due to the bonding distributions (Fig. 5). Adapted from J. F. Sadoc and J. Dixmier, Mat. Sci. Eng. **23**, 187 (1976).

to the case of a crystalline structure at the same composition, in which bond lengths and angles are single-valued. Hence, it will come as no surprise that there may in fact be a number of different amorphous states, corresponding to different minima in configurational space (Fig. 7). The existence and influence of medium range order [11], typically beyond the third or fourth nearest neighbor seen in neutron diffraction or EXAFS, is increasingly recognized.

2.4 What About the Electronic Properties?

The CSRO has a major incidence on the electronic properties of amorphous materials (e.g., see W. L. Johnson and M. Tenhover in [10], and [14]). This may be shown via a crudely simplified transport property analysis. Amorphous metallic alloys, for example, display resistivities in the range $(150 - 400)$ $\mu\Omega.cm$, i.e., corresponding to an electron mean free path roughly on the order of the interatomic distance. How does this come about ? For a free electron metal, the resistivity derived from the Boltzmann equation for incoherent scattering is

$$\rho \approx [e^2 v_F^2 N(E_F)\tau]^{-1} \tag{1}$$

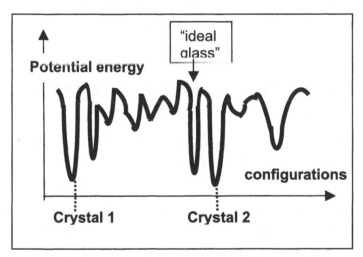

Figure 7. Potential energies corresponding to the different possible configurations (i.e., different amorphous phases) of a glass. Two crystalline phases are also shown, as well as the most stable (long-term annealed) glass. The different configurations are traditionally accessed by different cooling rates. Although a transition from one configuration to some other is possible via, e.g., annealing or applied pressure, appropriate irradiation conditions should provide a way to travel systematically from one minimum to any other. Figure adapted from C. A. Angell, Science 267, 1924 (1995).

where v_F is the Fermi velocity, $N(E_F)$ the density of states at the Fermi level and τ the electronic relaxation time. With some drastic approximations analogous to those made for the case of a liquid metal, it may be shown [9] that in simple amorphous metal alloys, precisely due to the existence of CSRO, conduction electrons are scattered by a *correlated ion distribution* leading to coherent diffracted waves. Equation (1) is then replaced by

$$\rho \approx [e^2 v_F^2 N(E_F)]^{-1} \int \left\{ [V(k)]^2 S(k) \right\}^4 (k/2k_F)^3 d(k/k_F) \qquad (2)$$

where the first term in the integral is the scattering matrix and $S(k)$ is the structure factor of the alloy as deduced from diffraction measurements (Fig. 6a). Another example is the influence of CSRO on the existence and size of semiconductor band gaps (e.g., [12]).

In summary, we emphasize that 1) glass formation is related to the interaction between free energy terms – the relative free energies of different phases, strain, interface energy differences, nucleation and/or growth enthalpies – and kinetic terms related to atomic movements (diffusion, precipitation). All of these features may be expected to affect the formation and stability of ion beam-synthesized amorphous compounds; and 2) an amorphous structure is by no means a random one – the existence of a CSRO is actually one of its main characteristics, not only from the structural viewpoint but also as regards the electronic properties. Understanding the effect of irradiation on the CSRO will be crucial to the analysis of the amorphization mechanism.

3   APPROACH TO IRRADIATION EFFECTS ON MATERIALS ("OPEN SYSTEMS")

How do irradiation or implantation affect the above considerations? Two aspects are important. First, irradiation enhances (or, at lower temperatures, induces) atomic movements, tilting the balance between free energy and kinetic terms in favor of the latter. Radiation-enhanced (or – induced) diffusion, for example, leads to drastic modifications of the phase diagram, and the departure of the latter from the equilibrium phase diagram depends on the degree of diffusion enhancement. A second, more fundamental modification is that of the very nature of the system under study. Under irradiation, elastic collisions produced by incoming particles lead to atomic displacement sequences that are local, temperature-independent and generally anisotropic; if ion implantation or ion beam mixing is involved, they are accompanied by a nonuniform compositional change. Such changes are due to externally produced forces, and the system becomes a non-ergodic, *open* system which is driven and generally kept away from thermodynamic equilibrium until the external force is interrupted by cutting off the ion beam. The system may then return to some equilibrium state, but the latter is by no means guaranteed to relate to the equilibrium thermodynamics state that would have prevailed at the same temperature in the absence of the external driving force. Similar remarks may apply to the relative stability of differing CSRO configurations under irradiation (as during a quench), with the additional problem of differences between "macroscopic" phase stability criteria and *local* stability criteria (e.g., in a glass or in a system subject to amorphization) in which small, differing local configurational entropy terms can add up to a large contribution to the Gibbs free energy. Two different but complementary approaches to ion-induced amorphization are legitimate: a "macroscopic" one that rests on considering the system as a whole, attempting to pursue the analogy with (and relation to) equilibrium thermodynamics as far as possible, and a "microscopic" approach, which considers lattice modifications on the atomic scale due to ion-induced displacements, including athermal or thermally activated defect recombinations and evolution, etc. The former approach touches on basic concepts and provides useful insight into possibly rather general criteria, whereas the latter relies more directly on material-specific or ion-beam properties (e.g., deposited energy densities, defect creation and recombination cross-sections, etc.), but may provide more precise information for a given system, especially with the confrontation of adequate numerical simulations with experimental results.

3.1 Nonequilibrium vs. Equilibrium Thermodynamics

A seminal 1975 paper by Adda [15] provided a basis for further work in this specific area. He pointed out that the irradiation-induced atomic displacement flux (which forces atomic mobility) affects not only the phase stability but also the phase transformations, and he drew a qualitative "phase diagram" in a temperature – ion flux plane (Fig. 8), displaying some possible transformations according to the relative contributions of thermally activated mobility and athermal displacement density. The latter two sometimes reinforce each other (e.g., radiation-enhanced diffusion or radiation-induced precipitation), whereas in other instances they tend to produce quite specific phenomena (e.g., amorphization or void formation). This difference in behavior is directly related to the relative influence of kinetics *versus* thermodynamics.

As opposed to that of a closed system, the evolution of a system under irradiation may be described by a series of equations containing an external source term. For such an "open system", stability does not require potential minimization, and there may well be no unique equilibrium state.

The source term contains at least a collisional jump term, which acts as an effective "collisional diffusion" term in parallel (or – at higher temperatures – interfering) with thermally activated diffusion, and possibly also (in the case of implantation) a chemical mass input and a concentration gradient.

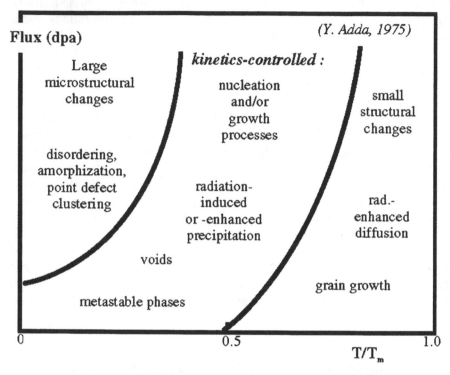

Figure 8. The heuristic phase diagram drawn by Adda for driven systems (here, particle flux versus temperature). T_m is the melting temperature. (see text for discussion).

This leads to the very characteristic features of Fig. 8. At low temperatures and at a speed depending on the ion flux, lattice disordering sets in progressively as the fluence is increased, leading to major phase transformations and possibly to amorphization (see Section 4). The transformation mechanism, a solid-solid transition, obviously differs greatly from the effects of rapid supercooling discussed above. The latter involves the entire system, whereas except at extremely high fluxes and fluences, irradiation-induced transformations are the result of a series of events occurring on a *local* (at most 10^2–10^3 nm, or less) scale. It is important to realize that the latter is the scale of atomic movements that allow the system to reach a quasi-equilibrium state *in the local volume*. It is by no means a uniquely determined dimension: it obviously depends on the size of a displacement cascade for heavy incident ions at keV-MeV energies (it may even be on the nm scale if light ion- or electron irradiation is involved), but at temperatures where defect motion occurs, it will also depend on their diffusion and possible interaction. It is thus a compound of these events that finally leads to the macroscopic transformation. Even at low temperatures, local (CSRO) stability criteria directly related to near-neighbor chemical bonding will come into play. As the temperature (or the ion flux) is raised, thermally activated (or collisional) diffusion becomes significant. This is illustrated by the onset of second-phase nucleation, for example, which may then compete with collisionally induced amorphization, just as in the near-equilibrium cases of Figs. 3 and 4.

The analysis of such open, "forced" or "driven" systems is a major, largely unsolved problem in nonequilibrium thermodynamics. Solutions have been offered [16] for a number of cases, in close similarity to the classical Cahn-Hilliard mean-field theory of diffusional transformations in solutions [17]. For example [18], one may take Cahn's theory of solution ordering or unmixing [19] as a starting point, and assume that collisional and thermodynamic driving forces operate in parallel. If collisional effects are large and isotropic but differ for the two components, the overall collisional (or "ballistic") pseudodiffusion coefficient $D_{coll}*$ for a two-component (i=1,2) system with respective concentrations (1-c) and c is

$$D_{coll}* = cD_1 + (1-c)D_2 , \text{ where } D_i = \tfrac{1}{2} \phi \, \sigma_r <r_i^2> ,$$

ϕ being the irradiation flux and σ_r the displacement (or replacement) cross-section, so the collisional interdiffusion flux is

$$J_{coll} = -D_{coll}*N_v \, \partial c/\partial x ,$$

formally analogous to chemical diffusion (N_v is the number of atoms per unit volume). The diffusion flux due to the chemical potential gradient is (by formal equivalence with Cahn's equation)

$$J_{th} = -M*N_v[f' \partial c/\partial x - 2K \partial^3 c/\partial x^3]$$

where M* is the (irradiation-enhanced) atomic mobility, f' the second derivative of the free energy, K the gradient energy term. Here,

$$M* = [c(1-c)/kT]D*$$

where D* is the irradiation-enhanced chemical diffusion coefficient. For a closed system in the absence of irradiation, J_{th} should be zero at equilibrium where the concentration profile minimizes the free energy F, which is then

$$F\{c(x)\} = N_v \int \{f[c(x)] + K(dc/dx)^2\}dx \qquad (3)$$

In the open system, under irradiation, the total flux is $J_{coll} + J_{th}$, and one can write

$$\psi(c(x)) = N_v \int \{\psi[c(x)] + K(dc/dx)^2\} dx \qquad (4)$$

where ψ is a formal equivalent, in an open system, of the free energy so that

$$\psi[c(x)] = U[c(x)] - T'.S[c(x)]$$

where U and S are the contributions to the internal energy and the entropy. By combining Equations (3) and (4),

$$\psi(c,T) = f(c,T'),$$

where

$$T_{eff} = T[1 + D_{coll}/D^*] \qquad (5)$$

in which D_{coll} is again the collisional pseudodiffusion coefficient and D^* the irradiation-enhanced thermal diffusion coefficient. Hence in this particularly simple case, the state ψ of the nonequilibrium system at temperature T (i.e., the phase one obtains) under irradiation "corresponds" to the free energy state of the system at a temperature $T_{eff} > T$ in the equilibrium phase diagram.

This approach is quite suggestive, as seen in Fig. 9: for large collisional effects and/or sufficiently low temperatures (so that the enhanced thermal diffusion is relatively weak), it leads to the prediction of an irradiation-induced amorphous state, the "corresponding state" being the liquid state at temperature T''' in the equilibrium phase diagram. At higher temperatures T' or T'' where significant diffusion sets in, nucleation and growth of one or more solid phase(s) may be expected to occur. A detailed account of such models and their extensions (including atomistic Monte Carlo simulations) is given in [16].

3.2   A Comment on Simulations

An unsurprising limitation is that when quantitative results are sought on specific systems, general mean field equations such as described above must be fed very detailed information. E.g., the collisional pseudodiffusion quantities depend on the features of the collision cascades, which in turn are ion beam- and target-specific (see Section 4 and Fig. 10). Also, modeling of the thermally activated diffusion properties requires

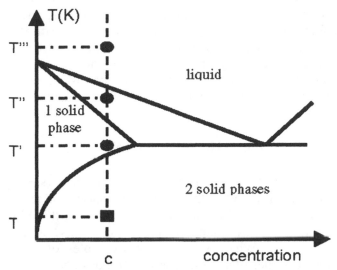

Figure 9. Example of a phase diagram for a driven alloy, adapted from Ref. 18. If the experiment is performed at concentration c* and at (low) temperature T, with a significant collision rate, the effective temperature T''' is high, see Eq. (5), and amorphization is predicted (i.e., the "liquid" phase). If the irradiation is performed at temperature T' or T", nucleation and growth of a single solid phase is predicted.

knowledge of jump energy barriers for different site environments, of the influence of outer shells (beyond near-neighbor) on jump frequencies in different directions, etc. This has led to the implementation of kinetic Monte Carlo (KMC) calculations, more or less "hybridized" by including quantities deduced from thermodynamic calculations; efforts are made to test their long-term evolution against the mean field results (e.g., [16, 20]). Such calculations are increasingly significant in view of their constantly improved capability of comparison to experiments; they generally make clear approximations, and operate on large numbers (several millions) of atoms. Limitations as regards our topic sometimes include the necessity of defining a fixed lattice (there are tricks to circumvent this limitation) and, more seriously, the fact that very often only Markovian processes are taken into account, whereas particularly in glasses the transition rate for an atom at a given time may well depend on its previous history (e.g., correlated jumps and jumps related to the local CSRO) when collisional pseudodiffusion is involved. Molecular dynamics (MD) calculations with detailed (material-dependent) atomic potentials (e.g., [21, 22]) are an alternative that, in principle, should provide exact results if adequate potentials are used. At present, a difficulty still lies in the rather small system size (from a few thousand to some 10^5) that such simulations can study, but this is progressively being overcome. For the moment, improved results are obtained by combining MD and KMC calculations, with the former providing inputs for critical parameters (jump barriers, succesive neighbor shell effects, etc.) and the latter allowing long-term studies. Simulations of much larger systems will be required in order to study such questions as the nucleation and growth of amorphous volumes in crystals, or the contribution of lattice strain to amorphization. Most questions regarding the diversity of amorphization mechanisms thus remain open.

4   BEAM-INDUCED AMORPHIZATION

Can an amorphous system form from the crystalline phase (see Fig. 1) ? If so, is there any relation between the amorphous phase thus produced and the criteria for amorphous phase formation from the melt? Why are most amorphous materials alloys rather than pure elements? Regarding amorphization mechanisms: What are the respective roles of chemical disorder versus lattice disorder (defects)? Is the amorphous phase produced by heterogeneous or homogeneous nucleation, does the amorphous phase grow from its crystalline counterpart, and if so, how? Is the crystalline-to-amorphous transition a phase transition? None of these rather fundamental questions are restricted to the realm of particle-induced amorphization. However, a brief review of the latter area shows that not only do the general features discussed above help to elucidate the effects of irradiation, but the reverse may occur, with irradiation (or implantation) experiments providing information on the static and dynamic properties of the amorphous state.

Let us first consider the apparently simplest possible case: that of ion-induced disordering at low temperatures, where defect (and constituent atom) mobility is absent or at least considerably reduced. It is in fact by no means simple at all. Irradiation by 1-3 MeV electrons leads to collisions with sufficient recoil energies to displace most elements, but the displacement cross sections are sufficiently low to induce mostly isolated defects (Frenkel pairs, a vacancy and an interstitial) and, to a far lesser extent, very small, isolated defect clusters (e.g., di- or tri-interstitials and their vacancy counterparts). Ion irradiation, on the other hand, generally (except for the lightest elements, protons or He ions) produces damage cascades whose deposited energy density – which depends both on the elements and on the incoming ion energy – is a crucial factor in determining the density and homogeneity of atomic displacements and stable damage. An example is shown in Fig. 10, which is a direct (field emission microscopy) measurement at 20 K of displaced target atoms in a W tip under irradiation at 30 keV by different ion masses. The differences are striking, and cannot be ignored when interpreting ion-induced disordering or amorphization effects, particularly if such effects appear in the ion fluence range where these damage cascades are just overlapping. The analysis of disorder in terms of point defect evolution is amply documented elsewhere (e.g., [23] and Chapter 10). Here, the reader is invited to meditate on the fact that when the proportion of vacant neighboring sites is very high (W cascade), the surrounding lattice may simply collapse athermally (minimizing its strain and potential energy) to form a small, stable vacancy

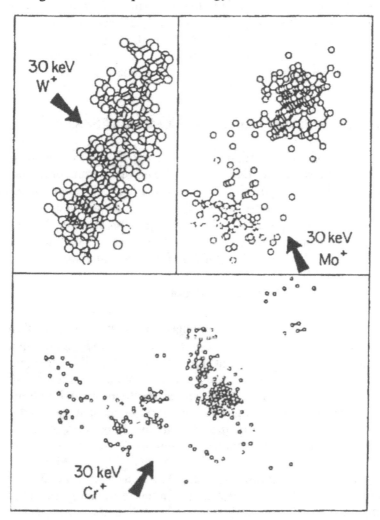

Figure 10. Experimental field ion microscopy: defect (vacancy) cascade in a W metal tip at 20 K by a single 30 keV ion impact of either W, Mo or Cr. Shows effect of deposited energy density on vacancy distribution. Figure from M. I. Current et al., Phil. Mag. **A47**, 407 (1983).

loop; when the proportion is a few percent (Cr cascade), the entire (larger) area which contains the damage sees its potential energy rise and its stability fall drastically. Hence, although the deposited energy density (and the defect production concentration) is higher in the former case, the resulting disorder may be less than in the latter case. If a phase transformation occurs at significantly higher fluences than those corresponding to cascade overlap, other effects (e.g., chemical effects in alloys) will clearly have to be taken into consideration. If it occurs at fluences well below this threshold, the influence of defects outside the cascade core – or of defect-induced nucleation *and growth* – will have to be considered. As discussed below for Si and metals, the ultimate fate of such a zone depends on the bonding properties of the material and on defect mobility: albeit in different ways, it always contributes to amorphous 'growth'.

4.1   Amorphization by Kinetics : The Case of Silicon

Now consider the particularly well-studied case of elemental Si, whose beam-induced crystalline-to-amorphous (c/a) transformation was first observed decades ago. Fig. 11 shows a typical room-temperature amorphization curve under Si ion irradiation, as determined via a RBS-channeling (RBS/C) experiment. The "damaged fraction" in the figure is actually the amorphous fraction α, as verified by transmission electron microscopy (TEM). The experimental curve may be fitted by

$$\alpha(c/a) = [1 - \exp(-\sigma \phi t)] \qquad (6)$$

where ϕ is the ion fluence and σ a "damage cross-section" to be determined. Essentially identical curves (with different values of σ, ϕ) are obtained for ifferent irradiating ions. It was tempting to interpret Fig. 11 as meaning that amorphization occurs inside individual cascades, whose ultimate superposition leads to overall amorphization in the implanted volume [24]. A crude but unfortunately long-lived assumption is that each individual cascade leads to a "thermal" (or rather collisional) spike, forming a liquid drop that is quenched in the amorphous phase. A wealth of experiments have proved that this analogy to Fig. 3 was carried too far. Perhaps the most convincing is the demonstration [25] that MeV electron irradiation (which precludes displacement cascade formation, albeit producing defects slightly more complex than Frenkel pairs) can induce Si amorphization at 25 K. Early on [26] it was suggested that the driving force for the transformation was due to the storing of lattice defects, which would locally raise the free energy of the sample above that of amorphous Si. Many experiments showed that amorphization became more difficult or impossible just as defect mobility set in, and recent MD simulations (e.g., [27]) concur with high resolution TEM experiments [25], showing that the most effective defects actually resemble "amorphous nuclei" composed of 5- and 7-member Si rings. Where are such defects created when the sample is ion-irradiated ? If they were only in the cascade core, ion-induced amorphization would be far less efficient. In-situ TEM (i.e., with an ion beam hitting the sample in the TEM chamber) [28] showed (Fig. 12) that there is actually a form of nucleation and growth in the amorphization process, mediated by the existence of defected zones outside of the cascade core, and the aforementioned MD simulations (see also [23, 29]) have given insight into the process.

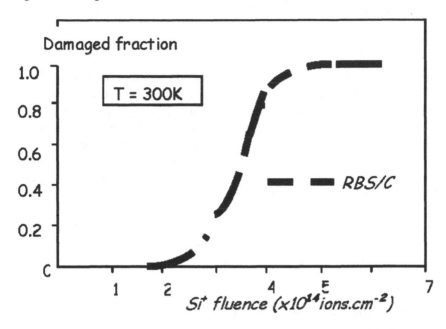

Figure 11. Amorphization curve of Si under Si ion irradiation at 300 K, as determined by a Rutherford Backscattering/channeling experiment. Adapted from O. W. Holland, S.J. Pennycook and G. L. Albert, Appl. Phys. Lett. **55**, 2503 (1989).

Thus, for elemental Si with covalent bonds, the transformation to the amorphous phase is neither unique nor a simple one. Heterogeneous amorphization may occur (e.g., via electron irradiation), but a form of "local" homogeneous nucleation and growth can also, and more frequently, occur in ion-irradiated samples. The structure of the corresponding "nuclei" (complex defects, odd-numbered rings...) is still a matter of debate, as is the experimental analysis of medium range order [11]. Moreover, the in situ TEM experiments show that the strained-induced contrasts due to the dislocations formed in the cascade core survive at least partially as amorphization proceeds, so it is quite possible that relatively long-range strain effects – indicated in early MD simulations [23] – may also contribute to lattice destabilization ("jamming effects", see below) and subsequent amorphization. The two mechanisms cannot be discriminated in Fig. 11 and Equation (1): this is not too surprising, since the latter just says that ion irradiation proceeds via a Poisson distribution of impacts with n=1, and provides no information on the microscopic process. In fact, it should be recognized that the agreement of results with Equation (6) (or with higher order Poisson distributions, for that matter) is only indicative. For example, fits of data very similar to Fig. 11 by the well-known Johnson-Mehl-Avrami phase transformation equation [30] were convincingly presented as evidence for locally "homogeneous" nucleation and growth of amorphous zones in Si [31]. Its form is

$$\alpha(c/a) = (1-\exp Kt^n),$$

in which $K =(\pi/3)R_N R_G^3$ (where R_N and R_G are the rates of nucleation and of growth, respectively), and n is a coefficient depending on the nucleation and growth mode (varying between 3/2 and 4, always superlinear – the values of this coefficient for different transformation modes are tabulated in metallurgy textbooks). This equation very likely applies to irradiation experiments performed at temperatures high enough for the amorphous lattice to relax (without recrystallizing), so that the system's free energy may travel from one minimum to another in Fig. 7. The superlinear fluence-dependence, and the corresponding value of the mode parameter n, change when the amorphous component's growth is thus enhanced. Note that the same behavior holds in the reverse case, when heterogeneous crystallization occurs in an initially amorphous sample.

4.2 Amorphization by Combining Kinetics and Chemistry: Metallic Alloys

Although they contain two or more components, the analysis of crystalline-to-amorphous transformations in metallic alloys actually provides more information on the process itself, because it relates more directly to the features described in Sections 2 and 3 of this chapter. Consider a simple ordered intermetallic compound (e.g., ABABABA... sequences along one or more low-index directions). Conceptually, it may be affected by irradiation in two different ways, chemical disordering and structural defect accumulation. Electron or ion irradiation at energies above a threshold can displace the atoms, leading to individual replacements or replacement sequences. Ultimately, the alloy is chemically disordered (e.g., ABAABABBB...). The initial crystal structure is preserved if the ordering energy is lower than the energy difference between the amorphous and crystalline states (in fact, a chemically disordered alloy may even undergo irradiation-induced ordering in this way [32]), or the reverse may occur and amorphization can take place. Actually, of course, both processes occur ; MD simulations indicate that in some cases (e.g., NiZr$_2$ [21]) chemical disorder suffices to amorphize, while in others (e.g., Cu-Ti intermetallics [22]) they may act sequentially (first chemical disorder, then defect accumulation in a destabilized lattice) to amorphize. The features (relative free energies, stability of the initial and final states, interfacial energies and strain) discussed in Section 2.2 and shown in Figs. 3 and 4 definitely play a role in determining the GFA of the irradiated intermetallic alloy. They serve as a guide for experiments and largely account for the fact that pure metals are practically not amendable to amorphization. Also, they indicate

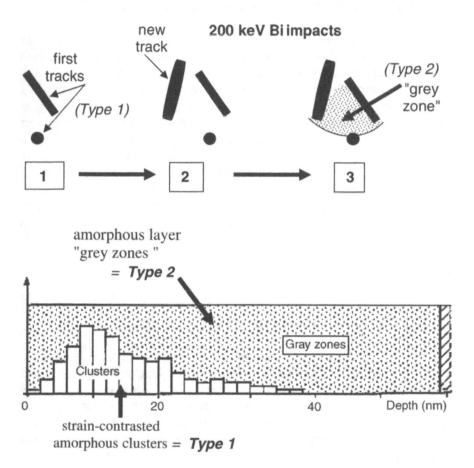

Figure 12. In situ TEM experiment showng 200 keV Bi$^+$ ion-induced amorphization process in Si at 300 K. Upper: (1) Two impacts within ca. 60 nm produce two high-contrast defected areas (dislocations, Type 1 damage); (2) a new impact occurs within ~60 nm, producing another Type 1-contrast area; (3) simultaneously, the entire area within the 3 impact tracks becomes "grey", identified by selected area diffraction as amorphous (Type 2 damage). Lower: comparing the defect cluster depth histogram as measured via TEM with the amorphous layer thickness as measured by RBS/channeling. Type 1 damage anneals at 500 K; Type 2 damage anneals at 800 K. For discussion, see text. Figure adapted from M. O. Ruault, J. Chaumont and H. Bernas, Nucl. Inst. Meth. **209/210**, 351 (1983).

when long-term irradiation might lead to formation of an amorphous compound, via a sufficiently broad available compositional range, sufficient structural complexity or structural differences between unmixed phases. This is even more important when the initial sample is a A_xB_{1-x} multilayer subject to ion beam mixing. In that case, the evolution of the interfaces goes through the entire composition range before reaching the aimed average composition, so that the GFA or phase separation characteristics described in Section 2.2 are to be taken into account over the whole range when attempting to anticipate the final compound structure [33].

But this is not the whole story by far, since it does not include the fact that the system is dynamic; its evolution is driven by a continuous flow of atomic displacements and (in the ion beam mixing or implantation case) a continuous compositional change. Hence, the guidelines provided by Figs. 2 – 4 are insufficient, and one must think in terms nearer to those of Fig. 9: depending on the composition, temperature and ion beam flux, the system may go directly to the amorphous state or to a 2-phase state. This is analogous to changing the quenching speed in the quasi-equilibrium case, with the proviso that here the temperature range depends on the collisional contribution. These dynamical features of amorphization were spectacularly illustrated [34] in the case of NiTi MeV electron-induced disordering (requiring both chemical and defect-induced contributions) where, due to atomic displacements, flux- and fluence-dependent structural fluctuations between the ordered and amorphous phases were monitored in situ under irradiation.

4.3   The Case of Metal-Metalloid Compounds

Let us now consider the case of the transition metal-metalloid (B, C, N, P, Si...) compounds around the deep eutectic composition (e.g. Fe_3B, $Ni_{80}P_{20}$, $Pd_{80}Si_{20}$,...). These were the first "metglasses" successfully produced by ultra-quenching, due to the presence of the covalent metalloid bonds and to the corresponding structural complexity of many of their ordered phases. The progressive amorphization under ion irradiation of such compounds produced in crystalline phases at compositions in the GFA range, was studied by several groups. Complete amorphization occurred at low fluences, in the 0.1 dpa (displacements per atom) range, and the amorphization fluence dependence resembled that of Equation (5) and Fig. 11. Under ion irradiation, the glass-forming compositional range and their CSRO – in those cases where it was determined – were identical to that of their quenched counterpart.

Ion implantation offers another means of producing alloys of the same nominal composition, by direct implantation of the metalloid into the initially pure crystalline metal. For the Ni-P system, a combined channeling [35] and in situ TEM study [36] of progressive amorphization as a function of the implantation-induced compositional change showed (Fig. 13) that at 80 K (precluding defect or atom movement) the amorphous fraction α varies as

$$\alpha = \sum_{N = N_c}^{\infty} \frac{(\langle N \rangle)^N}{N!} \exp(-\langle N \rangle) \qquad (7)$$

where it is assumed that amorphization proceeds by a build-up of implantation-induced elementary amorphous clusters of volume v_c, synthesized when $N \geq N_c$, and whose size may be determined from a single-parameter fit of the curve to Equation (7). The mean number $\langle N \rangle$ of P atoms in a cluster is proportional to the latter's volume and to the mean concentration c_c of P in the sample. The "amorphization threshold" (half-way point of the curve) corresponds to a critical concentration of $c_c = 12\%$, as opposed to the fast-quenching amorphization threshold of 18-19%, showing that irradiation has indeed broadened the GFA concentration range. Note that such implant concentrations correspond to a total dpa rate of about 100, so that defect density saturation has long been reached: amorphization is essentially due to chemical effects, just as it was due to chemical disordering when irradiating intermetallic alloys such as NiTi or the metal-metalloid compounds at the deep eutectic composition.

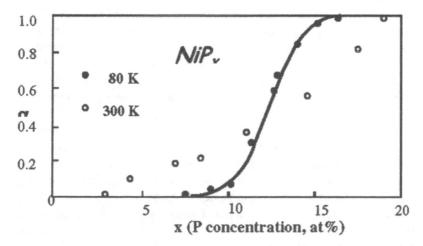

Figure 13.   Dependence of amorphous fraction α on concentration x of 100 keV P ions implanted into Ni at 80 K or 300 K. Solid line: best fit to 80 K results with Equation (7) and a critical volume v_c of radius ~1 nm. At 300 K, both collisional defects and P atoms move under implantation; it is likely that amorphous clusters grow, so that the percolation model is no longer appropriate. Adapted from Ref. 35.

The critical volume v_c (assumed to be a sphere), as deduced at 80 K, is found to have a radius corresponding to the average near-neighbor distance around a solute atom – just the distance over which the CSRO may be defined according to EXAFS, X-ray or neutron diffraction measurements. A room temperature implantation changes the curve significantly (open dots in Fig. 13), due to defect evolution and related metalloid atom short-range diffusion – i.e., the system relaxes, amorphous CSRO clusters rearrange and probably grow similarly to the observation of [28], so that the amorphization threshold increases to the value observed after fast quenching where analogous processes occur. All these features were also found in the metalloid implantation-induced amorphization behavior of Fe-B, Pd-Si, Ni-B, for which the elementary cluster size determined at low temperatures remained the same, or after D^+ irradiation-induced amorphization at 15 K of the Ni_3B and Fe_3B compounds (i.e., in the absence of defect cascades). As noted above, a Poisson distribution in itself does not provide information on the amorphization mechanism. However, the size deduced from the fit to Equation (7) is an interesting indication that at least for those metglasses whose CSRO is determined by covalent bonds, the amorphous lattice results from a progressive accumulation of nanometer-size elements, whose packing properties are presumably determined both by the CSRO around a solute atom and by the organization of the initially crystalline host. The heterogeneous character of this implantation-induced crystalline-to-amorphous transformation of NiP_x was confirmed by conduction and magnetic property measurements [37]. Its significance will be discussed in Section 4.4.

How does this relate to the discussion in Section 3? If, instead of a comparatively homogeneous system such as exists inside an implantation profile, one considers the implantation profile edges or the ion-beam mixing of multilayers, the chemical potential gradient combined with defect mobility is an effective driving force, and the references given in Section 3 are relevant. At room temperature (or above), for the metalloid-implanted metals considered here, defect mobility should play a role so that both the implantation-induced chemical potential change and the accompanying atomic displacement flux contribute to the driving force, and hence to the amorphization dynamics. The latter is correspondingly complicated, especially if – as is likely – atomic motion favors growth of the amorphous clusters in the destabilized crystalline lattice. At 80 K it is also a "driven system", but the external driving force is then largely provided by the compositional change more than by the related rate of atomic displacements. The total number of dpa is very large, so that there are typically 10^2 collision-induced rearrangements of each local nanometer-size CSRO configuration in the sample. Due to the high total collision rate, the initially "frozen" system approaches ergodicity. This should bring it into one of the secondary minima in Fig. 7, albeit not necessarily the lowest one.

These low-temperature ion implanted (or low-flux irradiated, for that matter) metal-metalloid systems are thus limiting cases which probably do not require the approach of Section 3. They are amendable to the percolation description below.

4.4 Nature of the Amorphization Threshold and the Glass Transition

Regarding the general features of the glass transition, we have so far emphasized the kinetic aspect. Discussions of the excess enthalpy temperature dependence in glasses, and experimental results such as the more-or-less λ-shaped peak in the specific heat at T_g, have suggested the possibility of an underlying thermodynamic glass phase transition (see Fig. 1b, and the discussions by J. Joffrin and P. W. Anderson in [38] or by C. A. Angell in [2]). This remains quite uncertain [1, 2, 7], and raises fundamental questions that are well worth delving into, but are outside the scope of this lecture.

A way to approach the glass transition without entering into the discussion on an underlying phase transition was the "free volume" theory [39], based on the notion that glasses (as liquids in J. Frenkel's classical theory) contain extra volume relative to the crystalline phase, with atomic transport being possible only between the corresponding "open cells". This projects the glass transition problem onto that of a percolation transition (see [1] for an excellent exposition), an approach that is particularly well-suited to studies of random systems. Percolation theory is a mathematical construction: its application does not require that the physical structure of the "open cells" be defined beyond a clarification of geometry (i.e., whether the connections between "lattice points" are determined by sites or by bonds). Introducing this method into the analysis of different experimentally determined physical properties of amorphous materials provides a useful guide for comparisons with MD simulations based on varying assumptions as to the nature of the elementary structures that might percolate in the glass.

Let us take the implantation-induced amorphization experiments referred to in Section 4.3 as an example. Can the amorphization process be analyzed by a percolation model ? According to the simple assumptions behind Equation (7), the Ni sample is made up of a crystalline (fcc) base which is progressively transformed at 80 K via P implantation and the accompanying atomic collisions, by turning elementary nm-size volumes from crystalline to amorphous as described above. As the concentration of these elementary building-blocks increases, they randomly connect to each other in the fcc lattice. Fig. 13 shows that the average metalloid concentration at which amorphization occurs at low temperature in these compounds is close to 12%, just the three-dimensional bond-percolation threshold. of the fcc percolating lattice. (When the implantation is performed at room temperature, static percolation no longer holds: as noted above, under the combination of irradiation and thermal activation, the system experiences time fluctuations, diffusion and growth that bring it closer to the "quenched glass" states). A very recent experimental, simulation and modeling study of atomic arrangements in both metalloid-metal (including Ni-P) and metal-metal metglasses [40] has shown that at comparatively low concentrations, solute atoms surround themselves with near-neighbor solvent atoms only, forming different types of icosahedra-like clusters which in turn tend to form "clusters of clusters" via symmetry and connectivity rules. The ion beam amorphization process suggested above is quite consistent with this picture.

In the case of intermetallics such as bct-structure $NiZr_2$, it was shown [41] that the onset of amorphization is accompanied by a drastic softening of the elastic properties. MD simulations [21] showed that this effect is directly related to the production of "distorted volumes", by accumulation of antisite defects, and that these volumes percolate at a threshold concentration of 15%, leading to an abrupt increase in the shear modulus, resembling a second-order phase transition. Because it includes covalent bonds, a heterogeneous system such as NiP_x should be expected to form even more highly distorted volumes leading to larger localized strain, inducing shearing at and above the percolation threshold. This is found indeed, as shown by in situ TEM experiments [36] on unsupported films. As the increase of implanted metalloid concentration in the metal leads the sample through the amorphization curve of Fig. 13, periodic stress first appears in the film [42]. When the metalloid concentration reaches about 12 %, shear thinning and shear softening lead to viscous flow and subsequent tearing of the film (a rather striking "chewing-gum" effect, Fig. 14). Recent MD simulations of the consequences of strain in amorphous materials [43] strongly support the existence of small volumes (estimated size typically on the nm, CSRO scale) in which stress is localized ("jammed"), and whose percolation corresponds to a threshold for plastic flow, leading to *progressive* "unjamming" by the creation and disappearance of such trapping volumes. In quenched glasses, the "unjamming" is usually provided by the strain-rate induced deformation (i.e., changing the quenching speed). The specific feature of ion beam experiments is that viscous flow sets in quite *suddenly* as soon as the percolation threshold is met. This is likely due to the constant creation and destruction of the "jamming volumes" by the successive collisional displacements accompanying metalloid implantation: at the percolation threshold there may be virtually no change in the metalloid concentration, but there is an onset of intense structural reorganization via atomic motion *on the nm scale*. This suggests that we give a novel turn to our outlook on ion beam experiments.

4.4 Ion Beams and Ergodicity

Let us compare samples prepared via the different irradiation or implantation experiments with the ultraquenched metglass in the same composition range. We have noted that the latter is non-ergodic: at any time, the free energy state of a sample depends on its history. Overall, the same is true for ion beam mixing, whose analogy with "classical" solid-state amorphization is rather well-established (see [33] and refs. therein). Suppose we wish to form a homogeneous alloy with a composition given by the ratio of the multilayer film components. As long as there is a chemical potential gradient, the final state of the multilayer will actually depend on the ion beam mixing conditions (ion flux, total deposited energy density and temperature), on the relative heats of mixing and nucleation rates of the compounds that may form on the way to uniform mixing, and on the way in which they form. Depending on these parameters and their interaction, stable intermediate phases may or may not nucleate and grow, and amorphization may (or may not) occur. This is amply exemplified in the ion beam mixing literature (cf. Chapter 10), and puts ion beam mixing-synthesized alloys in the same realm of complexity as ultraquenched metglasses.

What about the driven 'metglass' (such as NiP_x or $PdSi_x$) alloys, as they evolve towards the amorphous state either by irradiation of the crystalline metal-metalloid compound or by implantation of the metalloid atoms into the initially pure metal? If we consider such a system as a whole, its evolution is definitely non-ergodic, and increasingly so as it becomes totally amorphous. This is true whatever the time window we use to study them. But now – in the *same* samples – let us consider the subsystem defined by (i) the population of nm-sized amorphous clusters (inside the crystalline lattice), for any concentration up to the percolation threshold, and (ii) in the case of metalloid implantation into the metal, a time-window such that the average concentration increment is negligible (i.e., collisions are very numerous and effective, but there is no concentration change). Because of the collision-induced restructuring of the small entities, the evolution of this subsystem is no longer restricted to a limited region of phase space: *for the nm-scale population the evolution tends to become ergodic*, even to the point where some amorphous clusters revert to the crystalline state, while others are subject to the reverse transformation (e.g., the Watanabe experiment [34], Section 4.2). In the concentration range above the percolation threshold, the transformed clusters – which now

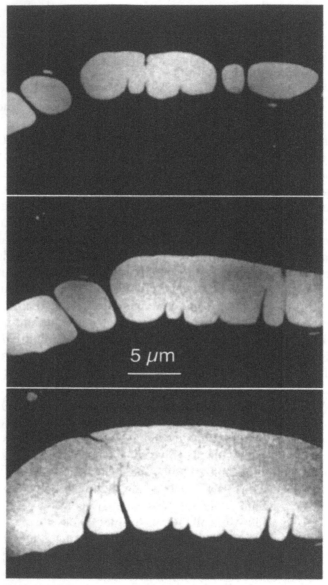

Figure 14 . In situ TEM observation of shear thinning, shear tearing and viscoplastic flow during Si ion implantation into Pd. The figure shows 3 stages of the same sample area (about 20% increase of the Si concentration between upper and lowest frame) just above the percolation threshold of the amorphous volumes (see text). From M. Schack, PhD. Thesis, University Paris XI, 1984.

contain more metalloid atoms – tend to stay amorphous and contribute to the branching of the infinite percolating clusters in the sample; the effect of collisions is to locally change their CSRO, hence the local configurational entropy (a very interesting result of Ref. 40 is the very large number – hundreds – of coordination polyhedra available in an amorphous structure), and the position (but not the average number) of "jamming sites". At higher concentrations, the amorphous clusters invade the whole sample. The latter is, so to speak, "uniformly amorphous", i.e., a non-ergodic metglass, but we may retain our definition of the subsystem by dividing up the metglass into nm-sized volumes as before, according to [40]. And on that scale, which determines the evolution of the jamming site population, it is essentially ergodic *because of the low temperature irradiation,* and it can fairly be described by equilibrium statistical mechanics with beam-induced collisions as the source of atomic motion. The existence in physical systems of components with differing ergodicities [8] is not exceptional (e.g., in magnetism). But particle irradiation is a physical tool that modifies the statistical, as well as the structural, behavior of the overall system's crucial (CSRO-scale) component, and irradiation can act *independently* on the structural and the dynamic mechanisms of a subsystem's evolution. Let us consider them in turn.

The structural modification induced by a particle depends on its deposited energy density (i.e., particle mass and energy – individual atomic displacements or displacement cascade). There have been experiments [44] in which ion irradiation and subsequent annealing provided evidence of anneal-driven transitions from a 'more distorted' to several different 'less-distorted' amorphous states. It was thus possible to find conditions that allowed the system's potential energy to jump from one minimum to a deeper one in Fig. 7, in direct correspondence to what is found by researchers performing annealing studies of the glass-to-liquid transition.. I argue that it should, in principle, be possible to travel by appropriate control of the irradiation conditions from an initial amorphous configuration in Fig. 7 either *up or down* the potential energy stream. The electron irradiation-induced transformations [45] of Co_3B shown in Fig. 4 of Chapter 10 may be viewed in this way, which is complementary to the description in terms of defect motion. Under irradiation there is, as the temperature increases, a crystal-to-amorphous phase transformation followed by an amorphous-to-crystalline transformation. By subjecting a non-ergodic system (the Co_3B metglass) to an appropriate combination of ion-induced and thermal atomic mobility, its nanoscale subsystem becomes ergodic and may explore the crystalline phase (this is somewhat analogous to the effect of adding noise onto a chaotic process in order to restore or obtain a regular behavior); the irradiation-induced amorphous-to-crystalline transformation thus loses some of its "unusual" character and may be analyzed in terms of the discussion in Section 3.

Regarding dynamics, the situation is quite intriguing. Many studies [2] of quenched glasses naturally focus on the temperature range around T_g, in which the viscosity (or relaxation time) changes by about twelve orders of magnitude when shifting the temperature by less than 10%. A major reason for this focus is obviously the existence of a reasonable experimental time window (relaxation phenomena are on the 100-second scale when the viscosity is around 10^{13} poise) in that region (Section 2.1). Here, irradiation experiments could open new vistas. At low temperature under irradiation, the nanoscale dynamics are ergodic, directly controlled by statistical collisions due to the particle beam rather than to the temperature – the "observational time window" (which is actually an average collision time in a nm-size volume of the sample) depends on the beam intensity, but this does not affect the ergodicity. The results of Fig. 14, performed at 15 K, suggest that there is indeed such a thing as "radiation-induced viscoelasticity", but they are only eye-openers at this stage. It is clear that experiments should be done in conditions where the temperature plays a more significant role, in the range nearer to (but still far below) T_g where thermal relaxation times are considerably shorter and so comparison and overlap with more familiar glass studies and models can be made.

5   CONCLUSION

Our purpose was to describe present knowledge on the relation between ion-induced amorphization and the near-equilibrium thermodynamic conditions for synthesizing "glasses" of different types, and to emphasize the specific contribution of ion beam experiments to the existing base of knowledge on glasses and the glass transition.

From this particular viewpoint, via a combination of irradiation, ion beam mixing and implantation experiments with theoretical approaches and simulations, many facets of the rather uncommon solid (crystal)-to-glass (as opposed to the usual liquid-glass) transition have been uncovered. However, it is probably fair to say that we have not yet exploited all the specific advantages of ion beam physics in controlling the structural and dynamic properties. The ability to (at least partially) control a hierarchy of characteristic scales with different statistics in a physical system is not common, and the possibilities given by irradiation in this regard should be explored in further detail.

ACKNOWLEDGMENTS

I am indebted to many colleagues, especially the members, post-docs and students of the Solid State Physics Group at CSNSM-Orsay and notably M.-O. Ruault and P. Nedellec, for their contribution to the topics of this chapter. Discussions with J. Friedel, J. F. Sadoc, F. Cyrot, C. Janot, P. Averbuch, R. Jullien and K.-H. Heinig have been of great value.

REFERENCES

[1]   R. Zallen, The Physics of Amorphous Solids (John Wiley, New York, 1983).

[2]   see "Frontiers in Materials Science", Science **267**, 1924-1953 (1995).

[3]   D.E. Porter and K.E. Easterling, Phase Transformation in Metals and Alloys (Van Nostrand Reinhold, Wokingham, England, 1981).

[4]   U. Mizutani and T.B. Massalski, Proc. AIME Symposium on Alloying Properties of Noble Metals, ed. by T.B. Massalski, L. H. Bennett, W. B. Pearson and Y. A. Chang (AIME, Warrendale, PA., 1986).

[5]   J. H. Perepezko, Progr. Mater. Sci.,**49**, 263 (2004).

[6]   L. Landau and I. M. Lifschitz, *Statistical physics*, Addison-Wesley (1958).

[7]   J. Jäckle, Rep. Prog. Phys.**49**, 172 (1986).

[8]   R. G. Palmer, Adv. Phys. **31**, 669 (1982).

[9]   N. E. Cusack, The Physics of Structurally Disordered Matter (Adam Hilger, Bristol, 1988).

[10]   Glassy Metals, Magnetic, Chemical and Structural Properties, ed. R. Hasegawa (CRC Press, Boca Raton, 1983). See article by D. Boudreaux.

[11]   Ju-Yin Cheng et al., J. Appl. Phys., **95**, 7779 (2004) and refs. therein.

[12]   G. T. Barkema and N. Mousseau, Phys. Rev. **B62**, 4985 (2000).

[13]   W. L. Johnson and M. Tenhover, in Ref. 10.

[14]   H. Bernas, in " Materials Modification by Ion Beams ", ed. R. Kelly and M. F. Da Silva, NATO ASI Series E, (Kluwer Academic, 1989).

[15]   Y. Adda, M. Beyeler and G. Brebec, Thin Solid Films **25**, 107 (1975).

[16]   G. Martin and P. Bellon, Sol. State Phys. 50, 189 (1997) and refs. therein.

[17]   J. W. Cahn & J. E. Hilliard, J. Chem. Phys. **28**, 258 (1958); **30**, 1121 (1959); and **31**, 688 (1959); and J. W. Cahn and J. E. Hilliard, Acta Metall. **19,** 151 (1971).

[18]   G. Martin, Phys. Rev. **B30**, 1424 (1984).

[19]   J. W. Cahn, Acta Met. **9,** 7 (1961).

[20]   M. Strobel, K. H. Heinig and W. Möller, Phys. Rev. **B64**, 24 5422 (2001), V. Borodin, K. H. Heinig and S. Reiss, Phys. Rev. **B56**, 5332 (1997).

[21]   C. Massobrio et al., Phys. Rev. **B41**, 10486 (1990).

[22]   M. J. Sabochick and Nghi Q. Lam, Phys. Rev. **B43**, 5243 (1991).

[23]   R. S. Averback and T. D. de la Rubia, Solid State Phys. **51**, 281 (1998) and refs. therein.

[24] F. F. Morehead & B. L. Crowder, Rad. Eff. Def. Sol. **6**, 27 (1970) ; J. F. Gibbons, Proc. IEEE **60**, 1062 (1972).

[25] J. Yamasaki, S. Takeda and K. Tsuda, Phys. Rev. **B65**, 115213 (2002).

[26] F. L. Vook & H. J. Stein, Radiat. Eff. **2**, 23 (1969) ; M. L. Swanson et al., Radiat. Eff. **9**, 249 (1971).

[27] Lourdes Pelaz et al., Nucl. Inst. Meth. Phys. Res. **B216**, 41 (2004) and refs. therein.

[28] M. O. Ruault, J. Chaumont and H. Bernas, Nucl. Inst. Methods **209/210**, 351 (1983) and H. Bernas, M.O. Ruault, and P. Zheng, p.459, Crucial Issues in Semiconductor Materials & Processing Technologies (Kluwer Academic, N.Y., 1992).

[29] Laurent J. Lewis and Risto M. Nieminen, Phys. Rev. **B54**, 1459 (1996).

[30] M. Avrami, J. Chem. Phys. **7**, 1103 (1937); **8**, 212 (1940); **9**, 177 (1941).

[31] S. U. Campisano et al., Nucl. Inst. Methods **B80/81**, 514 (1993).

[32] H. Bernas et al., Phys. Rev. Lett. **91**, 077203 (2003).

[33] B.X. Liu, W. S. Lai and Z. J. Zhang, Adv. Phys. **50**, 367 (2001).

[34] S. Watanabe et al., Phil. Mag. **83**, 2599 (2003).

[35] C. Cohen et al., Phys. Rev. **B31**, 5 (1985).

[36] M. Schack, PhD. Thesis, University of Paris XI (1984).

[37] A. Traverse et al., Phys. Rev. **B37**, 2495 (1988).

[38] see lectures by J. Joffrin and P. W. Anderson in *Ill-condensed Matter* , ed. R. Balian, R. Maynard, G. Toulouse (North Holland, 1979).

[39] M. L. Cohen and G. S. Grest, Phys. Rev. **B20**, 1077 (1979) and G. S. Grest and M. L. Cohen, Adv. Chem. **48**, 455 (1981).

[40] H. W. Sheng et al., Nature **439**, 419 (2006)

[41] L. E. Rehn et al., Phys. Rev. Lett. **59**, 2987 (1987).

[42] M. O. Ruault et al., Phil. Mag. A58, 397 (1988).

[43] Yufeng Shi and Michael L. Falk, Phys. Rev. Lett. **95**, 095502 (2005); M. L. Falk and J. S. Langer, Phys. Rev. **E57**, 7192 (1998).

[44] K. Laaziri et al., Phys. Rev. Lett. **82**, 3460 (1999) and subsequent studies by S. Roorda et al.

[45] T. Sakata, H. Mori and H. Fujita, Acta. Metall. Mater. **39**, 817 (1991).

Chapter 13

ION BEAM MIXING

Michael Nastasi[1] and James W. Mayer[2]
¹Los Alamos National Laboratory, Los Alamos, NM 87545 USA
²Arizona State University, Tempe, AZ 85287 USA

1   INTRODUCTION

Materials under ion irradiation undergo significant atomic rearrangement. The most obvious example of this phenomenon is the atomic intermixing and alloying that can occur at the interface separating two different materials during ion irradiation. This process is known as *ion beam mixing*. An early observation of the ion mixing phenomenon was made following the irradiation of a Si substrate coated with a thin Pd film. A reaction between Pd and Si was observed when the irradiating Ar ions had sufficient energy to penetrate through the Pd/Si interface [1]. This process is schematically displayed in Fig. 1 for a layer M on a substrate S for successively higher irradiation doses.

Early in the irradiation, when ion tracks are well isolated, each incident ion initiates a collision cascade surrounding the ion track. Atoms within the cascade volume will be mobile and undergo rearrangement for a short period of time, resulting in an intermixed region near the interface. At this stage of the ion mixing process, the interfacial reaction is considered to be composed of many localized volumes of reaction (Fig. 1a). As the irradiation dose is increased, overlap of localized regions occurs (Fig. 1b), and for higher doses a continuous reacted layer is formed at the interface (Fig. 1c).

A major driving force in the development of the ion beam mixing process is its ability to produce ion-modified materials with higher solute concentrations at lower irradiation doses than can be achieved with conventional high-dose implantation techniques. A case in point is the formation of Au-Cu alloys on Cu substrates (Fig. 2).

Figure 2 shows Au concentration as a function of ion dose for both Xe ion mixing of an Au layer on Cu and the direct implantation of Au into Cu. The ion mixing experiment was carried out by depositing a 20 nm Au film on Cu, then irradiating the sample with Xe ions at energy sufficient to penetrate the Au layer. As the data shows, the amount of Au that can be

387

K.E. Sickafus et al. (eds.), Radiation Effects in Solids, 387–400.
© 2007 *Springer.*

introduced into the Cu by ion mixing greatly exceeds the maximum concentration that can be achieved by direct implantation where sputtering effects generally set the concentration limits.

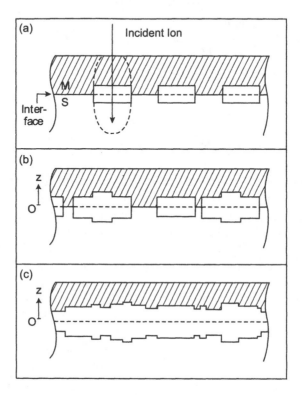

Figure 1. Schematic of the ion mixing process for a layer M on a substrate S for successively higher irradiation doses.

Several processes are responsible for the ion mixing effect, all of which are initiated by the interaction of an energetic ion with a solid. The ballistics or kinematics of the ion/target interaction plays a role, as does the formation of collision cascades and the total number of ions that have passed through the interface, i.e., the ion dose, ϕ. Both ballistic and cascade effects can be altered by changing the mass of the irradiating ion; increasing the mass of the ion increases the amount of energy deposited in nuclear collisions per unit length traveled by the ion. An example of mass and dose effects in ion mixing is clearly evident in Fig. 3. This shows that the average thickness of reaction at the Pt/Si interface, in units of Si atoms/cm^2, increases with both increasing mass of the incident ion and the increasing dose, ϕ, and that the mixing rate for all irradiating ions is proportional to $(\phi)^{1/2}$. For a given dose, the ratio of the number of atoms reacted, Q, scales with the ion mass and with the nuclear energy loss $(dE/dx)_n$ to the 1/2 power. These trends lead to a general condition for the amount of mixing, Q, at the interface between two different materials, which can be expressed as

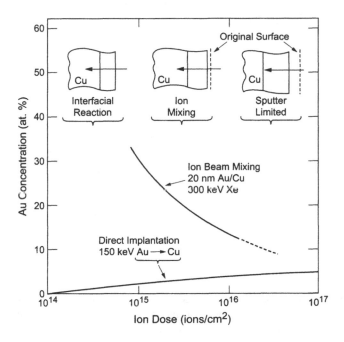

Figure 2. Concentration as a function of ion dose for the ion mixing of a 20 nm Au layer on Cu with 300 keV Xe and the direct implantation of 150 keV Au into Cu. (After Ref. [2].)

$$Q \propto \sqrt{\phi \left(\frac{dE}{dx} \right)_n} \qquad (1)$$

Since the dose rate, ions/cm2/sec., is nominally held constant during each ion mixing experiment and, hence, ion dose is proportional to time, the observation that mixing is proportional to $(dose)^{1/2}$ implies that ion mixing is also proportional to (ion mixing time)$^{1/2}$. This latter proportionality is very similar to that observed for a reaction layer formed between two materials by thermally-activated interatomic diffusion. The width of the reacted layer, W, in a thermal diffusion experiment has been observed to have the following trend:

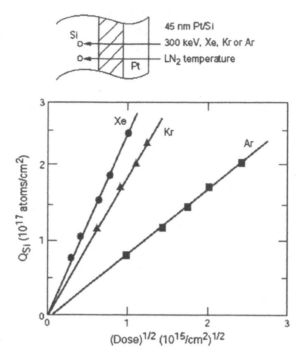

Figure 3. An example of mass and dose effects in the ion mixing of Pt/Si. Mixing increases with both increasing mass of the incident ion and increasing dose ϕ. The mixing rate for all irradiating ions is proportional to $\phi^{1/2}$. (After Ref. [3].)

$$W \propto \sqrt{\tilde{D} t} \qquad (2)$$

where $\tilde{D}$ is the interdiffusion coefficient. This observation has led to a general formulation that ion mixing has characteristics similar to a diffusion-like process.

In addition to the primary effects of ion/target atom collisions, external variables, such as the sample temperature during irradiation, can also influence mixing behavior. At low temperatures, the amount of mixing observed for a given ion dose is typically found to be insensitive to temperature variations, while above a critical temperature the mixing is very temperature-dependent. This behavior can be seen in Fig. 4 for the case of Cr/Si irradiated with 300 keV Xe ions to a dose of 1×10^{16}/cm^2. For temperatures below $0°$ C, the amount of mixing that occurs is relatively insensitive to the sample temperature during irradiation; this temperature interval is known as the temperature-independent ion mixing regime.

However, as the sample temperature is increased to above ~ 100 oC, the mixing rate changes rapidly with temperature and is observed to increase

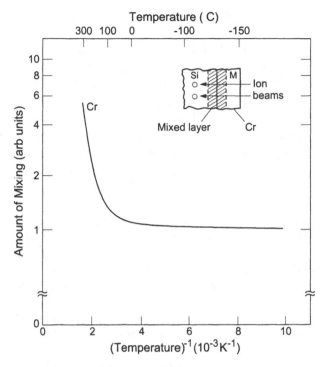

Figure 4. The amount of Si atoms contained in the Cr/Si mixed layer versus reciprocal ir-radiation temperature. (After Ref. [4].)

with increasing temperature; this temperature interval is known as the temperature-dependent ion mixing regime.

2   BALLISTIC MIXING

The interaction of an energetic ion with a solid involves several processes. As an ion penetrates a solid, it slows down by depositing energy both to the atoms and to the electrons of the solid. During the nuclear collision portion of this process, target atoms can be permanently displaced from their lattice sites and relocated several lattice sites away. When this process takes place at the boundary separating two different materials, interface mixing results. The displacement mechanism of atomic rearrangement is the fundamental principle governing ballistic mixing.

2.1   Recoil Mixing

When an incident ion strikes a metal target atom, M, near a metal/substrate interface, some of the incident ion's kinetic energy is transferred to the target atom. For high-energy collisions, the target atoms recoil far from their initial location. This process, which results in the transport of atoms through repeated single collision events between the incident ions and target atoms, is the simplest form of ballistic mixing. It is known as *recoil implantation* or *recoil mixing*. For mixing to be effective by this process, the recoil should travel the maximum range possible; a maximum range will result when the collision between the incident ion and the target atom is head-on ($\theta = 0°$). The probability of a head-on collision is very small, with most collisions being *soft* (i.e., $\theta > 0°$). The recoils produced in such soft collisions will possess significantly less energy, and head-on collisions and their trajectory will not be in the forward direction. As a result, the number of target atoms contributing to mixing by the mechanism of recoil implantation will be small.

Insight into the collisional mixing factors operating in ion beam mixing can be gained by the use of embedded marker studies. In these studies, a thin layer of a marker material, such as Ge, is placed between two layers of a matrix element, such as Si. The effective diffusion coefficient observed during marker ion mixing is proportional to the ion mixing dose, ϕ, and the damage energy deposited per unit length, F_D. These two relationships are shown in Fig. 5 for several different markers in amorphous Si. These and other similar results have led to the definition of an effective mixing parameter, Dt/ϕ, as indicated in Fig. 5b. The deposited energy, F_D, should be considered when defining the effective mixing parameter, casting it as $Dt/\phi F_D$. Strictly speaking, F_D is the total kinetic energy of the incident ion deposited into nuclear collisions per unit length, and it is obtained from the recoil energy by factoring in electronic losses. However, the damage energy at a given location in the sample is occasionally approximated by the nuclear stopping at that location, which provides an overestimate of F_D.

2.2   Cascade Mixing

In addition to recoil mixing, other ballistic phenomena are possible during ion irradiation and implantation. For example, enhanced atomic mixing can occur when multiple displacements of target atoms result from a single incident ion. In the multiple displacement process, an initially displaced target atom (primary recoil) continues the knock-on-atom processes, producing secondary recoil atom displacements which in turn displace additional atoms. The multiple displacement sequence of collision events is commonly referred to as a *collision cascade*.

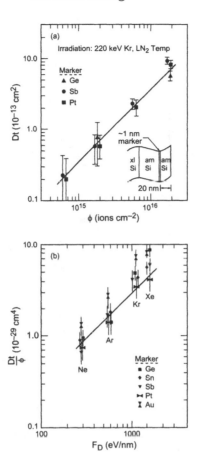

Figure 5. Marker mixing data of several different markers in amorphous Si. These data show the relationship between the effective ion mixing diffusion coefficient, Dt, and ion mixing dose, φ, (a), and the effective mixing parameter, Dt/φ, and the damage energy deposited per unit length, F_D (b). (After Ref. [5].)

Unlike the highly directed recoil implantation process, in which one atom receives a large amount of kinetic energy in a single displacement, atoms in a collision cascade undergo many multiple uncorrelated low-energy displacement and relocation events. Atomic mixing resulting from a series of uncorrelated low-energy atomic displacements is referred to as *cascade mixing*.

The ballistic interactions of an energetic ion with a solid are shown schematically in Fig. 6. The figure shows sputtering events at the surface, single-ion/single-atom recoil events, and the development of a collision cascade that involves many low-energy displaced atoms. The cascade is shown in the early displacement stage of development, where the displaced atoms occupy interstitial positions surrounding a core of vacant lattice sites.

Ballistic Processes

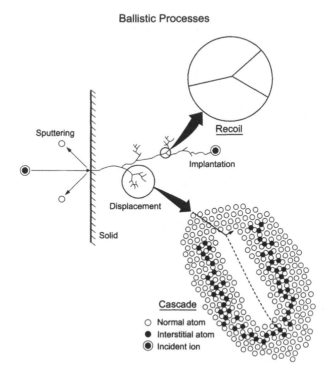

Figure 6. The ballistic interactions of an energetic ion with a solid. Graphically displayed are sputtering events at the surface, single-ion/single-atom recoil events, the development of a collision cascade which involves a large number of low energy displace atoms, and the ion implantation of the incident ion.

Calculations of the mean energy of atoms in a cascade show that most recoils are produced near the minimum energy necessary to displace atoms, E_d. Due to the low-energy stochastic nature of these displacement events, the initial momentum of the incident particle is soon lost, and the overall movement of the atoms in a collision cascade becomes isotropic. This isotropic motion gives rise to an atomic redistribution that can be modeled as a random-walk of step size defined by the mean range of an atom with energy near E_d. The effective diffusivity, D_{cas}, for a collision cascade-induced random-walk process is expressed in the diffusion equation as [6]

$$D_{cas}\, t \;=\; \frac{dpa(x)\, <r^2>}{6} \tag{3}$$

where dpa(x) is the number of cascade-induced displacements per atom at distance x, and $<r^2>$ is the mean squared range of the displaced target atoms. The dpa, resulting from a given dose of ions can be expressed as

$$dpa(x) = \frac{0.4 F_D(x) \phi}{E_d N} \tag{4}$$

where $F_D(x)$ is the damage energy per unit length at distance x, ϕ is the ion dose, and N is the atomic density. Combining Eqs. (3) and (4) gives the effective diffusion coefficient due to ballistic cascade mixing as

$$D_{cas} t = 0.067 \frac{F_D(x) < r^2 >}{N E_d} \phi \tag{5}$$

Sigmund and Gras-Marti [7] made a more detailed theoretical formulation of collisional mixing, based on linear transport theory. This formulation also accounts for mass difference between the ion, M_1, and the target atom, M_2. The effective diffusion coefficient obtained from calculations examining the ion irradiation induced spreading of an impurity profile in a homogeneous matrix is given by

$$D_{cas} t = \frac{\Gamma}{6} \xi \frac{F_D < r^2 >}{N E_d} \phi \tag{6}$$

where Γ is a dimensionless parameter with a value of 0.608, and ξ is a mass-sensitive kinematic factor given by $[4 M_1 M_2 / (M_1 + M_2)^2]^{1/2}$. In the case of $M_1 = M_2$, $\xi = 1$, and Eqs. (5) and (6) become very similar.

The primary features of Eqs. (5) and (6) are that the effective diffusion coefficient should scale with the dose, ϕ, and the damage energy, F_D, in good agreement with the trends observed in Fig. 5. An additional characteristic of this expression is that it does not contain any temperature-dependent terms. The effective diffusion coefficient described by Eq. (5) is independent of temperature and can be compared only with experiments in which mixing also is observed to be independent of temperature. The data in Fig. 3 shows the influence of temperature on the mixing between Cr and Si. This data shows that ion mixing remains independent of temperature up to about 0 °C.

Equation (5) and the marker data presented in Fig. 5 can be used to estimate the average atomic displacement distance of a marker atom in a collision cascade formed in a matrix of amorphous Si. For example, from the temperature-independent data in Fig. 5a, a typical value of $D_{irr} t / \phi$, for both Sn and Sb markers, is $4(10^{-29} \text{ cm}^4)$, or 0.4 nm^4. From Fig. 5b, the corresponding damage energy is 1500 eV/nm. Using the atomic density of crystalline Si, 50 atoms/nm^3, for the amorphous Si value of N, the ratio of F_D/N will be 30 eV/nm^4. This indicates that $<r^2>^{1/2}$ should be approximately 1.6 nm for a Si displacement energy of $E_d \cong 13$ eV.

3   THERMODYNAMIC EFFECTS IN ION MIXING

Chemical driving forces, which not considered in ballistic models, play an important role in ion mixing when concentrated alloys are formed. For example, Au on Cu and W on Cu should both have the same ballistic response to ion mixing because of their nearly identical parameters (atomic density, atomic number, and atomic mass) for ion-solid interactions. The Au/Cu data shows that Au is well intermixed, while the W/Cu data shows W is relatively unchanged after irradiation, showing only signs of material loss due to sputtering. An evaluation of these data reveals that ion mixing in the Au/Cu system is 10 times that observed in W/Cu [8]. These results were attributed to the miscibility differences in the two systems: Au and Cu are completely miscible in both the liquid and the solid states, while W and Cu are immiscible in both the liquid and the solid states.

Similar observations were also made in the different ion mixing responses of Hf/Ni and Hf/Ti bilayers [9]. Again, from a ballistic view, ion mixing for these two systems should be nearly identical. However, it was observed that the mixing rate of Hf/Ni was significantly higher than that of Hf/Ti. This difference was attributed to differences in the heat of mixing, ΔH_{mix}, for the two systems. In Table 1, the four bilayer systems — Au/Cu, W/Cu, Hf/Ni, and Hf/Ti — are listed, along with their values of ΔH_{mix} at the equiatomic alloy concentration, and the nuclear energy loss at the initial bilayer interface.

Table 1 shows that the bilayer systems that undergo extensive ion mixing (Au/Cu and Hf/Ni) possess a negative heat of mixing, while the systems that experienced little or no mixing (W/Cu and Hf/Ti) possess a zero or positive heat of mixing.

Table 1 Heat of Mixing and Stopping in Bilayer Systems

Bilayer System	ΔH_{mix} (kJ/g.at)	dE/dx (eV/nm)
Au/Cu	-9	31.0
W/Cu	+36	32.2
Hf/Ni	-62	32.9
Hf/Ti	0	23.5

dE/dx from SRIM, ΔH_{mix}. After Ref. [10].

ure of how attractive different elements are to each other relative to their attractiveness to themselves. The enthalpy difference, ΔH_{mix}, results from the chemical joining of A and B atoms and the formation of A-B bonds during alloying. The more negative the heat of mixing, the greater the tendency to form A-B alloys.

In bilayer systems with negative heats of mixing, there is a driving force to form interface alloys during ion irradiation. Although it would seem that that ion beam irradiation would intermix layered structures, thermodynamic effects can overwhelm ballistic processes. If heats of mixing are positive and the sample temperature is sufficiently low, ion irradiation can cause intermixing. However, when the sample temperature is increased, the mixed layer back-segregates into its components, a process known as *de-mixing*.

The mixing rate in bilayer systems is expressed as the derivative expression $d(4\tilde{D}t)/d\phi$, where the variable $\tilde{D}$ is the chemical interdiffusion coefficient. In Fig. 7, $d(4\tilde{D}t)/d\phi$ is plotted for several metal bilayer systems as a function of the systems heat of mixing (heat of mixing data is taken at the 50 atomic % composition). Fig. 7 shows that the mixing rate is thermodynamically biased. When thermodynamic driving forces are important, the mixing rate equation for bilayer mixing can be expressed as

$$\frac{4\tilde{D}t}{\phi} = \frac{4\tilde{D}_o t}{\phi}\left(1 - \frac{2\Delta H_{mix}}{k_B T}\right) \tag{7}$$

where $4\tilde{D}_o t/\phi$ is the ballistic-induced mixing term, defined as

$$\frac{4\tilde{D}_o t}{\phi} = 0.268\,\frac{F_D <r^2>}{NE_d} \tag{8}$$

Equation (7) is a linear expression with ΔH_{mix}, with a slope defined by

$$\frac{4\tilde{D}_o t}{\phi}\frac{2}{k_B T} = 0.536\,\frac{F_D <r^2>}{NE_d k_B T} \tag{9}$$

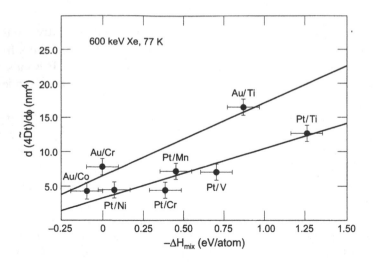

Figure 7. The experimentally observed mixing rates and ΔH_{mix} for several metallic alloy systems following ion irradiation with 600 keV Xe ions and a sample temperature of 77 K. (After Ref. [11]).

Reasonable average values for the ballistic terms in Eq. (9) are: $F_D = 5500$ eV/nm, $<r^2> = 2.25$ nm^2, $N_0 = 74$ atoms/nm^3, and $E_d = 30$ eV. These values give a calculated slope of $3.0/k_BT$ nm^4/eV. The experimental slopes (Fig. 10) for the Au and Pt bilayer mixing data are approximately 5 and 7.0 nm^4/eV, respectively. Setting the experimental and calculated values equal to each other and taking $k_B = 8.63 \times 10^{-5}$ eV/K gives an effective cascade temperature of 3310 K and 4966 K in the Au and Pt samples, respectively. The average kinetic energy of the atoms in the cascade can be estimated using the law of equipartion of energy, $E = 3/2k_BT_{eff}$, which gives 0.43 eV/atom and 0.64 eV/atom in the Au and Pt bilayers, respectively.

A more detailed analysis of the thermodynamic influence on ion mixing by Johnson *et al.* [12] resulted in the following phenomenological expression

$$\frac{d4\tilde{D}t}{d\phi} = \frac{K_1 F_D^2}{\bar{N}^{5/3}(\Delta H_{coh})^2}\left(1 + K_2 \frac{\Delta H_{mix}}{\Delta H_{coh}}\right) \qquad (10)$$

where ΔH_{coh} is the cohesive energy of the alloy system; F_D is the damage energy per unit length; $\bar{N}$ is the average atomic density; and K_1 and K_2 are fitting parameters. Based upon experimentally determined mixing rates for 600 keV Xe irradiation at 77 K, the values of the fitting parameters were found to be $K_1 = 0.0034$ nm and $K_2 = 27.4$. The cohesive energy of the alloy system composed of elements A and B are approximated by

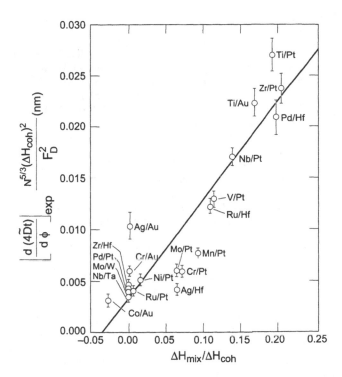

Figure 8. Experimental mixing data showing a linear relationship between the mixing rate versus the ratio $\Delta H_{mix}/\Delta H_{coh}$. (After Ref. [13].)

$$\Delta H_{coh} \cong (X_A \Delta H_A^o + X_B \Delta H_B^o) + \Delta H_{mix} \qquad (11)$$

where ΔH^0 are the cohesive energies of individual elements (see Ref. [14]), X is the atomic fraction of the elements and ΔH_{mix} alloy's heat of mixing.

Equation (10) is demonstrated in Fig. 8, which shows the experimental mixing rate $d(4\tilde{D}t)/d\phi$ (scaled by $\bar{N}^{5/3} \Delta H_{coh} F_D^{-2}$) versus the ratio $\Delta H_{mix}/\Delta H_{coh}$. The linear relationship indicates that the amount of mixing scales with $\Delta H_{mix}/\Delta H_{coh}$ and that Eq. (10) provides a reasonable prediction of the ion beam mixing rate in metal/metal bilayer systems irradiated with heavy ions at low temperatures.

Ballistic effects resulting from the early stages of cascade evolution are important factors in the ion mixing process, but materials properties, such as heat of mixing and cohesive energy, also influence the process. Such material properties dominate the rate of mixing and the phase formation possibilities in the ion-reacted layers, influencing the final structures that form.

REFERENCES

[1]   van der Weg, W.F., D. Sigurd, and J.W. Mayer (1974) "Ion Beam Induced Intermixing in the Pd/Si System," in *Applications of Ion Beams to Metals* eds. S.T. Picraux, E.P. EerNisse, and F.L. Vook (Plenum Press, New York) p. 209.

[2]   Mayer, J.W. and S.S. Lau (1983), "Ion Beam Mixing" in Surface Modification and Alloying by Laser, Ion, and Electron Beams, eds. J.M. Poate, G. Foti, and D.C. Jacobson (Plenum Press, New York, 1983) p. 241.

[3]   Tsaur, B.Y., S.S Lau, Z.L. Liau, and J.W. Mayer (1979) "Ion-Beam Induced Intermixing of Surface Layers," *Thin Solid Films* **63**, 31.

[4]   Mayer, J.W., S.S. Lau, B.Y. Tsaur, J.M. Poate, and J.K. Hirvonen, (1980) "High-Dose Implantation and Ion-Beam Mixing," in *Ion Implantation Metallurgy*, eds., C.M. Preece and J.K. Hirvonen (The Metallurgical Society of AIME, New York) p. 37.

[5]   Matteson, S., B.M. Paine, M.G. Grimaldi, G. Mezey, and M.-A. Nicolet (1981) "Ion Beam Mixing in Amorphous Silicon I. Experimental Investigation" *Nucl. Instr. Meth.* **182/183**,43.

[6]   Andersen, H.H. (1979) "The Depth Resolution of Sputter Profiling" *Appl. Phys.* **18**, 131.

[7]   Sigmund, P., and A. Gras-Marti (1981) "Theoretical Aspects of Atomic Mixing by Ion

[8]   Westendorp, H., Z-L. Wang, and F.W. Saris (1982) "Ion-Beam Mixing oc Cu-Au and Cu-W Systems" *Nucl. Instr. Meth.* **194**, 543.

[9]   van Rossum, M., U. Shreter, W.L. Johnson, and M-A. Nicolet (1984) "Amorphization of Thin Multilayer Films by Ion Mixing and Solid State Amorphization" *Mat. Res. Soc. Symp. Proc.* **Vol. 27**, 127.

[10]   de Boer, F.R., R. Boom, W.C.M. Mattens, A.R. Miedema, and A.K. Niessen (1989) *Cohesion in Metals* (North-Holland, Amsterdam).

[11]   Cheng, Y.-T., M. van Rossum, M-A. Nicolet, and W.L. Johnson (1984) "Influence of Chemical Driving Forces in Ion Mixing of Metallic Bilayers," *Appl. Phys. Lett.* **45**,185.

[12]   Johnson, W.L., Y.-T. Cheng, M. Van Rossum, and M-A. Nicolet (1985) "When is Thermodynamics Relevant to Ion-Induced Atomic Rearrangements in Metals?" *Nucl. Inst. Meth.* **B7/8**, 657.

[13]   Cheng, Y.-T. (1990) "Themodynamic and Fractal Geometric Aspects of Ion-Solid Interactions" *Materials Science Reports*, **5**, 45.

[14]   Kittel, C. (1976) *Introduction to Solid State Physics*, 5th edition (John Wiley & Son, Inc, New York).

Chapter 14

RADIATION EFFECTS IN NUCLEAR FUELS

Hans Matzke

Institute for Transuranium Elements, Karlsruhe, Germany

1   INTRODUCTION

Nuclear fuels have to operate reliably for years under extreme conditions of radiation damage. The main physical process for heat production, used to generate electricity, is to slow down and stop the high-energy heavy ions, i.e., the fission products (FPs). When an actinide atom is fissioned, about 200-MeV energy is dissipated in the fuel lattice. Most of this high energy is carried by the FP that cover the mass range from A = 75 to 160, hence elements between Ga and Dy. The FPs fall into two groups: the light ones, typically Kr, with about 100 MeV energy, and the heavy ones, typically Ba, with about 70-MeV energy. Intense neutron fluxes with energies ranging from eV to MeV interact with the fuel atoms, and there is an intense ß.γ-radiation field because most FPs are radioactive with different decay energies and very different half-lives.

An additional important damage source is the α-decay of the original actinides, and even more so, of large amounts of "minor actinides," e.g., Np, Am, and Cm that are formed by successive neutron capture during the operation of the fuel.

Figure 1 shows that the α-decay, in this case of Am^{241}, produces about 1700 displacements, most caused by the heavy recoil atom, here, Np^{237} with a very short range of about 20 nm forming a dense collision cascade.

The lower part presents the conditions during fission, the light fission product (LFP) that are being emitted to the right, and the heavy fission product (HFP) to the left. Together, they produce about 100,000 defects.

K.E. Sickafus et al. (eds.), Radiation Effects in Solids, 401–420.
© 2007 *Springer.*

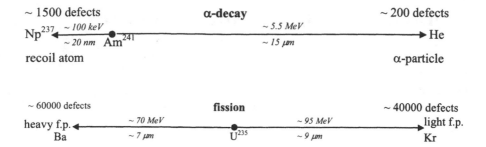

Figure 1. Schematic presentation of energies, ranges, and defects produced during α-decay and fission in the fuel of most currently operating electricity producing power reactors, i.e., UO_2.

Figure 2 shows the electronic energy loss (dashed lines) and range distributions (full lines) of the α-particle and the two FPs; the arrow indicates the dense collision cascade of the recoil atom of the α-decay. Its stopping is predominantly nuclear. The upper part of Fig. 2 shows the large differences between the high-defect density formed along the path of the FPs (fission spike) and the largely isolated defects formed along the (longer) path of the α-particles. Note that the energy-loss curves show only electronic stopping, with high values for FPs at the point of fission, i.e., full energy (18–22 keV/nm). The nuclear stopping part is small at the point of fission (~ 0.1 keV/nm), and it peaks at the end of the range with a value of ~ 1 keV/nm. The ratio of electronic-to-nuclear stopping is always high at those high energies. It amounts to 180:1 at the point of fission, and it decreases toward the end of the range to 3:1 or even 1:1 at the very end of the trajectory.

The contribution of neutrons to the total displacements per atom (dpa) level of the fuel at end of life is much smaller. Neutrons will occasionally collide (mean free path ~ 1 cm) with lattice atoms and impart some energy, depending on the impact parameter. The maximum energy transferred to the U-atoms of UO_2 is 17 keV, and for the O-atoms it is 250 keV. However, most interactions with neutrons transfer less energy. The minimum neutron energy to produce a displacement in UO_2, using the above E_d-values, is ~ 0.1 keV. Low-energy, "thermal" neutrons thus do not produce direct displacements.

High-energy electrons, hence β-decay, can also produce some displacements, both by direct energy transfer (very few isolated point defects) or by ionization, if the material is not resistant against radiolysis.

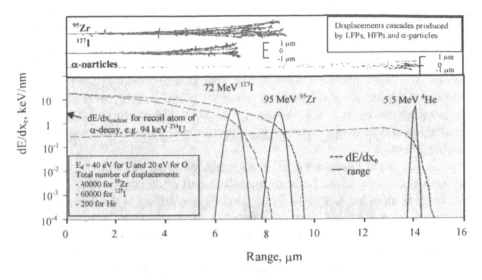

Figure 2. Electronic energy loss and ranges of α.-particles and of light and heavy fission products in UO_2. The displacement energies E_d used for the calculations with the SRIM code [1] were 20 eV for O- and 40 eV for U-atoms.

A well-known consequence of accumulated radiation damage is damage-induced phase changes, most notably amorphization (or metamictization) of originally crystalline matter, or polygonization, also named grain subdivision—a process that transforms a typical grain of an originally well-crystallized ceramic into thousands of small grains in the submicron range. Polygonization occurs in some nuclear fuels, including UO_2, and amorphization is of concern for the selection of suitable "inert matrices": it occurs in the candidate material spinel $MgAl_2O_4$.

When discussing damage accumulated by fission, or to a lesser extent, by radioactive decay, one has to consider the simultaneous change in chemistry. Each fission, besides producing the above 100,000 displacements, also produces two fission products: new chemical elements between Ga and Dy, including volatile elements such as Br, I, Kr, Xe, Cs, etc., accumulating to often more than 10 at% at the end of life of the nuclear fuels. Or in α-decay daughter atoms, e.g., Np in the decay of Am and He-atoms formed in addition to the displacements, the He atoms may precipitate into bubbles, thus causing the specimens to swell. Similarly, in ß-decay, new elements are formed. For example, Cs, a high-abundance fission product, decays into Ba with another valence state and a different chemical behavior. Therefore, we deal with complex phenomena that explain why interest is still high in understanding damage effects and mechanisms that are not only in new but also in conventional nuclear fuels, despite the large amount of work devoted to this subject in the past five decades.

An important aspect in understanding the complex damage consists of single-effect studies in which accelerators are used with ion beams of different elements and energies, also in combination with in situ transmission electron microscopy. Particularly useful are irradiations with beams of fission products of new but also in conventional nuclear fuels, despite the large amount of work devoted to this subject in the past five decades.fission energy, e.g., beams of 70-MeV iodine, to simulate reactor irradiation and in using high energies, up to GeV, to study the formation of visible ion tracks.

The techniques used to investigate damage accumulation and its consequences in nuclear fuels are manifold and often require work in glove boxes or in shielded hot cells. In the studies performed at the Transuranium Institute in Karlsruhe and at cooperating institutions, the following techniques were used: quantitative ceramography, transmission and scanning electron microscopy, measurements of volume changes and density, Rutherford back-scattering and channeling (RBS-C), elastic recoil detection analysis, x-ray diffraction including micro-x-ray diffraction, electromotive force measurements, differential scanning calorimetry, Vickers indentation to measure hardness and fracture toughness, atomic force microscopy, laser-flash techniques with thermogravimetry, and He-release measurements.

2   TODAY'S NUCLEAR FUELS UO_2 AND $(U,Pu)O_2$

Most electricity-producing nuclear power stations use uranium dioxide (UO_2) as fuel, enriched in U^{235} to about 5%, and some also use in MOX (mixed oxide) fuel, i.e., UO_2 with about 5% of PuO_2 replacing the U^{235}. The fuel consists of sintered pellets of about 1 cm in diameter. These pellets are used to produce long stacks that are sheathed in tubes (cladding) of a Zr alloy. For a typical 900-MW pressurized water reactor, 272 sintered pellets of UO_2 form a rod 3.85 m long. Of these rods, 264 are assembled into a fuel assembly, and 157 such assemblies form the reactor core that contains about 11.3 million sintered UO_2 pellets.

Uranium dioxide has a fluorite structure and a high melting point ($T_m = 3150$ K) and a rather low thermal conductivity, typically ~ 4 W/mK at 1000 K. Uranium dioxide oxidizes easily during heating, forming higher oxides such as U_3O_7 or U_3O_8, and it disintegrates into powder. It can also form hyper- and hypostoichiometric compositions, UO_{2+x} and UO_{2-x}. Uranium dioxide should therefore be stored under inert atmosphere; for any annealing, the proper oxygen potential has to be selected.

As described in the introduction, uranium dioxide fuel is subjected to all the radiation and damaging sources for about 5 years. During this period, fission products grow in and change both chemical and physical properties. For example, a fuel with a simulated burn-up of 8% (by adding nonradioactive fission product elements in the fabrication step; the product is called SIMFUEL [2]) shows an important decrease of the thermal conductivity of about 25% just because of the chemically added fission product elements. Any treatment of damage effects has to allow for such changes that occur even in the absence of damage buildup.

The most interesting damage effect in UO_2 fuel at high burn-up (expressed in GWd/t_U or in percent fissions of the heavy metal atoms) is polygonization—the formation of about 10^4 small submicrometer grains from each original UO_2 grain. Polygonization is the term used to describe the rearrangement of those dislocations formed in the earlier stage of irradiation that do not annihilate one another into walls of dislocations, forming low-energy "subboundaries" and perfect but slightly misaligned subgrains.

The onset burn-up for this process to take place is about 5 at %. The phenomenon was already observed in early test irradiations and was called grain subdivision [3]. It was rediscovered in the 1980s when power reactors increased the fuel burn-up. In power reactor fuel, the small grains, together with micron-sized pores containing the insoluble fission gases Kr and Xe, are first formed in the outer zone of the fuel pellets. The reason for this formation is that near the fuel surface, the local temperature is lower than in the center; the burn-up is significantly higher because of the resonance neutron capture of nonfissile U^{238} that eventually forms fissile Pu^{239}. This explains the term "rim effect," which is frequently used to refer to the fuel part showing the polygonized high burn-up structure. The experimental results are well documented; tailor-made test irradiations are performed and relevant measurements along the radii of the fuel pellets are being made [e.g., 4]. First modeling activities show promising results, and the details of the underlying mechanisms are studied in international projects.

The accumulated experience shows that the formation of the high burn-up structure is acceptable for safe reactor operations. The early concerns about a possible increased gas release and strongly decreased thermal conductivity could be shown to be not relevant, even in fuel with a wide polygonized rim. Up to burn-ups significantly higher than today's end-of-life burn-ups of 50 to 60 GWd/tU, fission gas release from the rim zone was not significantly increased, the thermal properties of the fuel matrix were not further deteriorated, and the fuel became softer and tougher.

Another interesting fission-related process is radiation- or, more specifically, fission-enhanced diffusion [e.g., 5, 6]. For diffusion of U and

Pu, a large series of experiments were performed, and the diffusion couples were placed in a furnace inside a nuclear research reactor. The temperature range covered was 13–1400 °C, and a variety of UO_2, $(U,PuO)_2$, UC, (U,Pu)C, UN, and (U,Pu)N specimens were investigated (for the carbides and nitrides see next the section). Between 130 and ~ 1000 °C, U, and Pu diffusion was completely athermal, hence, independent of temperature. For the oxide fuels, the fission-enhanced diffusion coefficients could be described by the relation

$$D^* = AF \text{ with } A = 1.2 \times 10^{-29} \text{ cm}^5 \text{ and } F = \text{fission rate in fissions cm}^{-3}\text{s}^{-1}.$$

D^* was independent of neutron flux and irradiation time but was directly proportional to the fission rate F. The results were explained by the formation of thermal spikes along the trajectory of the fission fragments in combination with a pressure gradient. Because of the high-energy deposition rate of typically 20 keV/nm (see Fig. 2), a locally (over)heated track (fission spike or thermal spike) may be formed. Such fission tracks are seen in transmission electron microscopy in many materials. An extreme case of fission spikes interacting with the UO_2 matrix is the destruction of preexisting fission gas bubbles by passing a fission spike. The phenomenon was called "re-solution" of fission gas and was known for about 40 years. It was explained by the above-mentioned hydrostatic pressure component of the thermoelastic stress field of the fission spike interacting with the bubbles [7]. The same pressure gradients serve to explain the surprisingly high D^*-values above. To decrease these gradients, the highly mobile uranium interstitials are pushed away from the spike axis, thus increasing the U-diffusion to values higher than those calculated for atomic mixing and thermal spike effects alone. Like diffusion, in-pile creep of UO_2 was shown to be athermal and fission-enhanced below ~ 1000 °C as well [8]. Another related phenomenon is the well-known and technologically important fuel densification during irradiation even at low temperatures, which is caused by destruction of small sintering pores in the wake of fission spikes.

The sequence of events in the fission spike is the following. After a primary or ballistic phase that produces high-density collision cascades, a second or quenching phase follows when the spike approaches thermal equilibrium, and an interstitial-rich outer zone and a vacancy-rich inner zone are formed. In the third or track-annealing phase, vacancy-interstitial recombination and formation of small vacancy clusters as nuclei for fission gas bubbles occur. For more details see [9].

Simulating fission tracks by implantation of high-energy heavy ions, including fission energy, is a powerful means to increase the knowledge of damage formation in UO_2. Heavy ions (Zn, Mo, Cd, Sn, I, Xe, Au, Pb, and also U) with energies between 173 MeV and 2.7 GeV were used, and the irradiated specimens were analyzed for track formation by transmission electron microscopy [10]. Fission tracks were never found in bulk-irradiated UO_2 (although they were seen at UO_2 surfaces [11] and explained by the interaction of shock waves with the surface). Clear evidence of the formation of visible tracks in bulk UO_2 was found when using 173 MeV Xe-I ns (dE/dx = 29 keV/nm) and other heavy ions at higher energy. This clearly shows that the threshold for visible track formation in UO_2 falls between 22 keV/nm (no tracks seen) and 29 keV/nm (clear tracks found). With the measured track radii and a thermal spike model [12], the temperature/time relations along the trajectory of fission fragments in UO_2 could be calculated. The peak value was 5000 K, indicating that a molten zone really exists in the fission spikes.

Additional results were obtained when a fission product beam of fission energy was used (here 70-MeV iodine from the TASCC accelerator at Chalk River Labs, AECL Canada) to measure fission-induced gas diffusion, bubble destruction, and formation of new bubbles. A typical example is shown in Fig. 3. The release of preimplanted (radioactive) Kr^{85} caused by the impact of the iodine ions was followed as a function of the iodine fluence, which can be taken as a time scale. While Kr is immobile at ambient temperature and at 500 °C in the absence of fission, the figure shows an athermal release due to the impact of iodine that yields, normalized to fission rate, a radiation-enhanced diffusion coefficient that is smaller by a factor of 6 than the above D* for U and Pu, as expected. The Kr diffusion is not increased by the additional contribution of high mobility of the U-interstitials. Additional experiments of this type confirmed the concept of fission-induced re-solution of rare gases from preexisting bubbles and also their precipitation into new bubbles. In parallel experiments, lattice parameter increases with saturation at $\Delta a_0/a_0$ at ~ 0.3% because the impact of the iodine ions of fission energy was found in these specimens in a very similar way to that observed in studies of self-radiation damage due to α-decay in many actinide oxides including UO_2.

More information on damage ingrowth, largely related to α-decay, can be obtained by irradiation with heavy ions in the 10^2-keV energy range and with He-ions of energy in the million electronvolt-energy range. By using elements that also occur as fission products, information on fuel-fission product chemistry can be obtained. In addition, damage recovery mechanisms during anneals can be investigated. Many results of this type

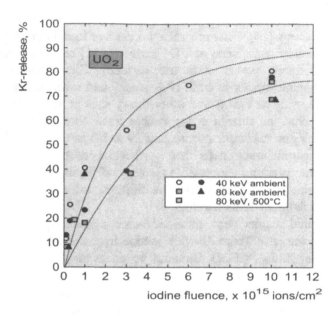

Figure 3. Kr-release from UO$_2$ preimplanted with 40 or 80 keV Kr-ions and subsequently irradiated with iodine ions of fission energy, either at ambient temperature or at 500 °C.

have been accumulated over the years. Rutherford backscattering, using typically 2 MeV He-ions, often combined with channeling was used to study the damage ingrowth and recovery in the U-sublattice; resonance scattering of high-energy He-ions was used to obtain this information for the oxygen sublattice.

Space does not allow a presentation and discussion of all of these often unpublished results. Only two figures are shown here. Figure 4 shows RBS-C results for self-damage in the UO$_2$ lattice, i.e., the recovery of UO$_2$ implanted with U-ions.

This avoids possible interference of damage with the implanted impurity, an example being UO$_2$ implanted with rare gas ions when, upon annealing, gas bubbles are formed affecting the surrounding lattice atoms and the RBS-C spectra and results. Figure 4 shows that damage recovery proceeds in two stages: one at ~ 550 °C can be attributed to U-vacancy mobility and the broader one around 800 °C to cluster annealing. A stage

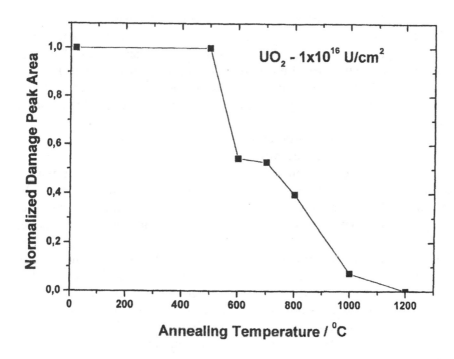

Figure 4. RBS-C results of defect recovery in the U-sublattice of a UO_2 single crystal implanted with $1x10^{16}$ U-ions/cm^2.

for mobility of U-interstitials is not seen because they are mobile below room temperature. Oxygen defects cannot be observed with RBS-C techniques using 2-MeV He-ions (see above for selectively measuring defects in the oxygen sublattice with resonance scattering of high-energy He-ions). In UO_2, the diffusion of the metal atoms is much slower than that of oxygen atoms. The mobility of the metal atoms is therefore rate-controlling for many processes of technological interest, like grain growth and sintering, creep, etc. In much of the early work on damage ingrowth and recovery, properties that are not specific for a given sublattice were used as a probe for damage, such as changes in lattice parameter in thermal and electrical conductivity, etc. This explains the interest in RBS-C techniques to selectively obtain information on the U-sublattice as shown in Figs. 4 and 5. Similar results for a large variety of ions, ion energies, and doses have provided a consistent picture on defect formation, saturation, annealing, and retention of the implanted impurity. An example is shown in Fig. 4 where RBS-C results on implanted UO_2 single crystals are presented. The U-damage peak of the spectra was evaluated to obtain the displacement efficiency, i.e., the number of measured U-defects per incoming 40-keV ion of Kr, Te, and Cs. Damage calculations on the basis of primary defect production due to nuclear energy deposition largely overestimate the lattice disorder.

The value of 250 U-defects for the conditions of Fig. 5, predicted by the theory of Kinchin-Pease [13], is not even observed for the lowest dose, which indicate that defect recovery in the collision cascades is very effective and that instantaneous defect recombination processes play the main role in the formation of the damage structure remaining in UO_2 after implantation at room temperature. The number of surviving U-defects per incoming ion decreases with ion dose to a value of less than 1, and the total damage approaches saturation (see inset in Fig. 5). These low values help to explain the good radiation stability of UO_2, which is in marked contrast to the ease of amorphization (metamictization) found in many other ceramics including some ceramics studied as inert matrix (see Section 2) or for final waste storage. This stability of UO_2 is predicted by the available models that are based on structural arguments—thermodynamical and bonding properties [14].

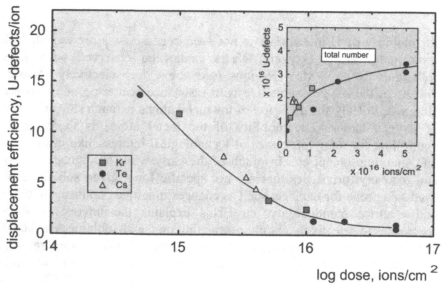

Figure 5. Displacement efficiency (U defects per incoming ion, surviving instantaneous annihilation) as a function of ion dose for implantation of 40 keV Kr-, Te-, and Cs-ions in UO_2 single crystals. The inset shows the total number of surviving defects.

The polygonization described above could also be produced with ion beams. These experiments [e.g.,15] yielded no effect of temperature between ambient and 500 °C and they showed that ion beams of an insoluble impurity (e.g., Xe, I, etc.) had to be used to produce polygonization. However, soluble impurities such as La were not causing polygonization, even when the dose was increased to very high levels.

At the end of this section, a few words on α–decay damage are needed although space does not allow details. Many actinide dioxides, old natural minerals like uraninite and UO_2 doped with a short-lived α-emitter such as Pu^{238}, were analyzed through the years, usually to measure the lattice parameter increase, its saturation (see above), and recovery. Thermal conductivity decreases were also followed, and recently [e.g., 16] specific heat, stored energy, and He-release were investigated; transmission electron microscopy was performed on the damaged specimens. Such information is important for the concepts of final or retrievable storage of spent nuclear fuel, and more information is needed, e.g., on the question of possible α-decay-enhanced diffusion during long-time storage.

3   ADVANCED FUELS: CARBIDES, NITRIDES, AND INERT MATRIX FUELS FOR TRANSMUTATION OF ACTINIDES

Despite the excellent performance of today's nuclear fuels, UO_2 and MOX (U, Pu)O_2, advanced reactor concepts need advanced fuels. Rock salt-structured carbides and nitrides MC and MN (M = U and/or Pu) offer a higher thermal conductivity (by approximately a factor of 7 to 8) and a larger metal atom density than the oxides (by ~ 30% for MC and ~40% for MN). The damage sources are identical to those for the oxide fuels. The effect of fission spikes are however different because of the different thermal properties. Basic properties of MC and MN were measured, and the irradiation behavior was widely tested (see monograph [17]), mainly in connection with their planned application in fast breeder reactors—liquid metal-cooled fast neutron breeder reactors (LMFBRs).

In the recent past, the concept of inert matrices was studied for the safe transmutation of existing actinides of either excess military Pu or actinide amounts (mainly Np, Am, and Cm) arising from reprocessing conventional oxide fuel. These actinides are incorporated into a nonfissile material with a low enough atomic number so that no new actinides can be formed by multiple neutron capture. This material, e.g., MgO, ZrO_2, or spinel $MgAl_2O_4$, therefore provides an inert matrix that does not form new actinides while transmuting incorporated actinides by fission. The damage source conditions are often very different, and they can be tailored according to the technological demands.

3.1   Carbide and Nitride Fuels

Rock salt-structured MC and MN are predicted to be stable against amorphization based on the criteria mentioned above, and this was confirmed in many experiments, both in studies of self-damage due to α-decay, in ion implantation work, and in reactor irradiations up to high burn-ups [17]. Both tailored capsule irradiations with controlled parameters and full-scale fuel pin irradiations were made. The total database for the advanced fuels is, however, just a small fraction of that for the oxides. Polygonization occurred at burn-ups in excess of ~ 5 at. %, as in UO_2. An increase in lattice parameter was observed because of damage ingrowth, both caused by self-damage in Pu-doped specimens as well as by reactor irradiation, again in similarity with UO_2. Figure 6 shows an example of the volume increase or swelling due to α-decay self-damage during storage at ambient conditions for 60 days in $(U_{0.8},Pu_{0.2})C$, the 20% Pu being short-lived ^{238}Pu. The increase is larger than that because of the increase in lattice parameter. No transmission electron microscopy has been done, but the probable reason for the difference will be formation of He-filled bubbles.

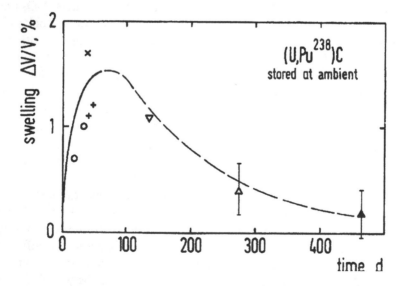

Figure 6. Volume increase of $(U_{0.8},^{238}Pu_{0.2})C$ due to self-radiation damage during storage at ambient temperature.

The group of Professor Kirihara at Nagoya University [e.g., 18] has measured different properties of UC and UN before and after reactor irradiation. The recovery of the observed changes during annealing was also measured. Similar data were obtained for UO_2 for comparison. The properties were lattice parameter, electrical resistivity, hardness, and different magnetic properties. In comparison to UO_2, less defect annealing occurred during fission, which was attributed to the higher thermal conductivity of the advanced fuels. Consequently, thermal and pressure spikes were expected to be less pronounced than in UO_2, and there was no indication of fast mobility of U-interstitials in UC and UN contributing to fission-enhanced mass transport. Compatible with this explanation are the results for fission-enhanced diffusion of U and Pu. The measured D*-values were temperature independent, athermal between 150 and ~ 1100 °C and were described by a relation similar to that for oxides (see Section 1), i.e., $D^* = AF$, F being the fission rate in fissions $cm^{-3}s^{-1}$. The A-values were, however, smaller by a factor of 5 for MC and a factor of 6.7 for MN than that for the oxides. These results are inversely proportional to the thermal conductivities which are, at 1000 °C, in the order UO_2:UC:UN, normalized for UC, 0.15 : 1 :1.2, again pointing to less defect mobility during fission in the advanced fuels. Athermal fission-enhanced creep and densification even at low temperatures were also found for both UC and UN, again less pronounced than in UO_2.

Irradiations with ion beams combined with RBS-C analysis were also performed with UC and UN single crystals, though again to a much smaller degree than with UO_2. Damage ingrowth and damage recovery were measured as a function of ion dose and for different ions, also in combination with transmission electron microscopy. The RBS-C spectra for implanted UN resembled those for metals more than those for ceramics: damaged and distorted layers were formed exhibiting continuously increasing dechanneling rather than a damage peak as found for more ionically bonded ceramics like UO_2 (and most candidates for inert matrices [see below])

Most work on these advanced fuels was done before about 1990 because the interest in fast breeder reactors decreased at that time. However, very recently, interest in nitrides for utilization in new reactor types is growing again. The existing data on radiation damage form a firm basis for future work.

3.2   Inert Matrix Fuels for Transmutation of Actinides

To reduce the radiotoxicity of radioactive high-level nuclear waste from reprocessing spent nuclear fuel from electricity-producing nuclear power stations, the long-lived actinides, mainly Am, but also Np and Cm, will be chemically separated (partitioned) and transmuted by fission in a nuclear facility (reactor or dedicated accelerator)). To achieve an effective and good performance, new fuels have to be developed. Adding the partitioned actinides to the fabrication process for UO_2 fuel, as it is presently done with the separated Pu, is in principle possible, although shielding would be needed because of the γ-radiation of Am. However, with this procedure, more actinides would be formed by multiple neutron capture and radioactive decay starting with the main isotope U^{238}. To avoid this drawback, so-called U-free fuels are considered. The actinides (Np, Am, Cm) have to be diluted for neutronic reasons, and this diluting support material is the inert matrix: it is inert with respect to the formation of new actinides. Several materials were proposed and studied as inert matrices, based on a critical evaluation of their properties. These activities started around 1990, grew into international cooperations, and the results were discussed at a sequence of workshops on inert matrices [e.g., 19]. A good example of international cooperation is the basic studies and reactor irradiations of inert matrix materials and concepts performed within the Experimental Feasibility of Targets for Transmutation (EFTTRA) cooperation for transmutation of the minor actinides, mainly Am. EFTTRA is a joint action of CEA and EdF in France, ECN in the Netherlands, FZK in Germany, and ITU, Karlsruhe and IAM, Petten of the Joint Research Centre of the European Commission. The philosophy underlying the choice of suitable materials was to find a compromise candidate with several desirable properties, resistance against radiation damage being a prime criterion for an acceptable matrix. The main requirements include ease of fabrication, availability, low cost of starting materials, neutron economy, good thermal properties (i.e., high melting point), good thermal conductivity and heat capacity, absence of phase changes and significant dissociation at high temperatures, good compatibility with cladding (Zircaloy, steel) and cooling (water, Na), good mechanical properties (elastic constants, hardness, fracture toughness), and finally, good stability against radiation (fission, α-decay, neutrons, β, γ-radiation) [e.g., 20].

Although the damage sources, the formation of He from α-decay, and the ingrowth of impurities (FP and α-decay products) are very similar to the processes occurring in conventional UO_2 fuel, the fact that only a fraction of the fuel is fissile suggests that tailor-made fuels may be produced to reduce the overall damage effects. The following possibilities exist and are tested for inert matrix fuels:

- Ceramic solid solution of the actinides in the matrix, e.g., $(Pu,Zr)O_2$
- CERCERs, a dispersion of a fissile ceramic in an inert ceramic matrix, i.e., a two-phased material in which the actinide compound

is present as a second phase in the inert matrix, either as fine particles or as of macroinclusions, e.g., AmO_2 in spinel $MgAl_2O_4$

- CERMETs, dispersion of a fissile ceramic in a metal, e.g., PuO_2 in W

These different possibilities can be used to produce tailor-made inert matrix fuels with regard to radiation damage, center temperatures, and temperature profiles during irradiation, etc. If a solid solution matrix is chosen, all of the matrix will be subjected to all radiation sources known from conventional UO_2 fuels. If, on the other hand, a dispersion-type matrix is chosen with inclusions of the fissile phase in the inert matrix, most damage will be produced in the inclusions. More exactly, all of the fuel will experience interactions with neutrons and β,γ-irradiation. In addition, the inclusions containing the actinides will be subjected to fission and α-decay damage, hence, to all damage sources. A very shallow shell around the inclusions (~ 20 nm) will experience all of the damage, a thicker shell of ~ 10 μm thickness will experience all damage except that of the recoil atoms, and a still thicker shell of about 20 μm thickness will experience damage of the α-particles in addition to that of the neutrons. These values of the thickness of the different shells correspond to the ranges of recoil atoms of the α-decay (~ 20 nm), the fission products (~ 10 μm), and the α-particles (~ 20 μm) in a matrix like e.g., spinel. The fraction of the undamaged matrix (with the exception of neutrons and β,γ radiation) can be chosen within a wide range. For small μm-sized inclusions at a typical actinide content in the range of 10 to 20 wt.%, the case of a solid solution is approached. For bigger inclusions, say of 100-μm diameter, more than 90 vol.% of the matrix remains undamaged in the above sense. Figure 7 shows the effect of particle size on the damaged fraction at constant actinide addition (here 20 vol.%).

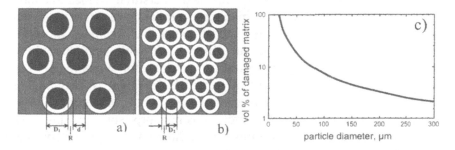

Figure 7. Parts a and b show projections of two particle sizes on a plane. Both have the same fractions of actinide inclusions, i.e., 20 vol.%. Part c shows the relation between particle size and fraction of damaged matrix, again for 20 vol.% actinide inclusions.

A series of potential inert matrix candidates were selected and tested, based on the criteria mentioned above. This series included the binary ceramics Al_2O_3, MgO, ZrO_2, CeO_2, Y_2O_3, ZrN, and TiN; the ternary ceramics spinel $MgAl_2O_4$; monazite $CePO_4$; zircon $ZrSiO_4$ and $Y_3Al_5O_{12}$; and the metals W and V, etc. In particular, extensive tests including neutron irradiations, ion bombardment, and reactor irradiation were performed with spinel $MgAl_2O_4$ and zirconia ZrO_2 [21–24], and the more complex inert matrix version was investigated in Japan [25], which was called ROX (for rock-like oxide fuel, a multiphase mixture of mineral-like compounds like yttria-stabilized zirconia, spinel, and corundum). For the reactor irradiations, the matrices themselves and matrices with Am, Pu, or U, either in solution or as inclusions, were used. Some of the above candidate materials were excluded because of their poor radiation resistance, e.g., Al_2O_3, since it amorphizes with a large volume increase under ion impact at unacceptably low doses (which actually was known already for decades [26]), or monazite $CePO_4$ and zircon $ZrSiO_4$, because these two materials show a Coulomb explosion due to the impact of high-energy heavy ions. Particularly worth mentioning is the EFTTRA-T4 irradiation of spinel with 10–12 wt.% Am_2O_3 microinclusions [21]. After irradiation in the HFR for 358 full powder days, the initial Am^{241} concentration was reduced to 4%; however, large swelling occurred (up to 18%). In the following section, the results with spinel and zirconia are briefly summarized.

Spinel: The most complete damage studies were performed with spinel $MgAl_2O_4$. Spinel is known to be resistant against neutron irradiation, including fast neutrons and very high doses. Its swelling under neutron irradiation is small, in contrast to that of Al_2O_3, and because of this stability, spinel is a candidate as an insulator in fusion applications, e.g., for radio frequency (RF) heating systems or beam injectors. The reasons can be qualitatively understood. In Al_2O_3, unfaulted interstitial dislocation loops can easily be formed, leading to a bias for precipitation of interstitials, thus leaving vacancies to form cavities which cause the swelling. In spinel, fewer loops form because loop nucleation is difficult, and these loops remain faulted, thus acting as less-effective sinks for interstitials. Also, achieving the stoichiometric ratio of vacancies in spinel is difficult, and efficient site exchange of different cations is possible, which explains the absence of voids and cavities. Therefore, it is not surprising that spinel was initially considered to be a promising candidate for inert matrix fuels because of its excellent stability against neutron damage. As the next test, the impact of heavy ions and He-ions simulating α-decay damage was tested in extensive parametric studies. The resistance against He-ion impact was found to be good and that against heavy ion (of the order of 10^2-keV energy) impact was acceptable: for room temperature implantation and for very high doses, spinel became only partially amorphous, i.e., RBS-C spectra showed absence of crystallinity in the Al-sublattice but the Mg-sublattice remained crystalline. At liquid nitrogen temperature, not relevant for technological applications, spinel became amorphous during high-dose impact of heavy ions.

The main deficiency of spinel was, however, discovered only when using beams of heavy ions at fission energy—swelling of up to 35% for irradiation with 70-MeV I-ions at ambient temperature, by ~10% at 500 °C and very small swelling for irradiation at 1200 °C [24, 27]. Irradiations with ions of Kr, I, Xe, Bi, and U of fission energy or higher energies yielded visible tracks and amorphization of spinel with a threshold (electronic) energy loss of 6–7 keV/nm [27, 28], and the application of the thermal spike model, already used for UO_2, allowed estimates of the temperatures along the tracks, The measured track radii (TEM) increased significantly with dE/dx for the low-velocity regime, whereas for energies beyond the Bragg-peak (in the high velocity regime), the track radii were always small (1–2 nm) and independent of the energy loss. Unfortunately, during reactor irradiation and fission, the low-energy regime is activated with much larger tracks, hence less favorable conditions for operational stability (metamictization, swelling). However, the amorphous spinel recrystallizes easily at ambient temperature under ionizing radiation, e.g., in the electron beam in an electron microscope and probably under the intense β,γ-radiation in a nuclear reactor.

Table 1 compares UO_2 with spinel $MgAl_2O_4$ and shows the superior properties of UO_2. Obviously, minimization of radiation damage in a spinel-supported inert matrix fuel can be achieved only in a dispersion-type target (see above) in which the advantages of spinel (high-thermal conductivity, good stability against neutrons) can be used while largely avoiding the drawbacks of its low stability against fission product impact. Spinel samples with different actinide inclusions have been produced and the irradiation performance is being tested and evaluated.

Table 1: Fission Damage in UO_2 Compared with That in Spinel (for T < 500 °C)

	UO_2	$MgAl_2O_4$
Threshold energy loss for track formation, dE/dx_e	22–29 keV/nm	6–7 keV/nm
Visible tracks for fission energy	No	Yes
Amorphization	No	Yes
Swelling	Very small	~ 25%

ZrO_2, Stabilized Zirconia: ZrO_2 is presently a main favored candidate for inert matrix fuels. Actinides are soluble in zirconia, in contrast to spinel, and the radiation stability of ZrO_2 is good [e.g., 22, 23]. It can therefore be used as a homogeneous fuel with the actinides transmuted in a solid solution. It is actually used as $(Zr,Pu)O_2$ in test irradiations. However, the same is true for its use as dispersion particles in a spinel matrix. Zirconia has a low thermal conductivity which is a problem for a homogenous fuel, but this is much less important for a dispersion-type fuel. Attention has also to be given to the structure of zirconia. ZrO_2 is a polymorphic oxide existing in three different crystal structures below its melting point of 2950 K. The high-temperature cubic form, which is of interest for inert matrix fuels, is isostructural with UO_2 and exists between 2650 K and the melting point. An intermediate-temperature form is tetragonal (1440–2640 K), and the low temperature, naturally occurring monoclinic form (distorted fluorite structure), exists below 1440 K. Each successive high temperature form is more symmetric and more dense. To avoid phase changes during heating or cooling, the dense, cubic high-temperature form is stabilized by adding yttria Y_2O_3 or calcia CaO. This causes the stabilized cubic form to be substoichiometric $Zr,Y)O_{2-x}$.

Less work has been devoted to study the radiation stability of ZrO_2 than was done for spinel. The stability of ZrO_2 against neutrons is well known. The first confirming tests were published between 1955 and 1960. In 1963, results of a large series of irradiations of $(Zr,^{235}U)O_2$ up to high burn-ups were published [29] that showed a good, overall performance. In the past decade, ion irradiation tests were performed by researchers at different laboratories. As expected based on criteria for structural stability [14], cubic zirconia never got amorphous (except when implanted with very high percentages of the large Cs-ions that produced a different material). As with UO_2, the impact of 70-MeV iodine ions caused no measurable swelling. However, RBS-C results clearly showed high degrees of disorder after the impact of heavy ions, much higher than those in UO_2 for the same irradiation conditions. Also polygonization occurred, again much earlier (by about a factor of 10 in dose) than with UO_2.

Space limitations do not allow a discussion of radition damage in other candidates for inert matrix fuels in detail, although there are interesting results. For instance, both MgO and CeO_2 show polygonization due to the impact of heavy ions etc.

SUMMARY

Nuclear fuels are subjected to a series of damage sources: neutrons of energy up to the MeV range, β,γ-radiation due to the decay of fission products, He-ions of ~5 MeV, daughter atoms of the α-decay of actinides with ~ 100 keV energy, and the fission products (heavy ions in the energy range of 70–100 MeV). And they have to operate reliably under these extreme conditions of radiation damage for years, accumulating up thousands of displacements per atom. Damage processes and mechanisms are discussed in this paper that shows that today's fuel—uranium dioxide (UO_2)—is the most radiation-resistant ceramic studied so far (with, or next to, ThO_2, which is not discussed here). Instantaneous damage recovery is very effective in UO_2; it never becomes amorphous and there are no visible tracks unless the energy of the damaging heavy ions is increased above fission energy. The threshold for track formation by heavy ion impact falls between 22 and 29 keV/nm. At very high burn-up, polygonization causes the fuel to develop a high burn-up structure characterized by sub-μm-sized grains with μm-sized pores. As shown here, this structural modification is acceptable for safe reactor irradiation.

Many results of relevance for radiation damage are also available for the advanced fuels MC and MN—carbides and nitrides of U and/or Pu. No important detrimental effect has been reported so far. For the new class of inert matrix fuels for the effective transmutation of actinides, much basic work and many reactor irradiations were performed that provide a good basis to decide on fuel strategies. Most data exist for spinel $MgAl_2O_4$ and stabilized zirconia ZrO_2, but a series of other materials have also been studied to some extent, including MgO, CeO_2, the Japanese ROX fuel, SiC, and others. The relevant results are presented in this paper.

REFERENCES

[1] J.F. Ziegler, J.P. Biersack, and U. Littmark, *The Stopping and Range of Ions in Solids*, (Pergamon, London) (1985)

[2] P.G. Lucuta, Hj. Matzke, and I. Hastings, J. Nucl. Mater. **232** (1996) 166

[3] M.L.Bleiberg, R.M Berman, and B.Lustman, in Radiation Damage in Reactor Materials, IAEA Vienna (1963) 319

[4] J. Spino, J. Cobos-Sabate, and F. Rousseau, J. Nucl. Mater. **322** (2003) 204

[5] Hj. Matzke, Radiation Effects **75** (1983) 317

[6] Hj., Matzke, T. Wiss, and P.G. Lucuta, Nucl. Instr. Methods in Phys. Research B **166–167** (2000) 920

[7] H. Blank, and Hj. Matzke, Radiation Effects **17** (1973) 57

[8] D. Brucklacher and W. Dienst, J. Nucl. Mater. **42** (1972) 285

[9] Hj. Matzke and T. Wiss, ITU Annual Report 2000, EC Report EUR 19812 (2000) 30

[10] T. Wiss, Hj. Matzke, C. Trautmann, M. Toulemonde, and Klaumünzer, S. Nucl. Instrum. Methods in Phys. Research, **B122** (1997) 583

[11]   C. Ronchi, J. Appl. Phys. **44** (1973) 3575

[12]   M.Toulemonde, C. Dufour, A. Meftah, and E. Paumier, Nucl. Instrum. Methods in Phys. Research **B166–167** (2000) 903

[13]   G.H. Kinchin and R.S., Pease Rep. Progr. Phys. **18** (1955) 1

[14]   Hj. Matzke, Radiation Effects **64** (1982) 3

[15]   Hj. Matzke, A. Turos, and G. Linker, Nucl. Instrum. Methods in Phys. Research **B91** (1994) 294

[16]   D. Staicu, T. Wiss, and C. Ronchi, Proc. Int. Meeting LWR Fuel Performance Orlando, USA September 19–22 (2004) paper 1087

[17]   Hj. Matzke, *Science and Technology of Advanced LMFBR Fuels, a Monograph on Solid State Physics, Chemistry and Technology of Carbides, Nitrides and Carbonitrides of Uranium and Plutonium* (North Holland, Amsterdam, 1986)

[18]   H. Matsui, M. Horiki, and T. Kirihara, J. Nucl. Sci. Technol. **18** (1981) 171

[19]   C .Degueldre and J., Porta, editors, Inert Matrix Fuel 6", Proc. 6th Workshop on Inert Matrix Fuels E-MRS Symposium 30.5. – 2.6.2000, Elsevier

[20]   Hj. Matzke, V.V, Rondinella, and T.Wiss, J. Nucl. Mater. **274** (1999) 47

[21]   R. Konings, et al., EC report EUR 19138 EN (2000)

[22]   K.E Sickafus et al., Amer. Ceram. Soc. Bulletin **78** (1999) 60

[23]   K.E. Sickafus et al., J. Nucl. Mater. **274** (1999) 66

[24]   T. Wiss and Hj. Matzke, Radiation Measurements **31** (1999) 507

[25]   T.Yamashita et al., in Proc. Ref. 19 327

[26]   Hj. Matzke and J.L. Whitton, Can. J. Physics **44** (1966) 995

[27]   T. Wiss and Hj. Matzke et al., Prog. Nucl. Energy **38** (2001) 281

[28]   S.J. Zinkle, Hj. Matzke, and V.A. Skuratov, Mater. Res. Soc., Warrendale, Pennsylvania, MRS, Symp. Proc. **540** (1999) 299

[29]   M.L. Bleiberg, R.M. Berman, and B. Lustman, in Radiation Damage in Reactor Materials, IAEA, Vienna (1963) 319

Chapter 15

ROLE OF IRRADIATION IN STRESS CORROSION CRACKING

Gary S. Was
University of Michigan, Ann Arbor, MI 48109 USA

1   INTRODUCTION

A growing concern for electric power utilities worldwide has been degradation in core components in nuclear power reactors, which make up ~17% of the world's electric power production. Service failures have occurred in boiling water reactor (BWR) core components and, to a somewhat lesser extent, in pressurized water reactor (PWR) core components consisting of iron- and nickel-base stainless alloys that have achieved a significant neutron fluence in environments that span oxygenated to hydrogenated water at 270–340°C. Because cracking susceptibility is a function of radiation, stress and environment, the failure mechanism has been termed irradiation-assisted stress corrosion cracking (IASCC). Initially, the affected components have been either relatively small (bolts, springs, etc.) or designed for replacement (fuel rods, control blades, or instrumentation tubes). In the last decade, there have been many more structural components (PWR baffle bolts and BWR core shrouds) that have been identified to be susceptible to IASCC. Recent reviews [1–5] describe the current knowledge related to IASCC service experience and laboratory investigations and highlight the limited amount of well-controlled experimentation that exists on well-characterized materials. This lack of critical experimentation and the large number of interdependent parameters make it imperative that underpinning science be used to guide mechanistic understanding and quantification of IASCC.

The importance of neutron fluence on IASCC has been well established, Fig. 1. Intergranular (IG) SCC is promoted in austenitic stainless steels as a critical "threshold" fluence is exceeded, (such "thresholds" appear only in some tests, as discussed later). Cracking is

K.E. Sickafus et al. (eds.), Radiation Effects in Solids, 421–447.
© 2007 Springer.

observed in BWR oxygenated water at fluences above 2 to 5 x 10^{20} n/cm^2 (E>1 MeV), which corresponds to about 0.3 to 0.7 displacements per atom (dpa). A comparable "threshold" fluence for IASCC susceptibility has been reported for high-stress, in-service BWR component cracking and during ex-situ, slow-strain-rate SCC testing of irradiated stainless steels. This indicates "persistent" radiation effects (material changes) are primarily responsible for IASCC susceptibility, although in-situ effects like radiation creep relaxation of weld residual stresses and increased stress from differential swelling can be important.

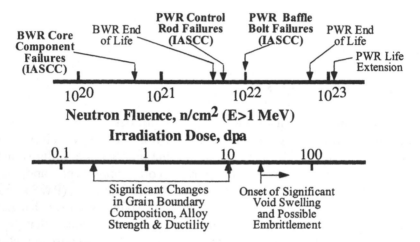

Figure 1. Neutron fluence effects on irradiation-assisted stress corrosion cracking susceptibility of type 304SS in BWR environments. [4]

Recent work has enabled many aspects of IASCC phenomenology to be explained (and predicted) based on the experience with IGSCC of non-irradiated stainless steel in reactor water environments. This continuum approach has successfully accounted for radiation effects on water chemistry and its influence on electrochemical corrosion potential. However, all radiation-induced microstructural and microchemical changes that promote IASCC are not fully known. Well-controlled data from properly irradiated and properly characterized materials are sorely lacking due to the inherent experimental difficulties and financial limitations. Many of the important metallurgical, mechanical and environmental aspects that are believed to play a role in the cracking process are illustrated in Fig. 2. Only persistent material changes are required for IASCC to occur, but in-core processes such as radiation creep and radiolysis also have an important effect on IASCC. The following section examines the current understanding of persistent material changes that are produced in stainless alloys during LWR irradiation based on the fundamentals of radiation damage and existing experimental measurements.

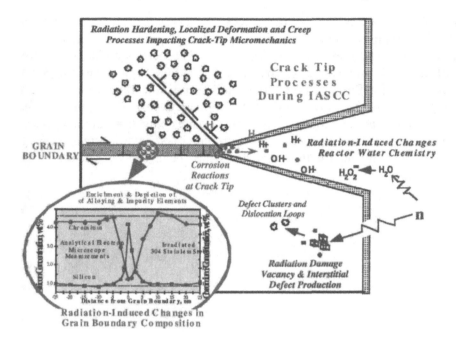

Figure 2. Schematic illustrating mechanistic issues believed to influence crack advance during IASCC of austenitic stainless steels. [4]

IASCC can be categorized into radiation effects on the water chemistry (radiolysis) and on the material properties. The cracking response to changes in water chemistry is similar for both irradiated and unirradiated materials. In both cases, there is a steep increase in environmental cracking kinetics with a rise in the corrosion potential above about 100 mV$_{SHE}$. [5–7] At high corrosion potential, the crack growth rate also increases sharply as impurities (especially chloride and sulfate) are added to pure water in either the irradiated or un-irradiated cases. In post-irradiation tests, the dominant radiation-related factors are microstructural and microchemical changes, which can be responsible for "threshold-like" behavior in much the same way as corrosion potential, impurities, degree of sensitization, stress, temperature, etc. Other radiation phenomena, like radiation creep relaxation and differential swelling, could also have "persistent" effects if we relied on the sources of stress present during radiation (e.g., weld residual stresses, or loading from differential swelling) during post-irradiation testing. The effects of radiation rapidly (in seconds) achieve a dynamic equilibrium in water, primarily because of the high mobility of species in water. In metals, dynamic equilibrium is achieved – if ever – only after many dpa, typically requiring years of exposure. While both radiation induced segregation (RIS) of major elements, and radiation hardening (RH) and the associated microstructural development asymptotically approach a dynamic equilibrium, other factors (e.g., RIS of Si, or precipitate formation or dissolution) may become important. Yet data on post-irradiation slow strain rate tests (SSRT) on stainless steels show that there is a distinct (although not invariant) "threshold" fluence at which IASCC is observed under LWR conditions. [8] Because this "threshold" occurs at a fraction to several dpa (depending on the alloy, stress, water chemistry, etc), Fig. 3, in-situ effects (corrosion potential, conductivity, temperature) may be important, but only "persistent" radiation effects (microstructural and microchemical changes) can be responsible for the "threshold-like" behavior vs. fluence in post-irradiation tests.

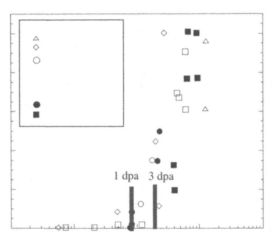

Figure 3. Dependence of cracking in neutron-irradiated high-purity 304SS on accumulated high-energy neutron fluence. [8]

2.   EFFECTS OF IRRADIATION ON MICROSTRUCTURE AND MICROCHEMISTRY

Having established the importance of persistent changes to the material in post-irradiation observations of IASCC, the challenge then becomes one of identifying the specific changes in the material that accelerate SCC. These material changes fall into three categories; (1) microcompositional effects; radiation-induced segregation of both impurities and major alloying elements, (2) microstructural changes; the formation of dislocation loops, voids, precipitates and the resulting hardening (increase in yield strength), and (3) deformation mode; formation of dislocation channeling and localized deformation in irradiated materials. As shown by the composite schematic diagram in Fig. 4, all of the observable effects of irradiation on the material increase with dose in much the same manner, making isolation of, and attribution to individual contributions difficult. The dislocation loop microstructure is closely tied to radiation hardening and both increase with dose until saturation occurs by ~5 dpa. RIS also increases with dose and tends to saturate by ~5 dpa. Although dependent on both metallurgical and environmental parameters, IASCC generally occurs at doses between 0.5 dpa (for BWRs) and 2–3 dpa (for PWRs), [3] which encompasses the steeply rising portion of the curves in Fig. 4 that describe the changes in materials properties with irradiation. Following a review of service experience, subsequent sections will focus on the possible mechanisms by which water chemistry, RIS, microstructure, hardening, deformation mode and irradiation creep – individually or on concert – may affect IASCC.

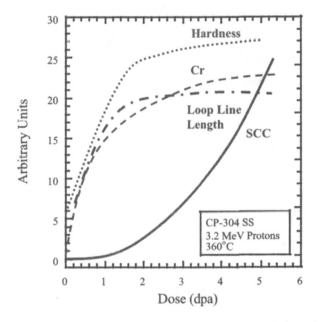

Figure 4. Schematic illustration of radiation induced segregation at grain boundaries.

2.1 Radiation Induced Segregation

Radiation induced segregation (RIS) occurs by the preferential association between solutes and of one type of point defect, giving rise to enrichment of some solutes and depletion of others at sinks. Figure 5 shows the composition profiles for several alloying elements in a 304 stainless steel, irradiated in a reactor to about 10^{22} n/cm^2 at 275°C. Note that chromium, iron and molybdenum deplete, while nickel, silicon and phosphorus enrich at the grain boundary. Since the depletion of chromium is of much interest, Fig. 6 shows the rate at which grain boundary chromium depletion occurs for several 300 series stainless steels irradiated around 300°C. Note that the grain boundary chromium level depletes initially very rapidly and then appears to approach saturation by 5-10 dpa. Silicon also undergoes significant segregation with irradiation, but it enriches at the grain boundary in very narrow profiles. Figure 7 shows the enrichment of Si with dose, which appears to continue up through 12 dpa without evidence of saturation. In fact, the segregation of silicon to grain boundaries is a strong function of the amount of silicon in the alloy. Figure 8 shows that the silicon segregation ratio is about 6, meaning that for the same dose, the grain boundary concentration is 6 times that of the bulk.

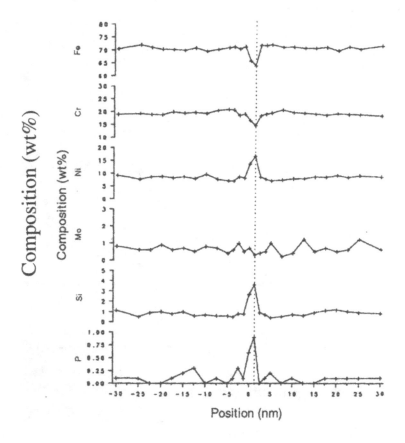

Figure 5. Typical segregation profiles in commercial purity type 304 SS irradiated to ~ 10^{22} n/cm² at 275°C. Composition profiles were measured using a field emission gun scanning transmission electron microscope, FEGSTEM. [9]

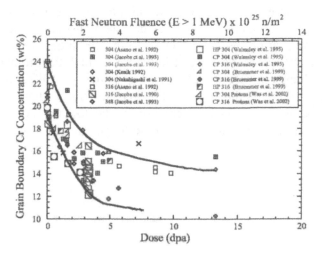

Figure 6. Dose dependence of grain boundary chromium concentration for several 300-series austenitic stainless steels irradiated at a temperature of about 300°C. [10-17]

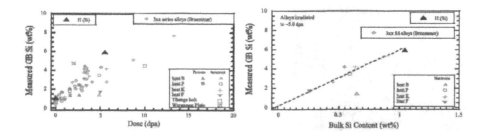

Figure 7. Grain boundary segregation of silicon in 300 series stainless steel alloys as a function of irradiation dose at temperatures near 300°C. [18]

Figure 8. Grain boundary silicon concentration as a function of the silicon content in the bulk alloy for 300 series stainless steels and for irradiations at about 5 dpa and near 300°C. [18]

One of the consequences of solute enrichment is the precipitation of second phases due to enrichment. Figure 9 shows a dark-field image and selected area diffraction pattern from a cold worked 316 stainless steel baffle bolt from a PWR that was irradiated to 7.2 dpa at 299°C. The spots in the image have been indexed as γ', Ni_3Si precipitates, that have presumably formed due to the local enrichment of silicon at defect sinks. When the concentration of silicon reaches the solubility limit, the formation of another phase is thermodynamically favored. As long as the phase is stable under irradiation, precipitation can occur and may result in significant property changes in the alloy.

Not all segregation follows the simple "V" shaped depletion or enrichment profiles. Figures 10 and 11 show how an initial enrichment of chromium at the grain boundary (due to processing) can lead to the formation of a "W" shaped profile at low doses, which then transforms to a classical "V" shaped profile at higher dose. Because the "W" shaped profile tends to occur at doses in the range of 1-3 dpa, and this is where the onset of IASCC occurs, there is much interest in understanding how this grain boundary chromium concentration profile forms and how it can impact IASCC.

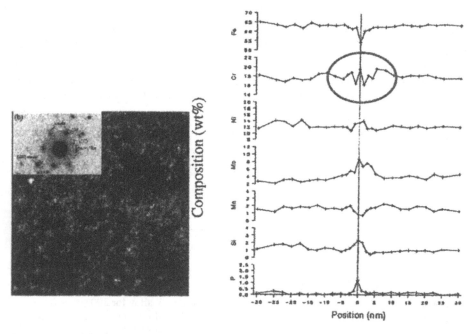

Figure 9. Precipitation of γ' (Ni₃Si) in cold-worked 316 stainless steel baffle bolts irradiated at 299°C to 7.2 dpa in reactor. [19]

Figure 10. The "W" shaped chromium concentration profile at the grain boundary in 304 stainless steel. [9]

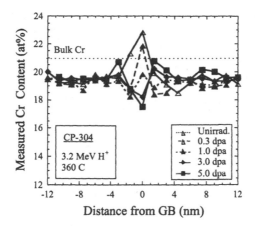

Figure 11. Development of the "W" shaped chromium concentration profile in 304 SS as a function of irradiation dose of 3.2 MeV protons at 360°C.

In addition to the elements discussed, small solutes such as B, C or N may also segregate and also may influence the segregation behavior of the major alloying elements. However, these elements are very difficult to detect. Figure 12 shows an atom probe field ion microscopy result of an unirradiated 304 stainless steel sample containing a grain boundary plane that is roughly parallel with and centered between the large top and bottom faces of the unit cell. Note that B, C and P appear to be concentrated near this plane, indicating enrichment the grain boundary plane. In contrast to small, interstitial solutes, the effect of oversize solute elements is shown in Fig. 13, which presents data from electron, proton and neutron irradiation showing that the presence of these oversized solutes reduces the amount of grain boundary chromium depletion under irradiation. These results are significant because if grain boundary chromium level is important in the IGSCC process, then IGSCC may be controllable through the addition of oversize solutes.

2.2 Irradiated Microstructure

The irradiated microstructure is a strong function of irradiation temperature and dose. Figure 14 shows the dose and temperature regimes where the various irradiated microstructure features dominate. At temperatures below about 300°C, dislocation loops dominate the irradiated microstructure. Above 300°C, cavities can nucleate and become increasingly important with temperature.

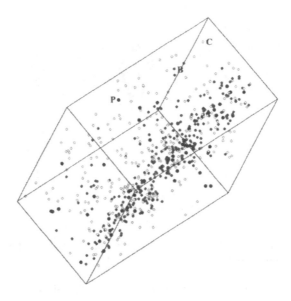

Figure 12. Atom probe – field ion microscopy (AP-FIM) image of a region in an unirradiated 304 stainless steel sample containing a grain boundary and exhibiting enrichment of B, C and P in the boundary region. [20]

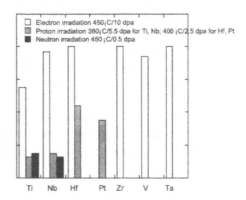

Figure 13. Effect of oversize solutes in the radiation induced segregation of chromium away from grain boundaries in stainless steel alloys. [21]

At high doses and high temperature, second phase formation becomes possible. For light water reactors, the environment is characterized by lower temperatures and low to moderate doses so the dislocation microstructure is the most important feature. Figure 15 shows how the dislocation loop density and size evolve with dose for irradiation of stainless steels around 300°C. The loop number density saturates by 2 dpa, while the loop size

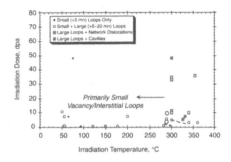

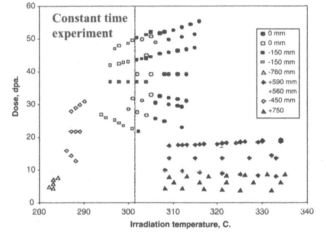

Figure 14. Summary of reported defect structures in 300-series stainless steels as a function of irradiation dose and temperature. [22]

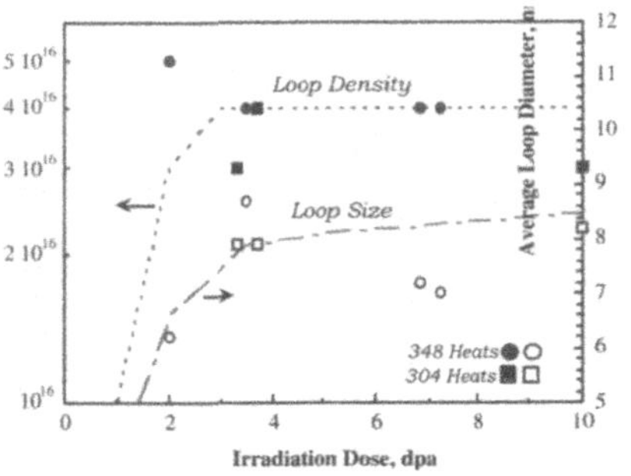

Figure 15. Measured change in density and size of interstitial loops as a function of dose during LWR irradiation of 300-series stainless steels at 275-290°C. [22]

Closed symbols denote voids observed

Void formation does not occur below ~300¼C in this steel when irradiated in a fast reactor at PWR-relevant dpa rates.

Figure 16. Temperature dependence of void formation under irradiation. [23]

continues to evolve until 5 to 7 dpa. Figure 16 shows the sensitivity of the void formation to temperature under irradiation.

The irradiated microstructure is influenced by many factors during irradiation. An illustration is provided in Figs. 17 and 18 in which the effect of oversize elements (Hf and Pt) on the dislocation microstructure is shown. Figure 17 shows that Hf significantly increases the dislocation loop density and decreases loop size following irradiation with Ni ions or protons. Hf also dramatically affects void nucleation as illustrated in Fig. 18, in which the formation of voids is completely suppressed in the optimized, Hf-containing 316 SS, while nucleation occurs by 2 dpa in the undoped alloy.

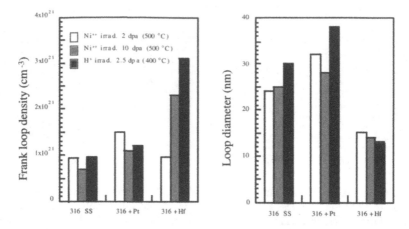

Figure 17. Effect of Hf and Pt additions on the dislocation loop size and number density in irradiated 316 stainless steel. [21]

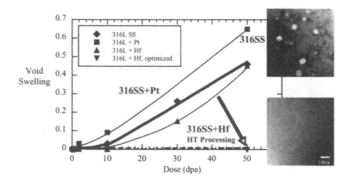

Figure 18. Effect of Hf and Pt additions on the void swelling behavior of 316 stainless steel.

2.3 Irradiation Hardening and Deformation

Irradiation generally results in hardening of the alloy, which can be attributed to the dislocation microstructure. The dislocation loops act as obstacles to the motion of dislocations, resulting in an increased yield strength. Figure 19 shows the yield strength as a function of irradiation dose for the 300 series stainless steels irradiated and tested near 300°C. Note that hardening rises rapidly and then saturates by about 5-10 dpa. Yield strength increase by a factor of 5 over the unirradiated value is not unusual for light water reactor irradiation conditions. But in addition to the hardening, the deformation mode changes, as evidence by a decrease in the work-hardening ability of the alloy. The loss in work hardening is due to the localization of deformation in the form of dislocation channels. Dislocation channels are a group of slip planes on which one or more leading dislocations have "swept out" the obstacles, clearing a path for subsequent dislocation passage. Figure 20 shows the drop in work hardening exponent and the corresponding increase in dislocation channel area fraction with dose. Figure 21 shows a transmission electron micrograph of a dislocation channel in a sample of 304 stainless steel, where the obstacles have been cleared along the path, and the impingement of the channels on the surface resulting in slip steps. Because the deformation mode changes dramatically in addition to the yield strength, the role of hardening on IASCC is difficult to separate from the change in deformation mode.

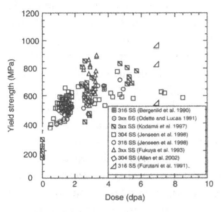

Figure 19. Yield strength of irradiated 300 series stainless steels irradiated and tested around 300°C. [24-27]

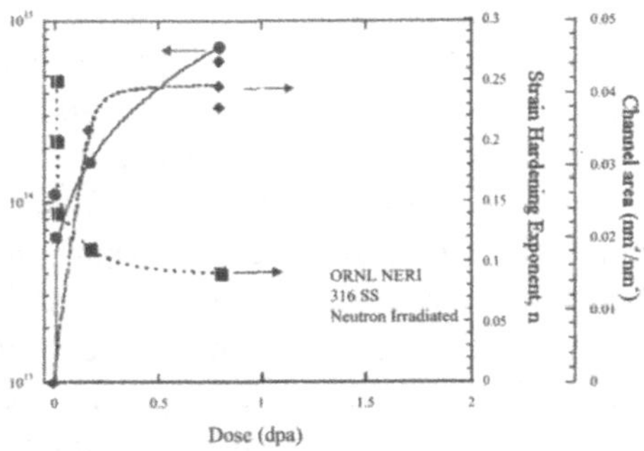

Figure 20. Increased localization of deformation with neutron irradiation in 316 stainless steel as measured by the channel area and the drop in the strain hardening exponent. [28]

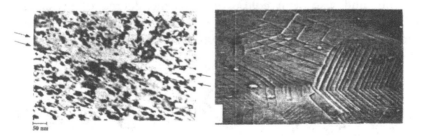

Figure 21. Dislocation channels in irradiated high purity 304 stainless steel as evidenced by a) transmission electron micrograph of a channel, and b) slip lines on the sample surface after straining. [29]

3   STRESS CORROSION CRACKING

In order to understand how the irradiated microstructure, microchemistry and hardening can affect stress corrosion cracking, it is first necessary to review the principle mechanisms of stress corrosion cracking. The slip-oxidation, hydrogen enhanced localized plasticity and selection internal oxidation mechanisms will be briefly summarized.

3.1   Mechanisms

Slip-oxidation refers to a mechanism in which cracking of the oxide film (due to the interaction of dislocations from the bulk with the oxide for transgranular cracking or dislocations in the grain boundary for intergranular cracking) results in rapid corrosion and then passivation of the exposed metal. [30] As the protective oxide film continues to grow, the oxidation rate drops with the one-half power of time. However, under the action of stress, dislocation sources are activated and result in localized increase in stress at the oxide film, which ruptures again to repeat the previous cycle. The schematic representation of the oxide rupture event, the dissolution of the exposed metal and the repassivation is shown in the schematic of Fig. 22 and the cyclic nature of the oxidation process is shown in Fig. 23. The rate of growth of a crack according to this mechanism is controlled by a number of factors including the corrosion potential, solution conductivity and the material condition (sensitized, cold-worked). Figure 24 shows some examples of the model predicted dependence of crack growth rate in sensitized 304 stainless steel in BWR normal water chemistry at 288°C as a function of corrosion potential for different values of solution conductivity. Crack growth rate increases with conductivity and solution conductivity and also with cold work and the degree of sensitization. According to this model, irradiation induced grain boundary chromium depletion and perhaps hardening (as with cold work) would be expected to increase the crack growth rate.

Figure 22. Schematic illustration of the slip-oxidation mechanism of stress corrosion cracking.

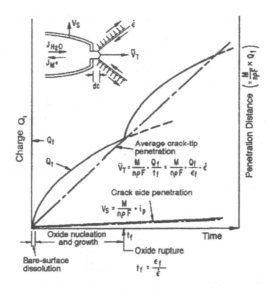

Figure 23. Cyclic behavior of the oxidation process in the slip-oxidation model for stress corrosion cracking.

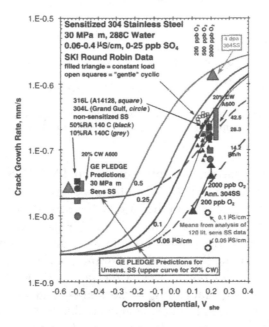

Figure 24. Effect of corrosion potential on crack growth for sensitized 304 stainless steel in 288°C BWR normal water chemistry. [31]

The hydrogen enhanced localized plasticity (HELP) model [32] describes a mechanism in which hydrogen may aggravate or induce intergranular stress corrosion cracking through an increase in plasticity. According to this model, hydrogen accumulation at dislocations shields the stress field of the dislocation, making plastic flow easier. Figure 25 shows images and an accompanying schematic of a dislocation pile-up in Ni-16Cr-9Fe in the TEM stage in the absence and presence (80 torr) of hydrogen. Note that the introduction of hydrogen resulted in a collapse of the dislocations on the slip plane, which is consistent with a reduced stress field. If such a process were to occur at grain boundaries, which are excellent traps for hydrogen, then the increased plasticity may result in additional grain boundary deformation and subsequent sliding, cracking and rupture of the oxide film.

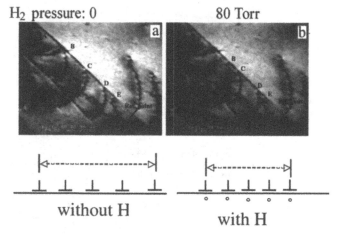

Figure 25. Transmission electron microscopy images and accompanying schematics of the collapse of a dislocation pile-up upon the introduction of 80 torr of hydrogen gas in the TEM stage. [33]

A third mechanism that is relevant to the IASCC phenomenon is selective internal oxidation, SIO. [34] In this mechanism, the alloy content of the film forming element (Cr in the case of stainless steel or alloy 600) is insufficient to form a protective surface film in the particular environment, allowing oxygen to diffusion down the grain boundaries. This process can result in the internal oxidation in which chromium oxides form in the bulk and near or on the grain boundaries. Upon application of a stress, the embrittled boundary can crack. Figure 26 shows schematically the SIO process and how it may lead to intergranular cracking. Continued supply of oxygen ahead of the crack tip will drive cracking deeper into the sample.

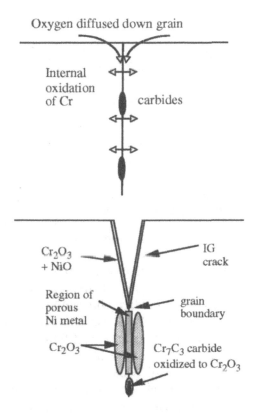

Figure 26. Selected internal oxidation mechanism of IGSCC in which oxygen penetration along grain boundaries can lead to embrittlement ahead of a crack tip and drive crack growth.

3.2   Irradiation Assisted Stress Corrosion Cracking

If any of the known stress corrosion cracking mechanisms applies to irradiated materials, then the IASCC data should be explained by the mechanism. For example, Fig. 27 shows the effect of grain boundary chromium content on IGSCC susceptibility in slow strain rate tests of both stainless steel and the nickel base alloy 600. In both cases, there is a clear dependence of cracking on grain boundary chromium content. The equivalent data for irradiated materials is given in Fig. 28, in which the dependence on grain boundary chromium content is no so clear. While the general trend is for increasing IGSCC with decreasing grain boundary chromium content, there is considerable scatter such that at any value of grain boundary chromium content, the %IG varies from zero to some maximum amount.

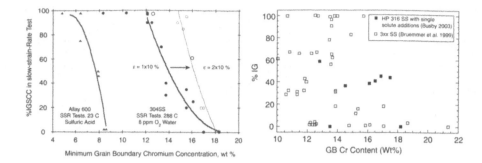

Figure 27. Effect of grain boundary chromium content on IGSCC in slow strain-rate tests. [35]

Figure 28. Effect of grain boundary chromium content on intergranular cracking in slow strain rate tests of irradiated materials. [22]

A similar situation exists with the effect of hardening on IASCC. Figure 29 shows the %IG in slow strain rate tests (SSRT) on the yield stress. As with the dependence on grain boundary chromium content, there is a very large amount of scatter such that the effect can vary from nearly zero to some maximum. Scatter in SSRT tests is well known and it is likely that some of the scatter in these two plots is inherent in the test technique, but the range of scatter; zero to full scale, is troubling and an alternative interpretation is that these features are not the controlling features of the observed cracking. In fact, the data in Fig. 30 on the effect of yield strength on crack growth rate provides better evidence that yield strength can affect cracking.

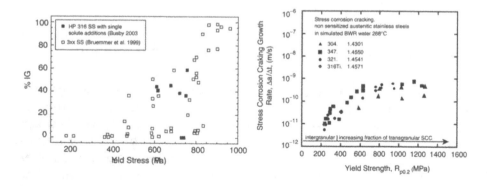

Figure 29. Effect of yield strength on intergranular cracking in irradiated materials. [22]

Figure 30. Effect of yield strength on crack growth rate in several 300 series stainless steels. [36]

The difficulty in attributing the observed IASCC behavior in irradiated materials to a single irradiated microstructure feature is illustrated earlier in Fig. 4, which shows that all of the effects of irradiation change rapidly and saturate according to the same dose dependence. Further, the SCC behavior displays an incubation period before cracking starts and the dose at which cracking starts does not identify any of the microstructure features as uniquely responsible for the effect. Another approach that has been taken is that of annealing irradiated materials to incrementally recover the unirradiated microstructure and then to track cracking as a function of recovery. Figure 31 shows how RIS, the irradiated microstructure (as measured by loop line length) and hardening recover with the extent of annealing, and also how the IASCC severity decreases with annealing. What is striking is that the IASCC susceptibility has dropped to zero before there is any detectible change in the grain boundary chromium content. Changes have started to occur to both in the dislocation loop structure and the hardening, but they are only about 15% recovered at the point where the IASCC susceptibility is completely removed. This may indicate that there is another parameter in addition to hardening that is important.

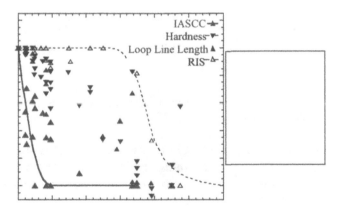

Figure 31. Percentage of remaining irradiation effect with extent of annealing and comparison to recover of IASCC susceptibility. [21]

Figure 32 provides evidence that the deformation mode may be the important parameter. This figure shows an experiment in which samples from a single alloy were subjected to combinations of cold work and irradiation so as to span the range from all cold work/no irradiation to all irradiation/no cold work while keeping the hardness constant. Subsequent SSRT tests in simulated BWR normal water chemistry showed that only the two samples with the highest irradiation doses (lowest levels of cold work) cracked. Since all of the samples had the same hardness, the implication is that the source of hardening is also important, and a distinction between hardening by cold work and by irradiation is that irradiation leads to localized deformation whereas cold work does not. In fact, observations of the surfaces of the samples revealed greater slip band (dislocation channel) activity in the more highly irradiated samples that cracked than in the other samples that didn't.

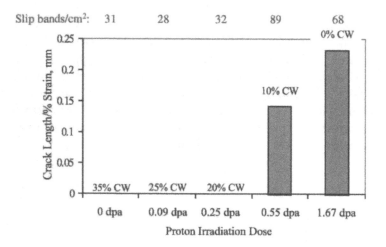

Figure 32. Effect of combinations cold work and irradiation on cracking susceptibility in simulated BWR normal water chemistry. [37]

The formation of localized deformation is not unique to irradiated microstructures as it is also controlled in unirradiated alloys by stacking fault energy. Alloys with high stacking fault energy result in wavy slip much cross slip whereas alloys with low stacking fault energy result in planar slip. The situation is illustrated in the three cases shown in Fig. 33. In the first, the high stacking fault energy results in the development of a

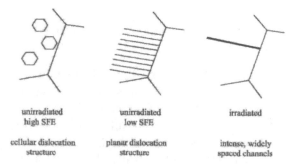

Figure 33. Schematic illustration of the effect of SFE and irradiation on the nature of slip.

dislocation cell structure. In unirradiated alloys with low SFE, a planar dislocation structure will form where the slip planes terminate at grain boundaries. In the third case, irradiation is responsible for the formation of dislocation channels, which are similar to planar slip but even more extreme in the localization of deformation.

Slip planarity has been linked to IGSCC and so it may be that irradiation-induced slip localization can be understood in the context of slip planarity. Figure 34 shows a correlation between stacking fault energy and %IGSCC in irradiated stainless steels. [38] It shows that as the stacking fault energy increases, the amount of IGSCC drops dramatically. But the question is why dislocation channels or planar slip should be a cause of IASCC? The way in which localized deformation can cause IASCC is through the interaction of the dislocation channels with the grain boundary. In irradiated materials, the channels always terminate at the grain boundary, and each channel can carry a tremendous amount of deformation; upwards of 200%! For the case of a grain boundary intersecting the free surface, large amounts of localized deformation in the grain boundary could rupture the oxide film and result in IGSCC by a mechanism similar to that in the slip-oxidation model, Fig. 35.

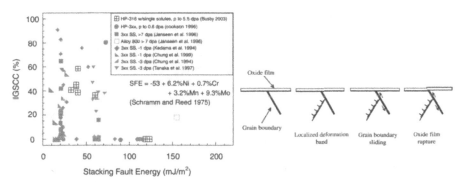

Figure 34. Correlation between IGSCC and stacking fault energy. [38]

Figure 35. Localized deformation leading to boundary sliding and the subsequent rupture of the protective oxide film.

There are limited options for this amount of deformation when it reaches the grain boundary. If lattice planes between adjacent grains are closely aligned, the deformation can cross from one grain to the next. But the alignment requirements are difficult to achieve and this is unlikely to account for the majority of strain transfer. Figure 36 shows the slip continuity as a function of the lattice mismatch across a boundary as described by the sigma number. Dislocations can also pile-up at the boundary and nucleate cracks or be absorbed into the boundary and result in grain boundary sliding. Either mechanism will likely cause the rupture of on oxide film, thus exposing the underlying metal to the water that will result in corrosion and further oxidation. Figure 37 shows the formation of wedge cracks due to the pile up of dislocations against a grain boundary. Figure 38 shows that the amount of IGSCC as measured by crack length per unit strain in SSRT tests in simulated BWR normal water chemistry is reduced for CSL-enhanced samples over annealed samples. Coincident site lattice (CSL) enhancement results in an increased fraction of special or low angle grain boundaries that are more resistant to grain boundary deformation than are high angle boundaries. Sliding in CSL boundaries is reduced in comparison to high angle boundaries, and as shown in Fig. 38, samples with an enhanced fraction of CSLs show less IG cracking in the irradiated state as well.

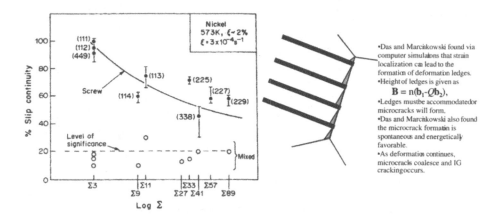

Figure 36. Slip continuity as a function of sigma number describing the mismatch of grains across a grain boundary. [39]

Figure 37. Microcrack formation due to the pile-up of dislocations in a dislocation channel where it intersects the grain boundary. [40]

Evidence for the role of localized deformation in promoting crack growth is shown in Fig. 39, which is a TEM micrograph of the tip of a growing crack, showing the intersection of deformation bands that coincide with the stepped appearance of the crack wall. Figure 40 summarizes how deformation can affect both crack nucleation and crack growth. By forcing slip of grain boundaries, the surface oxide film can rupture, causing exposure of the underlying metal to the solution resulting in corrosion and subsequent passivation. For a growing crack, the dislocation channels can feed dislocations to the crack tip, allowing for crack extension and oxide film rupture within the crack. For either crack initiation or crack growth, localized deformation can enhance the process of the irradiated material.

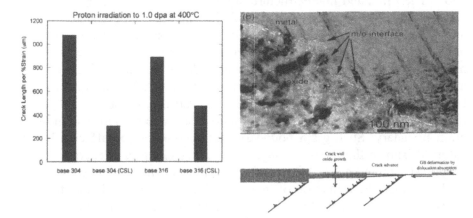

Figure 38. Reduction in IGSCC susceptibility in 304 and 316 alloys with enhanced fractions of coincident site lattice (CSL) boundaries. [41]

Figure 39. Possible role of localized deformation in the growth of a crack by IASCC.

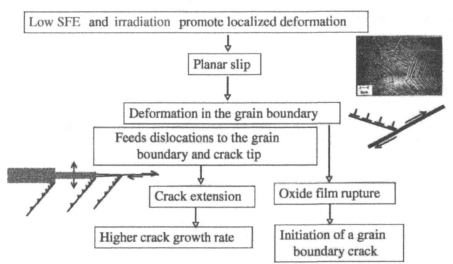

Figure 40. Process by which localized deformation can lead to either crack initiation or crack growth.

4 SUMMARY

Irradiation of stainless steels in reactor core components results in profound changes in the microstructure, microchemistry and hardness of the alloy. The irradiated microstructure, along with the alloy composition can significantly influence the mode of deformation as compared to an unirradiated sample. While the mechanistic foundation for IASCC has yet to be identified, it is likely that it depends on both the state of the irradiated material and the way in which the material deforms. RIS, dislocation loop microstructure and radiation hardening are all potential contributors to the IASCC mechanism in LWRs. Another potential mechanism for IASCC is rupture of the oxide film near the grain boundary due to deformation in the grain boundary. Such grain boundary deformation can be affected by the nature of slip in the matrix. Alloys with low stacking fault energy deform by planar slip and a low SFE combined with the irradiated microstructure could enhance the formation of intense, deformation channels that transmit dislocations to the grain boundary. The accommodation of dislocations in the boundary can result in localized slip or sliding of the grain boundary that can rupture the oxide film, and cause initiation or propagation of an IG crack. Further research is required that extends beyond the current bounds of irradiated material characterization and considers the role of deformation in the cracking process. Key experiments to measure SFE and to link crack initiation and growth to SFE and irradiation-induced localized deformation have yet to be performed.

REFERENCES

[1]   P.L. Andresen, F.P. Ford, S.M. Murphy and J.M. Perks, *Proc. Fourth International Symposium on Environmental Degradation of Materials in Nuclear Power Systems – Water Reactors*, D. Cubicciotti and G.J. Theus, Eds., NACE, Houston, 1990, p. 1-83.

[2]   G.S. Was and P.L. Andresen, *J. Metals* Vol. 44 (No. 4), 1992, p. 8.

[3]   P.M. Scott, *J. Nucl. Mat.*, Vol 211, 1994, p. 101.

[4]   S.M. Bruemmer, E.P. Simonen, P.M. Scott, P.L. Andresen, G.S. Was and J.L. Nelson, *J. Nucl. Mater.*, Vol 274, 1999, p. 299.

[5]   F.P. Ford and P.L. Andresen, "Corrosion in Nuclear Systems: Environmentally Assisted Cracking in Light Water Reactors", in *"Corrosion Mechanisms"*, Ed. P. Marcus and J. Ouder, Marcel Dekker, p. 501-546, 1994.

[6] F.P. Ford and P.L. Andresen, *Proc. Third International Symposium on Environmental Degradation of Materials in Nuclear Power Systems – Water Reactors*, G.J. Theus and J.R. Weeks, eds., The Metallurgical Society of AIME, Warrendale, PA, 1988. p. 789.

[7] P.L. Andresen and F.P. Ford, *Mat. Sci. Eng.*, Vol. A1103, 1988, p. 167.

[8] M. Kodama, R. Katsura, J. Morisawa, S. Nishimura, S. Suzuki, K. Asano, K. Fukuya and K. Nakata, *Proc. Sixth International Symposium on Environmental Degradation of Materials in Nuclear Power Systems – Water Reactors*, R.E. Gold and E.P. Simonen, eds., The Minerals, Metals & Materials Society, Warrendale, PA, 1993, p. 583.

[9] A. Jenssen, L.G. Ljungberg, J. Walmsley and S. Fisher, *Corrosion*, Vol 54, (No. 1), 1998, p. 48.

[10] K. Asano, K. Fukuya, K. Nakata and M. Kodama, "Changes in Grain Boundary Composition Induced by Neutron Irradiation on Austenitic Stainless Steels," *Proc. Fifth International Symposium on Environmental Degradation of Materials in Nuclear Power Systems – Water Reactors*, D. Cubicciotti, E.P. Simonen and R.E. Gold, Eds., American Nuclear Society, LaGrange Park, IL, 1992, p. 838.

[11] A.J. Jacobs, "Effects of Low Temperature Annealing on the Microstructure and Grain Boundary Chemistry of Irradiated Type 304SS and Correlations with IASCC Resistance," *Proc. Seventh International Conference on Environmental Degradation of Materials in Nuclear Power Systems – Water Reactors*, NACE, Houston, TX, 1995, p. 1021.

[12] A.J. Jacobs, G.P. Wozadlo, K. Nakata, S. Kasahara, T. Okada T, S. Kawano and S. Suzuki, "The Correlation of Grain Boundary Composition in Irradiated Stainless Steel with IASCC Resistance," *Proc. Sixth International Symposium on Environmental Degradation of Materials in Nuclear Power Systems – Water Reactors*, R.E. Gold and E.P. Simonen, Eds., TMS, Warrendale, PA, 1993, p. 597.

[13] E.A. Kenik, *J. Nucl. Mater.*, Vol 187, 1992, p. 239.

[14] S. Nakahigashi, M. Kodama, K. Fukuya, S. Nishimura, S. Yamamoto, K. Saito and T. Saito, *J. Nucl. Mater.*, Vol 179-181, 1992, p. 1061.

[15] A.J. Jacobs, R.E. Clausing, M.K. Miller and C.M. Shepherd, "Influence of Grain Boundary Composition on the IASCC Susceptibility of Type 348 Stainless Steel," *Proc. Fourth International Symposium on Environmental Degradation of Materials in Nuclear Power Systems – Water Reactors*, D. Cubicciotti, Ed., NACE, Houston, TX, 1990, p. 14-21.

[16] J. Walmsley, P. Spellward, S. Fisher, and A. Jenssen, "Microchemical Characterization of Grain Boundaries in Irradiated Steels," *Proc. Seventh International Symposium on Environment Degradation of Materials in Nuclear Power System – Water Reactors*, NACE, Houston, TX, 1997, p. 985.

[17] G.S. Was, J.T. Busby, J. Gan, E.A. Kenik. A. Jenssen, S.M. Bruemmer, P.M. Scott and P.L. Andresen, *J. Nucl. Mater.*, Vol. 300, 2002, p. 198.

[18] G. S. Was and J. T. Busby, "Use of Proton Irradiation to Determine IASCC Mechanisms in Light Water Reactors: Solute Addition Alloys," Final Report, EPRI Project EP-P3038/C1434, , Electric Power Research Institute, Palo Alto, CA, April, 2003.

[19] L. Thomas and S. Bruemmer, "High Resolution Characterization of SCC Cracks in LWR Core Components," Interim Report, EPRI Project EP-P2291/C1028, Electric Power Research Institute, Palo Alto, CA, May, 2002.

[20] E.A. Kenik, M.A. Miller, M. Thavander, J.T. Busby and G.S. Was, "Origin and Influence of Pre-existing Segregation on Radiation Induced Segregation in Austenitic Stainless Steel," *Proc. Materials Research Society*, Materials Research Society, Pittsburgh, Vol 54, 1999, p. 445.

[21] G.S. Was, "Recent Developments in Understanding Irradiation Assisted Stress Corrosion Cracking," Proc. 11th Int'l Conf. Environmental Degradation of Materials in Nuclear Power Systems – Water Reactors, American Nuclear Society, La Grange Park, IL, 2003, p. 96.

[22] S.M. Bruemmer, E.P. Simonen, P.M. Scott, P.L. Andresen, G.S. Was and J.L. Nelson, *J. Nucl. Mater.*, Vol 274, 1999, p. 299.

[23] F. A. Garner, S. I. Porollo, A. N. Vorobjev, V. Konobeev and A. M. Dvoriashin Proc. *Ninth International Symposium on Environmental Degradation of Materials in Nuclear Power Systems – Water Reactors*, F.P. Ford, S.M. Bruemmer and G.S. Was, Eds., The Metallurgical Society of AIME, Warrendale, PA, 1999, p. 1051.

[24] U. Bergenlid, Y. Haag and K. Pettersson, *The Studsvik MAT 1 Experiment. R2 Irradiations and Post-Irradiation Tensile Test.* Studsvik Report STUDSVIK/NS-90/13, 1990.

[25] M. Kodama, S. Suzuki, K. Nakata, S. Nishimura, K. Jukuya, T. Kato, Y. Tanaka and S. Shima, "Mechanical properties of Various Kinds of Irradiated Austenitic Stainless Steels," *Proc. Eighth International Symposium on Environment Degradation of Materials in Nuclear Power System – Water Reactor,* A.R. McIlree and S.M. Bruemmer, Eds., American Nuclear Society, La Grange Park, IL, 1997, p. 831.

[26] G.R. Odette and G. Lucas, *J. Nucl. Mater.,* Vol 179-181, 1991, p. 572.

[27] A. Jenssen, written communication, Studsvik Nuclear, Sweden (1998).

[28] K. Farrell, T.S. Byun and N. Hashimoto, "Mapping Flow Localization Processes in Deformation of Irradiated Reactor Structural Alloys," Oak Ridge National Laboratory, report ORNL/TM-2002/66, July 2002.

[29] S. Bruemmer, J. Cole, R. Carter and G. Was, "Radiation Hardening Effects on Localized Deformation and Stress Corrosion Cracking of Stainless Steels," Proc. Sixth International Symposium on Environmental Degradation of Materials in Nuclear Power Systems - Water Reactors, Minerals, Metals & Materials Society, R. E. Gold and E. P. Simonen, eds., Minerals, Metals & Materials Society, Warrendale, PA, 1993, pp. 537-546.

[30] F. P. Ford, Proc. Environment-Induced Cracking of Materials, NACE-10, NACE International, Houston, TX, 1988, p. 130.

[31] G. S. Was and P. L. Andresen, "Effect of Irradiation on Corrosion and SCC in LWRs," Chapter 5, ASM Handbook, in press.

[32] J. Eastman, T. Matsumoto, N. Narita, F. Heubaum, H. K., Birnbaum, Proc. 3rd Int. Conf. Hydrogen Effects in Metals, I. M. Bernstein and A. W. Thompson, eds,, The Metallurgical Society of AIME, New York, 1980, p. 397.

[33] D. Paraventi and G. S. Was, "Environmentally-Enhanced Deformation of Ultra High Purity Ni-16Cr-9Fe Alloys", Metall. Trans. A, 31A (2000) 2383-2388

[34] P. M. Scott, Proc. Ninth International Conference on Environmental Degradation of Materials in Nuclear Power Systems – Water Reactors, S. Bruemmer, P. Ford and G. S. Was, eds., The Minerals, Metals and Materials Society, Warrendale, PA, 1999, p. 3.

[35] S. Bruemmer and G.S. Was, "Stress Corrosion Cracking Fundamentals in Nuclear Power Systems," J. Nucl. Mater. 216 (1994) 326-347.

[36] M.O. Speidel, R. Magdowski, "Stress Corrosion Cracking of Stabilized Austenitic Stainless Steels in Various Types of Nuclear Power Plants," Proc. *Ninth International Symposium on Environmental Degradation of Materials in Nuclear Power Systems – Water Reactors*, F.P. Ford, S.M. Bruemmer and G.S. Was, Eds., The Metallurgical Society of AIME, Warrendale, PA, 1999, p. 325.

[37] M. C. Hash, L. M. Wang, J. T. Busby and G. S. Was, "The Effect of Hardening Source in Proton Irradiation-Assisted Stress Corrosion Cracking of Cold-Worked Type 304 Stainless Steel," Effects of Radiation on Materials: 21st International Symposium, ASTM STP, M. L. Grossbeck, T. R. Allen, R. G. Lott and A. S. Kumar, Eds., American Society for Testing and Materials, West Conshohocken, PA, in press.

[38] R. E. Schramm and R. P. Reed, *Met. Trans. A.*, 6A (1975) p. 1345.

[39] L. C. Lim and R. Raj, Acta Metall. 33 (8) (1985) p. 1577.

[40] E.S.P. Das and M.J. Marcinkowski, "J. Appl. Phys., 43, No. 11, (1972), p. 4425.

[41] R. B. Dropek, G. S. Was, J. Gan, J. I. Cole, T. R. Allen and E. A. Kenik, Proc.11th Int'l Conf. Environmental Degradation of Materials in Nuclear Power Systems – Water Reactors, American Nuclear Society, La Grange Park, IL (2004) pp. 1132-1141.

Chapter 16

ION BEAM SYNTHESIS AND TAILORING OF NANOSTRUCTURES

Harry Bernas and Roch Espiau de Lamaestre
CSNSM-CNRS, Université Paris, Orsay, France

1   INTRODUCTION

The basic electronic structure properties of materials are all related to some characteristic lengths whose scales depend on which property is considered. Examples are the Fermi wavelength or the electron mean free path for conductivity, the Debye wavelength for phonons, the dipolar interaction distance for electromagnetic interactions, the pair correlation length for superconductivity, etc., all of which vary typically from ∼ 0.1 to several tens of nanometers. A fundamental question in nanoscience is what happens when the physical size of the sample shrinks down to the characteristic length scale of one or another of its basic physical properties? The sample's electrical, optical, magnetic or mechanical properties will then be *radically* affected by its size and shape, by the symmetry of its environment, and by its coupling (chemical bonds, radiation, etc.) to the latter. Their hybrid "betwixt atom and bulk" matter nature often directly reflects their electrical, optical or mechanical properties. As we know, the quantum properties are extremely sensitive to the boundary conditions of wave functions in the nano-objects. Hence, not only do such studies require small samples, but in order to be meaningful, they require samples such as nanoclusters or ultrathin films and multilayers with well-controlled shapes, sizes, and interfaces.

Where possible, these objects are obtained by specific chemical techniques or by quasi-equilibrium thermodynamical methods, and remarkable successes have been recorded [1]. But there are cases where such methods fail because scientists attempt to form a thermodynamically "impossible" alloy or compound precipitates in a given host, or because the growth mechanisms in deposition techniques such as molecular beam

449

K.E. Sickafus et al. (eds.), Radiation Effects in Solids, 449–485.
© 2007 *Springer.*

deposition, molecular beam epitaxy (MBE), or sputter deposition for 2D films do not lead to homogeneous systems or to adequately shaped crystals or interfaces. Attempting to circumvent such difficulties is the main justification for using nonequilibrium techniques. Among the latter, ion beams are quite specific. A priori they usually induce disorder, and single ions often do this typically on the sample's nanometer scale (the collision cascade size, for example). These are of course generally regarded as unfavorable features, except in the case of ion beam mixing of thin layers, where the purpose is to form a metastable alloy with (hopefully) predesigned properties.

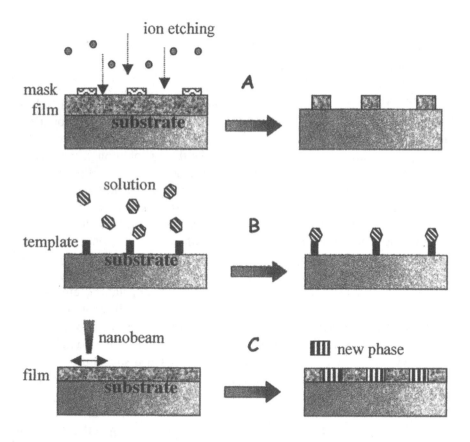

Figure 1. Generic nanostructuring techniques. Top-down (A) involves ion or photon (VIS, UV, x-ray) treatment through a mask (lithography). Bottom-up (B) involves template deposition, and self-organized deposition of a solution component on the template. The most recent and potentially efficient Third Way (C) involves treating a predeposited film or cluster array by an ion (or electron, or photon) nanobeam to induce localized phase changes without requiring further manipulation. The nanobeam may be produced by appropriate focusing or through a stencil mask. From [2].

But as is often the case, such "inconveniences" may turn into advantages as shown by two examples shown below and which epitomize what has been termed [2] the "Third Way" (Fig. 1) by comparison with—and in contrast to—the familiar "top-down" (involving lithography of some kind) and "bottom-up" (involving chemical, e.g., sol-gel) techniques.

As schematized in the figure, the corresponding deposition techniques involve a preparatory step for nanostructure formation, so that the latter is "revealed" by a simple (ion beam, electron, or photon) irradiation with little or no post-treatment. In Part 1, we consider deposited thin film multilayers overlayed with a fairly thick mask including an array of nanometer-sized lines or holes. Through ion beam mixing, ion irradiation through the mask will lead to the formation of a metastable alloy on the scale of the nanostructuring, thus forming an array of metastable alloy nanolines or nanodots. The crucial question is, of course, how to design the irradiation conditions so that excellent control is obtained over the nature and properties of the metastable alloy. Because of the size effects listed above, the answer will usually depend on a detailed understanding of the consequences of irradiation (implantation) on *both* the nanostructure *and* the physical properties to be tailored—a combination that is often nontrivial. This is also true as regards (Parts 2, 3, 4) nanocluster formation through forced nonequilibrium chemistry (supersaturation through ion implantation or irradiation-induced "milking out" of a specific impurity in a host), irradiation-modified nucleation and growth that allows us to form a decent concentration of nanocrystals (e.g., of Si, metals, or compound semiconductors in silica) with a reasonable size distribution.

Although both of the cases chosen technically involve the Third Way, the physics on which they are based differ. Among the important reasons for the difference is the dimensionality of the nano-objects to be produced; hence we chose to first study the 2D and then the 0D situation. Note that Part 1 shows an example of the use of ion beams as a means of nanopatterning *the required properties* (rather than the sample alone!).

Reviews of the nanosciences appear continuously, but much of the more perceptive work is to be found in primary papers and in partial summary articles (published especially in *Science*) over recent years. Our purpose here is to highlight selected aspects for which ion irradiation or implantation can provide specific advantages at some stage of nanostructure design, and our reference list is biased to exemplify this feature. One reason why ion beams are an asset is that typical ion impacts and displacement sequences are on the nanometer scale and that their size and density may be controlled by varying beam conditions. Another reason, of course, is forced solubility. But we argue that except in very few instances, the use of brute force ion implantation to obtain supersaturation and the taxidermal search of novel alloys by implantation or ion beam mixing are not the best methods in this field of research. Implementing ion beam treatments in nanostructure elaboration generally makes sense only within a "building strategy" that involves deliberate use of some specific finely tuned feature of ion beam interactions to perform a nonequilibrium modification of the starting material that could not be performed otherwise but which alone does not generally lead to the final nano-object. As in microelectronics, ion beams are often combined with other synthesis and fabrication methods to provide unique advantages.

2   ION BEAM MODIFICATION OF 2D (1D) OBJECTS

This is likely the most direct application of the Third Way method. In its apparently simplest form, a focused ion beam (FIB)—typically 30-keV Ga ions [3]—is used to literally draw lines or circumscribe dots of different sizes, potentially down to the 10-nm range, by the milling and disordering effects produced in a predeposited film or multilayer. A rather

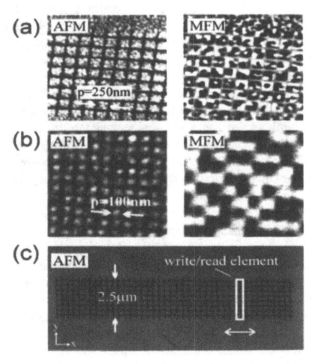

Figure 2. (a) and (b): Atomic (AFM) and magnetic (MFM) force microscopy images of a granular perpendicular recording CoPtCr alloy patterned at two different periods (250 and 100 nm); (c) : AFM image of a static write-read tester. From [4].

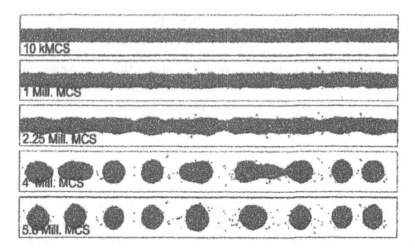

Figure 3. Kinetic lattice Monte Carlo (MC) simulation of the pearling instability for a 8.6-nm diameter wire. Decay proceeds through interface diffusion, forming a chain of nanoclusters with an average diameter 16.4 nm. Numbers of MC simulation steps are shown on lhs. From [8].

elaborate implementation of the technique for magnetic recording is shown in Fig. 2 [4]. Quantum wells or quantum dots for optics [5] have been designed in this way.

Alternatively, drawing an appropriate mask with an FIB and then filling the opened spaces with evaporated metal has led to the production of metallic rod or dot arrays for nonlinear optics applications [6]. Two more sophisticated uses of the FIB may be mentioned. First, the use of the FIB to draw nanometer-size defects on a surface (B. Prével et al. [6]) to induce nucleation of a surface-diffusing species at predetermined sites by interaction of the diffusing atoms with these artificial defects. This allows the buildup of a metallic dot array. Second, a beam of metal ions other than Ga may be produced by using specific intermetallic alloys in the FIB source, and highly localized ion implantation is performed in order to produce rods or dots directly in a host. An example [7] is the synthesis of $CoSi_2$ rods in an Si host by a sequence of Co nanoimplantation combined with annealing. Note that such applications, in spite of their interest, rest on nanocluster evolutions that are difficult to control as regards the final size distribution, at least if the average size is to remain small (a few nanometers). Attempts are made to complement this by a form of self-organized dot formation through [8] the Rayleigh instability decomposition (Fig. 3) of the rods by high-temperature annealing.

A significant difficulty in using FIB technology for many applications in nanostructure design, besides its relative slowness, is the need to limit and control the ion beam intensity and corresponding sputtering and redeposition rates. These problems are optimistically discussed in [3]. Below, we describe complementary ion beam techniques that allow novel applications by using ion beam irradiation as a control parameter of an appropriate physical quantity. We first show how to perform "magnetic patterning" of ultrathin magnetic films and multilayers. This involves modifying the magnetic properties with a minimal change in the structural properties.

2.1 "Magnetic Patterning" with Ion Beams

Many of the typical lengths involved in thin film magnetism (e.g., the exchange length, the domain wall width, the Barkhausen jump length) are nanometric, so that reducing the lateral sample size to that scale leads to new insights into nanomagnetism. Ultrahigh density recording through nanopatterned magnetic media is a major potential application (Fig. 4): the challenge is to obtain significant areal density enhancement by patterning without signal-to-noise ratio degradation and without the loss in sensitivity entailed by surface roughness due to physical etching. This may be done by controlled atomic-scale perturbations of heterogeneous magnetic materials.

Structural perturbations act on the magnetic properties at various scales. We focus on the nanometer scale. The magnetic free energy includes a term (the magnetocrystalline anisotropy, MA) that depends on the orientation of the magnetization relative to different crystal axes (hence, the magnetization curves differ along different crystal directions). It is most sensitive to the structural asymmetry around the atoms sitting at an interface. An example (Fig. 5) is that of a Pt/Co bilayer or multilayer. Such systems are of interest for ultrahigh density magnetic recording, notably because their strong perpendicular magnetocrystalline anisotropy (PMA) improves the spin-flip detection limit [9]. The magnetic state (spin alignment, domain structure) of these sandwiches is the result of a competition between the shape anisotropy due to the 2D-geometry of the films (dipolar term, which tends to bring the magnetization axis in-plane), the magnetoelastic anisotropy due to the heteroepitaxial sandwich structure, and the magnetocrystalline anisotropy due to the spin-orbit interaction which couples the spin direction to the lattice. The effect of symmetry-breaking at interfaces is to enhance the latter term and to induce PMA. In order to reduce the magnetic anisotropy of the layers—which is most sensitive to the asymmetry of the Co atoms' immediate environment—we modify the short-range order around the Co and Pt atoms at the interface and roughen the latter. In the case of ultrathin films, any change in the interface structure actually modifies the local environment around a large

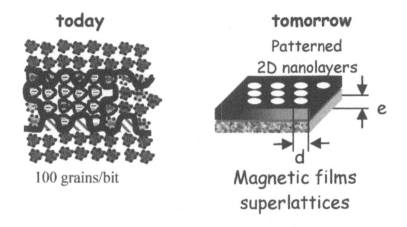

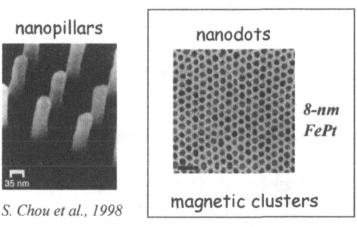

Figure 4. Prospective evolution of magnetic memory elements. In order to reduce signal-to-noise ratio, present elements comprised small grains separated by nonmagnetic material; magnetic anisotropy (MA) is in-plane. Patterned layers will likely have perpendicular MA (PMA) to ease field-induced read-write. Future media (terabit memories) should involve regular magnetic nanocluster arrays, likely with PMA. See text for discussion. Micrographs from [13] and [15].

fraction of the film's atom population. A means to design structural modifications that affect the magnetic anisotropy in a controlled way is the use of a very simple version of ion beam mixing that involves room-temperature He ion-induced mixing at energies typically ranging from 10 to

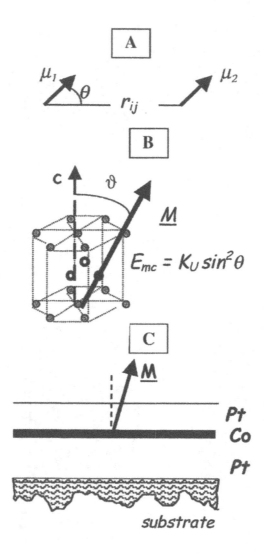

Figure 5. Magnetic anisotropies (MA). A: magnetic dipolar (shape) anisotropy (tends to bring magnetization in-plane); B: MA due to spin-orbit interaction (example of Co, with c-axis perpendicular to basal plane); C: MA in an "artificial crystal" (ultrathin film bilayer), in which the orientation of the MA is determined by the symmetry breaking at the interface.

150 keV. In that energy range, the average energy transferred by the He ion in a collision is very low (~ 100–200 eV), so that Pt or Co recoils only travel 1–2 interatomic distances on average. The temperature is sufficiently low to avoid diffusion (but relaxation occurs). The displacement cross sections are low (typically 1% per atomic layer). Hence, the irradiation parameters (including beam flux and fluence) may lead to very precise control over the 'mixing rate." Note also that this is an ideal case for Monte

Carlo simulations of the corresponding interface modifications. This is only a first step, however: what is the fate of the displaced atoms? There is clearly a temperature-dependence involved. Suppose the atoms can move to some extent: how does the interface relax? This was studied some time ago by neutron- and x-ray reflectometry experiments [10], precisely for systems analogous to the Pt/Co system (Fig. 6). For metallic multilayer systems with masses similar to the couple (Pt/Co) studied here, a linear superposition of ballistic and chemical mixing (depending linearly on the heat of mixing) was found to be effective. The Pt/Co system has a large negative heat of mixing so that the "mixed state" on the atomic scale is stable (i.e., Co displaced into the Pt layer or Pt displaced into Co remain where kinetics have led them). On the other hand, Fig. 6 predicts that the same treatment performed on systems such as Ag/Co or Au/Co would—if the atoms were allowed to move somewhat by weak heating—actually improve the interface by moving the components away from each other. This effect has now been seen experimentally. For Pt/Co layers, the corresponding He ion beam mixing conditions lead [11] to remarkably square hysteresis loops (Fig. 7, upper) and perfectly controllable changes in the coercive fields. They have allowed a complete study [12] of the phase diagram (Fig. 7, lower) for two-dimensional magnetism in this system.

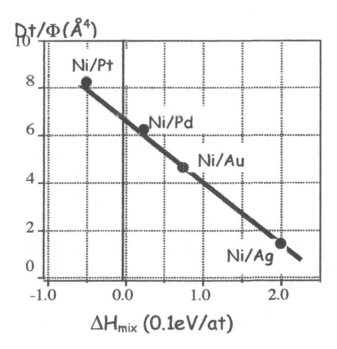

Figure 6. Initial stages of He ion beam mixing of metallic multilayers, studied through x-ray and neutron reflectometry (the ordinate describes the mixing rate; the abscissa is the mixing free energy ; see [10]).

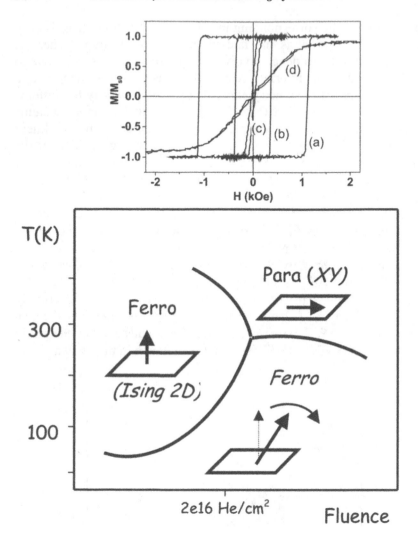

Figure 7. Upper: Hysteresis loops (measured through the Kerr ellipticity) of a Pt/[Pt(1.4nm)/Co(0.3 nm)]$_6$/Pt/Herasil multilayer versus irradiation fluence (He$^+$ ions, 30 keV): (a) as-grown sample; (b) 2 x 10^{15} ions/cm^2; (c) 6 x 10^{15} ions/cm^2; (d) 10^{16} ions/cm^2. The magnetization M of all curves has been normalized to the saturation magnetization M$_{S0}$ of the as grown sample. This figure is from [11]. Lower: Temperature-fluence magnetic phase diagram (see text) from [12].

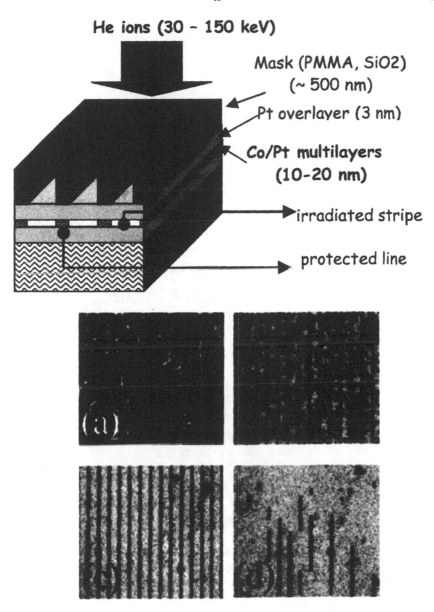

He ions (30 – 150 keV)

Mask (PMMA, SiO2)
(~ 500 nm)

Pt overlayer (3 nm)

Co/Pt multilayers
(10-20 nm)

irradiated stripe

protected line

Figure 8. Ion beam-induced magnetic patterning. Upper: schematics of the process, showing irradiation through a lithographed mask (MA is higher in protected areas). Lower: The vector magnetization of the system was probed through the changes it imparts to incident polarized light, in a polar magneto-optical Kerr effect experiment. Kerr microscopy images (size 30 μm) showing progressive reversal of irradiated line domains. 60 nm-wide lines are separated by 2.5 μm. The initial state is magnetically saturated. Images (a) to (d) are taken in zero field after applying short pulses of increasing fields to reverse magnetization (reversed domains appear white). From [11].

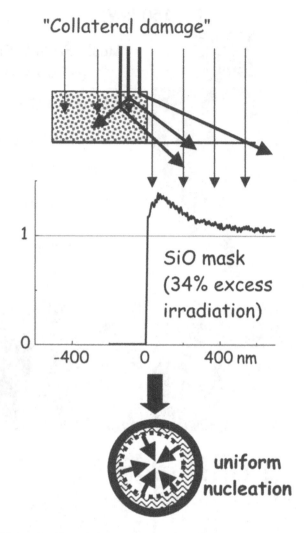

Figure 9. Amplitude of collateral damage (excess irradiation) due to keV He ion scattering outside of mask. It limits resolution to 20 nm but triggers magnetic reversal from a well-defined, uniform nucleation area in the immediate vicinity of the mask edge. As a result, all dots in an array are reversed at the same field. See [14].

The technique involves quite conventional ion irradiation at keV energies, so it is basically identical to the familiar planar technology for semiconductor production and easy to implement. The next step is to study domain wall propagation (i.e., magnetization reversal) in the same system after having nanostructured it not only in depth, but also *laterally*. Here again, ion beam patterning opens new studies in micromagnetics at very short length scales as well as new possibilities regarding magnetic data storage. As shown in Fig. 8, lithographic masking (in this case using the "nanoimprint" technique [13]) allowed (i) ultrahigh resolution magnetic patterning down to 20-nm line widths (Fig. 9) of continuous films without any significant change in their optical properties or surface roughness; (ii) excellent control over the magnetic anisotropy amplitude and orientation, thus also providing control over the coercivity [14]; and as shown in the figure, magnetic reversal triggering occurs in a well-defined geometry that is identical for all the dots in an array, for example. This is most useful for applications.

2.2   Ion Irradiation-Induced Chemical Ordering

This is a typical example of a feature that characterizes the nanosciences: a technical challenge related to applications poses a problem that impacts directly on the most basic physics. As mentioned above, efforts to produce high-density recording media led to the implementation of alloys that maximize magnetic anisotropy (and preferentially PMA). The tetragonal $L1_0$ phase of FePt is one of the best candidates for this. Potentially terabit-density ordered FePt (2–5 nm diameter) nanocluster arrays were produced by solution-phase syntheses [15], but after deposition, the Fe and Pt atoms are randomly distributed on the sites of a cubic lattice. The cubic (low, in-plane MA)-to-$L1_0$ phase transformation (Fig. 10) only occurs by annealing around 950 K. Unfortunately, this anneal leads to sintering and coalescence of the initial nanoclusters. Even for films, such temperatures lead to impractical thermal budgets. The challenge was

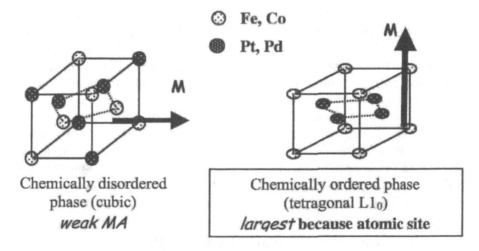

Chemically disordered
phase (cubic)
weak MA

Chemically ordered phase
(tetragonal $L1_0$)
largest because atomic site

Figure 10. Structure and MA orientation of cubic versus $L1_0$ phase of equiatomic Fe(or Co) Pt (or Pd) alloys. The cubic phase's MA is weak, and in-plane; because of symmetry-breaking in the $L1_0$ structure, its MA is perpendicular to the (100) planes. For FePt, this leads to the largest known MA for transition-metal alloys.

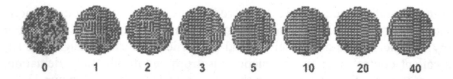

Figure 11. Kinetic lattice Monte Carlo simulation of chemical ordering in an FePd nanocluster. Vacancies are introduced one by one (experimentally, by low-energy He irradiation) and induce successive pairwise exchanges. Selection of a single c-axis variant is performed in a film by inducing directional short-range order during deposition [16]. The challenge for applications is to do the same in a nanocluster. Figure courtesy of K.H. Heinig.

therefore to reduce the transformation temperature significantly. Combining the same type of He ion irradiation as above with limited thermal mobility below 550 K, the cubic-to-$L1_0$ phase transformation (chemical ordering) was promoted in intermetallic FePt and FePd thin films [16]. In the previous paragraph, we saw that room temperature irradiation led to interface disordering by the high-energy barriers. But introducing mobile vacancies by irradiation at sufficiently high temperatures (typically 450–550 K) allows successive pairwise exchanges that are determined by the vacancy jump probability which in turn depends exponentially on the ratio of Fe-Fe, Fe-M, and M-M binding energies to kT (where M=Pt or Pd). In this way, the system explores the nonequilibrium paths toward the low-energy configurations that correspond to chemical ordering. KLMC simulations show that chemical order propagates along the directions that correspond to the lower energy barriers, thus favoring Fe-Fe and Pt-Pt plane formation. This has important consequences. For applications, one requires a unique, well-defined MA orientation, whereas the $L1_0$ phase has three equivalent variants along the three (100) directions. It was shown [16] that one can actually select a single variant (e.g., PMA to the film plane) by introducing a directional short-range order (e.g., in the film plane) during film deposition by layer-by-layer growth, before the ion irradiation: the corresponding symmetry-breaking favors the PMA variant (Fig. 11). This shows how ion irradiation may be combined with other features of the process (here, adequate initial film growth conditions) to obtain control over the ultimate physical properties. The remaining challenges in this area are related to the production of FePt nanocluster arrays and controlling the MA of an entire population. Whether ion beams can help in this endeavor is under investigation.

2.3   Ion Irradiation Control of Exchange-Bias Systems

This is another area of magnetism in which ion irradiation allows tailoring of significant magnetic properties through well-controlled structural modifications. The exchange-bias phenomenon [17] is observed in samples involving a ferromagnetic (FM) layer in contact with an antiferromagnetic (AFM) layer, when they have been prepared (or when the AFM has been run down below the Néel temperature) in an applied field. Under such conditions, the uncompensated spin density acts as an effective field: the FM hysteresis loop is shifted by the so-called exchange bias field, and the coercivity is enhanced. Conflicting models have been offered to explain the effect. In a recent version of one of these [18], it is ascribed to the existence of nonmagnetic atom sites or defects in the AFM, which are domain wall pinning sites (defect-induced pinning of the domain walls stabilizes the domains by reducing the domain wall energy) and which—because their populations are statistically unequal in the AFM sublattices—produce a remnant magnetization in the magnetic domains. Changing the overall population of nonmagnetic sites in the AFM is shown to affect both the exchange bias field amplitude and its orientation. Ion irradiation experiments [19] have shown that through a mechanism of this kind, these parameters can indeed be controlled in an FM permalloy/AFM FeMn system. As shown by simulations and recent irradiation experiments [20], however, the observed changes, including pinning, involve local changes (inside the AFM) in the uniaxial MA. There is nothing to prove that the MA is defined on the scale of a single atomic site. One may conjecture that irradiation defects act on the local symmetry in the AFM through Fe-Mn pairwise exchanges, leading to a local adjustment of the MA amplitude and orientation on a scale significantly larger than that of an atom. This would account for the fact that the comparatively weak magnetic interactions override all others (including those involved in defect creation) in determining exchange bias properties.

3. SOME PROPERTIES OF NANOCLUSTERS AND OF THEIR FORMATION

In the next parts we discuss the role of ion beams as a means of orienting the evolution of a solid solution toward nucleation and growth of a predetermined metallic or semiconducting nanocluster species. This is a far more complex endeavor than those listed above; it involves many of the nonequilibrium thermodynamics problems mentioned in Chapter 12, notably regarding "driven alloy" formation. Although it is by no means a great competitor to several other nanocluster-forming techniques, ion irradiation or implantation has led to novel results that could, with some refinements, provide unique applications. We begin by a brief reminder of characteristic nanocluster properties.

3.1   Some Basic Properties of Quantum-Confined Nanostructures

A simple example is provided by metallic nanoclusters on which light is shone and the optical absorption measured. For a bulk sample, because of the high electron density there is a quasi-continuum of states up to the Fermi energy: Very large clusters have "nearly bulk" optical properties. However, if they become smaller than about a tenth of the incoming wavelength, their optical absorption is drastically affected, as noted by Faraday in 1835 and quantified by Mie [21] in 1908. The electric field associated to light induces charge oscillations at the surface of the metal, generating a dipole electric field that acts on the electrons' motion. This is the surface plasmon resonance that assists both intraband interband transitions, which is related to conduction electrons and to the material's electronic density of states. When the surface-to-volume ratio of the sample becomes large (reduced sample size), this leads to a significant size dependence of the optical absorption and a nonlinear response of major importance for applications. Figure 12 [22] illustrates how nanocrystal behavior lies between that of the atom and the bulk solid. This can be seen more precisely in the semiconducting nanocrystals where, because of the comparatively low electron density, the dispersion curve $E(k)$ is discontinuous and wave function confinement may occur (an excellent tutorial on the entire field is in [23]).

Recall that for a single electron in a periodic crystal potential U, the Schrödinger equation states that

$$\{- (\underline{h}^2/2m_0)\nabla^2 + U(r)\}\psi = E\psi$$

(where $\underline{h}$ = "h-bar," Planck's constant divided by 2π), which by mass renormalization leads to the free-electronlike equation

$$- (\underline{h}^2/2m^*)\nabla^2\psi = E\psi$$

(where m* is the effective mass) and to its familiar dispersion curve $E(k)$ solution. In a system with a band structure, noninteracting quasi-particles are considered; when the system is excited, e.g., by photon absorption, an electron-hole pair (exciton) is created with the electrons (m_e^*) in the conduction band and the holes (m_h^*) in the valence band. The available energy spectrum of these excitations is given by the Hamiltonian

$$\mathbf{H} = - (\underline{h}^2/2m_e^*)\nabla_e^2 - (\underline{h}^2/2m_h^*)\nabla_h^2 - e^2/\varepsilon \,|\, r_e\text{-}\, r_h \,|$$

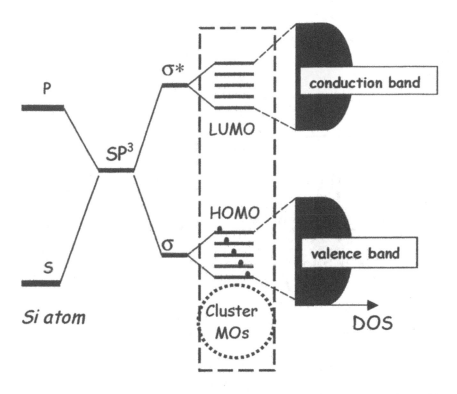

Figure 12. The evolution from Si atomic orbitals to the Si bulk. Atomic bonding orbitals progressively develop into molecular orbitals, into orbitals that cover the cluster (HOMO = highest occupied molecular orbital; LUMO = lowest unoccupied molecular orbital), and finally into valence and conduction bands. The intermediate state produced by the binding of a few tens of Si atoms in a cluster is highlighted. Adapted from [22].

which is analogous to that of the hydrogen atom, i.e., its features are characteristic of wave functions in a box, with correspondingly discrete energy levels. But here the Bohr radius for the exciton is $a_B = \varepsilon \underline{h}^2/\mu e^2$, where μ is the reduced mass of the electron-hole system, smaller than m_e, and ε is the nanocrystal's dielectric constant (significantly larger than that of vacuum). The corresponding exciton Rydberg energy is $R_y^* = e^2/2\varepsilon a_B =$ 13.6 $(1/\varepsilon^2)(\mu/m_0)$, and upon photon absorption (neglecting photon momentum) excitons are created at successive discrete energies

$$E_n(k) \approx E_g - R_y^*/n^2$$

where E_g is the ground state. The typical exciton Bohr radii for semiconducting nanocrystals range from 1 to about 10 nm. A comparison of the exciton's Bohr radius with that of the nanocrystal will tell us whether

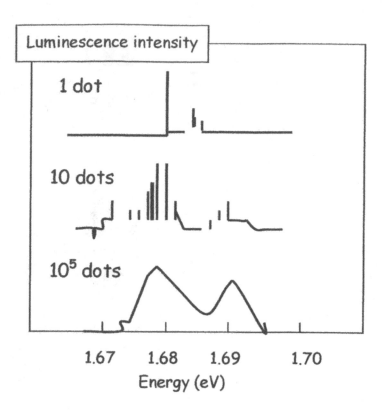

Figure 13. – Photoluminescence spectra from quantum dots excited and detected through apertures in a mask with diameters 0.2 μm to 25 μm, in order to identify the emission from individual dots. Schematically adapted from D.Gammon et al., Phys. Rev. Lett. 76, 3005.

the exciton's wave functions are weakly confined (a_B significantly smaller than the nanocrystal radius) or whether (a_B is larger than the nanocrystal radius) the nanocrystal constitutes a box whose walls determine the boundary conditions for the exciton wave functions—this is known as quantum confinement. We will restrict ourselves to this discussion, in which electron and hole motion (and the corresponding energy level spectra) are independent. Note that for dimensions D higher than 0, threshold energies are still sharp, but the shape of the density of states for electron and hole states is given by $E^{D/2-1}$, i.e., there are discrete subbands in the 1D and 2D cases corresponding to quantum wells. These properties show up clearly in the optical absorption and emission properties of the semiconducting nanoclusters. An important feature of the confinement property is the resulting sensitivity of the exciton state properties to the details of the potential barrier, viz., the size, of course, and the interface

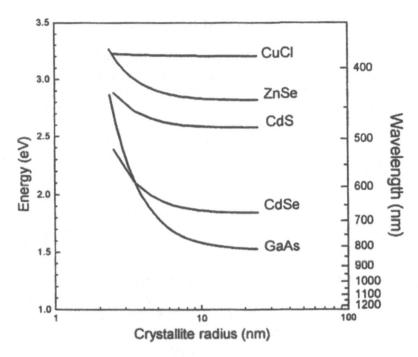

Figure 14. Calculated energy of absorption onset for different semiconductor nanocrystals versus crystallite size. From [23].

between the nanocluster and its host. A quantum-confined cluster behaves like an artificial atom, but not a totally isolated one. This is one of the crucial aspects to consider in elaborating nanocluster preparation techniques, and accounts for huge efforts to control the interface chemistry and obtain cluster surface passivation, adequate core-shell structures, etc. Even when the nanocrystal size and shape distribution are very narrow, size and interface effects may introduce differences in the emission spectra of quantum dots as shown in Fig. 13. Finally, note that semiconducting nanocrystals with a small effective electron and hole mass have a large Bohr radius (see above): when their size is small, such nanoclusters display strong wave function confinement, leading to potentially large blueshifts in the exciton energy spectra as the size is decreased (Fig. 14). As a result, controlling the size—and limiting the size distribution width—of such nanoclusters provides a means of choosing their emission or absorption wavelength in a broad range. Typically, the gap of PbS ($m^* = 0.12\ m_e$) in the bulk is 0.41 eV (corresponding to an emission at about 3 μm), whereas a 3 nm-radius PbS nanocluster emits at about 1.5 μm.

3.2   Nucleation and Growth of Nanoclusters: A Reminder

Many attempts have been made to synthesize metallic or semiconducting nanocrystals in a semiconducting or insulating medium by "brute force" ion implantation, i.e., forcing supersaturation of a solute at temperatures low enough to avoid unmixing and then obtaining a second phase by progressively annealing at increasing temperatures. This is based on the idea that one may somehow extend the physics expounded in thermodynamics textbooks [24] to cases in which miscibility can go down to zero, the solute profile is nonuniform, nonequilibrium defects play a (sometimes major) role, etc. This extension requires precautions, of course. Assuming that the solute incorporation has led to the metastable region of the phase diagram (in which two phases α and β can coexist) provides a guideline regarding the early stages of nucleation and growth. The first question concerns the very possibility of nucleation, i.e., the minimum size, if they exist, of equilibrium clusters at a given temperature. This is given by the driving force for nucleation and growth, i.e., the total Gibbs free energy for the formation of a cluster (radius r). The latter is the sum of a positive (destabilizing) surface energy term and of a negative (stabilizing) volume energy term. Figure 15 shows the corresponding variations as a function of n, the number of atoms in a cluster (i.e., the surface term is proportional to r^2: the volume term to r^3). If the cluster size is below a minimum

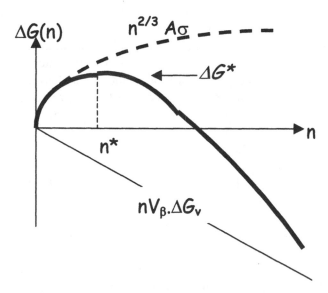

Figure 15. – The total Gibbs free energy for cluster formation is the sum of a surface term (σ is the surface tension) and of a volume term (ΔG_v is the volume free energy; V_β the atomic volume in the cluster). In order for the cluster to be stable, a minimum size is required, corresponding to n* atoms in the cluster. See Ref. 24.

(designated as n* in the figure), the cluster will redissolve. There is thus an energy barrier for nucleation.

In the same steady-state nucleation picture, the Gibbs-Thomson equation describes the relation between the nanocluster size and the average concentration c of solute atoms that it is in equilibrium with $c(r) = c_\infty \exp[r_c/r]$, where $r_c = 2\sigma V_\beta/kT$ is termed the capillary length and c_∞ is the (temperature-dependent) solubility corresponding to an infinitely large cluster. The equation shows that the smaller the cluster, the larger the solute concentration field and concentration gradient around it. The latter is the driving force for solute diffusion, so that the concentration gradients between clusters of differing sizes lead to growth of the larger ones at the expense of the smaller ones (Ostwald ripening). This process may be interface reaction-limited or (more often in our cases) diffusion-controlled,

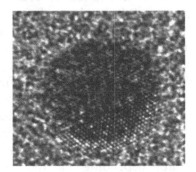

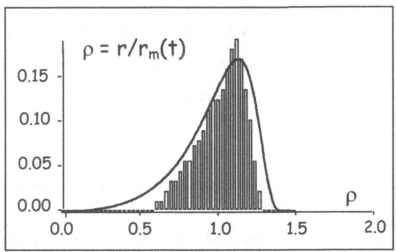

Figure 16. Full curve showing long-term limit of size distribution according to LSW [25]. It is compared with the MeV ion beam irradiation-induced nanocluster size distribution of [49], normalized to the average nanocrystal size. Upper: a typical nanocrystal (note facets) synthesized by MeV ion beam irradiation and annealing (from [49]).

a process that leads to an initially parabolic growth law. As ripening progresses, the shape of the nanocluster size distribution changes. This has been described analytically by Lifshitz, Slyosov, and Wagner [25] (LSW) in a mean field approach, assuming that the clusters are at very low concentration and only interact through the average concentration field, that the nanocluster radii are far larger than the capillary length, that the total number of clustered atoms is conserved and that the chemical potential gradient is near zero, i.e., only the long-term limit of ripening is considered. Under these conditions, LSW show that for diffusion-controlled ripening the average nanocluster size varies as $r^3 - r_0^3 = kt^{1/3}$, and their number as $N = k't^{-1}$, whereas for the interface-controlled ripening, the dependences are respectively $r^2 - r_0^2 = kt^{1/2}$ and $N = k't^{-3/2}$. The particle size distribution is shown to have a universal (analytical) shape when scaled by the average particle size (Fig. 16). This description is found to be adequate for experimental reults that are well beyond its relatively severe assumptions, notably as regards nanocluster concentrations and spatial uniformity, at least as long as the initial (preripening) distribution is narrow enough to justify the mean field description—a situation that is not so common in the real world, where nucleation and growth affect different fractions of the same nanocluster population in a host at a given temperature.

Finally, we note that as mentioned above, the random nucleation process corresponds to the metastable region of the phase diagram, in which the driving force is the supersaturation chemical potential. This is by no means the whole picture: depending on the implanted species-host combination, supersaturation may occur at near-zero concentrations, so that as soon as the temperature is high enough for atomic mobility to set in, the solution is unstable against very small concentration fluctuations and there is no nucleation barrier. This is described by the Cahn-Hilliard theory of spinodal decomposition and recent extensions (see Binder and Stauffer [24]). As noted in Chapter 12, Part 2, the description of nucleation and growth in driven alloys under irradiation or implantation is considerably more complicated.

4   TWO EXAMPLES OF ION-BEAM SYNTHESIZED NANOCLUSTERS

Size- and density-controlled nanoclusters are mostly produced through powerful and diverse chemical techniques (metallic or semiconducting nanoclusters in solutions or in various types of "cages"), and by molecular beam epitaxy (semiconducting quantum dots on or inside semiconductors) [1]. New approaches (e.g., physical UHV growth) methods are being implemented with some success. Ion implantation was initially

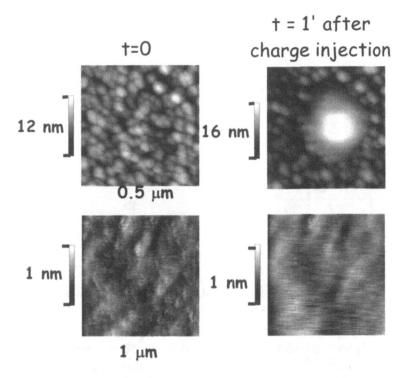

Figure 17. Charge injection into a (Si nanocrystal in silica) sample, performed and detected by AFM. Upper: AFM image of sample with Si nanocrystals, before (left) and after (right) charge injection. Lower silica sample without Si nanocrystals, before (left) and after (right) charge injection, shows no evidence of charging. Demonstrates that charges are localized on the Si nanocrystals. From [28].

introduced primarily to overcome the solubility limit of the desired element(s) in the required host; this may of course be obtained, but under conditions (temperature, irradiation damage and disorder, etc.) that are often stringent, empirically determined, and not necessarily compatible with, e.g., post-annealing conditions that allow manipulation of the size distribution. On the other hand, ion beam synthesis is presently either the easiest or the only known way to obtain and study some types of nanoclusters in appropriate hosts. In this section we briefly show an example pertaining to Si nanocrystals in SiO_2 and one related to anisotropic metal nanocluster production. Examples related to spintronics applications are given in [26].

Si nanocrystals in silica display intense photoluminescence in the visible range, as well as evidence for controllable electroluminescence [27]. This has led to many studies of the basic processes involved and of course to applications. Regarding the former, two major effects have been suggested and evidenced: quantum confinement, which contrary to early suggestions does not turn indirect-gap Si into a direct gap semiconductor,

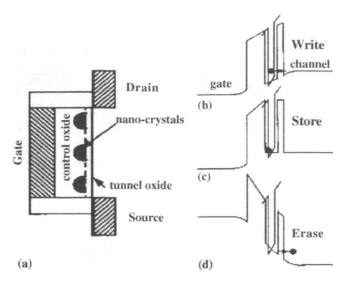

Figure 18. Schematic cross section (a) and band diagram during injection (b), storage (c), and removal (d) of an electron from a nanocrystal. From [29].

and the influence of radiative defect states at the nanocrystal surface. Quantum confinement increases the band-gap energy and the ratio of radiative to nonradiative transition probabilities. The latter is also obviously sensitive to the possibility of (often nonradiative) exciton recombination at the interface of the nanocrystal with the surrounding host. Besides controlling growth and the size distribution, finding appropriate conditions to minimize the interface defect density is therefore a difficult challenge for ion beam as well as other techniques, especially since large Si nanodot density is required to adjust electron transport through tunneling in SiO_2 for emission control purposes. A basic question is: Can one find charging conditions for which the injected electrons (hence the excitons) are localized on the Si nanoclusters, or are they delocalized in the oxide as well as on the clusters? It has been shown [28] (Fig. 17) that charge localization is indeed effective. The fabrication of nonvolatile semiconductor memories using Si nanoclusters in a tunneling oxide as discrete charge storage centers (floating gate) was demonstrated (Fig. 18) by Tiwari [29] using chemical vapor deposition. Physical deposition from the gas phase and excess Si precipitation in the oxide have also been used, the Si excess in the oxide being often produced by Si ion implantation [27]. In the latter case, control over the size distribution is practically impossible, interdot tunneling is significant, and ultralow implantation energies are needed in view of the very thin tunneling oxide layers that are required for fast, low-power memories. A combination of basic knowledge on nucleation and growth (Section 2.2) as well as on ion beam mixing with lattice Monte Carlo simulations led the authors of [30] to design a process in which Si implantation is performed through the completed poly-Si control gate so that ion beam mixing at the Si-SiO$_2$ interface leads to Si displacements into the oxide, hence to local Si supersaturation very close to the interface and, upon annealing, to the formation of a self-aligned Si nanodot array with a narrow zone (typically 2–3 nm) depleted in Si, very well adapted to controlled single particle tunneling (Fig. 19). Not all problems are solved. Lateral self-organization would be required to avoid the residual intralayer tunneling, but the results are encouraging and have led to an industrial 128-Mo memory demonstrator.

Anisotropic metal particles and their arrays are particularly interesting because of their optical properties: they produce lateral confinement below the diffraction limit, and their surface plasmon resonance bands experience a strong redshift when longitudinally polarized light is shown along their long axis. This leads to interesting applications in nano-optoelectronics, including polarized wave-guiding at the nanoscale and nonlinear effects. Anisotropic particles were fabricated [31] by irradiating initially spherical Au-core/silica-shell particles with 30-MeV medium-mass ions. The silica shell expands anisotropically, perpendicular to the beam, producing a large lateral stress around the ion track that makes the (relatively soft) metal core deform along the beam direction. Line arrays of small (2–15 nm), spherical particles in subdiffraction limit proximity were fabricated [32] by a somewhat similar technique. An initially random distribution of spherical Ag nanoparticles was produced in a supported silica film through a technique rather analogous to that described below in Section 4.2 (the interpretation of Ag nanoparticle nucleation in terms of collisional effects offered in [32] totally disagrees with that given here, however). Further irradiation by 30-MeV medium-mass ions led to the same beam-induced anisotropic strain effect as above, but acting here on the silica film, so that the initially randomly dispersed particles end up in approximate alignments along the beam direction. In both experiments, interpretations of stress creation and deformation are proposed in terms of thermal spike effects (see papers at Mat. Res. Soc. Meeting, San Francisco, April 2006). As in many other cases discussed in this chapter, it is fair to recall that these ion beam techniques are in competition with e-beam lithography (e.g., the experiments on Au nanorods of [33]) and FIB fabrication [6]. Other effects and applications related to MeV ion irradiation are presented, e.g., in. [34].

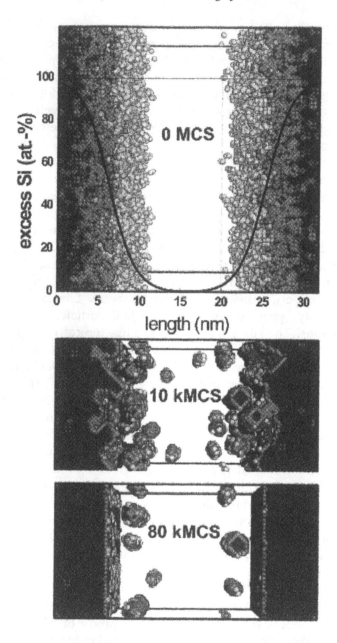

Figure 19. Kinetic lattice Monte Carlo simulation of self-aligned Si nanocrystal layer formation in SiO2 tunneling barrier. Upper: Initial, as-irradiated state (full line is Si density). Middle: Early phase separation stage – Si interfaces recover while excess Si nucleates clusters in oxide. Lower: Intermediate phase separation stage – interfaces have recovered completely, while Si nanoclusters form a delta-layer on each side (note depleted zones between nanocrystal layers and interfaces). From [30].

5 MECHANISMS OF ION BEAM SYNTHESIS OR MODIFICATION OF NANODOTS

5.1 Ion Implantation Synthesis

Regarding nanocrystal precipitation, the initial hope was that the quasi-athermal supersaturation produced by implantation would, upon appropriate thermal annealing, lead to a precipitation behavior adequately described by classical growth and coarsening (LSW) mechanisms, hence, with rather well controlled size distributions. Considerable work in very different directions [35] showed that nanoclusters of many different types (including core-shell or even empty-core-shell) could be ion beam-synthesized, but the results of ion beam induced growth usually differ significantly from the growth and coarsening mechanisms expected from the binary alloy scheme. In fact, control over nucleation and growth is generally difficult, broad (sometimes even bimodal) size distributions being the rule rather the exception; in the worst cases, annealing even at comparatively low temperatures causes a loss of control over the precipitation depth, which no longer corresponds to the implantation profile. Can one rationalize and improve ion beam synthesis of nanocrystals? The problem has been studied in some detail for the case of metallic and semiconducting nanoclusters in insulators and we discuss it here.

Among the reasons usually offered to account for observed discrepancies with the classical nucleation and growth scheme, some are related to physics, others to chemistry. The former are based on the effects of collisional defects and their interactions, sometimes including the electronic stopping contribution, whereas the latter have focused on the chemical interactions of implanted ions with their environment (particularly in glasses). We briefly review both approaches and show that at least in oxide glasses, they may be combined through the redox concepts developed to account for radiolysis phenomena in metal-containing solutions.

5.1.1 Physical Versus Chemical Approaches to Ion Implantation-Induced Precipitation

The physical approaches involve a description of the unmixing conditions for the implanted, supersaturated solution and account for solute diffusion and concentration changes. Ion stopping induces defects that affect solute mobility as soon as the sample temperature allows atomic movements. Diffusion may be enhanced, or (equivalently) the annealing temperature needed to trigger precipitation may be lowered [36]. In the spirit of driven alloy studies, Monte Carlo simulations [37] predict modifications of the precipitation mechanism under strongly nonequilibrium conditions with a source term (solute and/or defect input) in the evolution equations of the system. For example, varying the ion beam fluence allows control over the monomer flux during precipitation, providing an additional parameter (in addition to annealing time and duration) to reach controlled precipitation. Comparison of the predicted effects with experiments [38] showed that this case can still be analyzed in terms of classical nucleation and growth behavior, albeit at an effective temperature higher than the physical temperature (see Chapter 12, Part 2). On the other hand, ion beam mixing experiments revealed that precipitation parameters such as the capillary length and solubility are affected in a way that diverges from the picture of an effective (increased) temperature.

Another point, often ignored, is that precipitation in implanted samples occurs in inhomogeneous samples. Growth and especially coarsening then differ from the features expected in homogeneous systems [39]. Specifically, this has led to self-organization, an effect related to the diffusional screening length λ, describing monomer diffusion among a nanocrystal population. Regarding the nanocluster depth distribution, if the screening length is large compared with the implanted profile width Δ, the local solute concentration nonuniformity may be neglected, and LSW-type growth occurs inside a (narrowed) implant profile. On the other hand, if $\lambda \gg \Delta$, solute diffusion occurs because of the concentration inhomogeneity and clusters tend to dissolve by outdiffusion from the implant profile. This has been verified experimentally [40].

Although the physical parameters involved in ion beam assisted precipitation of nanocrystals provide a means of building up a qualitative description of observations, they have failed to support a more quantitative picture nor has their manipulation led to significantly improved precipitation control. As first emphasized by the Padova group [41], the implanted ions' interactions with the surrounding matrix cannot be ignored. Chemical interactions of implants with their surroundings were indeed detected very early on [42]. A striking example is that of iron implantation in pure silica [43], in which iron was found to occur in five different charge states.

Hosono [44] showed the importance of the implants' electronegativity in determining their post-implantation state as well as that of the related defects by using ion-solid chemistry to modify nanocrystal precipitation in silica. Likewise, the composition of the thermal annealing atmosphere was found to be crucial: depending on the element(s) comprising metallic nanocrystals to be formed in silica, the annealing atmosphere required to promote growth had to be either reducing or oxidizing [45]. The special role of ubiquitous oxygen has also been stressed in attempts to obtain a criterion for precipitation of various elements in oxide glasses. The use of thermodynamical data related to oxides [41] has met with some success, but it raises questions about the temperature at which the comparison of Gibbs free energies should be made, about the use of a high fictitious temperature based on the thermal spike model and, as noted by the authors themselves, about the nature of the oxides (or metals) to be compared. The latter point is especially significant: a given element in a glass is never in a single, uniquely defined chemical state because it interacts with different groups of molecular subunits. The following section suggests that redox chemistry can more readily account for this and provide a unified view of radiation-assisted precipitation regarding both the microscopic mechanisms and the thermodynamical features.

5.1.2 Role of Redox Interactions: An Example

The following example illustrates the influence of redox chemistry on nanocrystal formation and on the subsequent size distribution width in silica. Sequential implantation of both Pb and S into pure silica leads to PbS nanocrystal formation [35], but size distributions are broad: they are bimodal when high-fluence implantations are performed and practically lognormal (FWHM/<R> = 60%) when concentration gradients are low. A detailed study of the latter case showed [47] that both the –II and +VI redox states of sulfur were present in pure silica. A consequence of the –II/+VI sulfur equilibrium is that the PbS nanocrystal growth is coupled to that of $PbSO_4$ nanocrystals. This simultaneous growth rules out any simple description in terms of second-phase nucleation and growth. Moreover, the –II and +VI redox forms of sulfur have very different diffusion coefficients, leading to significant changes in the nanocluster depth distributions. The complex implant chemistry in silica therefore leads to multiple precipitation paths, and the very existence of such multiplicity has been shown [40] to broaden the size distribution and give it a lognormal shape.

By including the role of redox interactions in the nanocrystal design strategy, improved control over their evolution can be obtained. PbS nanocrystals are again a case in point. Large concentrations (up to 40%) of lead oxide may be introduced in silicate glasses so that Pb implantation is not required Large densities of PbS nanocrystals may potentially be obtained by simply implanting sulfur (which does not dissolve completely in the melt) into such glasses. Control over precipitation may then be obtained by tuning the remaining degree of freedom, i.e., the glass composition, which is adjusted so as to control the sulfur redox state populations. The width of the nanocluster size distribution was reduced by a factor of 4 relative to the initial, lognormal distribution in this way, by progressively replacing the calcium oxide in a standard soda-lime glass by zinc oxide [48]. As shown below, introducing the redox properties of implants clarifies a number of problems in ion beam-assisted precipitation, suggesting a reinterpretation of some previous results.

5.2   Ion Irradiation Synthesis

We have seen that using ion irradiation and post-annealing of a host to cause second-phase precipitation involves rather complex mechanisms, among these, the creation and thermal evolution of vacancies and interstitials and their compounds in metals and semiconductors or mostly of electronic defects in insulators. All of these mechanisms may interact with diffusing species (implanted, or belonging to the host). These effects are usually invoked in efforts to interpret nanocluster precipitation results. Defect mobility and interactions obviously play a role, but we stress that in insulating hosts, irradiation-induced defects also modify the redox potential, thus acting on the charge state, mobility, and chemical affinity of the components, and finally on their ability to form clusters. A case in point is that of metal cluster formation in silica and/or oxide glasses, which we show to be very similar to the early stages of photography.

5.2.1   "Ion Beam Photography"

Except in quite specific sol-gel and UHV nanocluster deposition techniques, controlling the size distribution and density of clusters in a host is rarely successful. This is mostly because annealing after deposition leads to clusters that usually nucleate and grow at different times, blurring and broadening the subsequent Ostwald ripening. If the nucleation and growth processes were entirely decoupled, one could anticipate that the initially monodisperse population would evolve according to the LSW prediction. Complete control over the density, the average size (from 2 to 8 nm) and the size distribution (FWHM typically 2 nm) of Cu metal clusters was indeed obtained in pure silica and a silica-based glass [49] by an MeV ion irradiation, in which the electronic energy loss of the incoming ions played the major role. The irradiation biased the cluster formation chemistry, producing only pure metal clusters, and led to a nearly perfect LSW-size distribution (Fig. 16). Nucleation was shown to require a threshold in the inelastic energy transfer to the electrons of the glass, and the nucleation probability (hence the cluster density) was modeled (Fig. 20) by noting that the electronic energy deposition is actually a statistical variable, and dividing the sample into elementary volumes and assuming that *separate,*

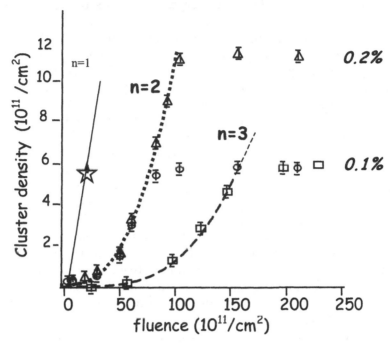

Figure 20. Comparison of experimental cluster densities with those deduced from the probability (Poisson distribution) of having more than n_c increments of deposited energy on the surface σ (full curves). F is the ion fluence per cm^2. The curves correspond to different deposited energy densities. For Br (11.9 MeV) irradiation (full triangles), n_c=2 increments within σ are required for nucleation to occur. For O (6 MeV) irradiation (open squares), n_c=3 increments are required. There is a unique "threshold" (5.0 keV/nm) for nucleation, which may be reached in independent increments. Nucleation occurred for n_c=1 with 580 MeV Ag ions (13 keV/nm). The saturation observed occurs when all the Cu atoms in the layer have clustered. This was checked on glasses with two different Cu_2O concentrations.

additive electronic energy depositions inside each volume induced Cu cluster nucleation. The cluster formation probability then depends on the total energy deposition inside each volume. This provides a simple model for the nucleation evolution, but it provides no information on the microscopic nucleation mechanism. The above results indicate that it is related to some form of charge neutralization, because this is the only way to "store" the consequences of energy deposition. This suggests that a close look should be taken at the photographic process, which operates in a similar way.

5.2.2   Redox Origin of Metal Nanocrystal Precipitation in Glasses

Ionizing radiation impinging on polarizable materials (glass, organic or aqueous solutions, etc.) leads to charge ejection and/or charge trapping. In oxide glasses, this produces a wealth of defects [50]. The ability of a site to trap or free a charge can usually be precisely quantified by the redox potential of this site. The latter is easier to measure in solutions, and redox chemistry has been successfully applied [51] to understand and improve photon and electron irradiation-triggered metal precipitation in aqueous and organic solutions. The redox conditions under which clusters nucleate and grow or dissolve are determined by the nature of the defects formed and modified by electron and hole traps. This approach may be extended to metal cluster precipitation in glasses, as shown by recent work performed on Ag clustering (the use of Ag ions opens more experimental possibilities than that of other metals, and eases comparisons between results on glasses and solutions). Experiments [52] first showed that Ag nanocrystal precipitation induced by *photon* irradiation in oxide glasses is very similar to that in solutions. The irradiation-induced defect configurations differ in glass (Fig. 21) and in aqueous solutions ($H°$ versus $HO°$), but in both cases they are either very oxidizing or very reducing, hence, their strong reactivity. Such centers all interact with the Ag species that form in the glass host or in solutions, leading to a correlated evolution of the reducing vs. oxidizing species concentrations. For example. reducing defects (E', and more generally trapped electrons in glass) reduce ionic silver present in the medium. At this point, one may argue that the neutralized metal species can diffuse, nucleate, and grow according to the classical precipitation scheme. The results (Fig. 16; see Section 2.2) on Cu in glasses [48] apparently agree with this simple view. However, they are likely an exception, since experiments conducted on Ag in both aqueous solutions [53] and in oxide glasses [52][54] show that some highly oxidizing defects may reoxidize neutralized silver and that the latter may also react with ionic silver to form charged molecular nuclei Ag_2^+, which interact in turn with reducing or oxidizing defects. Nucleation and growth processes thus involve not only Ag^0 but also Ag^+ and higher charge states. They are largely determined by the redox properties of metals in glasses and by irradiation-induced transient modifications of the medium's mean redox potential (defined as the average of defect redox potentials). This is indeed analogous to the photographic process [55], the "developer" in our case being the balance between the oxidizing and reducing defect populations instead of a specific organic molecule.

The ion beam control of precipitation being related to the redox behavior of induced defects, the relevant parameters are therefore those which adjust the defects' nature and population. Higher deposited energies lead to higher electron densities (reducing defects), enhancing precipitation. For example, switching from gamma photon irradiation to ion irradiation increases the deposited electron density by a factor of some 10^4, enhancing precipitation of noble metals such as Ag or Cu (Fig. 21).

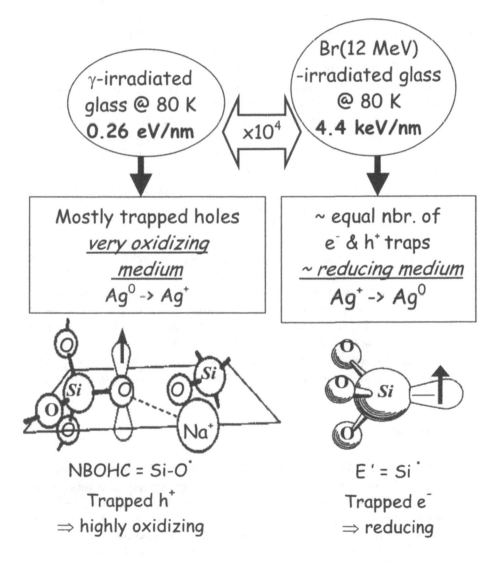

Figure 21. Defect production according to optical spectroscopy and ESR data. Different deposited energy densities (here, a factor ≈ 10^4) lead to the creation of differing defect reservoirs. Low deposited energy densities lead to a large majority of hole traps (hence a very oxidizing medium); high deposited energy densities lead to a similar number of electron and hole traps, and to a comparatively reducing medium. From [52].

Various ion beam experiments may be reconsidered in the light of such redox effects. Approaches to clustering efficiency [44] based on electronegativity involve the ability of an element to discard or attract electrons, obviously related to the redox potential. However, the latter measures this ability for a species in a given medium, whereas electronegativity, restricted to the isolated atom or ion—redox chemistry—is best adapted to describe charge exchange in condensed matter. The effects on precipitation of reducing or oxidizing annealing atmospheres [45] are similarly related to redox chemistry, a conclusion that is actually quantified by glass makers because, together with the annealing temperature and base glass composition, it determines the redox equilibrium of multivalent elements in glass [56].

The ion irradiation-induced adjustment of redox properties can also contribute to the nanocluster dealloying previously ascribed to vacancy diffusion [57]: Ag is more easily reduced than Cu. Irradiation produces reducing conditions that provide a nanoscale thermodynamic redox-driving force to extract and precipitate pure Au. The ion beam dealloying of AgAu and AuCu clusters [58] is caused by the higher redox potential of Au^+/Au^0 relative to both other (Ag^+/Ag^0 and Cu^+/Cu^0) redox potentials, implying the following reactions: $Au^+ + Ag^0 \rightarrow Au^0 + Ag^+$ and $Au^+ + Cu^0 \rightarrow Au^0 + Cu^+$. Collisional stopping efficiently assists the dealloying process, but the main driving force is thermodynamical and given by redox chemistry. In fact, ion beam-assisted dealloying experiments are steps toward a quantitative evaluation of metal redox properties in silica or glasses. From this rather consistent picture, we predict that an ion beam with a large electronic energy loss impinging on a binary alloy nanocrystal previously formed in an oxide glass will induce dealloying of the element which has the highest redox potential (the "most noble" metal). A similar discussion, including the additional influence of the deposited energy density flux, could account for the nucleation and growth of alloy versus core-shell nanoclusters.

Conclusions

Nanophysics and nanotechnology are fast-moving topics of interest. The purpose of this chapter was to be thought-provoking and reasonably close to the authors' areas of competence rather than to be complete. For these two reasons, many other exciting aspects are to be found in the references. We hope to have been convincing on two points. (1) In a field where novel techniques and technical refinements are constantly appearing, fabrication methods using ion beam interactions are not in a different world from other techniques—their specific features and complementarity to (and possibly integration with) others should be kept in mind. (2) On a size scale where the critical dimensions of the objects to be studied or modified are comparable to the characteristic dimensions of ion beam interactions, ion-matter interactions alone can never suffice to interpret results; they necessarily have to be integrated into the basic physics (and chemistry) of the system under study. The emphasis on redox effects in the latter part of the chapter was meant to illustrate this point. Finally, we hope that some of our examples will tempt the reader to participate in one of several fascinating areas where new basic properties and applications appear at this time.

Acknowledgements

Collaboration and discussions with many colleagues from Orsay (especially T. Devolder, C. Chappert, J. Ferré, H. Bea, J. Belloni, and J.L. Marignier), Grenoble (especially A. Marty and Y. Samson), Ecole Polytechnique (B. Boizot and G. Petite), FZ-Rossendorf (K.H. Heinig), and the University of Catania (D. Pacifici, G. Franzo, and F. Priolo) are gratefully acknowledged, as well as critical discussions with L. Bischoff (FZ-Rossendorf), G. Mattei (Padova), and J. Pannetier (Corning Research).

REFERENCES

[1] *Nanoelectronics and Information Technology*, ed. R. Wäser, Wiley-VCH (2003)

[2] N. Mathur and P. Littlewood, Nature Materials **3** (2005) 207

[3] J. Gierak et al., Microelectron. Eng. **78–79** (2005) 266 and **73–74** (2004) 610

[4] J. Lohau et al., Appl. Phys. Lett. **78** (2001) 990; see also G. Xiong et al., Appl. Phys. Lett. **79** (2001) 3461

[5] "Semiconductor Quantum Dots," MRS Bull. **23**(2), Feb. 1998

[6] B. Prével et al., Appl. Surf. Sci. **226** (2004) 173; M.D. McMahon et al., Phys. Rev. **73** (2006) 041401(R)

[7] S. Haussman et al., Jap. J. Appl. Phys. (Part 1) **38** (1999) 7148

[8] T. Müller et al., Mat. Sci. Eng. **C19** (2002) 209

[9] D. Weller and A. Moser, IEEE Trans. Magn. **35** (1999) 4423

[10] M.G. LeBoité et al., Mater. Lett. **6** (1988) 1089; A. Traverse et al., Europhys. Lett. **8** (1989) 633

[11] C. Chappert et al., Science **280** (1998) 1919

[12] J. Ferré et al., J. Phys. **D36** (2003) 1

[13]   S. Chou et al., Science **272** (1996) 85

[14]   T. Devolder et al., Appl. Phys. Lett. **74** (1999) 3383; T. Devolder et al., J. Magn. Mag. Mat. **249** (2002) 452

[15]   S.Sun et al., Science **287** (2000) 1989; S. Sun, Adv. Mat. **18** (2006) 396

[16]   H. Bernas et al., Phys. Rev. Lett. **91** (2003) 077203

[17]   J. Noguès and I. K. Schuller, J. Magn. Mag. Mat. **192** (1999) 203

[18]   Miltenyi et al., Phys. Rev. Lett. **84** (2000) 4224; Misra et al., J. Appl. Phys. **93** (2003) 8593

[19]   Mougin et al., Phys. Rev. **B63** (2001) 060409

[20]   Poppe et al., Europhys. Lett. **66** (2004) 430; J.V. Kim and R.L. Stamps, Appl. Phys. Lett. **79** (2001) 2785

[21]   G. Mie, Ann. Physik, **25** (1908) 377

[22]   M.Bawendi et al., Ann. Rev. Phys. Chem. **41** (1990) 477

[23]   S.V. Gaponenko, *Optical Properties of Semiconductor Nanocrystals* (Cambridge Univ. Press. Cambridge, 1998)

[24]   D.E. Porter and K.E. Easterling, *Phase Transformations in Metals and Alloys* (Van Nostrand Reinhold, Wokingham, England, 1981); K. Binder and D. Stauffer, Adv. Phys. **25** (1976) 343

[25]   I.M. Lifshitz and Slyozov, J. Phys. Chem. Solids **19** (1961) 35 and C. Wagner, Z. Elektrochem. **65** (1961) 581. See also [24].

[26]   A.F. Hebard et al., J. Phys. **D37** (2004) 511

[27]   S. Coffa and A. Polman, ed., *Materials and Devices for Si-based Optoelectronics*, Vol. 468, Materials Research Soc., Warrendale, Pa., (1998)

[28]   E. Boer, Ph.D. thesis, California Institute of Technology, Pasenda, California (2001)

[29]   S. Tiwari et al., Appl. Phys. Lett. **68** (1996) 1377

[30]   K.H. Heinig et al., Appl. Phys. **A77** (2003) 17

[31]   S. Roorda et al., Adv. Mat. **16**, (2004) 235

[32]   J. Penninkhof et al., Appl. Phys. Lett **83** (2003) 4137

[33]   A. Bouhelier et al., Phys. Rev. Lett **95** (2006) 267405

[34]   M. Toulemonde et al., Nucl. Inst. Meth. Phys. Res. **B216** (2004) 1

[35]   A. Meldrum et al., J. Mater. Res. **14** (1999) 4489

[36]   J. Philibert, *Atom Movements: Diffusion and Mass Transport in Solids*, Ed. Physique, Les Ulis (1991)

[37] K.H. Heinig et al., Appl. Phys. **77** (2003) 17; M. Strobel, Ph.D. Thesis, Forschungzentrum Rossendorf, Germany (1999)

[38] M. Strobel et al., Nucl. Instr. Meth. in Phys. Res. **B147** (1999) 343

[39] H. Trinkaus and S. Mantl, Nucl. Inst. Meth. Phys. Res. **B 80/81** (1993) 862; V.A. Borodin et al., Phys. Rev. **B56** (1997) 5332

[40] R. Espiau de Lamaestre and H. Bernas, Phys. Rev. **B73** (2006) 125317

[41] P. Mazzoldi et al., Nucl. Instr. Meth. Phys. Res. **91** (1994) 478 and reference therein

[42] see review by R. A. Weeks, in *Optical and Magnetic Properties of Ion Implanted Glasses*, Materials Science and Technology Vol. 9, ed. J. Zarzycki, (VCH, Weinheim, 1991) 331

[43] A. Perez et al., J. Mater. Res. **2** (1987) 910

[44] H. Hosono, J. Non-Cryst. Solids **187** (1995) 457

[45] A. Miotello et al., Phys. Rev. **B63**, (2001) 075409

[46] P. Mazzoldi et al., Nucl. Instr. Meth. Phys. Res. **B80/81** (1993) 1192

[47] R. Espiau de Lamaestre et al., Nucl. Instr. Meth. Phys. Res. **242** (2006) 214;
R. Espiau de Lamaestre et al., J. Phys. Chem. **B109** (2005) 19148;
R. Espiau de Lamaestre et al., J. Non-Cryst. Solids **351** (2005) 3031

[48] R. Espiau de Lamaestre and H. Bernas, J. Appl. Phys. **98** (2005) 104310

[49] E. Valentin et al., Phys. Rev. Lett. **86** (2001) 99

[50] Noriaki Itoh and Marshall Stoneham, *Materials Modification by Electronic Excitation* (Cambridge University Press, Cambridge, 2000)

[51] J. Belloni et al., Nature **402** (1999) 865 and refs. therein

[52] R. Espiau de Lamaestre, Ph.D. Thesis, University of Paris (2005) 11

[53] J. Belloni and M. Mostafavi, Metal Clusters in Chemistry **2** (1999) 1213

[54] A. Barkatt et al., J. Phys. Chem. **76** (1972) 203

[55] R.W. Gurney and N.F. Mott, Proc. Roy. Soc. **A164** (1938) 151

[56] A. Paul, *Chemistry of Glasses* (Chapman & Hall, New York, 1982); H.D. Schreiber et al., Glastechn. Ber. **60** (1987) 389

[57] D. Manikandan et al., Nucl. Instr. Phys. Res. **B198** (2002) 73

[58] G. Mattei et al., Phys. Rev. Lett. **90** (2003) 085502

Chapter 17

RESIDUAL STRESS EVOLUTION DURING ENERGETIC PARTICLE BOMBARDMENT OF THIN FILMS

Amit Misra[1] and Michael Nastasi[1],

Los Alamos National Laboratory, Los Alamos,NM 87545 USA

1   INTRODUCTION

A variety of surface modification and surface coating techniques are used in industry to modify the near-surface properties of the substrate materials. In the surface modification by ion implantation process, a surface alloy composed of a combination of the substrate elements and the implanted ions is formed. In the ion implantation process the substrate not only provides a backing for the surface alloy but also contributes the material that makes up part of the surface alloy. In this process, there is a slow transition between the surface modified zone and the substrate. The range of applications for such ion implantation-based surface engineering includes automotive and aerospace components, orthopedic implants, textile-manufacturing components, cutting and machining tools (e.g., punches, tapes, scoring dies, and extrusion dies), etc [1].

In the surface coatings area we consider the physical vapor deposition (PVD) processes such as sputtering and e-beam evaporation used to deposit thin films in a variety of applications such as microelectronics, microelectromechanical systems (MEMS), nanoelectromechanical systems (NEMS), integrated optoelectronics, etc. The significance of these applications can be appreciated from the simple fact that the PVD equipment market alone is over \$150B/year [2]. The primary function of the PVD synthesized thin films in engineering applications is often non-load bearing, e.g., electrical interconnects in integrated circuits, diffusion barriers, adhesion layers, seed layers to promote texture or epitaxy, coatings in magnetic recording media and heads, optical coatings, wear-resistant coatings, etc. However, the presence of residual stresses can adversely affect the properties, performance and long-term reliability of PVD thin

K.E. Sickafus et al. (eds.), Radiation Effects in Solids, 487–534.
© 2007 *Springer.*

films or ion-synthesized surfaces. Some common deleterious effects of residual stresses are film cracking, delamination from the substrate, stress-induced voiding (e.g., in the electrical interconnect lines in integrated circuits) and undesired modifications of physical properties (e.g. increases in the electrical resistivity). Hence, the need to understand the fundamental origins of residual stresses in these far-from-equilibrium materials and the ability to tailor residual stresses is critical to the integrity and performance of these materials and devices.

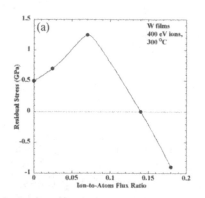 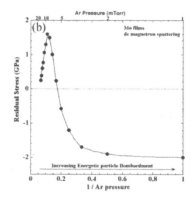

Figure 1. Typical trends in the evolution of thin film residual stresses during energetic particle bombardment; (a) IBAD W films [4] and (b) sputtered Mo films [5]. Positive values of stress indicate tensile stress and negative values compressive stress.

One reason for the catastrophic effects of residual stress on the film failure is the large magnitudes of stresses typically encountered, as shown in Fig. 1. It should be pointed out that the residual stresses are elastic and can build up to the level of the film yield strength, provided the yield strength is higher than the film fracture or delamination stress. It is well established that thin films with nano-scale grain sizes have much higher yield strengths as compared to bulk materials [3]. Note from Fig. 1, taken from old literature [4,5], (a great deal of the earlier work is reviewed by Koch [6]), that maximum residual stresses in metal films can be on the order of 1-2 GPa. Specifically, for the case of Mo films shown in Fig. 1(b), the peak residual stress is $\approx$ 2 GPa. If we assume that this peak residual stress is of the same order as the film yield strength, we infer a yield strength of $\approx$ 2 GPa for nanocrystalline Mo thin films which is equivalent to $Y/150$ where Y is the Young's modulus, and hence, only a factor of 5 lower than the theoretical limit of strength of materials of $\approx Y/30$. Typically, bulk metals have strengths that are lower than the theoretical strength limit by two to three orders of magnitude. Besides the high magnitudes of stresses, Fig. 1 also shows the strong dependence of stress on the synthesis parameters. For e-beam evaporated W films (Fig. 1(a)), bombardment with 400 eV Ar ions resulted in a rapid build up of tensile stress and transition to compressive stress with increasing ion-to-atom flux ratio. For sputtered Mo films (Fig. 1(b)), a similar transition in residual stress from tensile to compressive with reducing Ar pressure during deposition is observed. As discussed later in this article, decreasing Ar pressure results in an increase in bombarding particle energy. The effect of an ion beam on film growth and its resultant physical properties will depend on the ion species, energy, and the relative flux ratio of the ions, J_I, and deposited atoms, J_A, customarily defined as R_i. Data showing the effect of ion bombardment on film properties is either expressed simply as ion flux (assuming constant deposited atom flux), as a relative ion/metal atom flux at the substrate (i.e., R_i), or as the average energy deposited per atom, E_{ave}, in eV/atom, which is simply the product of the relative ion/atom flux and the average ion energy, E_{ion}. The mathematical description of these two ion-beam assisted deposition (IBAD) parameters is given by [1]

$$R_i = \frac{\text{ion flux}}{\text{flux of deposited atom}} = \frac{J_I}{J_A} \tag{1}$$

and

$$E_{ave} = E_{ion} \cdot \frac{J_I}{J_A} = E_{ion} R_i \tag{2}$$

In this chapter, we present a review of the physical mechanisms of residual stress evolution in thin films subjected to energetic particle bombardment either during or post-deposition. Following equations (1) and (2), energetic particle bombardment may be achieved by increasing ion (or atom) energy at a constant R_i, or increasing R_i at a constant E_{ion}. Before describing the models of stress evolution, we present some related general concepts as background.

2   MECHANICS OF BIAXIAL STRESS IN THIN FILMS RIGIDLY BONDED TO A SUBSTRATE

Residual stresses exist in thin films because of the constraint from the massive substrate, i.e., the film is rigidly bonded to the substrate. Thus, any change in the in-plane dimensions of the film that is not matched exactly by an equal change in the substrate dimensions will result in a stress in the plane of the film, with no stress in the direction normal to the film plane (i.e., biaxial stress state). The growth stresses discussed here arise due to micro-structural relaxation processes that tend to change the stress-free length of the film during deposition or during post-deposition ion bombardment. The stresses that occur because of external factors that tend to change the stress-free length of the film subsequent to deposition, e.g., thermal expansion or contraction, or application of an external force to the film, are not discussed. This article is focused on the physical mechanisms of growth stress evolution. Needless to say, the basic mechanics and methods of measurement of thin film stresses are the same, irrespective of the origins of the residual stress.

To express the biaxial stress in thin films [7], consider a 3-D isotropic system in which the stress σ and strain ε are related by the following elasticity equations (x and y are the in-plane directions, and z is the out-of-plane direction):

$$\varepsilon_x = [\sigma_x - \upsilon(\sigma_y + \sigma_z)] / Y$$
$$\varepsilon_y = [\sigma_y - \upsilon(\sigma_x + \sigma_z)] / Y$$
$$\varepsilon_z = [\sigma_z - \upsilon(\sigma_x + \sigma_y)] / Y \tag{3}$$

For thin films in plane stress, there is stress in the plane directions of the film (x and y) but not out of the plane ($\sigma_z = 0$). Therefore,

$$\varepsilon_x = (\sigma_x - \upsilon\sigma_y) / Y$$
$$\varepsilon_y = (\sigma_y - \upsilon\sigma_x) / Y$$
$$\varepsilon_z = (-\upsilon / Y)(\sigma_x + \sigma_y) \tag{4}$$

Rearranging eq. (4), we get,

$$\varepsilon_x + \varepsilon_y = (1-\upsilon / Y)(\sigma_x + \sigma_y)$$

and,

$$\varepsilon_z = (-\upsilon / 1-\upsilon)(\varepsilon_x + \varepsilon_y) \tag{5}$$

For isotropic systems where $\varepsilon_x = \varepsilon_y$,

$$\varepsilon_z = (-2\upsilon / 1-\upsilon)(\varepsilon_x) \tag{6}$$

Eq. (6) gives the simple relation between the in-plane (ε_x) and out-of-plane (ε_z) strains for isotropic case.

The effect of in-plane residual stress on wafer substrate is schematically illustrated in Fig. 2 [8]. Consider the case shown in Fig. 2(a) where a film if detached from the substrate would have smaller lateral dimensions than the substrate. To maintain the same in-plane dimensions as the substrate, the film will need to be stretched biaxially, resulting in an induced curvature of the wafer substrate. Similarly, for the case shown in

Fig. 2(b), a film with larger in-plane dimensions than the substrate, if unattached to the substrate, would need to be compressed biaxially to remain rigidly bonded to the substrate, resulting in a concave curvature of the substrate.

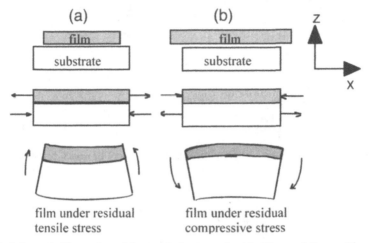

Figure 2. Schematic illustration of the residual stresses in thin films and the resulting substrate curvatures. A film that tends to shrink develops tensile stress and a film that tends to expand develops compressive stress due to substrate constraint [8].

The bending in the substrate can be used not only to determine the nature of the stress, compressive or tensile (Fig. 2), it can also be used to quantify the amount of stress. Consider a film of thickness t_f on a substrate of thickness t_s where the in-plane film stress (σ_f) resulted in a curvature of the substrate. A cross-section detail of the film/substrate couple, and the resulting stress distribution for this condition is shown in Fig. 3. From the theory of elasticity for the bending of plates and beams, there is always a location in the beam called the neutral plane where the stress is zero. For the condition $t_f \ll t_s$ the neutral plane can be taken as the middle of the substrate. Expressions for the stress in the film and substrate can be derived from the equilibrium requirement that the sum of the moments produced by the stress in the film, M_f, and in the substrate, M_s, be zero. For a stress σ_f, which is uniform across the film thickness, the moment (force times the perpendicular distance) due to the stress in the film with respect to the neutral plane is (Fig. 3):

$$M_f = (\sigma_f W t_f)\frac{t_s}{2} \tag{7}$$

where W is the width of the film normal to t_f. The moment produced by the stress in the substrate is given by

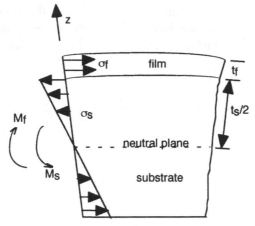

Figure 3. Schematic diagram showing the stress distribution in film and substrate and the corresponding bending moments [7].

$$M_s = W \int_{-t_s/2}^{t_s/2} z\sigma_s(z)dz \qquad (8)$$

Assuming biaxial stress in the substrate

$$\sigma_s(z) = \frac{Y_s}{1 - \nu_s} \, \varepsilon_s(z) \qquad (9)$$

From Fig. 3 and geometry it can be shown that

$$\varepsilon_s(z) = z/r \qquad (10)$$

where r is the radius of curvature of the substrate. From (10) and (9), we get:

$$\sigma_s(z) = \frac{Y_s}{1 - \nu_s} \frac{z}{r} \qquad (11)$$

The moment from this stress is now written as (substituting eq. (11) in (8)):

$$M_s = W \frac{Y_s}{(1 - \nu_s)r} \int_{-t_s/2}^{t_s/2} z^2 dz = \frac{WY_s t_s^3}{12r(1 - \nu_s)} \qquad (12)$$

Imposing the equilibrium condition, $\Sigma\, M = 0$, gives the stress in the film

$$\sigma_f = \left(\frac{Y_s}{1 - \upsilon_s}\right)\frac{t_s^2}{6rt_f} \tag{13}$$

which is also known as Stoney's equation [9]. This equation can also be derived using an energy minimization approach [10].

3   METHODS OF MEASUREMENT OF RESIDUAL STRESSES AND STRAINS IN THIN FILMS

The Stoney equation derived above is the basis for the most commonly used methods for residual stress measurements in thin films on wafer substrates. In the first type of measurement, the wafer is shaped as a cantilever beam with one end clamped (Fig. 4), and the deflection of the free end (δ in Fig. 4) is used to infer the film stress. The deflection is typically measured using a position-sensitive photo-detector that monitors a laser beam focused on the backside of the free end of the substrate [11, 12], although early measurements used a light microscope equipped with an ocular micrometer for measuring δ [13]. Stress in the film is calculated from the following equation which is the Stoney equation modified for the geometry shown in Fig. 4:

$$\sigma_f = \left(\frac{Y_s}{1 - \upsilon_s}\right)\frac{\delta t_s^2}{3t_f l^2} \tag{14}$$

where l is the length of the cantilever beam. This geometry, with suitable modifications as needed, has been used for *in situ* measurements of film stress during deposition [13] or post-deposition ion irradiation [11,14,15].

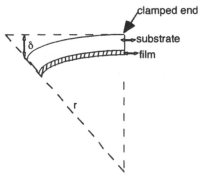

Figure 4. Schematic illustration of the principle of film stress measurement from bending of a cantilever beam substrate induced by film residual stress [8].

The other widely used approach is to measure the radius of curvature (r) using a laser scanning set up and deduce stress from eq. (13) [12]. The basic principle of the laser scanning technique to measure wafer curvature is quite simple, and is schematically illustrated in Fig. 5. A laser beam is reflected from the surface of the wafer, and the displacement of the reflected beam is determined as the wafer is scanned. The change in displacement of the reflected beam is proportional to the change in the angle between the incident laser beam and the wafer surface. In a typical set up, the incident laser beam passes through a galvanometer mirror that is oscillated to have the laser beam scan vertically across a lens. The light that comes through the lens remains parallel to the optic axis as it hits the wafer specimen (Fig. 5). By symmetry, if the sample is perfectly flat, the laser beam is reflected back to a single spot on the photo-detector throughout the scan. This reflected spot is displaced by a distance L (=2fθ) on the detector, where f is the focal length of the lens and θ is the angle by which the flat wafer is tilted (Fig. 5(a)). If the sample is curved (Fig. 5(b)), the reflected laser spot on the detector (L) moves as the laser is scanned along the sample, s. Mathematically,

$$\frac{\Delta L}{\Delta s} = 2f\frac{\Delta \theta}{\Delta s} = \frac{2f}{r} \qquad (15)$$

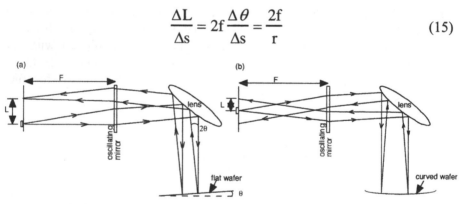

Figure 5. Optical ray diagram showing (a) a parallel incident beam on a flat wafer is reflected back to the same point on the photo detector, and (b) for a curved wafer, the position of the reflected spot varies linearly with the scan position on the sample with a slope that is inversely proportional to the radius of curvature of the sample [10-12].

From eq. (15), it follows that a plot of the reflected beam displacement (L) as a function of the scan distance (s) will be a straight line with a slope proportional to the inverse of the radius of curvature (r), f being a known constant [10, 12]. Since commercially available wafer substrates typically have a finite curvature in the as-received state, equation (13) is modified as follows:

$$\sigma_f = \left(\frac{Y_s}{1 - \upsilon_s}\right)\frac{t_s^2}{6t_f}\left(\frac{1}{r_d} - \frac{1}{r_s}\right) \tag{16}$$

where r_s is the curvature of the bare substrate and r_d is the curvature subsequent to film deposition. For *ex situ* stress measurements, r_s is measured first, then the substrate is loaded in the vacuum chamber for film deposition, and r_d is measured after the deposited substrate is taken out of the vacuum chamber. Thus, the film is exposed to ambient environment prior to curvature measurement. Several investigators have developed *in situ* measurement capability where the laser scanning set up is coupled to a film deposition chamber [16,17]. Hence, r_s is measured at time zero and r_d is measured in real time, thereby measuring the stress evolution as a function of film thickness, while the film is still in vacuum. Besides the real time measurement capability in vacuum (almost instantaneous measurement of curvature using lasers) that avoids any stress relaxation issues, the other obvious advantage of this technique is that the modulus of the film is not needed to know the stress in the film. However, this technique always gives the total stress in the film. Thus, if the film is multilayered, the respective residual stresses in the constituent layers and the interface are not measured, only the average is measured (for *ex situ* measurements). With *in situ* measurements, stresses in the individual layers can be measured [16].

X-ray diffraction (XRD) methods have also been used widely for measuring the residual strains in the thin films. Elastic stresses change the spacing of crystallographic planes in crystals by amounts easily measured by XRD [18]. Having measured the elastic strains by XRD, the stress can be calculated using Hooke's law provided the elastic constants of the thin film are known. In symmetric reflection geometry, (Fig. 6(a)), the scattering vector, which is the vector difference between the incident and diffracted beams, is held normal to the film surface while its length is scanned by changing the scattering angle 2θ. Thus, the lattice spacings measured are from planes parallel to the sample surface, i.e., the out-of-plane strain is measured. Assuming a biaxial strain state and an isotropic material, the in-plane strain may be calculated from eq (6), provided the Poisson's ratio of the film is known and the out-of-plane strain has been measured. To measure in-plane strains, the lattice spacings of planes that are approximately vertical, i.e., normal to the film surface, needs to be measured. In grazing incidence x-ray scattering (GIXS), Fig. 6(b), the incident and diffracted x-ray beams are at a very small angle $\square$ to the film surface so that the planes that diffract are nearly perpendicular to the film surface [19]. The XRD techniques have been applied to epitaxial as well as films with a fiber texture (i.e., preferred growth direction but random in-plane orientations). However, due to low diffracted intensities from thin film specimens, strong x-ray beams (rotating anode x-ray source or synchrotron) are often needed. For more details on the XRD technique for residual strain measurement in thin films, the reader is referred to review articles and books on this topic [18-21].

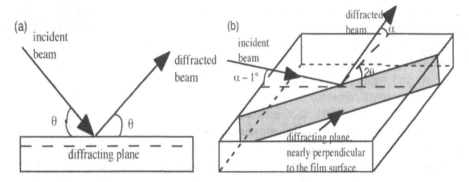

Figure 6 Schematic illustration of diffraction geometries used for (a) out-of plane, and (b) in-plane lattice spacing measurements in thin films [18, 20].

4 SOURCES OF ENERGETIC PARTICLE BOMBARDMENT IN PVD

Before we address the evolution of growth residual stress in thin films during energetic particle bombardment, it is important to briefly describe the sources of hyper-thermal particles that may bombard a growing film during PVD. The two common PVD processes we will consider are e-beam evaporation and magnetron sputtering.

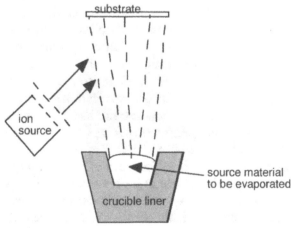

Figure 7. Schematic of the geometry employed in energetic ion bombardment of thin films deposited by e-beam evaporation [1].

In evaporation (Figure 7), atoms are removed from the source by thermal means, i.e., the source is heated, typically with a focused e-beam, until it melts and starts vaporizing. For some materials, semi-melting or sublimation are sufficient for evaporation as opposed to complete melting. In evaporation, the flux of atoms condensing onto the substrate from the vapor phase is typically "thermalized", i.e., atoms have energies on the order of kT where k is the Boltzman constant and T is the substrate temperature. At low homologous temperatures (T/T_m, where T_m is the melting point), the evaporated films are often *under-dense* resulting in tensile stress as discussed in the subsequent sections. Energetic ion bombardment of evaporated films during deposition is achieved via an ion source, as schematically illustrated in Fig. 7, and the process is commonly referred to as ion-beam assisted deposition (IBAD). The ions are typically produced by a low energy (0.2 – 2 keV) broad-beam gridded ion source producing beam currents up to 1-2 mA/cm^2 ($\approx 10^{16}$ ions/cm^2). For more details on the design of IBAD systems, the reader is referred to [1, 22-24].

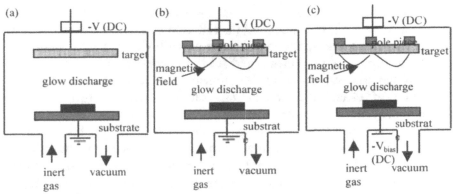

Figure 8. Schematic of the geometry used in sputtering: (a) dc sputtering, (b) dc magnetron sputtering and (c) dc magnetron sputtering with negative substrate bias.

In sputtering (Figure 8), the atoms to be deposited are dislodged from a solid target (source) surface through surface bombardment by energetic ions. Consider the simplified geometry shown in Fig. 8(a) for dc sputtering. The target (cathode) is a plate of the material to be deposited onto the substrate. A dc voltage is applied to the cathode while the substrate is at ground potential. After evacuation of the chamber, an inert gas (e.g., Ar) is introduced in the chamber. The electrons emitted from the cathode ionize the inert gas and the Ar+ ions are accelerated to the negatively charged cathode. The collision of the Ar+ ions with the cathode knocks-off (sputters) the target atoms that are then condensed onto the substrate. This process leads to a visible glow discharge between the electrodes. During sputtering, some of the Ar+ ions also get back reflected from the cathode as neutral atoms and travel towards the substrate. In the simple dc sputtering case shown in Fig. 8(a), a large fraction of the electrons emitted from the cathode may escape towards the walls of the chamber; hence, a high base pressure of inert gas (~50-100 mTorr) is needed to initiate the glow discharge. The sputtered atoms from the target and the reflected neutral Ar atoms lose most of their energies during transport from the cathode to anode by collisions with the gas particles in the chamber. Higher gas pressure implies higher density of gas particles and therefore, shorter mean free path between collisions. Thus, the flux of atoms condensing on the substrate is essentially "thermalized" (similar to e-beam evaporation). The dc sputtering case is, therefore, a non-energetic deposition. Bombardment of the growing film by energetic ions may be achieved by adding an ion source, similar to the IBAD case shown in Fig. 7. The development of magnetron sputtering in 1970s allowed the operation of the glow discharge at low Ar base pressures (typically, 1-10 mTorr). This is achieved by adding magnets behind the cathode (Fig. 8(b)) such that the magnetic field will keep the emitted electrons confined to the cathode region, thereby making the ionization process more efficient. The low inert gas pressure implies that sputtered atoms from the target and the reflected neutral Ar atoms lose little energy via collisions in the transport from cathode to anode. Thus, the depositing flux is hyper-thermal and lead to energetic particle deposition. Most of the modern sputtering equipment use dc magnetron guns for deposition of electrically conducting materials. The energetic particle bombardment during sputter deposition may be further enhanced by applying a negative dc bias voltage to the substrate, (Fig. 8(c)), that is otherwise kept at ground potential. Typically, dc bias to the substrate is much less than the dc voltage applied to the cathode. The negative dc bias attracts some of the Ar+ ions from the plasma to the anode leading to energetic ion bombardment of the film. In summary, there are three sources of energetic particle bombardment during magnetron sputtering:

(a) Sputtered atoms that are ejected from the target with average energies of a few to a few tens of eV with a significant high-energy tail in the energy distribution

(b) Energetic neutral atoms of inert gas that are back reflected from the target with average energies that scale with the ratio of target atomic mass to incident ion mass and to the energy of the incident ions.

(c) Energetic ions bombarding the film due to application of a negative substrate bias.

For more quantitative details on the energies of particles bombarding the film during magnetron sputtering, the reader is referred to the literature on measurement of such energy distributions by energy-resolved mass spectrometry [26-28] and theoretical calculations [29].

Finally, thin films on substrates may be subjected to post-deposition ion-irradiation in a high-energy ion implanter. Typically, ion energies on the order of hundreds of KeV may be needed to penetrate the entire film thickness. While the low-energy bombardment during growth is preferred, there may be situations where subsequent processing after film deposition (e.g., thermal anneal) introduces undesirable film stress that may be modified by a high-energy ion-irradiation of the film.

5   REVIEW OF KEY CONCEPTS RELEVANT TO THE ATOMIC ORIGINS OF INTRINSIC RESIDUAL STRESS IN THIN FILMS

In the next section, the physical mechanisms of the growth residual stress evolution in thin films during energetic particle bombardment will be discussed, and models will be developed to interpret these stresses in terms of atomic scale defects. Some basic concepts relevant to the description of atomic origins of stress are briefly reviewed here. First, we review the universal binding energy relation that describes the interatomic potential energy (and hence, the interatomic force) as a function of interatomic distances in solids. Secondly, we review the ion-solid interactions concepts, particularly the formation of lattice defects during ion irradiation of solids.

5.1   Universal Binding Energy Relation

Smith and co-workers [30-32] have developed a universal description of the atomic binding energy as a function of atomic separation in solids. The shape of the curve can be obtained from a simple two-parameter scaling of a universal function $E^*(a^*)$,

$$E(r_{ws}) = \Delta E \cdot E^*(a^*) \tag{17}$$

where E is the binding energy per atom, r_{ws} is the radius of the Wigner-Seitz sphere containing an average volume per atom, and ΔE is the equilibrium binding energy. The parameter a^* is a scaled length given by:

$$a^* = \frac{r_{ws} - r_{wse}}{\lambda} \tag{18}$$

The parameter r_{WSE} is the radius of the equilibrium Wigner-Seitz sphere and can be calculated from the expression:

$$r_{WSE} = \left(\frac{3}{4\pi N_o}\right)^{1/3} = \left(\frac{3V_o}{4\pi}\right)^{1/3} \tag{19}$$

where N_o and V_o are the equilibrium atomic density and volume, respectively. The length scale λ describes the width of the binding energy curve and is dependent on the isothermal bulk modulus B and the equilibrium binding energy ΔE, and can be expressed as:

$$\lambda = \left(\frac{\Delta E}{12B\pi r_{WSE}}\right)^{1/2} \tag{20}$$

Finally, the form of the universal function $E^*(a^*)$ is given as:

$$E*(a*) = -(1+a*)\exp(-a*) \tag{21}$$

Substituting eq. (21) into (17) gives the final form of the universal binding energy relation:

$$E(r_{WS}) = -\Delta E(1+a*)\exp(-a*) \tag{22}$$

The values of λ, r_{WSE}, and ΔE are listed in Table 1 for several elements.

Table 1. Values of parameters λ, r_{WSE}, and ΔE used in the universal binding energy relation (see equations 17-22 in the text for details)

Metal	λ (nm)	r_{WSE} (nm)	ΔE (eV)
Mg	0.0316	0.177	1.53
Al	0.0336	0.158	3.34
Si	0.0344	0.168	4.64
Ti	0.0340	0.162	4.86
V	0.0310	0.149	5.3
Cr	0.0254	0.142	4.10
Fe	0.0274	0.141	4.29
Co	0.0262	0.139	4.39
Ni	0.027	0.138	4.435
Cu	0.0272	0.141	3.50
Zn	0.0215	0.154	1.35
Ge	0.0348	0.176	3.87
Y	0.047	0.199	4.39
Zr	0.0395	0.177	6.32

Nb	0.0336	0.163	7.47
Mo	0.0265	0.155	6.810
Ru	0.0245	0.148	6.62
Pd	0.0237	0.152	3.936
Ag	0.0269	0.160	2.96
Cd	0.0214	0.173	1.16
In	0.0360	0.184	2.6
Hf	0.0373	0.174	6.35
Ta	0.0330	0.163	8.089
W	0.0274	0.156	8.66
Re	0.0247	0.152	8.10
Ir	0.0230	0.150	6.93
Pt	0.0237	0.153	5.852
Au	0.0236	0.159	3.78
Pb	0.0331	0.193	2.04

Essentially, the potential can be obtained using the bulk modulus, lattice parameter and cohesive energy of elements. Figure 9(a) shows the interatomic energy-distance curve for Cr, plotted using equation 22 and the characteristic material parameters listed in Table 1. While the exact features of this curve are model dependent, the general shape and trends are universal for all materials. The minimum energy in this curve corresponds to the most stable configuration for Cr atom spacing and represents the maximum energy that must be applied to pull a Cr atom free from the crystal, i.e., the cohesive energy of Cr. The interatomic distance corresponding to the minimum in the interatomic potential energy corresponds to the equilibrium distance between Cr near neighbors. As evident from Fig. 9(a), any departure from the equilibrium interatomic distance results in an increase of the energy of the material (less negative) and makes the material less stable. An alternate way to look at figure 9(a) is to recall that $-dE/dr = F$, where F is the interatomic force and r is the interatomic distance. Note that under this convention, a positive force occurs when dr is negative and a negative force results when dr is positive, which is opposite to the stress convention applied in materials science. To maintain the stress convention, i.e., increasing the distance between atoms produces a positive restoring force and compressing the spacing produces a repulsive force we plot $F = dE/dr$ in Fig 9(b). These data show that there are no forces acting on Cr atoms that have the equilibrium spacing of ≈ 0.249 nm. However, if the interatomic distance is greater than equilibrium, a positive or tensile force results. On the contrary, if the interatomic distance is decreased below equilibrium, repulsive or compressive force tries to return the atoms to their equilibrium position.

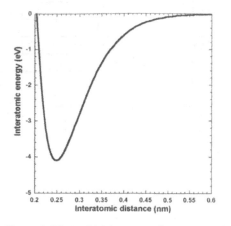 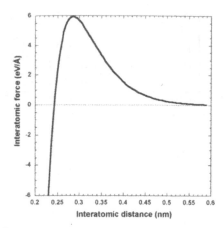

Figure 9. Plots of (a) inter-atomic energy and (b) inter-atomic force as a function of inter-atomic distance for Cr using the universal binding energy relation [32].

5.2 Fundamentals of Ion-Bombardment Induced Lattice Defects

It has been known for many years that bombardment of a crystal with energetic ions produces regions of lattice disorder. This lattice disorder results from the physical processes responsible for slowing the ion down and allowing it to come to rest in the crystal. There are two energy loss mechanisms that contribute to the slowing down processes; nuclear collisions, in which energy is transmitted as translatory motion to a target atom as a whole, and electronic collisions, in which the moving particle excites or ejects atomic electrons. The total energy-loss rate dE/dx can be expressed as:

$$\frac{dE}{dx} = \frac{dE}{dx}\bigg|_{n} + \frac{dE}{dx}\bigg|_{e}$$

where the subscripts n and e denote nuclear and electronic collisions, respectively. Nuclear collisions can involve large discrete energy losses and significant angular deflection of the trajectory of the ion. It is this process that is responsible for the production of lattice disorder by the displacement of atoms from their positions in the lattice. Electronic collisions involve much smaller energy losses per collision, negligible deflection of the ion trajectory, and negligible lattice disorder. The relative importance of the two energy-loss mechanisms changes rapidly with the energy E and atomic number, Z_1, of the ion: nuclear stopping predominates for low E and high Z_1, whereas electronic stopping takes over for high E and low Z_1.

The nuclear stopping power of an ion of mass M_1 and atomic number Z_1 incident on target atoms of mass M_2 and atomic number Z_2 can also be calculated with a high level of accuracy using the Ziegler, Biersack and Littmark (ZBL) formula [33]:

$$S_n(E_0) = \frac{8.462 \times 10^{-15} \, Z_1 \, Z_2 \, M_1 \, S_n(\varepsilon)}{(M_1 + M_2)(Z_1^{0.23} + Z_2^{0.23})} \quad \frac{eV \, cm^2}{atom} \tag{23}$$

where the reduced nuclear stopping cross-section is calculated using

$$S_n(\varepsilon) = \frac{0.5 \ln(1 + 1.1383\varepsilon)}{(\varepsilon + 0.01321 \, \varepsilon^{0.21226} + 0.19593 \, \varepsilon^{0.5})} \quad \text{for } \varepsilon_{ZBL} \leq 30 \tag{24}$$

and

$$S_n(\varepsilon) = \frac{\ln(\varepsilon)}{2\varepsilon} \quad \text{for } \varepsilon_{ZBL} > 30 \tag{25}$$

The term ε_{ZBL} is the ZBL reduced energy and is given by

$$\varepsilon_{ZBL} = \frac{32.53 \, M_2 \, E_0}{Z_1 \, Z_2 \, (M_1 + M_2)(Z_1^{0.23} + Z_2^{0.23})} \tag{26}$$

where E_0 is in keV. The advantage of the reduced energy notation is that a single expression can define nuclear stopping cross-section for all ion/target atom combinations.

For most ion implantation and ion assisted deposition processes, nuclear collisions dominate the energy losses experienced by the ion. As an ion slows down and comes to rest in a crystal, it makes a number of collisions with the lattice atoms. In these collisions, sufficient energy may be transferred from the ion to displace an atom from its lattice site. Lattice atoms that are displaced by the incident ion are called *primary knock-on atoms* or PKA. If the PKA has been given sufficient energy it can in turn displace other atoms, secondary knock-on atoms, and so on, tertiary knock-ons, etc. -- thus creating a cascade of atomic collisions, Fig 10. This process leads to a distribution of vacancies, interstitial atoms, and other types of lattice disorder in the region around the ion track. The vacancy-interstitial defect formed by this process is commonly referred to as a *Frenkel-pair* or a *Frenkel-defect*.

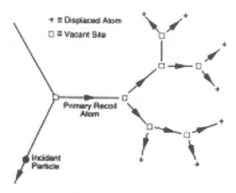

*Figure 10.*Schematic of a collision cascade [1].

In the collision process it is also possible that the PKA has sufficient energy to displace an atom off its lattice site, but after the collision is left with insufficient energy to continue its flight and falls into the vacated site it just created. This process is called a *replacement collision*. While such events have little influence on monatomic materials, replacement collisions can produce considerable disorder in ordered polyatomic materials. One of the principle effects of introducing point defects into a crystal is the long-range displacement field they produce around themselves. An example of this effect is schematically presented in Fig. 11 for a vacancy and an interstitial. As can be seen from this example, the defects displacement field results in a change in the crystal's volume, which is commonly referred to as the defects *relaxation volume*, V^{rel}. In the example presented in Fig. 11 the relaxation volume of the vacancy, V_v^{rel}, is negative and the relation volume of the interstitial, V_i^{rel}, is positive.

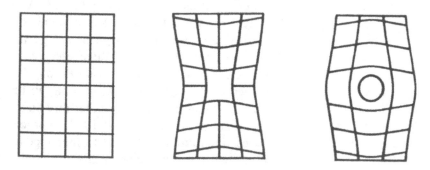

Figure 11. Schematic of lattice strains around point defects.

The relaxation volume for vacancies and interstitials has been experimentally determined for a number of metals [34]. A summary of these data is presented in Table 2. The relaxation volumes presented in Table 2 are given in terms of the atomic volume of one atom in the perfect crystal, V_{atom}, and the relaxation volume of a Frenkel-defect, v_{FD}^{rel}, is taken as the sum of vacancy and interstitial relaxation volumes. Also presented in Table 2 are values of the material's Young's modulus, Y, and Poisson's ratio, v. Examining Table 2, we see that for the metals listed, the relaxation volumes of the interstitials, v_i^{rel}, is always positive and greater that one atomic volume while the relaxation volumes of the vacancies, v_v^{rel}, is always negative and a small fraction of an atomic volume. Therefore, when vacancies and interstitials are created in equal numbers by an irradiation process, positive volume will be added to the crystal causing it to swell.

6   STRESS EVOLUTION MODELS

6.1   Stress Evolution as a Function of Film Thickness in non-Energetic Deposition

As discussed in earlier sections, non-energetic deposition primarily applies to the deposition conditions when the depositing flux is essentially thermalized. Since the focus of this article is on stress evolution during energetic particle deposition conditions, only a brief description of stress evolution during non-energetic deposition will be presented here. For more details, the reader is referred to other literature [16, 17, 35-45].

Table 2. Relaxation volumes for interstitials (V_i^{rel}), vacancies (V_v^{rel}) and Frenkel defect (V_{FD}^{rel}) and Young's modulus (Y) and Poisson's ratio (v) for several elements

Crystal Structure	Element	V_i^{rel}	V_v^{rel}	V_{FD}^{rel}	Y (GPa)	v
fcc	Cu	1.55	-0.25	1.30	123.6	0.35
	Al	1.9	-0.05	1.85	71.0	0.34
	Ni	1.8	-0.2	1.6	193.2	0.30

	Pt	1.8	-0.2	1.6	170.6	0.38
bcc	Mo	1.1	-0.1	1.0	327.6	0.30
	Fe	1.1	-0.05	1.05	209.9	0.28
hcp	Zn	3.5	-0.6	2.9	397.2	0.28
	Co	1.5	-0.05	1.45	104.9	0.35

In the absence of bombardment from hyperthermal particles, the general trend of the evolution of growth stress in island growth thin films as a function of film thickness is as shown in Fig. 12 [38,39]. For materials with high atomic mobility at room temperature [16, 17, 38-42] such as Ag, Al, Au, Cu, etc, stress evolution with increasing film thickness has three stages: (i) initial compressive stress, (ii) rapid build up of tensile stress to a peak, and (iii) relaxation of tensile stress and eventual build up of compressive stress (Fig. 12(a)). Often these measurements are performed *in situ* and some relaxation of the compressive stress is observed when the deposition flux is stopped [44]. For refractory metals with high melting points (e.g., Cr, Mo, W), diffusion at room temperature is limited and the transition from tensile to compressive stress with increasing film thickness is typically not observed during deposition, Fig. 12(b)), [38, 45].

The initial compressive stress that occurs when the film is discontinuous has been attributed to surface stress [39]. First, we need to define surface stress. Due to the differences in the nature of bonding between surface and bulk atoms in a solid, surface atoms tend to have an equilibrium interatomic spacing different from bulk. However, the surface atoms must remain in atomic registry with the bulk, and this constraint results in a force per unit length acting on a solid surface, referred to as surface stress. For low index surfaces of metals, the atoms would prefer to adopt a lower equilibrium spacing than bulk in order to increase the local electron density. In this case the atomic registry between the surface and underlying bulk atoms would result in stretching of the atomic bonds at the surface (i.e., surface stress is tensile) and shrinkage of the bonds in the underlying bulk (i.e., induced compressive stress in the bulk of the island). It should be emphasized that the surface stress represents the work done to elastically *strain* the surface atoms, and hence is different from surface energy that is the work done in *creating* a free surface [47-49]. Similarly, there is interface stress related to the elastic straining of atoms at interfaces between two solids. The compressive stress induced in early stages of island growth is thus attributed to the surface stress. For fcc metals with high surface mobility (Fig. 12(a)), island coalescence occurs early on in the deposition process and hence, large compressive stresses typically do not develop. However, in bcc refractory metals (Fig. 12(b)) such as Mo, compressive stress on the order of 1 GPa has been observed in early stages of deposition [50].

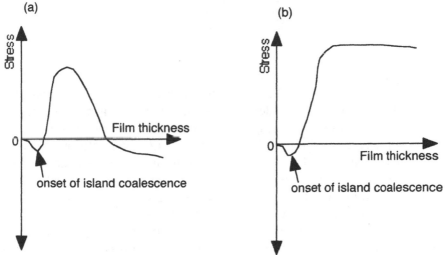

Figure 12. Typical trends of growth stress evolution as a function of film thickness for non-energetic deposition: (a) materials with high atomic mobility, and (b) materials with low atomic mobility at the deposition temperature [38,39].

Next we consider the evolution of tensile stress as the islands coalesce to form a continuous film. A model for this has recently been proposed by Nix and Clemens [36]. Consider the film has square cross-section islands of height t and width d. The driving force for coalescence is the removal of two free surfaces at the expense of forming a grain boundary (note that grain boundary energy is typically about one-half to one-third of surface energy in fcc metals) and elastic strain energy that results from the biaxial tensile straining of the film as the islands coalesce. This energy balance can be used to estimate the maximum stress that could result from the island coalescence. Let the surface energy per unit area be γ_s, grain boundary energy per unit area be γ_{gb}, and the elastic strain energy per unit volume be $\sigma\varepsilon$ where σ is the stress and ε is the strain in the film. Note that

$$\varepsilon = \frac{\sigma}{\left(\dfrac{Y}{1-\upsilon}\right)} \qquad 27)$$

where $Y/(1-\upsilon)$ is the biaxial modulus.

The energy balance gives the following relation:

$$2\gamma_s(\text{td}) = \gamma_{gb}(\text{td}) + \frac{\sigma^2(d^2t)}{(Y/(1-\upsilon))} \tag{28}$$

Therefore,

$$\sigma = \sqrt{\frac{(2\gamma_s - \gamma_{gb})Y}{d(1-\upsilon)}} \tag{29}$$

Substituting typical values for low melting point fcc metals such as Au, Ag, Al etc, Nix-Clemens showed that eq. (29) gives an upper bound stress of 5-6 GPa which is at least an order of magnitude higher than the peak tensile stress typically observed during island coalescence of these metals. The reason for overestimation of stress from eq. (29) was attributed to the assumption in the model that all crystallites are imagined to coalesce at the same time, with the consequence that no shear stress can be developed on the film/substrate interface. A more realistic picture would be to allow different crystallites to coalesce at different times and some sliding at the interface to occur. Furthermore, any strain relaxation by defect motion within the crystallites was also not considered. Phillips *et al* [40] described the gradual build up of tensile stress during film growth by considering cracks in the film (equivalent to incompletely coalesced islands) that would tend to reduce the curvature of the substrate. As the area fraction of the cracks decreased (i.e., extent of coalescence increased) the tensile stress would increase. This hypothesis was supported by TEM observation on Pt films deposited on SiO_2 where the area coverage of the substrate surface by the Pt islands was observed to increase gradually with film thickness [40]. However, the maximum stress was still significantly lower than what equation (29) would predict.

Several investigators have attempted to extend the Nix-Clemens work to come up with quantitative predictions of tensile stress that compare reasonably with experimentally measured stresses [41-43]. The basic idea is to examine the island coalescence process in more detail, as was first suggested by Nix-Clemens. If the islands are assumed to have an elliptical shape, then the initial contact between growing islands will be at a single point. The contact between islands could then be visualized as an elastic crack edge, and the subsequent coalescence process equivalent to crack closing or "zipping". Following the idea that coalescence follows a crack zipping process, Seel *et al* [42] used a finite element method to model it. By minimizing the sum of the positive strain energy and associated reduction in the boundary energy, the equilibrium configuration resulting from island coalescence was determined as a function of island radius. For a given island radius at impingement, this approach yielded average stress values that were an order of magnitude lower than that predicted by eq. (29). Freund and Chason [43] modeled the zipping process as a Hertzian contact mechanics problem. Their model is based on the theory of elastic contact of solids with rounded surfaces (Hertz contact theory). This analytical approach also predicted stresses on the order of a couple hundred MPa (as opposed to several GPa from eq. 29) from island coalescence. Floro *et al* [41] and Seel *et al* [42] also considered the effect of stress relaxation during growth, pointing out that the observed stress is due to a dynamic competition between stress generation due to island coalescence and stress relaxation due to surface diffusion.

Finally, we consider the evolution of compressive stresses during film growth for conditions of non-energetic deposition. Chason *et al* [44] have developed a model for compressive stress generation based on the idea that an increase in the surface chemical potential caused by the deposition of atoms from the vapor drives excess atoms into the grain boundaries. Plating of extra atoms at grain boundaries would lead to an in-plane expansion of the film were it not rigidly bonded to the substrate, and the constraint from the substrate leads to compressive stress in the film. Since the compressive stress raises the chemical potential of atoms in the grain boundary, the driving force for additional flow of atoms decreases with increasing stress and eventually a steady state is reached. The model also explains the relaxation of compressive stress, often observed during *in situ* measurements of stress of metals such as Ag with high mobility of surface atoms at room temperature, as the reverse flow of excess atoms from boundaries to surfaces. The mechanisms of stress evolution and relaxation under conditions of non-energetic deposition described above are schematically shown in Fig. 13.

6.2   Stress Evolution During Energetic Particle Bombardment

Typical trends of residual stress evolution during energetic particle bombardment were shown in Fig.1. In this section, we present experimental data of residual stress evolution in thin metal films subjected to energetic particle bombardment in different ways. The corresponding film microstructures are also shown and then models are presented for generation and relaxation of film residual stresses and limits to these stresses.

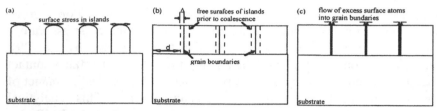

Figure 13. Schematic of the mechanisms of stress evolution as a function of film thickness during non-energetic (corresponding to the trend shown in Fig.12a): (a) compressive stress arises in the film due to constraint from the positive surface stresses, in the initial stages of growth when the islands are not coalesced [39], (b) generation of tensile stress as free surfaces of adjoining islands coalesce to form grain boundaries, tensile strain = Δ/d, [36], (c) relaxation of tensile stress and build up of compressive stress due to flow of atoms from surface to grain boundaries [44].

6.2.1   Correlation Between Stresses and Microstructures

For simplicity, we will take the example of Cr films deposited on Si substrates, investigated in detail by the authors [14, 46, 51-56], to show how tailoring of stress in thin films may be achieved by different types of energetic particle bombardment. In these studies, stress evolution is studied as a function of energetic particle bombardment for a constant thickness, as opposed to the non-energetic deposition case discussed in the previous section where stress evolution as a function of film thickness was reported. The thickness of the films in these studies will be on the order of a couple hundred nm or more. Hence, the films will be continuous (i.e., islands have coalesced). Since Cr represents a low atom mobility case (Fig.12 (b)), the intrinsic stress is not expected to be compressive in the absence of any energetic deposition. The mechanisms of stress generation developed later in this section will apply to high atom mobility case as well (Fig. 12 (a)). As discussed in section 4, sputtered films have two sources of hyper-thermal (tens to hundreds of eV) particle bombardment: (i) varying the inert gas pressure (lower pressure results in higher particle bombardment energies) and (ii) negative substrate bias. A third kind of bombardment considered here is post-deposition ion irradiation with high-energy ions (hundreds of keV).

The evolution of tensile stresses in sputtered 150 nm nominal thickness Cr films as a function of Ar gas pressure in the 1-7.5 mTorr range is shown in Fig. 14(a). With decreasing Ar pressure, the biaxial tensile residual stress increases to a maximum of ~1.7 GPa and then rapidly decreases. At the lowest Ar pressure of ~ 1 mTorr used, the stress was still tensile. With increasing substrate bias in the 150 nm thick film (Fig. 14(b)), we observe an increase in tensile stress to a maximum of ~1.6 GPa, followed by a complete relaxation of tensile stress, and finally the rapid build up of compressive stress to a saturation value of ~2.1 GPa. The stress evolution with substrate bias for 1 μm thick Cr films is shown in Fig. 14(c). The Ar pressure was kept constant at 2.5 mTorr. Note from Fig. 14(a) that 2.5 mTorr is to the left of the peak in tensile stress in Fig. 14(a) and hence, no initial increase in the tensile stress is observed with increasing bias at 2.5 mTorr deposition (Fig. 14(c)). However, with increasing bias at 5 mTorr deposition (which is to the right of the peak in Fig. 14(a), an initial increase in tensile stress to a maximum is observed (Fig. 14(b)). The evolution of stress in 150 nm thick Cr films, deposited at 5 mTorr Ar pressure without any substrate bias, with post-deposition ion irradiation is shown in Fig. 14(d). The irradiation was done with 110 keV Ar ions that penetrate about 80 nm of film thickness. With constant ion energy and increasing dose, the stress evolution is similar to that shown in Figs. 14 (a) and (b), i.e., tensile stress initially increases, then decreases to zero, and then compressive stress builds up. No saturation of compressive stress was observed for the dose range considered here. Irradiation with 300 keV Ar ions, where the depth of ion penetration was greater than the film thickness, gave the same results. This indicates that factors such as modification of interface stress and the non-uniform stress due to irradiation of only a fraction of the film thickness have little influence on the observed stress evolution.

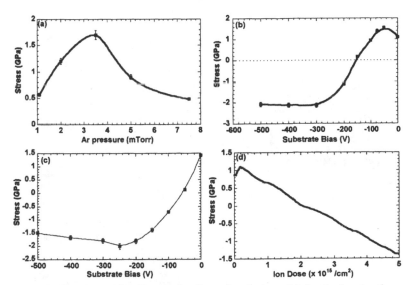

Figure 14. Experimental data showing the effects of energetic particle bombardment on the intrinsic residual stress in thin sputtered Cr films: (a) as a function of Ar pressure for 150 nm thick films, (b) as a function of negative substrate bias for 150 nm thick Cr films, (c) as a function of negative substrate bias for 1 μm thick Cr films, and (d) as a function of post-deposition ion-irradiation in 150 nm thick Cr films irradiated with 110 keV Ar ions (deposition was at 5 mTorr Ar pressure with no substrate bias).

The evolution of stress was correlated with the microstructures of the films through transmission electron microscopy (TEM) observations. The microstructures were investigated for four processing conditions: (a) films deposited at high Ar pressures without any substrate bias (Fig. 15(a)); (b) films deposited under conditions that produced the maximum tensile stress (Fig. 15(b)); films deposited under conditions that produced almost no stress (Fig. 15(c)), and (d) films deposited under conditions that produced the maximum compressive stress (Fig. 15(d)). Typically, nanocrystalline columnar microstructures are observed as shown in the cross-sectional TEM micrograph in Fig. 15(a) from a 150 nm thick Cr film deposited at 7.5 mTorr Ar pressure without substrate bias. The inter-columnar regions exhibited Fresnel fringes in through-focus imaging: bright fringes for under focused images (Fig.15(a), dark fringes for over focused images and no contrast for exact focus images [51,52]. While an exact focus image gives the impression that the film is fully dense and all islands have completely coalesced to form grain boundaries, the observation of Fresnel fringes in over and under focused images indicate voided inter-columnar regions that may only be a few atomic layers thick. Such Fresnel contrast imaging (phase contrast due to differences in atomic numbers) technique has been used to highlight nanoscale voids or films at grain boundaries in bulk materials [57].

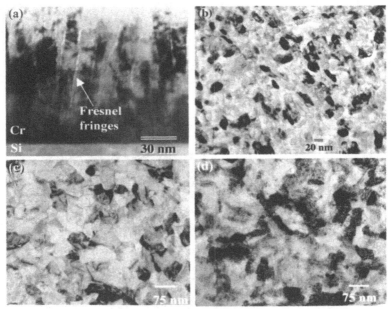

Figure 15. TEM micrograph showing (a) cross-sectional view of a 150 nm thick Cr film deposited at 7.5 mTorr Ar pressure without substrate bias, (b) plan view of a 150 nm thick Cr film deposited at 5 mTorr Ar pressure and −50 V bias, (c) plan view of a 1 μm thick Cr film deposited at −50 V bias at 2.5 mTorr Ar pressure, (d) plan view of a 1 μm thick Cr film deposited at −200 V bias at 2.5 mTorr Ar pressure.

As the bombardment energy is increased, the inter-columnar voids tend to shrink and at the tensile stress maximum, no Fresnel fringes were observed (Fig. 15(b)). At near-zero stress, clear grains are observed with no evidence of open volume or radiation-induced defects in the film (Fig. 15(c)). At compressive stress maximum, clear evidence of radiation-damage-type defect structures is observed (Fig. 15(d)), although the details of the radiation-induced point defects cannot be analyzed from these TEM images. However, by comparing figures 15(c) and (d), it is obvious that the major fraction of the grains is covered with a high density of radiation damage in (d) that may include vacancies (either isolated or clustered), self-interstitials (either isolated or clustered), Frenkel defects and entrapped Ar. Finally, it should be emphasized that the microstructure evolution with bombardment correlates well with the stress evolution irrespective of the film thickness. For a given stress, the primary difference in the microstructure of the 150 nm and 1 µm thick Cr films was in the grain size: ~16 nm in the former and ~50 nm in the latter.

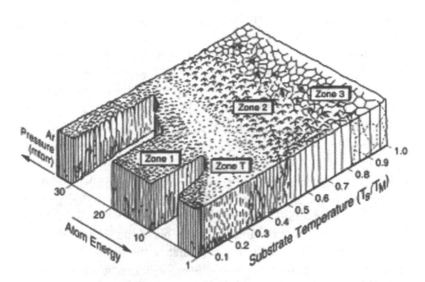

Figure 16. Schematic illustration of the relationship between the film microstructure and the energetic particle bombardment conditions [1].

Figure 16 shows the relationship between the film microstructure and the ion beam conditions could be represented schematically as shown in the structure zone diagram. The diagram is similar to the evaporation Zone diagram Grovenor *et al.* [58], but includes bombarding ion energy and the sputtering gas pressure as a processing parameter in addition to the substrate temperature (expressed as a fraction of the melting temperature in degrees Kelvin) [1]. The sputtering gas pressure is included because the energetic particle bombardment of films increases with decreasing gas pressure as discussed in section 4. Columnar grains with an open structure along the grain boundaries characterize zone 1 in Fig 16. The Zone 1 structure results when the deposited atoms diffusion is too limited to overcome the effects of shadowing, and therefore forms at low T_s/T_m where T_s is the temperature of the substrate and T_m is the melting point of the film material. Shadowing is the process whereby high points on the growing surface receive a higher coating flux than valleys do, and is most prevalent when an oblique deposition flux is present. Shadowing introduces open boundaries that reduce the film density. The addition of bombarding ions during film growth lowers both the temperature window over which Zone 1 structures are stable, and the transition temperature to the formation of Zone T microstructures.

The microstructures in Zone T, or the transition Zone, are also dominated by shadowing effects but have finer structures consisting of a dense array of poorly defined fibrous grains with boundaries that are sufficiently dense to yield respectable mechanical properties. The microstructure in this Zone is attributed to the onset of surface diffusion that allows the deposited atoms to migrate on the deposit surface before being covered by the arrival of further material. The boundary of this zone is strongly influenced by substrate temperature and bombarding energy. Intense substrate ion bombardment during deposition can suppress the development of the open Zone 1 structures at low T_s/T_m and ion bombardment yields structures typical of high T_s/T_m depositions. Zone 2 microstructures nominally form at $T_s/T_m = 0\cdot3 - 0\cdot5$, where the growth process is dominated by the surface diffusion of the deposited atoms. There is little sensitivity of the microstructure to the ion bombardment energy since temperatures are already sufficiently high enough to dominate kinetic processes. Columnar grains separated by distinct dense inter-crystalline boundaries dominate the structure in this zone. Zone 3 structures form at $T_s/T_m \geq 0\cdot5$, where bulk diffusion dominates the final structure of the coating. There is little sensitivity of the microstructure to increasing T or to the ion bombardment energy. Recovery and re-crystallization processes typically occur in this temperature regime, driven by the minimization of strain energy and surface energy of the grains. The effects of energetic particle bombardment on the intrinsic stress evolution and the film microstructures have been described above. In the following sections, we discuss the physical mechanisms that lead to this stress evolution.

6.2.2   Tensile Stress

As discussed in section 5, the interatomic forces acting between adjoining islands separated by a small gap (typically, < 1 nm) would lead to an attractive force between islands tending to close the intercolumnar gap. The driving force is the reduction of potential energy of the solid (Fig. 9) by reducing the interatomic distances in the voided intercolumnar region, at the expense of elastic strain energy from stretching the film in biaxial tension. The elastic strain continues to increase until the islands coalesce completely to form a grain boundary. In other words, significant tensile stress may be generated even before the two free surfaces are replaced by a grain boundary.

This basic concept was first proposed by Hoffman [59] referred to as the grain boundary relaxation model, and discussed in detail recently by Machlin [35] and Nastasi *et al* [1]. The idea that energetic particle bombardment leads to the generation of tensile stress via a "densification" process where inter-columnar voids are closed is supported by two kinds of observations. First, TEM observations (Fig. 15 and references [51,52]) that show closing of gaps on the order of a few atomic layers thick corresponding to an increasing tensile stress. Second, molecular dynamics simulation by Müller [60-62] that show the same phenomena at the atomic scale.

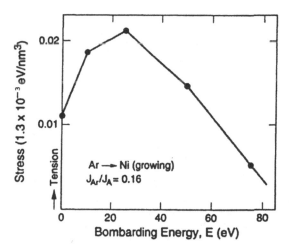

Figure 17. Intrinsic tensile stress from MD simulations of an IBAD Ni film as a function of the Ar ion bombarding energy where ion-to-atom flux is 0.16 [59].

Fig. 17 presents the calculated intrinsic stress of a Ni film versus the Ar ion bombarding energy for an ion-to-atom flux ratio of $J_I/J_A = 0.16$. In these simulations it is assumed that the vapor deposited Ni atoms arrived at the substrate with a kinetic energy of 0.13 eV. The data in Fig.17 show that stress state of the Ni film deposited without ion beam assistance is slightly tensile and that the tensile stress initially increases with low energy Ar bombardment, passes over a maximum at an Ar ion energy of about 25 eV and then decreases with increasing bombarding energy. Fig. 18, contains MD generated microstructure for the Ni film grown under the ion bombardment conditions. These microstructures show that the Ni film grown without Ar ion assist contains large micro-pores and open voids which close (Fig. 18(a), (b) and (c)) with increasing ion bombardment resulting in higher packing density of the film. Since these are very simplistic 2-D simulations, it is really the trends of stress evolution that should be compared with experimental data (e.g., compare Figs. 17 and 14(a)). The peak value of stress from the MD simulation (Fig. 17) is about 4 GPa while experiments show about 1-1.5 GPa peak tensile stress in Ni [59, 63].

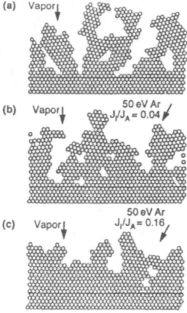

Figure 18. Evolution of thin film microstructure in 2-D molecular dynamics simulation for evaporated Ni films: (a) without ion bombardment, (b) with Ar ion bombardment of E = 50 eV, angle of incidence of ions (α) = 30° and ion-to-atom flux of 0.04, and (c) with Ar ion bombardment of E = 50 eV, α = 30° and ion-to-atom flux of 0.16 [61].

Using an appropriate interatomic potential, the tensile stress in a thin film can be computed analytically as a function of the gap distance between crystallites as described below. Consider fig. 19 that schematically shows interatomic forces between two crystallites separated by a distance Δ. As a lower bound, consider only the nearest neighbor interatomic forces. In other words, each atom on the surface of crystallite X facing the crystallite Y experiences an attractive force (F) that is readily known from a plot such as Fig. 9(b) for a given distance Δ. The tensile stress is then obtained as the net force on each atom times the number of atoms per unit area:

$$\sigma_T = F\, (N_{film})^{2/3} \tag{30}$$

where N_{film} is the atomic density. For bcc refractory metals, Morse potential has been found to give a reasonable fit to the properties [64, 65, 55]. Using this potential and $N_{film} = 8.33 \times 10^{22}$ atoms / cm^3 for Cr, σ_T can be plotted as a function of Δ, as shown in Fig. 20(a). The upper bound calculation would include the second and third nearest neighbors interatomic forces as well [55], but is only slightly higher than the lower bound due to rapidly decaying F with Δ relation.

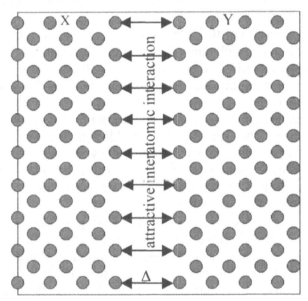

Figure 19. Schematic of the attractive interatomic forces between two adjoining grains labeled X and Y separated by a distance Δ.

We now compare the model prediction (Fig. 20(a)) with the experimental data, Fig. 14(a) and (b). With increasing bombardment, as the intercolumnar gaps close (TEM data in Fig. 15), the tensile stress increases to a peak and then rapidly relaxes. This trend is well predicted by the model except that the peak stress is much higher than the experimental observation. We hypothesize that the yield strength for plastic flow sets a limit to the maximum elastic residual stress that can be accumulated. This is discussed further in the next section. The initial rapid build up of the tensile stress (segment OA in Fig. 20(a)) matches with the tensile stress generation in Fig. 14(a) and (b). The horizontal segment AB in Fig. 20 (a) shows the film yield strength. Hence, the point labeled A in Fig. 20(a) marks the stress and corresponding Δ values at which islands are close enough to spontaneously snap together and in the process relieve some of the stress by dislocation motion (diffusive relaxation may also be possible in high atom mobility materials). This is consistent with recent MD simulation [66]. Reduction in Δ from A to B to C result in relaxation of the stress as Δ approaches the equilibrium value (lowest energy point in the interatomic potential in Fig. 9(a)).

To further illustrate the correlation between the model prediction and experimental data, a zoom-in of the OA segment of Fig. 20 (a) is shown in Fig. 20(b). The experimentally measured stress values (Fig. 14(a)) are used to read the corresponding Δ values in Fig. 20(b). For the Cr films deposited with Ar pressures of 7.5, 5 and 3.5 mTorr respectively, we get Δ of $\sim$ 0.56, 0.52 and 0.48 nm respectively. Two possible approaches for measuring Δ in these films are:

(i)   by measuring the film density [55]. Here film density is measured by obtaining the number of atoms per cm^2 by ion-beam analysis methods and the film thickness by profilometry. By knowing the average grain size from TEM and assuming a regular shape (e.g., cylinder or square) of all grains, an average Δ value can be obtained corresponding to the measured film density. This approach [55] yielded good agreement between the Δ values calculated from density measurements and those obtained from Fig. 20 (b), considering the crudeness of estimating Δ from density measurements and the fact that the interatomic potentials are approximate.

(ii) Δ is measured directly from TEM using the approach described by Page and co-workers [57]. In this approach, Fresnel fringe thickness is measured from the through focus images and plotted as a function of positive (over) and negative (under) defocus. Extrapolating the Fresnel fringe thickness to zero focus gives the gap distance at the boundary. We attempted this approach for the Cr film with 1 GPa tensile stress exhibiting Fresnel fringes at intercolumnar boundaries and the results are shown in Fig. 21. This approach shows that Δ is $\sim$ 0.5 –0.6 nm for this sample consistent with the model prediction of 0.52 nm from Fig. 20(b).

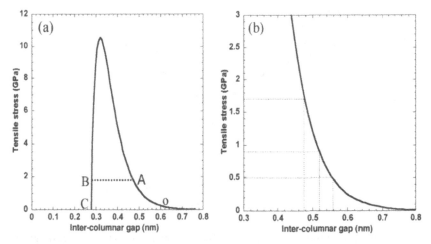

Figure 20. Interatomic tensile stress as a function of the intercolumnar gap distance. The line AB in (a) indicates the maximum tensile residual stress limit set by the film yield strength. Further reduction in the gap distance below the point A lead to relaxation of film stress (segment ABC). The segment OA of stress generation matches with the experimental data, shown in detail in (b).

In summary, starting with non-energetic deposition as the particle bombardment is increased, the generation of tensile stress can be estimated as follows: (i) obtain the interaomic force-distance curve for the material using accurate interatomic potentials (analogous to Fig. 9(b)), (ii) multiplying the interatomic force with the number of atoms per unit area in the material (eq. (30)), gives the stress as a function of interatomic distance (analogous to Fig. 9(b)), (iii) obtain the evolution of the intercolumnar gap distance from TEM measurements (Fig. 21) or from density and grain size measurements; the stress corresponding to a given intercolumnar gap distance can be read from data such as Fig. 20(b) to the point of peak in tensile stress. As discussed above, the peak in tensile stress may be set by residual stress increasing to the level of the film yield strength, at which point dislocation motion and spontaneous closure of the intercolumnar gaps relax the tensile stress to zero. More detailed studies of using MD simulation to study tensile stress evolution from intercolumnar gap closure have recently been initiated [67]. The generation of compressive stress at higher bombardment energy/flux is discussed next.

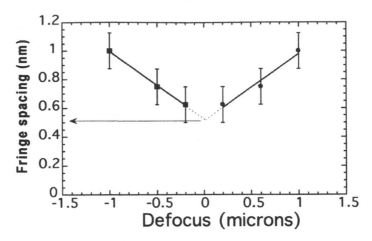

Figure 21. Estimation of the intercolumnar gap distance from measured thickness of the grain boundary Fresnel fringes in defocused plan view TEM images.

6.2.3 Compressive Stress

The hypothesis here is that compressive stress generation during energetic particle deposition is due to the production of irradiation-induced point defects that add positive volume to the film causing it to swell [53,55, 68-70]. However, the constraint that the film must remain rigidly bonded to the substrate prevents any in-plane expansion of the film leading to compressive stress generation in the plane of the film (Fig. 2).

Developing a model for compressive stress generation, based on the above hypothesis, requires knowledge of defect concentrations and the magnitude of local expansion in the lattice introduced by the defects. As described in section 5, interstitials add positive volume to the material and vacancies add a negative volume. However, an interstitial-vacancy pair (Frenkel defect) has a net positive relaxation volume. Thus, compressive stress can be generated if the number fraction of interstitials induced by irradiation either equals or exceeds the number fraction of vacancies generated.

Consider the case of ion-assisted growth of films, the ion bombarding energies used are typically low, of the order of a few100 eV to a few keV. Under these conditions, interstitials are created by displacing atoms from the surface and driving them into the bulk, leaving vacancies behind at the surface. The maximum amount of energy that can be transferred to a surface atom by the bombarding ion, T_M, occurs for a head-on collision, and can be calculated by [1]:

$$T_M = \frac{4 M_1 M_2}{(M_1 + M_2)^2} E_0 \tag{31}$$

where M_1 and E_0 are the mass and energy of the ion, respectively, and M_2 is the mass of the target atom.

Equation 31 shows that $T_M = E_0$ only for the condition when $M_1 = M_2$ and for all other combinations where $M_1 \neq M_2$, less energy than E_0 will be transferred. The more energy transferred to the surface atom the greater will be its range and the further the interstitial will reside from the surface. Monte Carlo simulations using the TRIM code show that surface atoms that recoiling with energies in the 100 eV to 2 keV range will be implanted to depths between 0.5 and 5 nm (TRIM) [1]. However, the probability of a head-on collision is extremely low with the majority of the collisions occurring at glancing angles where significantly less energy is transferred, which in turn will result in shallower recoil implantation depths. Therefore the interstitials created by this process will reside within a few monolayers of the surface, while the vacant site left at the surface will most probably be filled in by a subsequent depositing atom. Therefore it is expected that there will be an imbalance in the number of vacancies and interstitial formed by the bombardment process during ion beam enhanced deposition, with interstitials and interstitial derived extended defects being dominant. However it should be noted that the high stress associated with these interstitials, along with their close proximity to the surface, can result in the athermal migration of the interstitial back to the surface, thereby reducing the number of interstitials that are actually retained in the growing film. Therefore, for any given processing condition, it will be difficult to know the exact number fractions of interstitials and vacancies retained in the film. We proceed with developing a model with the assumption that all defects induced by irradiation are Frenkel pairs, to obtain a limit of the stress that can be developed by Frenkel pairs.

In general, the biaxial stress equation for the addition of a defect that changes the volume of the crystal by ΔV_{def} is given by [3]:

$$\sigma_{xx}^{defect} = -\frac{Y}{1-v} F_{defect} \frac{\Delta V_{defect}}{3} \tag{32}$$

where Y is the Young's modulus, v is Poisson's ratio of the material the defect resides in and F_{defect} is the number of defects per unit volume in the material. Generalizing this equation to the Frenkel- defect yields

$$\sigma_{xx}^{FD} = -\frac{Y}{1-v}F_{FD}\frac{V_{FD}^{rel}}{3} \tag{33}$$

where V_{FD}^{rel} is the relaxation volume of the Frenkel-defect given in units of atomic volumes and F_{FD} is the number fraction of Frenkel-defects. The data in Table 1 shows that V_{FD}^{rel} is always greater than zero, which when applied to eq. 33 shows that the sign of stress will be negative. Therefore, the formation of Frenkel-defects adds compressive stress to the irradiated material.

Given that V_{FD}^{rel} is known or can be estimated, the ultimate magnitude of the stress that results from Frenkel-defects will depend on the atomic fraction of Frenkel-defects in the material, F_{FD}. The average number of displaced atoms (N_d) produced by an ion of energy E can be estimated using Monte Carlo computer simulations, such as TRIM [33] or can be calculated using the Kinchin-Pease displacement damage function [71–74], which is given by:

$$\langle N_d(E) \rangle = \begin{matrix} \mathbf{0} & \textit{for } 0 < E < \mathbf{E_d} \\ 1 & \textit{for } \mathbf{E_d} \leq E < 2\mathbf{E_d}/\xi \\ \dfrac{\xi v(E)}{2E_d} & \textit{for } 2\mathbf{E_d}/\xi \leq E < \infty \end{matrix} \tag{34}$$

where E_d is called the *displacement energy,* and is the energy that a target atom has to receive in order to leave its lattice site and form a stable interstitial, thereby making a Frenkel-defect. This energy depends on the direction of the momentum of the target atom. Therefore, a range of displacement energies exists for the creation of a Frenkel pair in a given material. Typical average values of E_d are in the range of 15 to 40 eV [71]. The parameter (E) in Eq. 34 is called the *damage energy* and represents the fraction of the ion energy that goes into displacement (nuclear) processes. The damage energy is closely related to the PKA's total nuclear stopping, but is always ($\approx 20\text{–}30\%$) smaller [1]. Therefore (E) can be estimated according to:

$$v(E) \cong 0.8 \int_0^{E_0} \frac{dE}{dx}\bigg|_n dx \tag{35}$$

where $(dE/dx)_n$ is the nuclear energy loss rate, which can calculated from the following relation

$$S \equiv \frac{dE/dx}{N} \qquad (36)$$

where N is the atomic density (atoms/cm^3) and S is the nuclear stopping. At very low ion or PKA energies, such as those used in ion assisted deposition, $v(E) \cong E$

Equation (34) describes the case for one ion. For a flux of ions that can penetrate the target to a distance R, the ion range, the atomic fraction of Frenkel-defects formed can be estimated from

$$F_{FD} = \frac{\langle N_d \rangle}{N} \frac{0.4 v(E)}{NRE_d} \phi \qquad (37)$$

where ϕ is ion dose (ions/cm^2). It is important to note that when the above equations are used to calculate $\langle N_d \rangle$ and F_{FD}, or if these values are determined using Monte Carlo simulation, that the values obtained represent the number or fraction of Frenkel-defect generated by an irradiation process as opposed to actual number of defects that ultimately remain in the material. This difference exists because these defect estimates do not take into account defect diffusion, that lead to vacancy-interstitial recombination or the collapse of point defects into other defect structures such as dislocation loops.

The implementation of Eq. 37 into eq. 33 to calculate the stress requires a value for $v(E)$, which in turn requires that we can calculate $(dE/dx)_n$ (Eq. 35). The nuclear energy loss rate can be calculated using the Ziegler, Biersack and Littmark (ZBL) universal nuclear stopping formula [33], which for an ion with energy E_0 in the laboratory reference frame is given by eq. (23).

Eqs. 34–37 show how F_{FD} can be calculated using simple ion-solid interactions theory. For the data shown in Fig. 14 (b) and (c), where substrate bias was used to produce energetic particle bombardment, some parameters such as ion dose and range, etc are not known. Hence, we apply the eq. (33) to the experimental data as described below.

The maximum compressive stress was ≈ 2 GPa in the experimental data shown in Fig. 14(b). Ion-beam analysis revealed the presence of entrapped Ar in the compressively stressed Cr film. Hence, we rewrite eq. (33) as:

$$\sigma_{comp}^{max} = \left(\frac{1}{3}\right)\left(\frac{Y}{1-\upsilon}\right)\left(F_{FD}V_{FD}^{rel} + F_{Ar}V_{Ar}^{rel}\right) \tag{38}$$

where F_{Ar} and V_{Ar}^{rel} are the atomic concentrations and relaxation volumes of Ar in Cr respectively. F_{Ar} was measured by ion-beam analysis to be ~ 0.005 for the Cr films with maximum compressive stress [53], and V_{Ar}^{rel} was estimated to be 1.75 from literature values [34], as described in [53]. The relaxation volume for Frenkel-defects in Cr (V_{FD}^{rel}) is reported to be 1 [34]. Using $(Y/1-\upsilon) \approx 300$ GPa for Cr [53] and $\sigma_{comp}^{max} \approx 2$ GPa as experimentally measured, we calculate F_{FD} from eq. (38) as 0.011. Thus, a Frenkel defect concentration of ~1% could explain the compressive stress observed in Cr films.

Consider a Frenkel-defect concentration of 1%. This would mean that a defect pair must exist every 100 atoms. However, when vacancies and interstitials are in such close proximity to each other athermal recombination is highly probable. Both experiment (at 4.2 K) and computer calculations have found the spontaneous recombination volume for vacancy-interstitial pairs is between 50 and 100 atomic volumes, and that the Frenkel defect concentration saturates somewhere between 1 and 2% [1]. Since the athermal recombination volume limits F_{FD}, it sets the upper limit to the maximum compressive stress (Eq. 33) that a material can experience due to the presence of Frenkel-defects.

The assumption that all defects produced in thin films during energetic particle deposition are Frenkel-defects provides an estimate of the compressive stress generated. For post-deposition ion irradiation at high energies and to high doses, defects may form clusters or even collapse into dislocation loops [74]. These loops preferentially align normal to the tensile stress direction [75] to relax the tensile stress and continued growth in the size of the loops eventually leads to the development of compressive stress [14, 76]. The compressive stress buildup in these high energy ion-irradiated films has been shown to be proportional to $(dpa)^{2/3}$ where dpa is displacements per atom [14, 76]. The above description demonstrates that the generation of compressive stress in ion-synthesized films can be interpreted in terms of point defect structures. The correlation of processing parameters with defect structures produced in the film is not described here, and the reader is referred to other texts [1, 34].

6.2.4 Limits of Residual Stress

The mechanisms of generation of intrinsic elastic residual stresses (tensile or compressive) in thin films are discussed above. Here we show that the maximum values of these elastic residual stresses are set by the plastic yield strength of the film (diffusive relaxation is not considered).

In thin films with the microstructural length scales (in-plane grain size and film thickness) on the order of a few tens of nanometers, continuum-scale dislocation pile-ups cannot be supported and hence, Hall-Petch type models for strengthening due to grain refinement break down [77, 78]. The plastic flow in these fine-scale structures is accomplished by the motion of single dislocations by bowing between the boundaries (or interfaces) and creating interface dislocations, as shown schematically in Fig. 22 [79–82]. The yield strength as determined by this Orowan bowing stress is given as, for the case when the in-plane grain size (d) << film thickness (h), [82]:

$$\sigma_{Orowan} = M \frac{Gb}{4\pi(1-\upsilon)} d^{-1} \ln\left(\frac{d}{b}\right) \qquad (39)$$

where M is the Taylor factor (≈ 2.75 for bcc metals), G is the shear modulus (95 GPa for Cr films) and b is Burgers vector (≈ 0.25 nm for Cr) [53]. In general, grain size and film thickness are additive when applying the Orowan model to estimate the yield strength of thin films [82]. However, if h << d (Fig. 22(a)), then h is used in eq. (39) instead of d. For the data shown here for columnar-grain Cr films (Fig. 14), d << h (Fig. 22 (b)) and hence, eq. (39) applies with d instead of h. Substituting these values we obtain $\sigma_{Orowan} = 1.4 \pm 0.1$ GPa, for d = 21 ± 2 nm, in good agreement with the maximum tensile stress measured (Fig. 14(a) and (b)).

The compressive saturation stress could also be calculated using eq. (39), except that an additional term needs to be added to account for hardening due to the radiation-induced point defects (Frenkel pairs or interstitial clusters) in ion-bombarded films having large compressive stress (Fig. 15 (d)). Hardening due to interstitial solutes is more rapid as compared to substitutional solutes since the strain field around interstitial solutes is usually elliptical while substitutional solutes have spherically symmetric strain field [83]. This rapid hardening ($\Delta\sigma_{defects}$) due to interstitial solutes is described by the following model [83]:

$$\Delta\sigma_{defects} = M \frac{G\Delta\varepsilon\sqrt{c}}{\alpha} \qquad (40)$$

where Δ is an empirical parameter, c is the solute concentration and $\Delta\varepsilon$ is the difference between the longitudinal and transverse strains in the elliptical strain field of interstitials. The $\Delta\sigma \propto \Delta\varepsilon$ dependence implies a diffuse interaction between solutes and dislocations, whereas a localized interaction is more applicable for impenetrable solute clusters leading to a $\Delta\sigma \propto \Delta\varepsilon^{3/2}$ type dependence [83].

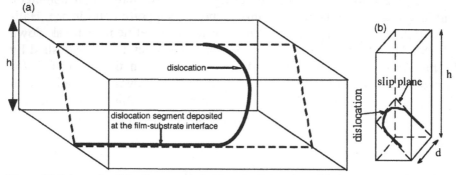

Figure 22. Schematic of the Orowan mechanism for dislocation glide in polycrystalline thin films (only one grain is shown for simplicity). The process involves dislocation bowing between the obstacles, film-substrate interface as in (a) or grain boundary as in (b), and creating interface dislocations as the bowed segment advances on the slip plane shown by dotted line. In (a), the film thickness (h) is much smaller than the grain size (d) and hence, strength is primarily determined by h. In (b), d << h and hence, d is used to determine the film yield strength.

We now apply eq. (40) to the case of Cr films. From the parameters used in eq. (38), we estimate the solute concentration as, $c = f_{FD} + f_{Ar} = 0.011 + 0.005 = 0.016$. $\Delta\varepsilon = 0.40$ and $\alpha = 20$ for typical interstitial hardening in bcc metals at room temperature [83]. Using these values, we obtain $\Delta\sigma_{defects}$ from eq. (40) as 0.66 GPa. The new yield strength of the film (at high bias deposition), σ_y, is given as:

$$\sigma_y = \sigma_{yield}^{gb} + \Delta\sigma_{defects} \tag{41}$$

With $\sigma_{yield}^{gb} = 1.56$ GPa (using grain size of ~ 19 nm for films with maximum compressive stress) and $\Delta\sigma_{defects} = 0.66$ GPa, it follows from eq. (41) that $\sigma_y = 2.22$ GPa. This is in good agreement with the experimentally measured saturation in compressive stress, indicating that the biaxial yield strength sets a limit to the maximum elastic intrinsic residual stress in films. The relaxation of compressive stress with increasing bias (Fig. 14c) has also been observed in other investigation [84]. Since we have shown that the maximum in compressive stress is correlated with point defect build up and saturation, it is likely that plastic flow in these materials is accomplished, in part, by irradiation enhanced creep [76].

7   OTHER PHENOMENA INDUCED BY ENERGETIC PARTICLE BOMBARDMENT THAT MAY INFLUENCE THE INTRINSIC FILM STRESS

7.1   Materials Transport and Ion Mixing

It is clear that significant atomic rearrangement occurs in films grown under ion bombardment. In fact it is well known that most materials subjected to ion irradiation will experience significant atomic rearrangement. The most obvious example of this phenomenon is the atomic intermixing and alloying that can occur at the interface separating two different materials during ion irradiation. This process is known as ion beam [85-91] mixing and is discussed in detail in another chapter by Nastasi in this volume. While many of the examples of ion mixing deal with ion irradiation stimulated interface alloying the fundamental physics that control this phenomena are also responsible for the evolving microstructure, atomic density, and stress that evolve in films grown under ion bombardment.

In addition to modifying stress, the ion mixing process is also influenced by stress, which has been examined by Nastasi and co-workers [85]. In their analysis the change in chemical potential that atom would experience upon a change in biaxial stress of $\Delta\sigma$ is given by

$$\Delta\mu_\sigma = -\frac{2}{3}\Omega\Delta\sigma \qquad (45)$$

where Ω is the molar volume. This equation indicates that there is a driving force for an atom to move towards regions of more positive stress (e.g., from regions of compressive stress to region of tensile stress). Given that interstitial atoms in most metals are in a state of compressive stress and that voids, pores and grain boundaries are in a state of tensile stress, cascades that form at the boundaries of open volumes will result in atomic transport into these volumes.

The influence of cascades on reducing tensile has been examined and modeled by Brighton and Hubler [86] who postulated that stress relief occurs if each atom in the growing film is involved with at least on cascade. If the average volume affected by a cascade is V_{cas}, and the average atomic density in the film is N, then the average number of atoms affected per cascade is NV_{cas}. Therefore, if one ignores cascade overlap, a lower limit for the critical ion-to-atom flux ratio, $\left(J_I/J_A\right)_c$, for stress relief can be expressed as:

$$\left(J_I/J_A\right)_c = \left(NV_{cas}\right)^{-1} \tag{46}$$

With cascade volumes derived from Monte Carlo simulations Brighton and Hubler found excellent agreement with experimental data of Hirsch and Varga [87]. Experiments have also shown that cascades and radiation damage can also lead to a reduction in compressive stress. The relief of compressive stress in ion bombarded amorphous materials has been extensively studied, with the bulk of the work focused on high-energy ion bombardment [12, 88-90]. This work shows that under these conditions stress relief is accomplished by radiation-induced plastic/viscous flow. Similarly, plastic flow is also expected [91] and observed [92, 14, 51] in irradiated metal films. In this case the plastic flow occurs by point defect rearrangement into dislocation loops that plate out in sympathy with existing stress. The effects of ion irradiation induced interfacial mixing on the intrinsic stress have been observed in metallic multilayers such as Ag-Fe [93] and Cu-W [94, 95]. Ion-beam induced intermixing at the interfaces, in the otherwise immiscible Cu-W system, was inferred from X-ray diffraction measurements [94]. In the case of Ag-Fe [93], ion beam mixing occurred to the extent that the crystal structure of the Fe layer changed from bcc to fcc (Fe-Ag alloy).

7.2   Bombardment-induced Phase Transformation

Another effect of energetic particle bombardment that has a consequence on intrinsic film stress is phase transformation in the film induced by energetic particle bombardment. The effect of phase transformation on stress originates from the volume difference in the product vs the parent phase. If the phase transformation tends to produce in-plane expansion, then the constraint from the substrate would lead to compressive stress. However, if the phase transformation would lead to an in-plane shrinkage of the film were it not attached to the substrate, then the constraint from the substrate produces tensile stress. A well-known example is amorphization (e.g., in Si [12]) at high doses of ion irradiation. The amorphization and the related radiation-enhanced plastic flow may lead to relaxation of compressive stress [12, 96, 97]. Similar effects have been observed in ion-irradiated metal silicide films on Si substrates [98, 99]. Other examples are discussed in ref. 35.

8   SUMMARY

The intrinsic stress evolution during energetic particle deposition of thin films is summarized in Fig. 23. Films deposited at low homologous temperatures without energetic ion bombardment, exhibit a low-density porous structure and will in general have a density that is lower than the nominal density and be in a state of tensile stress. As the energetic ion bombardment is increased, the densification of the film starts which leads to tensile stress generation due to attractive interatomic forces across the sub-nanoscale intercolumnar gaps in the film. The film is nearly stress free when the intercolumnar gap distance approaches the equilibrium interatomic distance for minimum potential energy. Under conditions where diffusion is limited and ion bombardment during growth is intensified, the implantation of the assisting ions and/or the formation of subsurface interstitial atoms that result from surface collision process that push deposited atoms into the bulk, can create an excess density and a state of compressive stress that scales with the concentration and relaxation volumes of these point defects. At high bombardment energies and/or dose, compressive stress may relax. While this work focused on correlating stresses to atomic-scale defects in thin films, more work is needed in correlating the processing conditions to the defect structure of ion-bombarded films or surfaces. This is a rich area for further research particular simulations (e.g., molecular dynamics, Monte Carlo, etc) that can accurately account for the time and length scales of the phenomena.

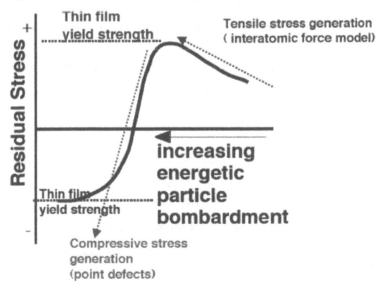

Figure 23. Schematic illustration of the stress evolution trend with energetic particle bombardment and the related physical mechanisms of residual stress generation and stress limits.

ACKNOWLEDGEMENTS

This research is funded by US Department of Energy, Office of Basic Energy Sciences. The authors acknowledge discussions with J.P. Hirth, H. Kung, R.G. Hoagland, J. Sprague.

REFERENCES

[1] Nastasi M., Mayer J.W., and Hirvonen J.K., *Ion-Solid Interactions: Fundamentals and Applications*. Cambridge University Press, Cambridge, 1996.

[2] Powell R.A. and Rossnagel, S., *PVD for Microelectronics*, **26**, Academic Press (1999).

[3] Doerner M.F. and Nix W.D., Stress and Deformation Processes in Thin Films on Substrates, CRC Critical Reviews in Solid State Materials Sciences, 1988; 14:225.

[4] Roy R.A., Cuomo J.J., and Yee D.S., J. Vac. Sci. Tech., 1988; A6:1621.

[5] Hoffman D.W., Thin Solid Films, 1983; 107: 353.

[6] Koch R., J. Phys.: Condens. Matter, 1994; 6:9519.

[7] Tu K.N., Mayer J.W. and Feldman L.C., *Electronic Thin Film Science for Electrical Engineers and Materials Scientists*, Macmillan Publishing Company, New York, 1992.

[8] Ohring M., *The Materials Science of Thin Films*, Academic Press, New York (1992).

[9] Stoney G.G., Proc. Roy. Soc. (London), 1909; A82:172.

[10] Flinn P.A., Gardner D.S. and Nix W.D., IEEE Transactions on Electron Devices, 1987; ED-34:689.

[11] Van Sambeek A.I. and Averback R.S., Mat. Res. Soc. Symp. Proc., 1996; 396:137.

[12] Volkert C.A., J. Appl. Phys., 1991; 70:3521.

[13] Kinosita K., Maki K., Nakamizo K. and Takeuchi K., Japanese Journal of Applied Physics, 1967; 6:42.

[14] Misra A., Fayeulle S., Kung H., Mitchell T.E. and Nastasi M., Nucl. Inst. And Methods B, 1999; 148:211.

[15] Kim Y.S. and Shin S.C., Mat. Res. Soc. Symp. Proc., 1995; 382:285.

[16] Shull A.L. and Spaepen F., J. Appl. Phys., 1996; 80:6243.

[17] Ramaswamy V., Clemens B.M. and Nix W.D., Mat. Res. Soc. Symp. Proc., 1998; 528:161.

[18] Clemens B.M. and Bain J.A., MRS Bulletin, 1992; July:46.

[19] Dosch H., Phys. Rev.B, 1987; 35:2137.

[20] Noyan I.C. and Cohen J.B., *Residual Stress: Measurement by Diffraction and Interpretation.* Springer-Verlag, New York, 1987.

[21] Segmuller A. and Murakami M., "X-ray Diffraction Analysis of Strains and Stresses in Thin Films", in *Treatise on Materials Science and Technology*, 27, 143, edited by H. Herbert, Academic Press, New York, (1988).

[22] Cuomo J.J. and Rossnagel S.M., Nucl. Instrum. & Meth., 1987; B19/20: 963.

[23] Hirvonen J.K., Mater. Sci. Reports, 1991; 6:215.

[24] Smidt F.A., Int. Mater. Rev., 1990; 35:61.

[25] Windishmann H., Crit. Rev. Sol. St. & Mat. Sci., 1992; 19:547.

[26] Jouan P.Y. and Lemperiere G., Vacuum, 1991; 42:927.

[27] Kadlec S., Quacyhaegens C., Knuyt G. and Stals L.M., Surf. Coat. Tech., 1997; 89:177.

[28] Petrov I., Myers A., Greene J.E. and Abelson J.R., J. Vac. Sci. Technol. A, 1994; 12:2846.

[29] Somekh R.E., J. Vac. Sci. Technol. A, 1984; 2:1285.

[30] Rose J.H., Smith J.R., and Ferrante J., Phys. Rev. B, 1983; 28:1835.

[31] Rose J.H., Smith J.R., Guines F., and Ferrante J., Phys. Rev. B, 1984; 29: 2963.

[32] Banerjea A. and Smith J.R., Phys. Rev. B, 1988; 37:6632.

[33] Ziegler J.F., Biersack J.P., and Littmark U., The Stopping and Range of Ions in Solids (Pergamon Press, New York, 1985).

[34] Ehrhart P., Jung P., Schultz H., and Ullmaier H., in *Atomic Defects in Metals*, Landolt_Bornstein, Group III, 25 (Springer-Verlag, Berlin, 1992) Chapter 2.

[35] Machlin E.S., *Materials Science in Microelectronics – The Relationships between Thin Film Processing and Structure*, vol. 1, pp 157-184, GIRO press, NY, (1995).

[36] Nix W.D., and Clemens B.M., J. Mater. Res., 1999; 14:3467.

[37] Thompson C.V., J. Mater. Res., 1999; 14:3164.

[38] Thompson C.V. and Carel R., J. Mech. Phys. Solids, 1996; 44:657.

[39] Cammarata R.C., Trimble T.M. and Srolovitz D.J., J. Mater. Res., 2000; 15:2468.

[40] Phillips M.A., Ramaswamy V., Clemens B.M. and Nix W.D., J. Mater. Res., 2000; 15:2540.

[41] Floro J.A., Hearne S.J., Hunter J.A., Kotula P., Chason E., Seal S.C. and Thompson C.V., J. Appl. Phys., 2001; 89:4886.

[42] Seel S.C., Thompson C.V., Hearne S.J. and Floro J.A., J. Appl. Phys., 2000; 88:7079.

[43] Freund L.B. and Chason E., J. Appl. Phys., 2001; 89:4866.

[44] Chason E., Sheldon B.W., Freund L.B., Floro J.A. and Hearne S.J., Phys. Rev. Lett., 2002; 88:156103.

[45] Sheldon, B.W., Lau, K.H.A., and Rajamani, A, Journal of Applied Physics, 2001; 90:5097.

[46] Misra A., Kung H., Mitchell T.E., and Nastasi M., Journal of Materials Research, 2000; 15:756.

[47] Cammarata R.C. and Sieradzki K., Ann. Rev. Mat. Sci., 1994; 24:215.

[48] Spaepen F., Acta Mater., 2000; 48:31.

[49] Cammarata R.C., Sieradzki K. and Spaepen F., J. Appl. Phys., 2000; 87: 1227.

[50] Adams D.P., Parfitt L.J., Billelo J.C., Yalisove S.M. and Rek Z.U., Thin Solid Films, 1995; 266: 52.

[51] Misra A., Fayeulle S., Kung H., Mitchell T.E. and Nastasi M., Applied Physics Letters, 1998; 73:891.

[52] Misra A. and Nastasi M., Journal of Materials Research, 1999; 14:4466.

[53] Misra A. and Nastasi M., Applied Physics Letters, 1999; 75:3123.

[54] Misra A. and Nastasi M., Journal of Vacuum Science and Technology-A, 2000; 18:2517.

[55] Misra A. and Nastasi M., Nucl. Inst. Methods B, 2001; 175/177:688.

[56] Misra A. and Nastasi M., Adhesion Aspects of Thin Films, 2001; 1:17.

[57] Ness J.N., Stobbs W.M. and Page T.F., Phil. Mag. A, 1986; 54:679.

[58] Grovenor C.R.M., Hentzell H.T.G., and Smith D.A., Acta Metall.,1984; 32:773.

[59] Hoffman R.W., Thin Solid Films, 1976; 34:185.

[60] Müller K.H., J. Appl. Phys., 1987; 62:1796.

[61] Müller K.H., J. Appl. Phys., 1986; 59:2803.

[62] Müller K.H., Phys. Rev. B, 1987; 35:7906.

[63] Mitra R., Hoffman R.A., Madan A. and Weertman J.R., J. Mater. Res., 2001; 16:1010.

[64] Girifalco L.A. and Weizer V.G., Phys. Rev., 1959; 114:687.

[65] Itoh M., Hori M. and Nadahara S., J. Vac. Sci. Tech. B, 1991; 9:149.

[66] Hoagland R.G., LANL, unpublished work.

[67] Sprague J.A., NRL, unpublished work.

[68] Windischmann H., J. Appl. Phys., 1987; 62:1800.

[69] Davis C.A., Thin Solid Films, 1993; 226:30.

[70] Knuyt G., Lauwerens W., Stals L.M., Thin Solid Films, 2000; 370:232.

[71] Robinson M.T., Phil. Mag., 1965; 12:741.

[72] Robinson M.T. and Oen O.S., J. Nucl. Mater., 1982; 110:147.

[73] Sigmund P., Radiation Effects, 1969; 1:15.

[74] Bacon D.J., Calder A.F. and Cao F., Radiation Effects and Defects in Solids, 1997; 141:283.

[75] Lidiard A.B. and Perrin R., Phil. Mag., 1973; 14:49.

[76] Jain A., Loganathan S. and Jain U., Nucl. Inst. and Methods B, 1997; 127/128:43.

[77] Clemens B.M., Kung H. and Barnett S.A., MRS Bulletin, 1999; 24:20.

[78] Misra A. and Kung H., Advanced Engineering Materials, 2001; 3:217.

[79. Embury J.D. and Hirth J.P., Acta Met., 1994; 42:2051.

[80] Nix W.D., Scripta Mat., 1998; 39:545.

[81] Misra A., Hirth J.P. and Kung H., Phil. Mag. A, 2002; 82:2935.

[82] Thompson C.V., J. Mater. Res., 1993; 8:237.

[83] Fleischer R.L., in *The strengthening of Metals*, (edited by D. Peckner), pp. 93-162, Reinhold Press, NY (1964).

[84] Window B., Sharples F. and Savvides N., J. Vac. Sci. Tech. A, 1988; 6:2333.

[85] Nastasi M., Fayeulle S., Lu Y.-C., and Kung H., Mat Sci. and Engr. 1998; A253:202.

[86] Brighton D.R. and Hubler G.K., Nucl. Instr. Meth., 1987; B28:527.

[87] Hirsch E.H. and Varga I.K., Thin Solid Films, 1978; 52:445.

[88] Trinkaus H., J. Nucl. Materials, 1995; 223:196.

[89] Brongersma M.L., Snoeks E., van Dillen T., and Polman A., J. Appl. Phys., 2000; 88:59.

[90] Mayr S.G. and Averback R.S., Phys. Rev. Lett., 2001; 87:6106.

[91] Jain A. and Jain U., Thin Solid Films, 1995; 256:116.

[92] Jain A., Surface and Coatings Technology, 1998; 104:20.

[93] Krebs H.U., Luo Y., Stormer M., Crespo A., Schaaf P. and Bolse W., Appl. Phys. A, 1995; 61:591.

[94] Gladyszewski G., Goudeau Ph., Naudon A., Jaouen C. and Pacaud J., Appl. Surf. Sci., 1993; 65/66:28.

[95] Pranevicius L., Badawi K.F., Durand N., Delafond J., Goudeau Ph., Surf. Coatings and Technology, 1995; 71: 254.

[96] Tamulevicius S., Pozela I. and Jankauskas J., J. Phys. D: Appl. Phys., 1998; 31:2991

[97] Snoeks E., Boutros K.S. and Barone J., Appl. Phys. Lett., 1997; 71:267.

[98] Hardtke C., Schilling W. and Ullmaier H., Nucl. Instr. Meth. B, 1991; 59/60: 377.

[99] Hardtke C., Ullmaier H., Schilling W. and Gebauer M., Thin Solid Films, 1989; 175:61.

Chapter 18

PEROVSKITE-BASED COLOSSAL MAGNETORESISTANCE MATERIALS AND THEIR IRRADIATION STUDIES: A REVIEW

Ravi Kumar,[1] Ram Janay Choudhary,[1,2] and Shankar I. Patil[2]
[1]Nuclear Science Center, New Delhi, India
[2]Department of Physics, University of Pune, India

1 INTRODUCTION

Recent technologies have developed appreciably over the last few years to maneuver the materials properties in the desirable range of performances. Swift heavy ion (SHI) irradiation has been utilized for modifying the properties of various materials to a great degree; SHI is used in various systems such as in mixing two different materials and in high-temperature superconductors, ferrites, manganites, etc. In this chapter, we focus our attention to the manganite materials. We will first provide an overview of the electrical and magnetic properties of manganites, then discuss the various modes of interaction of swift heavy ions with materials, and finally provide information about available reports on the effects of SHI irradiation on the structural, electrical, and magnetic properties of manganites.

Ever since people have realized the impact of magnetic materials in their day-to-day lives, natural inquisitiveness has driven individuals to probe into these materials in depth to better appreciate how these applications have improved. This has led to numerous classifications of magnetic materials, each of which demonstrates significant magnetic properties. In this context, the perovskite-based $LaMnO_3$ system is one such class of material, which has been a topic of extensive interest in recent years [1].

The exploration of perovskite-based transition metal oxides has a long history. These materials demonstrate their own diverse characteristics, such as magnetic, electrical, optical, and dielectric properties [2, 3]. However, after the breakthrough of superconductivity in cuprate-based perovskites [4], the perovskite-based transition metal oxides attracted

K.E. Sickafus et al. (eds.), Radiation Effects in Solids, 535–574.
© 2007 Springer.

enhanced consideration, which led to a profound understanding of these materials. Innovation of colossal magnetoresistance (CMR) in hole-doped $LaMnO_3$ compound was the consequence of one such effort [5]. Magnetoresistance is defined as

$$MR\% = \left| \frac{R(0) - R(H)}{R(0)} \right| \times 100$$

where R(0) and R(H) are the resistance values without and with magnetic field, respectively. This hole-doped $LaMnO_3$ compound not only presents many applications for futuristic devices, but it also offers many interesting topics in physics to analyze.

1.1   Perovskite Materials

Perovskite ABO_3-based materials (A is a rare earth element; B is a transition metal element) have been recognized to have diverse motivating properties, such as magnetic, electrical, optical, and dielectric properties. A typical example of this family is $LaMnO_3$. The unit cell of $LaMnO_3$ is a pseudocubic, as shown in Fig. 1.

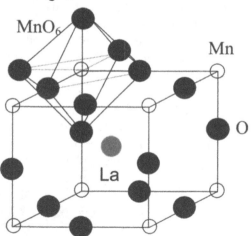

Figure 1. Perovskite structure of $LaMnO_3$.

The stability of a perovskite structure (ABO_3) depends on the Goldscmidt tolerance factor [2], formulated as

$$t = (R_A + R_O)/[(\sqrt{2}) (R_B + R_O)]$$

where R_A, R_B, and R_O are the radii of the A cation, B cation, and oxygen ion, respectively. The perovskite structure is supposed to be stabilized in the range of $0.75 \leq t \leq 1$; however, for an ideal perovskite, t is 1. For t < 1, the cubic structure transforms to the orthorhombic structure, which leads to deviation in the Mn – O – Mn bond angle from 180° (a case for ideal perovskite), resulting in the distortion in MnO_6 octahedra.

As shown in Fig. 2, $LaMnO_3$ or divalent atom (Ca, Ba, Sr, Pb, etc.) substituted $LaMnO_3$ has an Mn ion octahedrally surrounded by oxygen ions. Mn^{+3} ions with $3d^4$ outer electronic configuration are influenced by an octahedral electrical field due to oxygen ions, which splits the five degenerate d orbitals into t_{2g} (triply degenerate) and e_g (doubly degenerate) orbitals as shown in Fig. 2. Thus, out of four electrons of $3d^4$ in the Mn^{+3} ion, three electrons are arranged in the t_{2g} levels, whereas one electron goes to the e_g level due to strong Hund's coupling which favors high-spin configuration.

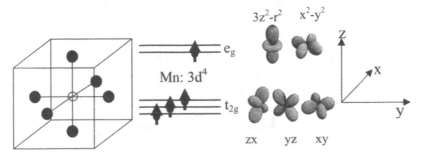

Figure 2. Crystal field splitting of Mn ion due to MnO_6 octahedra.

Undoped $LaMnO_3$ is an antiferromagnetic (AFM) charge-transfer insulator wherein Mn is in +3 valence state [6]. If any divalent atom (such as Ca, Sr, Ba, Pb, etc.) is doped at the La site, for instance, $La_{1-x}Ca_xMnO_3$, it converts Mn^{+3} to the Mn^{+4} state in equal proportion to the doping concentration. This is equivalent to the hole doping in the system. The hole doping illustrates several dramatic changes in electric and magnetic properties from the parent $LaMnO_3$ compound, such as insulator-metal transition, paramagnetic-ferromagnetic transition, CMR, charge-ordered state, phase separation, etc. [7]. In light of these rich physical properties, hole-doped $LaMnO_3$ has a potential for promising device applications such as magnetic sensors, magnetic valves, read head technology, bolometric application, etc. [8–15]. However, these properties strongly depend on ionic radii of the divalent cation and concentration of doping that govern the distortion of the MnO_6 octahedra. Zener's double-exchange model [16] and Jahn–Teller's distortion of MnO_6 octahedra [17] attempt to explain the diversity of the electrical and magnetic properties of doped $LaMnO_3$.

1.1.1 Zener's Double-Exchange Model

The co-occurrence of metallic and ferromagnetic properties in optimally hole-doped $LaMnO_3$ was explained by Zener [16]. He proposed a double-exchange mechanism between the Mn^{+3} and Mn^{+4} ions mediated through oxygen. In undoped $LaMnO_3$, the Mn ion is in +3 valence state with the electronic configuration of $t_{2g}^3 e_g^1$. All the 3d electrons are under the influence of a Coulomb repulsion interaction and are highly correlated. The e_g state electrons, which are highly hybridized with oxygen 2p states, have a tendency to localize the conduction electrons, making $LaMnO_3$ a Mott insulator. However, in hole-doped $LaMnO_3$, substitution of a divalent atom at La site drives an Mn^{+3} ion to an Mn^{+4} ion, equal in proportion with the divalent atom by creating a hole in the e_g level. So for 30% of Ca doping in $LaMnO_3$ (i.e., $La_{0.7}Ca_{0.3}MnO_3$), 70% of Mn^{+3} ions and 30% of Mn^{+4} ions are present. Zener suggested that an electron from the e_g level of Mn^{+3} ion hops to O^{-2} ion, and concurrently, an electron from O^{-2} hops to Mn^{+4} ion, giving rise to metallicity.

In this way in the Mn^{+3}–O–Mn^{+4} arrangement, Mn ions interchange their respective positions; however, the transfer of electron is controlled by the overlap between Mn 3d orbital and O 2p orbital. A schematic diagram of DE interaction is shown in Fig 3.

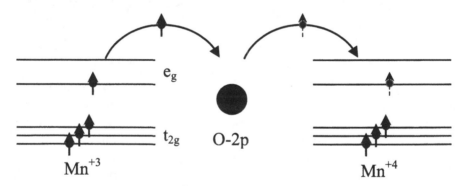

Figure 3. A schematic diagram of Zener's double-exchange mechanism.

The t_{2g} electrons are highly localized because of the strong correlation effect, and they form a net local spin of 3/2 and do not play a part in conduction. The conduction e_g electrons are strongly coupled to the t_{2g}^3 electrons due to Hund's coupling (J_H).[18] Because of this coupling, the effective hopping parameter (t_{ij}) of e_g electrons between the neighboring sites (i and j) depends on the relative angle (θ_{ij}) between the local spin (t_{2g}) of neighboring Mn sites and the formulation of this transfer integral in the strong coupling limit $J_H \rightarrow \infty$, as given by [19]

$$t_{ij} = t_0 \cos (\theta_{ij} / 2)$$

where t_0 is the intersite-hopping interaction of the e_g electron between the neighboring sites, i and j. Note that the hopping probability of e_g electron is maximum when local spins are parallel and minimum when antiparallel. The ferromagnetic interaction through exchange of the conduction electron, whose spin is influenced by the on-site Hund's coupling with the local spin, is called "double exchange (DE) interaction"—it involves the exchange of spin twice during the electron transfer from Mn^{+3} to Mn^{+4} through the oxygen ion. This double-exchange mechanism ensures the co-occurrence of metallicity as well as ferromagnetism.

Generally, the Mn^{+4}–O–Mn^{+4} set of connections shows superexchange AFM interaction, and the Mn^{+3}–O–Mn^{+3} network may exhibit ferromagnetic or AFM interaction depending on the orbital configuration [20]. In a lightly hole-doped $LaMnO_3$ system, due to competition of cos $(\theta/2)$ angular-dependent DE interaction between Mn^{+3}–O–Mn^{+4} of ferromagnetism and superexchange AFM interaction between Mn^{+4}–O–Mn^{+4} and / or Mn^{+3}–O–Mn^{+3}, the magnetic phase diagram of manganites is complex.

As apparent from the DE interaction formulation, the probability of electron transfer is maximum when all the spins are aligned parallel and favored at a low temperature due to a decrease in thermal fluctuation of spins at temperatures $(T < T_C)$. As the temperature is increased, the alignment of spins is disturbed, which causes a concomitant decrease in the hopping probability of electrons enhancing the resistivity. However, the application of the magnetic field can align the spins, which results in a decrease in resistivity. Therefore, the maximum change in resistivity due to external applied magnetic field is observed near the T_C. CMR originates as a consequence of a sharp drop in resistivity at higher temperature in the presence of magnetic field.

1.1.2 Jahn–Teller Distortion

At an early stage of research, it was believed that Zener's double-exchange interaction accounts for the electrical and magnetic properties of manganites. Recently, Millis et al. [17] argued that DE interaction was not sufficient to elucidate the CMR phenomenon, in particular, the large resistivity at $T > T_P$ and abrupt drop in resistivity at T_P. They and Röder et al. [21, 22] recommended that other effects resulting from Jahn–Teller distortion of Mn^{+3} ion such as electron phonon coupling, polaron effects, etc., might be additional contributing factors. According to the Jahn–Teller theorem, "Any nonlinear molecular system in a degenerate electronic state will be unstable and will undergo distortion to form a system of lower symmetry and lower energy thereby removing the degeneracy." Therefore, the doubly degenerate e_g level occupying a single electron in the Mn^{+3} ion lifts its degeneracy, and its energy is lowered, distorting the MnO_6 octahedra [23] as shown in Figs. 4 and 5. However, Mn^{+4} or Mn^{+2} ion is not a Jahn–Teller ion because there is no occupancy of an electron in the e_g level, or there are two electrons in the e_g levels, respectively; hence for these ions, the MnO_6 octahedra is not distorted.

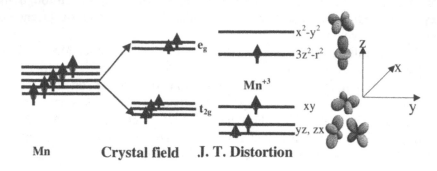

Figure 4. A schematic view of Jahn–Teller distortion of an Mn^{+3} ion.

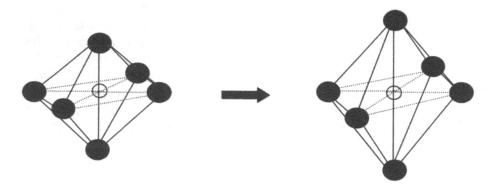

Figure 5. Distortion of an MnO_6 octahedra due to Jahn–Teller distortion.

1.1.3 Electrical and Magnetic Properties

For the $La_{0.67}Ca_{0.33}MnO_3$ compound, the electrical and magnetic properties are shown in Fig 6 [24]. It is evident from Fig. 6 that the insulator-metal transition temperature (T_P) is close to the paramagnetic to ferromagnetic transition Curie temperature (T_C). The observed magnetoresistance also is maximum near that temperature. Equally evident from the figure is that the application of the magnetic field results in an

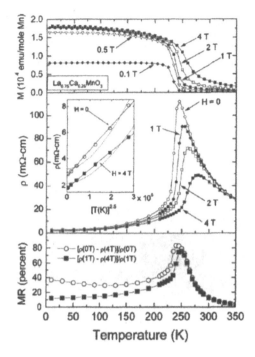

Figure 6. Electrical and magnetic transport properties of the compound $La_{0.75}Ca_{0.25}MnO_3$ (Schiffer et al. [24]).

increase in the T_P; clearly, the application of the magnetic field is further aligning the spins of Mn ions and it decreases the resistivity of the compound.

LaMnO₃ is an orbital-ordered antiferromagnetic insulator due to the presence of all Mn^{+3} ions only, and it is strongly affected by the collective Jahn–Teller distortion of Mn^{+3} ions [6]. This compound undergoes antiferromagnetic transition at 120 K. In $La_{1-x}D_xMnO_3$ (D = Ca, Sr, Ba, etc.) compounds, due to hole doping, the static Jahn–Teller distortion decreases with an increase in the doping level due to a decrease in the number of Mn^{+3} species. It has been observed that in a certain range of doping (x ~ 0.2-0.4), the ground state becomes ferromagnetic [24]. This is recognized as the creation of Mn^{+4} ions that tends to delocalize the e_g electron of Mn^{+3} ion under the influence of DE interaction. Thus, in this range of doping, a paramagnetic to ferromagnetic phase transition takes placewhich is accompanied by a sharp drop in resistivity. Electronic and magnetic phase diagrams of $La_{1-x}Ca_xMnO_3$ series are shown in Fig. 7 [24].

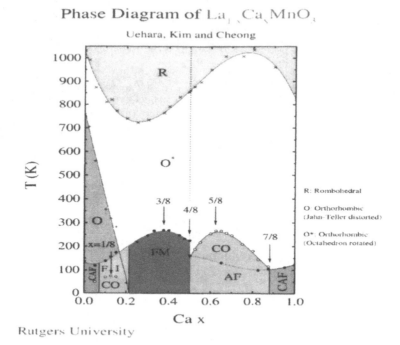

Figure 7. Electrical/magnetic phase diagram for La$_{1-x}$Ca$_x$MnO$_3$ [24].

For nearly 50% of the Ca doped manganites (at x ≈ 0.5), the Mn^{+3} and Mn^{+4} ions are equal in proportion, a real space charge ordering takes place when their long-range Coulomb interaction overcomes the kinetic energy [25–27]. The charge-ordered insulating state is favored by lower cation size at the La site as observed in Pr$_{1-x}$Ca$_x$MnO$_3$ [28, 29]. The transition to charge- order phase is a first-order transition accompanied by an abrupt change in resistivity at charge-order temperature as well as an abrupt change in crystal lattice parameters. Charge ordering is also reported to be closely related to the orbital ordering through the Jahn–Teller effect of Mn^{+3} ion [30].

At x > 0.5, the number of DE interaction links giving rise to ferromagnetism decreases, and the number of Mn^{+4} ion responsible for AF interaction increases. This arrangement gives rise to AF or canted AF, which hinders the motion of charge carriers.

There are also some reports on electron doping in the parent LaMnO$_3$ compound felicitated by substituting tetravalent atom like Ce at La site vig. La$_{1-x}$Ce$_x$MnO$_3$ [31–36]. This electron doping drives the Mn^{+3} to Mn^{+2} state. In this scenario, double-exchange interaction would take place in the Mn^{+3}–O–Mn^{+2} network while electron hopping from Mn^{+2} to oxygen ion and simultaneously from oxygen ion to Mn^{+3} ion. This can facilitate the co-occurrence of ferromagnetism and metallicity even in the electron-doped system. Both the electron-doped and hole-doped systems are intrinsically symmetrical as both Mn^{+4} and Mn^{+2} are non-Jahn–Teller ions and Mn^{+3} ion is a Jahn–Teller ion.

1.1.4 Substitution Effect at the *A* Site

As discussed in the previous section, doping level x in $La_{1-x}A_xMnO_3$ (A = divalent atom) controls the electrical and magnetic transport behavior. There is another evenly vital parameter, average *A* site ionic radii $<r_A>$. Parameter $<r_A>$ also joins in shaping the character of the compound because manganites are very susceptible to any pressure either intrinsic chemical or external physical pressure. Hwang et al. [37] analytically studied the evolution of T_C in the compounds $La_{0.7-y}Pr_yCa_{0.3}MnO_3$; $La_{0.7-y}Y_yCa_{0.3}MnO_3$ while keeping the Ca concentration fixed. They established that a decreasing $<r_A>$ corresponded to a decreasing T_C. Roughly, ferromagnetic DE interaction is maximum for x ~ 1/3 for an optimum value of $<r_A>$ ~ 1.24 Å. Departure from this value of $<r_A>$ leads to increase in a distortion of the crystallographic structure while keeping the Mn–O–Mn bond angle away from 180°. This weakens the DE interaction and also decreases the bandwidth which in combination tends to localize the conduction electrons, leading to increase in the resistivity and decrease in T_P.

As discussed above, it is evident that the magnetotransport properties of doped manganese oxides are very susceptible to the lattice strain and disorder. It is an entrenched reality that the irradiation of solids with energetic particle ions crafts a wide range of defect states in the system. In the following section we will briefly discuss some of the aspects of swift heavy ion irradiation.

2   SWIFT HEAVY ION IRRADIATION

Besides numerous prospective applications such as magnetic recording sensors, switching devices, bolometer, etc., that this compound offers [8–15], it also brings abundant issues in physics. Until now, because of the enthusiasm that this system has fashioned, there have been numerous publications at theoretical and experimental levels [38–44]. However, we still do not have an inclusive understanding of the physics involved in the process or of the optimum conditions needed to obtain high CMR values and the high temperature coefficient of resistance (TCR) values in the preferred choice of temperature and magnetic field. In order to learn the fundamentals of the mechanisms accountable for this consequence and to provide the optimum conditions to project this CMR-based compound as a potential futuristic device, new experiments and explorations of new materials that display comparable performance are being studied. The swift heavy ion irradiation of manganites is one such new adventure.

The SHI irradiation fashions controlled defect states in the material and modifies the properties of materials [45–52]. Depending on the choice of energy and type of ions, SHI irradiation can generate various kinds of defects such as point defects and columnar amorphization in the target material. These defects can create structural disorder or localized strain in the lattice of the target material. The results of SHI irradiation of hole-doped manganite system indicate significant modifications in their electrical, transport and magnetic properties [53–56].

2.1   Interaction of Swift Heavy Ions with Materials

There are basically two modes of energy deposition of an ion in a material medium: inelastic energy loss due to electronic excitations (i.e., electronic energy loss S_e or $(dE/dx)_e$) and elastic collision (i.e., direct nuclear collision—nuclear energy loss S_n or $[dE/dx]_n$). When energy of the incident ion is small enough ($\sim$ keV) so that the ion's velocity is smaller than the Fermi velocity of the electrons of the target material, the incident ion collides and loses its energy directly with the nucleus. In this situation, S_n dominates over S_e. However, for SHI (energy $\sim$ MeV), the incident ion has the order of Fermi velocity of the electrons of the target material, and it suffers collision with the electronic system of the target material. In such a situation, S_e is much higher than the S_n value. Therefore, almost all the energy loss or the energy deposited in the material can be considered because of the electronic energy loss if the thickness of the material is less than the range of the ion. Various models explain the energy loss process of SHI irradiation on a target material [57–60]. In Fig. 8, we show the dependence of S_e and S_n on the energy of the [107]Ag ion impinging on $La_{0.7}Ce_{0.3}MnO_3$ target. This type of curve is usually called the Bragg curve.

The two specific effects, in terms of materials modification that arises because of the S_e mode of energy deposition by the ion beam, are annealing and phase change. Depending upon the magnitude of the $(dE/dx)_e$ and the characteristics of the material (insulating, semiconducting, metallic, etc.), the energy deposited by the ion beam to the electronic subsystem of the material can either anneal out preexisting defects, or it can lead to the creation of defect or amorphized latent tracks along the ion path.

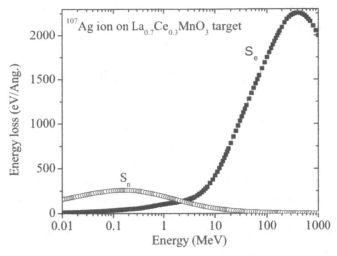

Figure 8. Variation of electronic energy loss S_e and nuclear energy lose S_n, of [107]Ag ion in $La_{0.7}Ce_{0.3}MnO_3$ target with respect to ion energy.

It should be emphasized that as the highly charged energetic ion passes through the material, the neighbor target atoms get positively charged (through electronic interactions) and repel each other. The time energetic ion spends passing through the material at a particular site is very short (10^{-17} sec) in comparison to the recombination time (10^{-12} sec). Thus SHI produces a long cylindrical channel of charged ions along its path. This channel of charged ions will explode radially because of the conversion of electrostatic energy to radial movement of atoms under the Coulomb repulsion force until the conduction electrons screen the ions. This results in the shock wave generation in the materials. This explanation is termed the Coulomb explosion model [61]. Another model used is the thermal spike model [62, 63]. The thermal spike is a two-step thermodynamic process described by two differential equations that give a time and space dependence of T_e and T_l (T_e and T_l are temperatures of electronic and lattice systems, respectively). According to this process, the energy locked into the electron excitation is transmitted to the lattice through electron-phonon interaction to increase the local temperature of the lattice above its melting temperature. The increase of temperature is followed by a rapid quenching ($10^{13} - 10^{14}$ K/sec), resulting in an amorphized columnar structure when the melt solidifies. To explain the track formation due to irradiation, the thermal spike model has been applied to metals and insulators [64–66].

Both processes mentioned above (Coulombic explosion and thermal spike) are inseparable. In a study, it has been shown that both of the effects are present simultaneously [54]. The cylindrical shock wave generates strain in the lattice around the ion track, and presence of the thermal spike creates the amorphized columnar tracks.

2.2   Swift Heavy Ions Irradiation on CMR Materials

External pressure has been acknowledged to influence the semiconductor (paramagnetic) to metal (ferromagnetic) transition in the manganite system because the magnetotransport properties of doped manganese oxides are very susceptible to the lattice strain and disorder [67, 68]. It has been known that the irradiation of solids with energetic particle ions can produce a wide range of defect states in the system. The SHI irradiation is an established technique in this context to fashion controlled defect states in the material, and it proposes the prospect of modifying the properties of materials [45–47].

Today, the facilities provide a means to study the irradiation effects in matter due to swift heavy ions whose main characteristic is to deposit a huge amount of energy on the target. During the last two decades, these effects have been widely reported on various types of materials: metals, semiconductors, and insulators [48–52]. There are also a few reports on irradiated samples of CMR materials [53–56].

In manganese perovskite-based $La_{1-x}D_xMnO_3$ (D = divalent metal) materials, the Curie temperature (T_c) increases when the bandwidth of itinerant e_g electrons broadens [25]. The bandwidth W is controlled by both the Mn–O bond length and bond angle of the Mn–O–Mn. The fine-tuning of both W and T_c can be obtained by an appropriate selection of the size of the ionic species to be substituted at rare-earth sites. However, the substitution of different ionic radii species can affect the bond angle but not the bond length due to the local disorder in the lattice [72]. On the other hand, the application of external hydrostatic pressure can modify both the bond length and the bond angle. It has been established that on the application of hydrostatic pressure, the bond length compresses, and the bond angle opens up which enhances the bandwidth [73]. Experimentally, it has been observed that under hydrostatic pressure, positive dT_c/dP reflects the contribution of the bond length, and the positive dW/dP reflects the contribution of the bond angle [37, 74].

From the application point of view, there is a need to enhance the metal insulator transition temperature (Tp), the temperature coefficient of resistance (TCR) (which is an important parameter for bolometric applications [15]), and magnetoresistance (MR) per unit magnetic field (which is important for magnetic sensors). It is speculated that the enhancement in T_c with hydrostatic pressure can be simulated by ion irradiation by introducing a type of permanent strain in the lattice [55, 56].

The SHI irradiation has also been projected to be a good tool to improve the TCR values of Ca doped LaMnO$_3$ samples [53].

2.3 Effect of Irradiation on Hole-Doped LaMnO$_3$

2.3.1 90-MeV O Ion

Ogale et al. [55] studied the influence of 90-MeV ^{16}O ion irradiation on the structural, magnetization, and magnetotransport properties of the pulsed laser deposited (PLD) epitaxial thin film of La$_{0.7}$Ca$_{0.3}$MnO$_3$ as a function of the ion dose value in the range of 10^{11}–10^{14} ions/cm^2. For 90-MeV oxygen ion, the typical values of Se and Sn as calculated by TRIM/SRIM [75] are 1.5 keV/nm and 1 eV/nm, respectively, indicating that most of the energy deposition is caused by Se. Also, these values are much smaller than the threshold value of electronic energy loss, (Se)$_{th}$ ($\approx$ 12 keV/nm) , as calculated using Szene's model [63], required for creating columnar amorphization along the track. The possible effects are point/clusters of defects and oxygen disorder in the present system. From

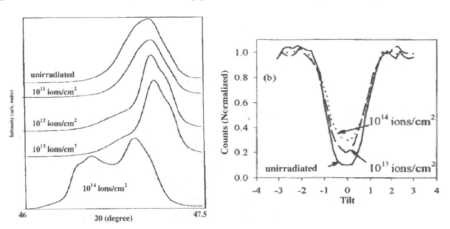

Figure 9. (a) X-ray diffraction patterns on expanded scale for the unirradiated La$_{0.7}$Ca$_{0.3}$MnO$_3$ film; the films irradiated at 10^{11}, 10^{12}, 10^{13}, and 10^{14} ions/cm^2; (b) RBS angular scans along the (100) channeling direction for the unirradiated and the irradiated films at fluence value of 10^{13} and 10^{14} ions/cm^2 (Ogale et al.[55]).

the x-ray diffraction (XRD) pattern of (004) peak (Fig. 9 (a)), it was observed that with an increase in dose from 10^{11} to 10^{13} ions/cm^2, the (004) peak shifts toward higher 2θ value, indicating a vertical compression or in-plane expansion.

It was observed that the film irradiated at 10^{14} ions/cm^2 exhibited a doublet feature on the LCMO peak, implying the existence of a two-phase constitution. However, the presence of other major peaks suggested that the crystal structure, as a whole, remains intact. This was also confirmed by the ion channeling experiment in the angular scan mode along the channeling direction as shown in Fig 9 (b). From Fig. 9 (b) it is evident that channeling occurs in all the samples, indicating that crystallinity is maintained in these samples even after irradiation with dose value of 10^{14} ions/cm^2, but the

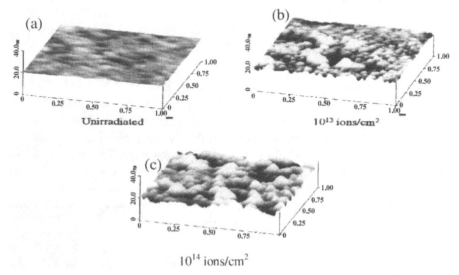

Figure 10. AFM images for the (a) unirradiated La$_{0.7}$Ca$_{0.3}$MnO$_3$ film, and the films irradiated at (b) 10^{13} and (c) 10^{14} ions/cm^2 (Ogale et al. [55]).

minimum yield (χ_{min}) of channeling increases. The increase in the χ_{min} with irradiation suggested the presence of point defects in the form of interstitials or vacancies-induced bond distortion in the Mn–O–Mn bond. Also the surface roughness was observed to increase with fluence, as shown by atomic force microscopy measurements (Fig. 10).

The effect of 90-MeV oxygen ion irradiation on the electrical resistivity and magnetoresistance (MR%, measured at 8.5 T) is shown in Fig 11. The metal-insulator transition temperature shifts toward the lower temperature region, and the resistivity increases for irradiated samples as compared with the unirradiated sample. Interestingly, MR% also increases with fluence; for the highest dose value of 10^{14} ions/cm^2, it remains constant at a lower temperature.

The effect of the irradiation on magnetization behavior at 5 K is shown in Fig. 12. The magnetization hysteresis loop is similar for the unirradiated film and the irradiated film with 10^{13} ions/cm^2 dose value. However, the film irradiated with a dose of 10^{14} ions/cm^2 shows a higher coercive field as compared with other films with two distinct coercive fields as shown in the inset of Fig. 12 (a), signifying the two-phase characters of

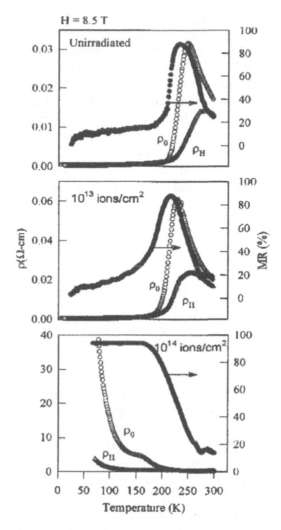

Figure 11. The effect of 90-MeV oxygen ion irradiation on the electrical resistivity and magnetoresistance (MR%, measured at 8.5 T) of $La_{0.7}Ca_{0.3}MnO_3$ film (Ogale et al. [55]).

the film. The Curie temperature follows the same trend (Fig. 12 (b)) as that of metal insulator transition. This study redefined the significance of the dependence of electrical transport or magnetotransport properties with the microstructure of the system.

In another study, Bathe et al. [76] studied the effect of 90-MeV oxygen ion irradiation on the electrical and magnetic properties of the PLD-grown epitaxial thin film of $La_{0.75}Ca_{0.25}MnO_3$. In Fig. 13 (a), the resistivity response of unirradiated and irradiated samples is shown. The corresponding variations of T_P and activation energy (E_a) for hopping transport in the semiconducting region ($T > T_P$) are shown in Fig. 13 (b). Irradiation with the initial fluence of 10^{11} ions/cm^2, T_p increases from 230 K of pristine film to 236 K; the activation energy of this film decreases as compared with the pristine film. However, further irradiation leads to a decrease in T_P and a increase in activation energy. Significantly, the film irradiated with the highest dose value showed the insulating behavior in the studied temperature range of 300 –60 K.

It was suggested that the lattice of the film relaxed at the lower irradiation dose, releasing the deposition-induced microscopic strain in the film that occurred during the film deposition. However, at high-dose values, irradiation-induced disorder introduced compositional inhomogeneity and related strain in the films which apparently affected the angle and length of the $Mn^{+3} - O - Mn^{+4}$ bond and hence the exchange interaction between the cations.

2.3.2 250-MeV Ag Ion

Until now, the energetics of the above-discussed irradiation studies were such that they could create only the point defects in the system because their electronic energy loss (S_e) was much below the threshold value of $(S_e)_{th}$ required for creating columnar amorphization along the ion track in the system. Now we will focus our attention to understanding the effects of columnar tracks to the various structural, electrical, and magnetic properties of the system. In another study, the effects of 250-MeV Ag ion irradiation on the structural, magnetization, and magnetotransport properties of the pulsed laser-deposited epitaxial thin film of $La_{0.7}Ca_{0.3}MnO_3$ grown on $SrTiO_3$ (STO) substrate as a function of the ion dose value in the range of $3 \times 10^{10} - 5 \times 10^{12}$ ions/cm^2 has been studied [56]. The S_e and S_n values, as calculated using TRIM97 simulation program [75] for 250-MeV Ag ion in LCMO system, is 24 KeV/nm and 48 eV/nm, suggesting that almost whole energy gets deposited in terms of electronic energy loss. Also in this case, Se >> (Se)$_{th}$, suggesting the possibility of creation of columnar amorphization zone in this system. Indeed, the columnar defects are clearly seen as dark spots in the AFM images of the irradiated sample. In Fig. 14, a comparision of an AFM study on the unirradiated and irradiated films is shown. The top view and side view of the amorphized track, as a consequence of irradiation, are shown in Fig. 14 (c) and (d), respectively. The white annulus correspond to the materials redeposited on the surface, whereas the dark region is the actual columnar amorphized defect with a typical diameter of ~ 10 nm. The interior of the defect was observed to be amorphous/highly disordered from the transmission electron micrograph (TEM) image (Fig. 15), suggesting that this region would be highly resistive as compared with other regions of the film.

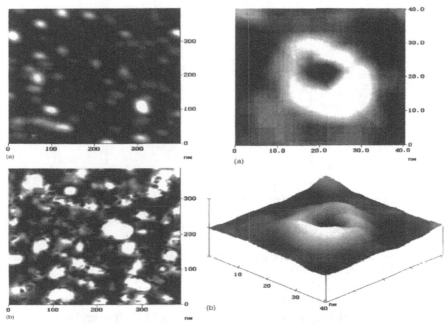

Figure 14. AFM images of (a) the unirradiated $La_{0.7}Ca_{0.3}MnO_3$ film and (b) the film irradiated at 4×10^{11} ions/cm^2 with 250-MeV Ag ions. The columnar defects are seen as dark spots in (b); (c) and (d) images are the top and side views of a single columnar track respectively (Ogale et al. [56]).

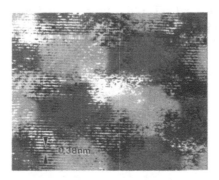

Figure 15. High-resolution transmission electron micrograph for the $La_{0.7}Ca_{0.3}MnO_3$ film irradiated with 250-MeV Ag ions showing a single columnar defect (Ogale et al. [56]).

It is worth mentioning here that the range of the 250-MeV Ag ion is much larger than the thickness of the LCMO film ($\sim$ 3000 Å), implying that the ions pass through the film deep into the substrate and do not make any chemical contamination during the irradiation.

The variation of $(dE/dx)_e$ and $(dE/dx)_n$ for the 250-MeV Ag^{+17} ion beam in LCMO target is shown in Fig. 16 as a function of depth, which has

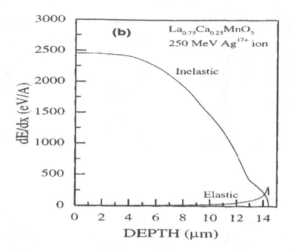

Figure 16. Elastic and inelastic stopping power as a function of depth for 250-MeV [107]Ag ion beam in LCMO (Arora et al. [53]).

a plateau of about 4-µm width, indicating uniform transfer of the energy due to irradiation throughout the film thickness [53]. To understand the effect of fluence values of irradiation on the structural, electrical, or magnetic properties, we divide the irradiation studies in two regimes: lower fluence value from 3×10^{10} to 7×10^{11} ions/cm^2 and higher fluence value. Figure 17 shows the normalized resistance (R/R$_{300K}$) as a function of normalized temperature (T/T$_p$) for unirradiated and 250-MeV Ag^{+17} ion irradiated LCMO thin films at various ion fluences in the range of 3×10^{10} to 7×10^{11} ions/cm^2 [77].

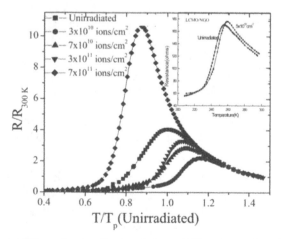

Figure 17. The variation of normalized resistance (R/R$_{300K}$) as a function of normalized temperature (T/T$_p$) for unirradiated and 250MeV Ag^{+17} ion irradiated LCMO thin films (Ravi Kumar et al.[77]).

Irradiation enhances the T_p to 242 K for irradiated film with dose value of 3×10^{10} ions/cm^2 as compared with the T_P value of 204 K for the pristine film. The peak resistivity is also found to be decreased by a factor of 2. Further increase in fluence values decreased the T_p values but remain in upper side than that of the unirradiated films up to the fluence value 3×10^{11} ions/cm^2. At 7×10^{11} ions/cm^2 fluence the T_p decreased to 177 K, and its peak resistance increased by a factor of 2.5 as compared with the unirradiated sample. Figure 18 shows the variation of T_p and normalized resistance as a function of ion fluence in LCMO.

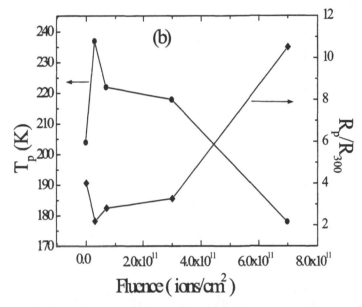

Figure 18. The variation of T_p and normalized resistance as a function of 250-MeV Ag ion fluence in LCMO thin films (Ravi Kumar et al. [77]).

One aspect clear from these observations is that the strain generated by the SHI irradiation provides a similar effect as the external hydrostatic pressure does. Second, there is a competition between the effect of structural strain and columnar defect generation. Up to 3×10^{11} ion/cm^2 of the dominant effect is caused by the structural strain, whereas at higher fluence values the columnar defects are dominant. At this juncture it is essential to point out that this effect is not caused by the starting sample which has lower T_p (~204 K). The enhancement in T_P has also been observed for LCMO films grown on NdGaO$_3$ substrates, which are known as strain free and have a higher T_p value (254K), where after irradiation with 250-MeV Ag^{+17} ions with fluence of 5×10^{10} ions/cm^2, the T_p value is increased to 260 K (i.e., the enhancement by 6 K; see inset in Fig. 17). This fact supports the observations of Fontcuberta et al. [74] who have shown that on a specimen with higher T_p, the response of external pressure is less. This implies that the SHI irradiation with optimized fluence value and energy generates a permanent structural strain in the specimen that is similar to that of external hydrostatic pressure, which is temporary.

Interestingly, the effect of SHI irradiation-induced defects has a different effect on the electrical transport behavior as compared with the defects induced by other means such as creating oxygen vacancies during the thin film deposition or doping some other element with different ionic radii. Figure 19 illustrates one of the basic differences in irradiation-induced defects and defects induced because of other means and the figure

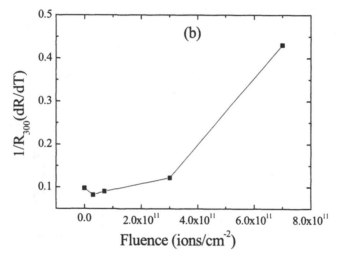

Figure 19. Variation of temperature coefficient of resistance values with ion fluence for the as-deposited $La_{0.7}Ca_{0.3}MnO_3$ film and films irradiated at different ion doses with 250-MeV silver ion (Kumar et al. [77]).

also shows the variation of maximum temperature coefficient of resistance values with ion fluence.

Initially, TCR value decreases with ion irradiation for fluence value of 3×10^{10} ions/cm^2. With the further increase in fluence value, TCR value increases and at 3×10^{11} ions/cm^2, the TCR value is higher by 25% than that of unirradiated films. It is recalled that except for irradiation with the first fluence value of 3×10^{10} ions/cm^2, irradiation at higher fluence values decrease the T_p of the film. Generally it is believed that a TCR value of a sample with a higher T_p is lower than that which has low T_p values [78]. From these observations we would like to suggest that the SHI irradiation is a good tool to improve the TCR values of LCMO specimens, which is an important parameter for bolometric applications [15].

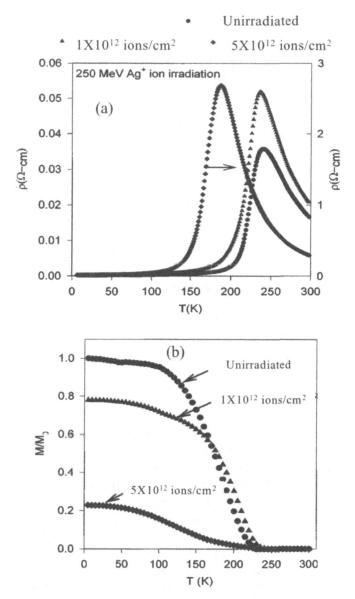

Figure 20. 25-MeV Ag ion irradiation-induced modifications on electrical resistivity (a) and magnetization data (b) of La$_{0.7}$Ca$_{0.3}$MnO$_3$ film (Ogale et al. [56]).

At higher fluence values (> 5 × 10^{11} ions/cm^2), 250-MeV Ag ion irradiation-induced modifications on electrical resistivity and magnetization data are shown in Figs. 20 (a) and (b), respectively [56]. The sample irradiated at the fluence of 1 × 10^{12} ions/cm^2 the resistivity is slightly higher, and the saturation magnetization is slightly lower than in the unirradiated sample, whereas neither the T$_P$ nor the T$_C$ defined from the

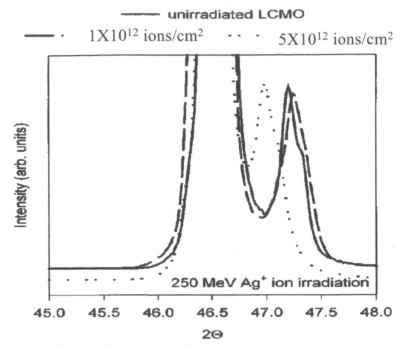

Figure 21. The XRD signature of (004) peak of the unirradiated and irradiated LCMO samples (Ogale et al. [56]).

magnetization curve changes significantly. Although the sample irradiated with the maximum dose value of 5×10^{12} ions/cm^2 exhibits a much higher resistivity, lower values of the saturation magnetic moment T_P and T_C compared with any other unirradiated or irradiated samples.

The possible explanation for this type of fluence-dependent electrical or magnetic behavior is that the amorphized region in the film created as a consequence of irradiation is highly disordered; hence, a region of high resistivity and the local moments in the region may be freezed or glassy because of the muddle structure in the region. Nevertheless, the unirradiated region in the film irradiated at highest dose value is sufficient enough to percolate through the film, tendering the typical metal-insulator transition and FM-PM transition, although the resistivity increases and saturation moment, T_P and T_C decrease.

In Fig. 21, the x-ray diffraction signature of the (004) peak of the unirradiated and irradiated LCMO samples is shown to investigate the strain and structure of the film. It is worth mentioning here that the LCMO films (a = 3.865 Å) deposited on STO (a = 3.91 Å) substrates are subjected to a tensile strain that causes a decrease of the out-of-plane lattice parameter. This strain is partially released as the film thickness grows. After irradiation at the low fluences up to 1×10^{12} ions/cm^2, there is hardly

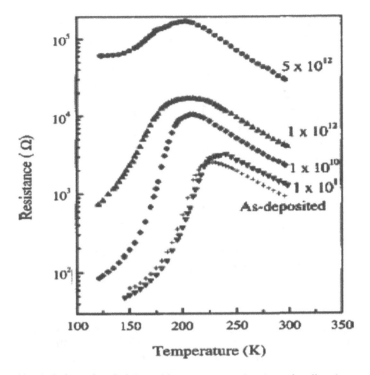

Figure 22. Variation of resistivity with temperature for the unirradiated as well as 250-MeV Ag ion irradiated thin films of $La_{0.75}Ce_{0.25}MnO_3$ (Arora et al. [53]).

any discernable change in the x-ray feature. However, in the sample irradiated at 5×10^{12} ions/cm^2 a dramatic downward shift in the position of the (004) peak is observed. However, the crystal structure is maintained even at such a high-dose value of irradiation.

In another study, the role of the 250-MeV [107]Ag ion irradiation-induced defects (fluence in the range of 1×10^{10} to 5×10^{12} ions/cm^2) has been investigated on the charge-transport process, $1/f$ conduction noise properties of the pulsed laser-deposited epitaxial thin films of $La_{0.75}Ca_{0.25}MnO_3$ grown on (100) oriented $LaAlO_3$ substrate [53]. Apart from providing the information about the conduction process, $1/f$ noise also provides a useful estimate of the signal-to-noise ratio, which determines the ultimate resolution possible with the devices made from these materials. The 1/f noise is known to be very susceptible to any fluctuation or transition whether electrical or magnetic. In this way, the measurement of 1/f noise gives input regarding the conduction mechanism in the system. Also, 1/f noise measurement is supposed to be an indirect method to probe the consequences of irradiation-induced modifications and formation of defects on the charge transport process in thin films.

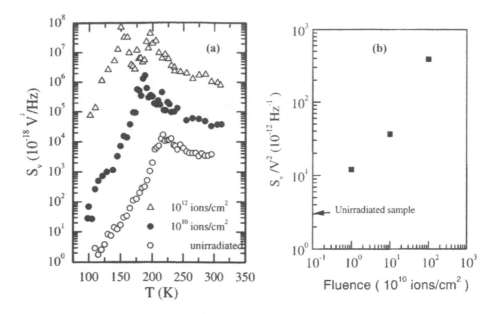

Figure 23. (a) The variation of S_V as a function of temperature for $La_{0.75}Ca_{0.25}MnO_3$ thin films measured before and after irradiation with 250-MeV Ag ion; (b) dependence of normalized spectral density of noise with irradiation dose value (Arora et al. [53]).

In Fig. 22, the effects of 250-MeV[107]Ag ion irradiation on the electrical resistivity of the thin films of $La_{0.75}Ca_{0.25}MnO_3$ are shown. It is clear that for the sample irradiated with 10^{11} ions/cm^2, T_P increased by 10 K, and the resistivity is decreased in the ferromagnetic metallic state from that of an unirradiated sample; however, at higher dose values T_P decreases and resistivity increases, similar to what we discussed above for the $La_{0.75}Ca_{0.25}MnO_3$ thin films.

Figure 23 (a) shows the variation of S_V as a function of temperature for $La_{0.75}Ca_{0.25}MnO_3$ thin films measured before and after irradiation. It is apparent from the Fig. 23 (a) that for unirradiated film, the magnitude of S_V is maximum near the T_P (225 K) of the film. At this temperature, the magnitude of S_V is about an order of magnitude higher as compared with its value at 300 K. As the sample is cooled down from this temperature, the noise magnitude first decreases sharply and then gradually. The sharp decrease in S_V occurs because of the transition of the material from a randomly aligned spin state to an aligned spin state. Also, the temperature dependence of the noise of irradiated samples remains similar as that of the unirradiated case. However, for the sample irradiated with fluences of 10^{10} and 10^{12} ions/cm^2, the magnitude of noise is higher by 1 and 2 orders of magnitude, respectively, at room temperature than that of the unirradiated film, although the peak in noise value follows the corresponding T_P value of the individual film. Higher noise values in the irradiated samples, both in paramagnetic-insulating phase as well as in ferromagnetic-metallic phase, is believed to be caused by the presence of irradiation-induced columnar defects and the structural strain surrounding the amorphized zone. The normalized spectral density of noise is also observed to increase with fluence value of 250-MeV Ag ion (Fig. 23 (b)).

The Hooge's parameter γ was determined for these unirradiated and irradiated films, using the empirical relation

$$S_v = (\gamma V^2/N_c f^\alpha)$$

where V is the dc voltage across the thin film, N_c is the carrier density in the thin films, and α is a constant close to unity. Table 1 provides a brief account of transport and noise studies performed on these unirradiated and 250-MeV Ag ion-irradiated $La_{0.75}Ca_{0.25}MnO_3$ thin films.

Table 1. Peak temperature Tp magnetoresistance at peak (MR%), resistivity $\rho_{300 K}$, ρ_{150K} and activation energy *Ea* extracted from the electrical transport measurements and spectral density of voltage noise *Sv* (10 Hz) at 300 and 150 K and the value of Hooge's parameter γ (300 K) estimated from the noise measurements as a function of ion fluence for the LCMO thin films irradiated with 250-MeV 107Ag ions.

Fluence ion/cm^2	Tp (K)	Peak MR %	Ea meV	$\rho_{300 K}$ mΩcm	$\rho_{150 K}$ mΩcm	Sv (10 Hz) (V^2/Hz) 300 K	Sv (10 Hz) (V^2/Hz) 150 K	γ 300 K
0	225	57	100	45	3.2	4×10^{-15}	34×10^{-18}	2×10^4
10^{10}	210	68	106	114.4	8.1	75×10^{-15}	14×10^{-15}	2×10^4
10^{11}	242	55	98	66.3	2.5	30×10^{-15}	1×10^{-15}	6×10^4
10^{12}	210	55	109	198.2	75.2	8×10^{-13}	33×10^{-12}	3×10^5

As clear from Table 1, the value of the Hooge's parameter increases with the fluence. The observed increase in γ value suggests that the lattice was damaged with the increasing fluence and contributed to the enhanced noise. The room temperature γ value is several orders of magnitude higher in comparison to the γ values reported for semiconductors and high *Tc* superconductors [79]. The suggest cause for the large γ values in the paramagnetic region (above T$_P$) in these films is from a rather complicated conduction behavior in the region because of localization of electrons participating in the conduction process or polaronic conduction-like behavior. Also it was reported that in the ferromagnetic-metal (FMM) state for the unirradiated sample, the value of Hooge's parameter was 1.9×10^4, several orders of magnitude higher than the values observed for the metals (typically 10^{-3}) or magnetoresistive multilayers [80]. The anomalously large γ value observed for manganites was caused by the local magnetization fluctuations coupled to the resistivity. This is also expected because the resistivity of the manganites in the metallic state itself is larger than the metals due to the presence of the strong electron-phonon coupling arising from the Jahn–Teller distortions.

In conclusion, we observed how SHI irradiation of hole-doped manganite system produced significant modifications in their structural, electrical transport, and magnetic properties. Now we will discuss some of the primary outcomes of SHI irradiation on electron-doped manganite system because there are not many studies performed concerning this topic.

2.4   200-MeV Ag Ion Irradiation of $La_{0.7}Ce_{0.3}MnO_3$ Thin Films

It is known that Mn exhibits three valence states: Mn^{+2}, Mn^{+3}, and Mn^{+4} with $3d^5$, $3d^4$, and $3d^3$ electronic configuration, respectively. Analgous to the divalent atom-doped $LaMnO_3$, where substitution of the divalent atom for the La atom results in hole doping and drives the system to the ferromagnetic metallic state, there are reports that partial substitution of some tetravalent ions such as Ce for the La^{+3} ion to form the composition as $La_{1-x}Ce_xMnO_3$ also drives the parent $LaMnO_3$ compound to a half metallic ferromagnetic state. In this scenario, if Ce exists in a tetravalent state, the equivalent amount of Mn^{+3} is believed to be converted into Mn^{+2} state (where the extra electron occupies the upper e_g level, making the system electron doped). However, quite a few essential issues remain unanswered [32–36]: (1) Is the Ce-doped $LaMnO_3$ a single phase? (2) Does all the Ce go to the La site? (3) How are they different from the divalent cation-doped CMR materials? To recognize some features of the above-mentioned issues, Kumar et al. and co-worker [36, 85, 86] studied thin films of pulsed laser-deposited thin films of a $La_{0.7}Ce_{0.3}MnO_3$ compound and examined their structural, electrical, and magnetic transport properties. In addition, thin films of this system were irradiated with a 200-MeV Ag ion to observe the effect of swift heavy ion irradiation on their structural, electrical and magnetic properties.

The x-ray diffraction patterns of the unirradiated and irradiated samples are shown in Fig. 24. A small peak of CeO_2 phase is detected in the XRD pattern of the unirradiated film, which has very small intensity around 4% of the intensity of the major peak (004) of the primary phase of LaCeMnO. Nevertheless, it is interesting to notice that on 200-MeV Ag ion irradiation, the intensity of all diffraction peaks decreases with ion fluence,

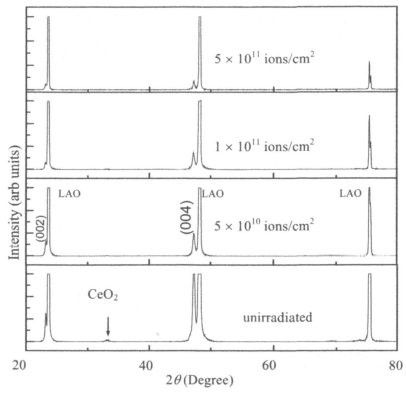

Figure 24. X-ray diffraction pattern for unirradiated and 200-MeV Ag ion-irradiated $La_{0.7}Ce_{0.3}MnO_3$ thin films (Kumar et al. [36]).

and at the highest fluence, there is no CeO_2 peak. This observation indicates that the SHI irradiation up to a certain fluence can be a good methodology to drive the materials to impurity-free single-phase forms.

The electrical transport properties of these films are shown in Fig. 25. The unirradiated film shows the insulator metal transition at 281 K, although a relatively broader insulator-metal transition is possibly caused by the presence of CeO_2 unreacted phase in the unirradiated sample. However, the disappearance of the CeO_2 phase in the irradiated samples brings in some remarkable changes in the electrical transport properties where the T_P increases from 281 K (for the pristine sample) to 294 K (for the film irradiated with fluence value of 5×10^{10} ions/cm^2). At a higher dose value (up to 5×10^{11} ions/cm^2), T_P decreases even higher than 281 K except for the film irradiated with highest dose of 1×10^{12} ions/cm^2 where T_P decreases to 254 K.

In Fig. 26, the variation of the resistivity peak temperature (T_P) and the activation energy (E_p) are shown as a function of ion fluence for

$La_{0.7}Ce_{0.3}MnO_3$ thin films. As the system tends to become impurity free after its first irradiation dose of 5×10^{10} ions/cm^2, its E_ρ value is lowered. However, with the increase in structural strain and lattice disorder caused by SHI irradiation, the value of E_ρ is raised. Indeed, this is also noticeable in the change in T_P with the irradiation dose as shown in the Fig. 26.

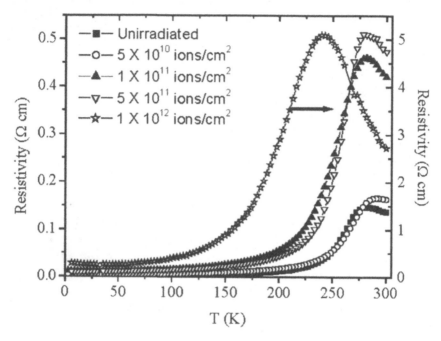

Figure 25. Resistivity vs. temperature plot for the unirradiated and 200-MeV Ag ion irradiated samples of $La_{0.7}Ce_{0.3}MnO_3$ thin films (Kumar et al. [36]).

The modifications on the structural and electrical properties have a curious impact on the TCR behavior with temperature. The variation of the temperature coefficient of resistance as a function of temperature is shown in Fig. 27. For the unirradiated sample, the maximum TCR value is 3% K^{-1} at 262 K; whereas for the sample irradiated with the initial fluence of 5×10^{10} ions/cm^2, the TCR remains almost the same (however, a bit higher ~3.2% K^{-1}). But significantly, the maximum TCR value shifts toward a higher temperature to 270 K. For samples irradiated with higher fluence value of 5×10^{11} ions/cm^2, the TCR value increases to 6.3% K^{-1}, but the corresponding temperature also shifts to a lower temperature of 258 K. At room temperature, the TCR value for the pristine film is –0.73 % K^{-1} which increases to –1.74 % K^{-1} upon irradiation with the fluence value of 5×10^{11} ions/cm^2. It is useful to mention here that in commercially available bolometers based on VO$_x$ or semiconducting YBCO, room temperature

TCR values vary from ~1.5% K^{-1} to 3% K^{-1} at room temperature [81–84]. For uncooled bolometric application, high negative TCR values are desired at room temperature.

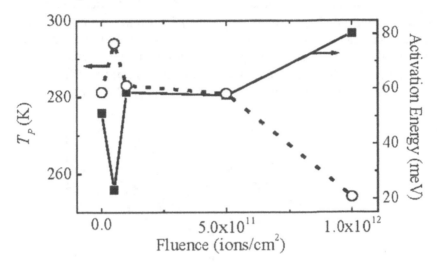

Figure 26. Variation resistivity peak temperature; activation energy of La$_{0.7}$Ce$_{0.3}$MnO$_3$ thin films with the 200-MeV Ag ion fluence values (Choudhary et al. [85]).

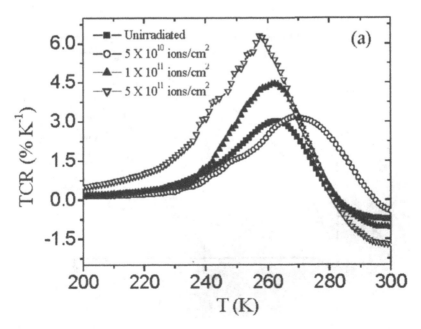

Figure 27. Variation of temperature coefficient of resistance with temperature of unirradiated and 200-MeV Ag ion irradiated samples of La$_{0.7}$Ce$_{0.3}$MnO$_3$ thin films (Choudhary et al. [85])

Figure 28 (a) and (b) presents the variation of magnetoresistance
(MR) data with magnetic field for the unirradiated as well as irradiated thin
films of $La_{0.7}Ce_{0.3}MnO_3$ at room temperature (300K) and 200K,
respectively. From Fig. 28 (a), it is evident that the MR% increases for the
film irradiated with the fluence value of 5×10^{10} ions/cm^2 as compared with
the pristine sample at 300 K. However, further irradiation results in the
decrease of the MR% value. The possible reason for the increase in the
MR% value at the initial fluence may be that with this fluence, the T_P value
increases toward room temperature as compared with 282 K for the pristine

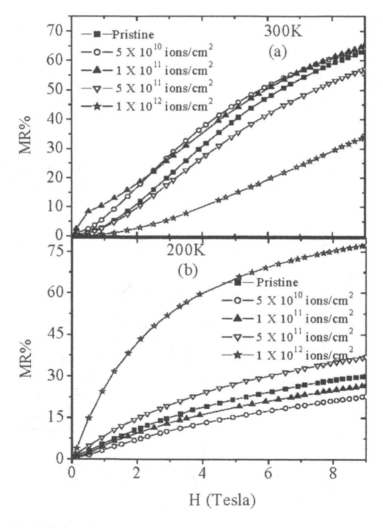

Figure 28. Variation of magnetoresistance with respect to the magnetic field for the
unirradiated as well as 200-MeV Ag ion-irradiated thin films of $La_{0.7}Ce_{0.3}MnO_3$ measured at
(a) 300 K and (b) 200 K (Choudhary et al. [85]).

sample, and it is known that the MR% is maximum near the transition temperature region. The irradiation with higher fluence values results in a decrease in the T_P; hence MR% is decreased at room temperature with an increase in the fluence value. Interestingly, the data at 200 K (Fig. 28 (b)) reflects a reciprocal behavior as that of Fig. 28 (a). The MR% of the film irradiated with 1×10^{12} ions/cm^2 fluence value exhibits the maximum value as compared with other samples. This could be understood in terms of the SHI irradiation-induced modifications on the structural, electrical, and magnetic transport properties. Because the film irradiated with the lowest fluence value is more chemically homogeneous and ordered as compared with the pristine sample, it shows a higher T_P value and is more ordered in the metallic regime, say at 200 K, as compared with other samples. Therefore, the MR% value at this temperature in this sample is the least among all the films. However, the film irradiated with the highest fluence value is tremendously under strain due to an increase in the columnar tracks density induced because of irradiation. This leads to a decrease in the T_P to 254 K and an increase in the resistivity as compared with the pristine sample. In this sample, the application of the magnetic field will significantly reduce the resistance as compared with other samples; hence, the maximum MR% is observed.

From an application point of view of the magnetic sensor, it is desirable that the response of resistance with the magnetic field should be swift. Figures 29 (a) and (b) show the graph of normalized derivative of resistance with respect to the magnetic field [field coefficient of resistance, FCR = $(1/R) \times (dR/dH) \times 100$, (% Tesla^{-1})] versus magnetic field at temperatures of 300K and 200 K, respectively.

Interestingly, it turns out that the film irradiated with the fluence value of 5×10^{10} ions/cm^2 shows the maximum FCR value among all the samples over the applied magnetic field at 300 K in the present study. More importantly, the maximum FCR value is attained at a relatively lower magnetic field value as compared with the pristine sample. This would be of high relevance from an application point of view. These observations indicate that the film irradiated with the fluence value of 5×10^{10} ions/cm^2 not only enhances the T_P, TCR and MR% value close to room temperature, but it also increases the FCR value as compared with the unirradiated sample. However, the film irradiated with the highest fluence value of $1 \times 10_{12}$ ions/cm2 shows a maximum FCR value at 200 K (Fig. 29 (b)). Because at 200 K all the samples, except one irradiated with the highest fluence, are in ordered ferromagnetic states, their FCR values are small and remain almost constant. This study reveals that by controlling the fluence value, the system from a device point of view can be tailored in a different temperature range of interest.

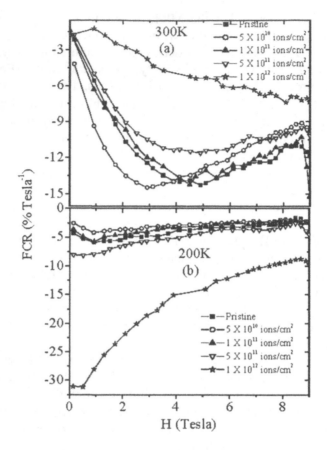

Figure 29. Variation of FCR with respect to magnetic field at (a) 300 K and (b) 200 K for the unirradiated as well as 200-MeV Ag ion-irradiated thin films of $La_{0.7}Ce_{0.3}MnO_3$ (Choudhary et al. [85]).

Figure 30 shows the S_v versus frequency plot for the unirradiated and irradiated $La_{0.7}Ce_{0.3}MnO_3$ thin film with fluence of 5×10^{10} ions/cm^2 at room temperature. The figure demonstrates that the conduction noise follows *1/f$^\alpha$* behavior with $\alpha \sim 1$. Such a behavior is observed for all the samples irradiated with higher fluence values. However, the noise level in all the samples increased with the fluence, similar to hole-doped manganites, as expected because of the increased number of columnar defects produced in the sample with irradiation dose values. The inset in Fig. 30 shows that S_v (measured at 10 Hz) follows the quadratic behavior with the bias current for unirradiated thin film. These features are followed throughout the temperature range (77–300 K) in unirradiated as well as in irradiated thin films. These observations suggest a conductance fluctuation type of noise.

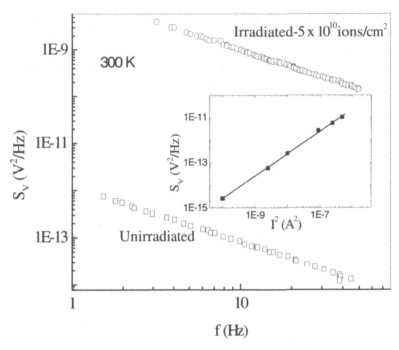

Figure 30. Spectral density S_v as a function of frequency with bias current 0.1 mA for pristine and the 200- MeV Ag ion irradiated $La_{0.7}Ce_{0.3}MnO_3$ thin film for fluence value of 5 $\times 10^{10}$ ions/cm^2. The inset shows the quadratic dependence of the spectral density with bias current for the unirradiated film (Kumar et al. [36]).

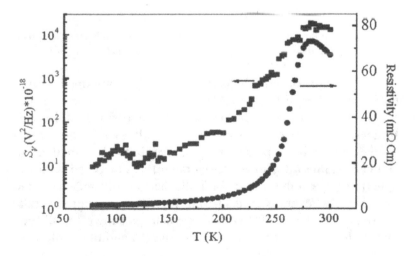

Figure 31. Dependence of spectral noise value (S_V) and resistivity with respect to temperature for the unirradiated sample (Choudhary et al. [86]).

The dependence of spectral noise value (S_V) for the unirradiated sample with temperature is shown in Fig. 31. It is evident that S_V follows the trend that resistivity exhibits. The noise value seems to be the maximum in the transition temperature region by the virtue of sensitivity of *1/f* noise to any electrical or magnetic fluctuation.

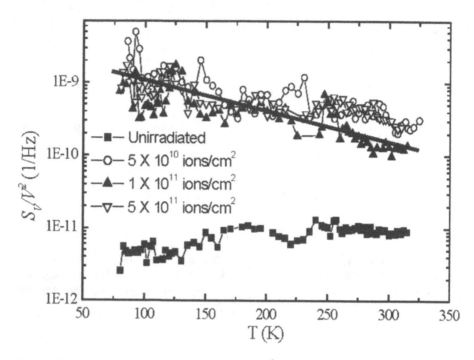

Figure 32. Variation of normalized noise value (S_v/V^2) with the temperature for pristine and the 200-MeV Ag ion-irradiated La$_{0.7}$Ce$_{0.3}$MnO$_3$ thin film (Choudhary et al. [86]).

These features of noise behavior in this electron-doped unirradiated or irradiated thin films of La$_{0.7}$Ce$_{0.3}$MnO$_3$ are similar to those of hole-doped samples. However, there is a difference in the normalized noise behavior between these two types of manganites. Recall that in the hole-doped sample, the normalized noise was observed to be smaller in the ferromagnetic regime (at a lower temperature) as compared with those in paramagnetic regime (above T_P) for both the unirradiated and irradiated thin films [53]. However, as evident from Fig. 32 in which the variation of normalized noise value (S_v/V^2) with temperature is shown for all the films, surprisingly for irradiated thin films, the magnitude of S_v/V^2 in ferromagnetic metallic state is higher than that in the paramagnetic insulating state. To guide the eyes, a straight line is drawn on the plot for irradiated films. One possibility of this occurrence could be the dominating contribution of magnon fluctuation to the noise in the irradiated samples as compared with the unirradiated sample, because electron-doping derives Mn^{+3} to Mn^{+2} which has a higher spin magnetic moment 5 μ_B as compared with 3 μ_B of Mn^{+4} that occurs because of hole doping.

Figs. 33 (a) and (b) show the magnetization and susceptibility, respectively, as a function of temperature. Similar to T_P, the Curie temperature T_C of the film moves up upon irradiation. However, unlike the resistive transition, higher fluences do not lower T_C; rather, the

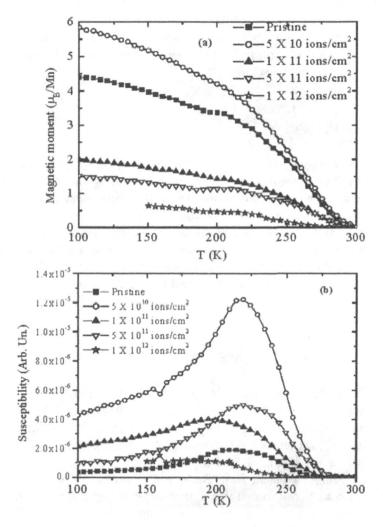

Figure 33. (a) Field-cooled dc magnetization versus temperature behavior at an applied field of 5000 Oe. (b) AC susceptibility of films measured at 15 Oe and 10 kHz for pristine and the 200-MeV Ag ion-irradiated $La_{0.7}Ce_{0.3}MnO_3$ thin film (Kumar et al. [36]).

susceptibility decreases. On the other hand, the saturation magnetization does track with T_P. It is surprising to detect rather unusual characteristics of the magnetization data. The unirradiated film attains a magnetization of 4.4 μ_B/Mn. This value is close to the projected spin-only value of 4.3 μ_B/Mn for the stoichiometric compound.

It must be mentioned that because there are Ce deficiencies in the unirradiated film, the average valence of Mn must be greater than 2.7, suggesting that the expected magnetization of the film is less than 4.3 μ_B/Mn. Nonetheless, it is remarkable that the film irradiated at the lowest fluence displays a magnetization that is significantly larger. The decrease in the magnetic moment with the larger fluence is consistent with the reported data on the hole-doped system.

3   SUMMARY

To summarize, we have presented a review of swift heavy ion irradiation-induced modifications in electrical transport, $1/f$ conduction noise, magnetic and structural properties of hole-doped and electron-doped colossal magnetoresistance (CMR) materials. SHI irradiation is a tool to tune the important parameters such as metal-insulator transition temperature T_p, temperature coefficient of resistance (TCR), and field coefficient of resistance (FCR), which are useful for applications. Finally, we conclude that such types of modifications are possibly made by controlled irradiation of swift heavy ions.

4   ACKNOWLEDGEMENTS

The authors are sincerely thankful to Dr. Amit Roy, Director, Nuclear Science Center, New Delhi, India, and Professor G.K. Mehta for their encouragement and support in pursuing this work. The authors are also thankful to DST for financing the project. Finally, the authors are grateful to Professors S. B. Ogale, T. Venkatesan, J.P. Srivastava, and to Doctors S.K. Arora, Shahid Husain, Ravi Bathe and to other colleagues who have participated in and contributed to this problem from time to time. Our group performed all studies presented here using 15UD Tandem Pelletron Accelerator at the Nuclear Science Center, New Delhi.

5   REFERENCES

[1]   *Colossal Magnetoresistance Oxides,* ed. Y. Tokura (Gordon and Breach, London, 1999)

[2]   G.H Jonker and J.H. Van Santen *Physica* **16** (1950) 337

[3]   J.H. Van Santen and G.H. Jonker *Physica* **16** (1950) 599

[4]   J. C. Bendorz and K.A. Muller, Z. Phys. B **64** (1986) 189

[5]   K. Chahara, T. Ohno, M. Kasai, and Y. Kozono, Appl. Phys. Lett. **63** (1990) 362

[6]   E. O. Wollan and W. C. Koehler, Phys. Rev. **100** (1955) 545

[7]   A. P. Ramirez, J. Phys. Conds. Mat. **9** (1997) 8171

[8]   J. Z. Sun, Phil. Trans. R. Soc. Lond. A**356** (1998) 1693

[9]   T. Venkatesan, M. Rajeswari, Zi-Wen Dong, S. B. Ogale, and R. Ramesh, Phil. Trans. R. Soc. Lond. A**356** (1998) 1661

[10] K. Steenbeck, T. Eick, K. Kirsch, K. O'Donnel, and E. Steinbeisse, Appl. Phys. Lett. **71** (1997) 968

[11] J. Z. Sun and A. Gupta, Annu. Rev. Matt. Sci. **28** (1998) 45

[12] J. Z. Sun, L. Krusin-Elbaum, P. R. Duncombe, A. Gupta, and R.B. Laibowitwitzpl, Phys. Lett. **70** (1997) 1769

[13] N.D. Mathur, G. Burnell, S. P. Isaac, T.J. Jackson, B.S. Teo, J.L. MacManus-Dricoll, L.F. Cohen, J.E. Evetts, and M. G. Blamire, Nature **387** (1997) 266

[14] M. Rajeswari, C.H. Chen, A. Goyal, C. Kwon, M.C. Robson, R. Ramesh, T. Venkatesan, and S. Lakeou, Appl. Phys. Lett. **68** (1996) 3555

[15] R.J. Choudhary, Anjali S. Ogale, S.R. Shinde, S. Hullavarad, S.B. Ogale, and T. Venkatesan, Appl. Phys. Lett. **84**, (2004) 3846

[16] C. Zener, Phys. Rev. **82** (1951) 403

[17] A.J. Millis, P.B. Littlewood , and B.I. Shraiman, Phys. Rev. Lett. **74**, 5144 (1995)

[18] P.A. Cox, *Transition Metal Oxides* (Clarendon, Oxford, 1992)

[19] P.W. Anderson and H. Hasegawa, Phys. Rev. **100** (1955) 675

[20] J.B. Goodenough, Phys. Rev. **100** (1955) 564

[21] A.J. Millis, B.I. Shraiman, and R. Mueller, Phys. Rev. Lett. **77** (1996) 175

[22] H. Röder, J. Zhang and A.R. Bishop, Phys. Rev. Lett. **76**, (1996) 1356

[23] J.B. Goodenough, Annu. Rev. Mater. Sci. **28** (1998) 1

[24] P. Schiffer, A.P. Ramirez, W. Bao, and S.W. Cheong, Phys. Rev. Lett. **75** (1995) 3336

[25] *Colossal Magnetoresistance, Charge Ordering and Related Properties of Manganese Oxides,* ed. by C.N.R. Rao and B. Raveau (World Scientific, Singapore, 1998)

[26] C.N.R. Rao and A.K. Cheetham, Science **276** (1997) 911

[27] H. Kuwahara, Y. Tomioka, A. Asamitsu, Y. Moritomo, and Y. Tokura, Science **270** (1995) 961

[28] Y. Tokura, Y. Tomioka, H. Kuwahara, A. Asamitsu, Y. Morimoto, and M. Kasai, J. Appl. Phys. **79** (1996) 5288

[29] C.N.R. Rao, Anthony Arulraj, A. K. Cheetham, and Bernard Raveau, J. Phys. Condens. Matter **12** (2000) R83

[30] Yamamoto, T. Kimura, T. Ishikawa, T. Katsufuji, and Y. Tokura, Phys. Rev. B **61** (2000) 14706

[31] P. Mandal and S. Das, Phys. Rev. B **56** (1997) 1507

[32] P. Raychaudhuri, S. Mukherji, A.K. Nigam, J. John, U.D. Vaisnav, R. Pinto and P. Mandal, J. Appl. Phys. **86** (1999) 5718

[33] Y.G. Zhao, R.C. Srivastava, P. Fournier, V. Smolyaninova, M. Rajeswari, T. Wu, Z.Y. Li, R.L. Greene, and T. Venkatesan, J. Magn. Magn. Mater. **220** (2000) 161

[34] C. Mitra, P. Raychaudhuri, K. Dörr, K.-H. Müller, L. Schultz, P.M. Oppeneer, and S. Wirth, Phys. Rev. Lett. **90** (2003) 17202

[35] K. Asokan, K.V. Rao, J.C. Jan, W.F. Pong, Ravi Kumar, Shahid Husain, and J.P. Srivastava, Surf. Rev. Lett. **9** (2002) 1053

[36] Ravi Kumar, R.J. Choudhary, S.I. Patil, Shahid Husain, J.P. Srivastava, S.P. Sanyal, and S.E. Lofland, J. Appl. Phys. **96** (2004) 7383

[37] H.Y. Hwang, S-W. Cheong, P.G. Radaelli, M. Marezzio, and B. Batlogg, Phys. Rev. Lett. **75** (1995) 914

[38] P.W. Anderson and H. Hasegawa, Phys. Rev. **100** (1955) 675

[39] K. Kubo and A. Ohata, J. Phys. Soc. Jpn. **33** (1972) 21

[40] J. Zang, A. R. Bishop, and H. Roder, Phys. Rev. B **53** (1996) 8840

[41] A.J. Millis, B.I. Shraiman, and R. Mueller, Phys. Rev. Lett. **77** (1996) 175

[42] A.J. Millis, R. Mueller, and B. I. Shraiman, Phys. Rev. B **54** (1996) 5405

[43] A.J. Millis, T. Darling, and A. Migliori, J. Appl. Phys. **83** (1998) 1588

[44] A. J. Millis, Nature **392** (1998) 147

[45] J. Asher, Nucl. Instr. Meth. Phys. Rev. B **89** (1993) 315

[46] D.M. Ruck, D. Boss, and I.G. Brown, Nucl. Instr. Meth. Phys. Rev. B **80–81** (1993) 233

[47] M. –A. Hasan, J. Knall, S. A. Barnett, J. E. Sundgren, L. C. Market, A. Rackett, and J. E. Greene, J. Appl. Phys. **65** (1989) 172

[48] A. Dunlop and D. Lesueur, Radiat. Eff. Def. Sol. **126** (1993) 123

[49] E. Balanzat, Radiat. Eff. Def. Sol. **126** (1993) 97

[50] M. Levalois, P. Bogdanski, and M.Toulemonde, Nucl. Instr. Meth. Phys. Rev. B **63** (1992) 14

[51] H. Matzke, Radiat. Eff. Def. Sol. **64** (1983) 3

[52] M.Toulemonde, S. Bouffard, and F. Studer, Nucl. Instr. Meth. Phys. Rev. B **91** (1994) 108

[53] S.K. Arora, Ravi Kumar, R. Singh, D. Kanjilal, G.K. Mehta, Ravi Bathe, S.I. Patil, and S.B. Ogale, J. Appl. Phys. **86** (1999) 4452

[54] Ravi Kumar, S.B. Samanta, S.K. Arora, A. Gupta, D. Kanjilal, R. Pinto, and A.V. Narlikar, Solid State Commun. **106** (1998) 805

[55] S.B. Ogale, K. Ghosh, J.Y. Gu, R. Shreekala, S.R. Shinde, M. Downes, M. Rajeswari, R.P. Sharma, R.L. Green, T. Venkatesan, R. Ramesh, Ravi Bathe, S.I. Patil, Ravi Kumar, S.K. Arora, and G.K. Mehta, J. Appl. Phys. **84** (1998) 6255

[56] S.B.Ogale, Y.H. Li, M.Rajeswari, L. Salamanca Riba, R. Ramesh, T. Venkatesan, A.J. Millis, Ravi Kumar, G.K. Mehta, Ravi Bathe, and S.I. Patil, J. Appl. Phys. **87** (2000) 4210

[57] J.H. Ormord, J.R. MacDonald, and H.E. Duckworth, Can. J. Phys **43** (1965) 275

[58] C.D. Moak and M. D. Brown, Phys. Rev. **149** (1966) 244

[59] *Handbook of Stopping Cross Sections for Energetic Ions in All Elements*, ed. by J.F. Ziegler (Pergamon Press, New York, 1990)

[60] W. Brandt and M. Kitagawa, Phys. Rev. B **25** (1982) 5631

[61] D. Luesuer and A. Dunlop, Radiat. Eff. Defects Solids **126** (1993) 163

[62] L.T. Chadderton and H.M. Montagu-Pollock, Proc. R. Soc. Lond. A **274** (1986) 239

[63] G. Szenes, Phys.Rev. B **51** (1995) 8026

[64] Z.G. Wang, Ch. Dufaour, E. Paumier, and M.Toulemonde, J. Phys. Condens. Matter **6** (1994) 6733

[65] Z.G. Wang, Ch. Dufaour, E. Paumier, and M.Toulemonde, J. Phys. Condens. Matter **7** (1995) 2525

[66] A. Meftah, F. Brisard, J.M. Costantini, E. Dooryhee, M. Hage-Ali, M.Y. Hervieu, J.P. Stoquert, F.Studer, and M. Toulemonde, Phys. Rev. B **49** (1994) 12457

[67] J.J. Neumeier, M.F. Hundley, J.D. Thompson, and R.H. Heffner, Phys. Rev. B52 (1995) R7006

[68] W Prellier, Ph. Lecoeur, and B. Mercey, J. Phys. Condens. Matter 13 R915 (2001)

[69] J. Asher, Nucl. Instr. Meth. Phys. Rev. B 89 (1993) 315

[70] D.M. Ruck, D. Boss, and I. G. Brown, Nucl. Instr. Meth. Phys. Rev. B 80–81 (1993) 233

[71] M. –A. Hasan, J. Knall, S. A. Barnett, J. E. Sundgren, L. C. Market, A. Rackett, and J.E. Greene, J. Appl. Phys. 65 (1989) 172

[72] J.L. Garcia-Munoj, J Fontcuberta, B. Martinez, A Seffar, S. Pinol, and X. Obradors, Phys. Rev. B55 (1997) 668

[73] P.G. Randaelli, G. Lannon, M. Marezio, H.Y. Hwang, S.W. Cheong, J.D. Jorgensen, and D. N. Argyriou, Phys. Rev. B56 (1997) 8265

[74] J. Fontcuberta, V. Laukhin, and X. Obradors, Appl. Phys. Lett. 72 (1998) 2607

[75] Improved version of the TRIM code originally developed by J.F. Ziegler, J.P. Biersack, and U. Littmark, *Stopping Power and Ranges of Ions in Matter* (Pergamon, New York, 1985)

[76] Ravi Bathe, S.K. Date, S.R. Shinde, L.V. Saraf, S.B. Ogale, S.I. Patil, Ravi Kumar, S.K. Arora , and G.K. Mehta, J. Appl. Phys. 83 (1998) 7174

[77] Ravi Kumar, F. Singh, R.J. Choudhary, and S.I. Patil, Nucl. Instrum. Meth. B, (2005), communicated

[78] A. Goyal, M. Rajeshwari, R. Shreekala, S.E. Lofland, S.M. Bhagat, T. Boettcher, C. Kwon, R. Ramesh, and T. Venkatesan, Appl. Phys. Lett. 71 (1997) 2535

[79] L.B. Kiss and P. Svedlindh, IEEE Trans. Electron Devices 41 (1994) 2112

[80] H.T. Hardner, M.B. Weissman, M.B. Salamon, and S.S P. Parkin, Phys. Rev. B 48 (1993) 16156

[81] C. Chen, X. Yi, X. Zhao, and B. Xiong, Sensors and Acuators A90 (2001) 212

[82] P.C. Shan, Z. Celik-Butler, D.P. Butler, and A. Jahanzeb, J. Appl. Phys. 78 (1995) 7334

[83] A. Jahanzeb, C.M. Travers, Z.C. Butler, D.P. Butler, and S.G. Tan, IEEE Trans. Electron Devices 44 (1997) 1795

[84] P.C. Shan, Z. Celik-Butler, D.P. Butler, A. Jahanzeb, C.M. Travers, W. Kula, and Roman Sobolewski, J. Appl. Phys. 80 (1996) 7118

[85] R.J. Choudhary, Ravi Kumar, Shahid Husain, J.P. Srivastava and S.I. Patil, Appl. Phys. Lett. (2005), communicated

[86] R.J. Choudhary, Ravi Kumar, Shahid Husain, J.P. Srivastava, S.K. Malik, and S.I. Patil. Nucl. Instrum. Meth. B, (2005), communicated

Chapter 19
EXPOSURE OF BONE TO IONIZING RADIATION

Leszek Kubisz

University of Medical Sciences in Poznan, Poland

Interest in the effects of the action of electromagnetic ionizing radiation on bone arises from medical applications of ionizing radiation. Gamma irradiation is commonly used for preservation and sterilization in bone banking. It is a most popular preservation means in tissue banking. The selection of a sterilization dose is a compromise between a dose that is low enough to preserve important biological properties of tissue allografts and high enough to inactivate as many microorganisms as possible. The problem is additionally complicated by the possible presence of pathogenic viruses such as the human immunodeficiency virus (HIV), hepatitis viruses and others. The currently recommended dose in bone banking is 25-35 k Gy, which inactivates bacteria, but it is ineffective in the case of viruses. The advantage of such sterilization over autoclaving and chemical sterilization is that irradiation neither reduces osteogenic activity nor makes the toxic chemical agent applied in the sterilization remain in bone. Often, prior to irradiation, bone undergoes other treatments such as freeze-drying and lipid extraction, which enables shelf storage of years.

Irradiation changes the physical properties of bone, mainly due to cross-linking and degradation of collagen, the organic component of bone. The above-mentioned additional treatment also influences the final effect of irradiation. Irradiation affects all levels of the structure of collagen. The chemical analysis has revealed damage of amino-acid side-chains, production of new groups, chemical transformation of amino acids, breakdown of peptide bonds, hydrogen bridges and cross-linking. There is also evidence that formation of inter- and intramolecular cross-linking takes place. Some of these effects must influence the tertiary structure of collagen

K.E. Sickafus et al. (eds.), Radiation Effects in Solids, 575–588.
© 2007 Springer.

and lead to conformational changes. Structural changes caused by cross-linking and degradation correspond to changes in macroscopic parameters of irradiated bone. Degradation leads to reduction of molecular weight and an increase in mobility. Thus degradation and cross-linking are responsible for the structural changes in the irradiated materials. The degradation process in collagen is expected to be dominant in solid state materials irradiated with doses higher than 100 kGy [1-4].

Studies on irradiated collagen suggest that the cross-linking process requires doses up to 30kGy [5,6]. Both processes, degradation and cross-linking of collagen, lead to its structural changes, which in turn are manifested in changes in such macroscopic properties as electrical, thermal and mechanical. In collagen, like in other polymers, the effect of irradiation is connected with the formation of radicals at the points of cleavage in the polypeptide chain. Radicals may finally react and disappear. However, in the solid state, two highly reactive radicals of relatively long lifetime, held for a long time in proximity, do not react with each other to repair a fracture, which is possible because of the structure of the collagen macromolecule. Its large side-groups, which are attached to the backbone of collagen, produce a state of molecular strain for steric reasons. Once the chain is broken the fracture allows the two fragments to separate. Free radicals may be trapped in collagen in regions of different states at the same temperature. Radicals or ions in the crystalline region may be unable to react there but be capable of migrating to the crystalline surface where, due to increased chain mobility, reaction takes place. Likewise, in the amorphous region free radicals may react faster. Free radicals are also generated in the mineral component of bone, which will be discussed later. Physical properties of biological materials irradiated with doses from 0 to 50 kGy are relatively well known. It has been established that some viruses and prions are destroyed only with doses as high as 200kGy, which raises the question of the effect of doses higher than 200kGy [7,8]. Because important chemical and conformational changes in the collagen of bone are associated with the phenomena of swelling, straining, heating and irradiation, the application of the solid –state sampling techniques enables us to study them.

1   WHAT IS BONE?

In biological terms, bone is described as connective tissue, which binds together and supports various structures of the body. At a macroscopic level there are two major forms of bone tissue: compact or cortical bone and cancellous or trabecular bone. Compact bone is a dense bone. Cancellous bone is composed of short struts of bone material called trabeculae. The connected trabeculea give the cancellous its bone spongy appearance. In physical terms, bone is natural composite material which consists of two phases. The combination of visco-elastic collagen and stiff, brittle HAP forms the anisotropic nanocomposite structure of bone. The physico-chemical and biological properties of the bone differ from those of each constituent considered separately. The collagen framework has a fibrous structure and supports mineral crystals, which reinforce its structure. Although the interplay of these two components is not fully understood [9,10], the interface plays a critical role in governing specific properties such as compressive and shear behavior, fracture modes, toughness and stress transfer from the applied load to the bone reinforcement. It is known, for example, that the presence of HAP brings about the increase in the denaturation temperature of collagen [11,12]. About 70% of mature bone is made up of the inorganic phase, with the balance consisting of organic matrix and water. The mineral fraction of bone is similar to hydroxyapatite $[Ca_{10}(PO)_4)OH)_2]$ and is often referred to as HAP. However, it is heterogeneous material and cannot be described as a single mineral [10]. Nevertheless HAP is thermodynamically the most stable mineral in bone. Yet other ions such as fluorine, chloride, and magnesium can also be incorporated into the crystal lattice. The substitution of monovalent OH sites and trivalent phosphate PO_4^{-3} sites by carbonate ions – CO_3 is particularly interesting. As a result, carbonated apatite, known also as dahllite $(Ca_5(PO_4,CO_3)_3$, is obtained. The organic matrix is 85-90% type I collagen fibres, which provide a supporting matrix upon which the mineral crystals grow. Minor proteins provide additional structural strength as well as regulatory and signalling functions. The remaining 10% of bone is made up of glycosoaminogycans, glycoproteins, lipids, peptides and enzymes. Collagen itself can be considered to be anisotropic material as it contains regions of different degrees of crystallinity [12].

According to many structural studies collagen triple helix is a unique protein defined by the supercoiling/superhelix of three polypeptide chains. It is a right – handed triple helix, which is generally composed of the amino-acid sequence repeat $(X-Y-Glycine)_n$, with proline and hydroxyproline often present at positions X and Y. Other frequently occurring amino acids in collagen include alanine, lysine, arginine, leucine, valine, serine, phenylalanine and threonine. The sequence of amino acids in polypeptide chains of collagen is called the primary structure. An individual chain has a left-handed helical structure, known as α-helix, which is not stable alone. The presence of glycine in every third residue is an absolute requirement for the triple-helix formation because glycine is a smallest amino acid; it is the only amino acid without a side chain. In this way we define the secondary structure of collagen. The triple helix forms the tertiary structure of collagen. The super-helix is stabilized by hydrogen bonds among the chains. The high content of hydroxyproline, which is also involved in interaction with water, is characteristic for collagen. Hydroxyproline residues play a critical role in stabilizing the triple-helical conformation of collagen because it participates in direct interchain hydrogen bonding. The triple-helical molecule is cylindrically shaped with a diameter of 1.5 nm and a length of 300 nm. The triple-helical molecules are all parallel, but their ends are separated by holes of about 35 nm. Neighboring molecules are staggered by 68 nm. The quaternary structure, microfibrils are formed by about five units of tropocollagen. The microfibrils are packed in a tetragonal lattice. The diameter of fibril is several angstroms long. Each fibril is made up of three polypeptide chains about 1000 amino acids long. These are wound together in a triple helix. Fibrils are assembled in fibers. Collagen itself is insoluble. Collagen fibrous tertiary structure combines with other macromolecules to form structural tissues, which support the mechanical stresses in organisms. In tendons, collagen fibers run parallel to the major stress axis. The gross mechanical properties of collagenous tissue must be derived from the conformation of the collagen molecule.

Water is the third component of bone. Traditionally water associated with proteins is divided into three types: structural, bound and free water. The structural water, about 0-0.07 g/g, is incorporated in the collagen structure. Its liberation is possible when collagen undergoes thermal denaturation. Bound water, 0.07-0.25 g/g, means water molecules which are tightly bound to specific sites in collagen chains filling in the spaces between molecules. The bound water -protein interaction is not strong as in the case of structural water. Free water is water of content higher than 0.45 g/g Between bound water and free water Nomura introduced a transition region in which both bound and free water are sorbed, that is water content belongs to the range 0.25-0.45 g/g [13]. Water sorption is possible due to both fibrous structure and chemical compositions; there is a large concentration of hydrophilic groups: C=O, N-H, COOH, OH.

2   MECHANICAL PROPERTIES OF IRRADIATED BONE

Bone sterilization aroused interest in the mechanical properties of irradiated bone because bone auto- and allografts are used in order to fill a bone gap or reinforce mechanical resistance. Under increasing applied force, bone maintains its rigidity to the point of fracture. The change in elastic modulus serves for initial softening but if the stress is too high, the mineral phase begins to fracture. Irradiated bone loses its mechanical resistance, which means the loss of its capacity to absorb energy, and results in bone's embrittlement. The loss depends on the condition of irradiation and is particularly high if prior to irradiation, bone is subjected to both dehydration by freeze-drying, and lipid extraction. Studies on the mechanical properties of bone have been undertaken up to the currently recommended dose of 30 ± 5kGy, which is sufficient to reduce a survived fraction of bacteria to 10^{-9}, and in doses which are high enough to kill viruses. Although in the case of HIV it is reported to be 60kGy, for other viruses the dose of 90kGy is necessary. In mechanical studies Young modulus, bending strength, work of fracture, and impact energy have been measured. According to the studies of Currey et al. [14], irradiation with the dose of about 95kGy produces unacceptable reduction in the mechanical integrity of bone. Their studies with doses of 17kGy, 29.5kGy and 94.7kGy showed that Young's modulus was unchanged by any level of radiation. However, radiation significantly reduced the bending strength, work of fracture and impact energy. In each case the severity of the effect increased with dosage. Irradiated samples did not usually reach such a high load as non-irradiated. Specimens irradiated with the dose of 94.5kGy absorbed only 5% of the energy of the controls.

Radiation, even in relatively small doses, makes the bone brittle. An increase in the embrittlement of bone reduces the bone's capacity to absorb mechanical energy. Similar results for the dose of 60kGy were reported in earlier studies of Komender, who found [15] the decrease in compressive, bending and torsional strength of 25, 35 and 40% respectively in human bone. However he did not find significant effect with the dose of 30kGy. Also Triantafyllou and Sotiropoulos discovered about 50% reduction in the bending strength of bovine bone for the dose of 50kGy [16]. Opposite results were reported by Anderson and Keyak [17] testing human cancellous bone. They found no effect of radiation on failure stress for doses lower than 60kGy. For the dose of 60kGy, they found reduction in both failure stress and Young's modulus by about 60%. Simonian and Conrad [18] observed no effect on human tibiae for the dose of 17kGy. Hamer and Strachan [19] found that irradiation with the dose of 28kGy did not affect the stiffness of bone but reduced the load to failure and the work to fracture in bending, decreasing monotonically with increasing dose.

The crucial question is why the mechanical properties are affected by radiation. According to the suggestion of Currey et al. [14] it is probably the strength properties of bone that are reduced as radiation partially denaturates the collagen matrix. Stiffness properties – Young's modulus of bone – remain unaltered because stiffness is mainly determined by the mineral which is less affected by radiation. Such speculations should be supported by studies on chemical changes occurring in mineral and organic phases of bone.

3   DSC TECHNIQUE AND IRRADIATED BONES

The differential scanning calorimetry (DSC) has recently been successfully applied to study thermal properties of such biological materials as leather, dentin, DNA, bones and collagen [20-23]. Moreover, this method has been used to study the thermal denaturation of proteins [23-25], water removal and thermal decomposition of amino acids [23,25]. In material containing collagen, which easily absorbs water, the large endothermic peak appears, usually in the temperature range of 290-440K [22]. Because the process of water release was the fastest in the temperature range of 353-373K [26], pre-heating at 385K led to liberation of free and bound water so that only structural water was left in the sample [8]. Heating of bone leads to thermal denaturation of collagen which according to Kronick and Usha [11,27,28] appears as endothermic peak at 429K. The thermally activated irreversible process of denaturation begins at 400K and involves uncoupling of the alpha chains leading to the helix-random coil transition and ends at 455K [24,29,30]. The DSC studies on bone exposed to the dose of 1 MGy, which brings about the domination of degradation of collagen, have shown that irradiation decreases the temperature of denaturation to 410K and shifts the onset temperature toward lower temperatures. The latter effect seems to be typical for other proteins because the decrease in the onset temperature with the increasing dose was observed also for globular protein irradiated with doses up to 30kGy and studied as liquid suspension [31]. Irradiation also leads to the higher enthalpy and entropy of the bone collagen denaturation process. They change respectively from -210 to $-337Jg^{-1}$ and from 0.49 to $0.82JK^{-1}g^{-1}$ [3]. The increase in entropy could be explained by a higher disorder due to smaller molecular mass of the scissed collagen polypeptide chain. Cleavage of the polypeptide chain also leads to solubility in collagen, which is usually insoluble.

4   CHANGES IN ELECTRICAL CONDUCTIVITY IN BONE

Electric experimental methods have been applied in studies on thermal and structural transition in bone. They provide information on important processes such as denaturation, thermal decomposition and water liberation. Electric properties are described by electric conductivity and electric permittivity. Dielectric relaxation studied as the effect of irradiation of the main organic component of bone, that is collagen, provides information on its thermal denaturation. Temperature and frequency dependencies of the $\varepsilon^{`}$ and $\varepsilon^{``}$ reflect the structural changes which take place upon irradiation. In heterogeneous systems dielectric polarisation is based on the Maxwell-Wagner-Sillars mechanism in the α-dispersion range [32]. In the case of collagen, conduction mechanism is explained in terms of the localized hopping of charge carriers, free protons, between neighbouring sites [32,33]. Such sites are formed by bound water attached strongly to the main chain of collagen [34]. The number of these sites increases during irradiation, due for example to water radiolysis. Generated free radicals and their molecular products H_2, H_2O_2 and H_2O produce hydrogen bonds with newly formed carbonyl and amide groups. Amide groups are formed during degradation of the polypeptide chain of collagen. In the case of irradiated collagen the release of structural water is completed below the temperature of 480K. It is manifested by a shift of a maximum of $\varepsilon`$ toward lower temperatures with increasing radiation doses. It indirectly shows that the irradiation decreases the denaturation temperature because the release of structural water takes place during thermal denaturation of collagen [35]. Temperature variation of dielectric constant of collagen also has shown that degradation processes dominate for doses higher than 100 kGy [4]. Dielectric measurements make it possible to determine relaxation time, which is inversely proportional to the conductivity at the low-frequency limit of the α-dispersion range, and proportional to the permittivity at the high-frequency limit of the α-dispersion range. However, relaxation time decreases with increasing temperature of collagen; the significant decrease was observed from the temperature of about 450K where the change in the slope of the log τ vs. temperature was observed (τ-relaxation time). Irradiation with doses of 100, 300, 500 and 1000kGy, leads to the decrease in the relaxation time at the temperature range up to about 500K, and the doses 300, 500 and 1000 kGy shifted the elbow of the mentioned curve toward lower temperatures, as the effect of degradation processes [4].

Studies on solid-state collagen enable us to predict the behaviour of bone collagen because the presence of HAP not only reinforces the structure of collagen but also increases its denaturation temperature [11,12,36]. Dielectric spectroscopy applied in studies on irradiated bone in the α-dispersion range indicated a domination of cross-linking for doses below 100kGy, and a domination of the degradation processes for higher doses. The latter process led to the decrease in the denaturation temperature. The effect of irradiation of bone is explained in terms of water free radicals, which were generated during irradiation. Initial small differences in the conductivity between irradiated and non-irradiated bone, which were observed in the temperature up to 373K, suggest that the radiolysis products of loosely bound water have no influence on dielectric polarization. Further heating of bone, which led to the increased conductivity, is explained as the indirect effect of water free radicals. Changes in the values of the dielectric constant $\varepsilon`$ and the loss factor $\varepsilon"$ of bone are, as in collagen, explained by a change in the number of jumps performed by protons (H^+) between sites formed by water molecules bound to collagen molecules in bone. With increasing temperature the number of sites among which protons can jump and the mobility of these protons is expected to increase as a result of release of water molecules. The conductivity arises from proton transfer through the intra- and intermolecular hydrogen bonds in bone [37]. The process of water release includes hydrogen bonds breaking, out of the bone sample water diffusion and mass loss [3,12]. Electrical conductivity, measured in d.c. fields in the range of voltage-current linearity [38], have shown that the degradation process also prevailed for doses higher than 100 kGy, and that the dose of 500kGy decreases the denaturation temperature from about 510K to about 500K [39], while the dose of 1000kGy decreases it to about 410K [3]. Electrical conductivity measured over the denaturation temperature showed the exponential increase. The fitting carried out in the dose range of 0-100kGy with the following equation $\sigma = A\exp(T/t)$ (σ-electrical conductivity), enabled us to find the relationship between pre-exponential factor A and t parameter: $lnA = a + b \cdot 1/t$, which was based on the Meyer – Neldel rule [40]. The lnA and $1/t$ were well correlated (the correlation coefficient of about 0.97) and lnA decreased linearly with increasing dose [41,42]. The effect is probably due to the fact that irradiation leads to structural defects in HAP that trap free radicals. The effect was observed up to the dose of 100kGy as it is clear that the concentration of both defects and free radicals in HAP depends on the absorbed dose of ionizing radiation and obtains saturation for the doses of 100 kGy [43]. The above relationship can be applied to determinate the dose of the ionizing radiation absorbed in bone.

5   EFFECTS ON AMINO ACID COMPOSITION

From the microscopic point of view, macroscopic changes arise from free radicals generated during irradiation. Also, chemical changes in proteins are produced through intermediate stages involving free radicals. It is known that free radicals take part in biochemical processes and they have been found in high concentration in tissues showing metabolic activity. On the other hand, ionizing radiation produces in biological systems paramagnetic species of high concentrations, which sometimes are stable. Damages observed during irradiation of bone occur due to radiation generated free-radicals such as hydroxyl radicals, superoxide ions, radicals derived from amino acids and inorganic radicals CO_3^{-3} and CO_2^{-1}. Collagen free radicals gradually disappear because they react with oxygen, which diffuses into tissue. Free radicals derived from the mineral part of bone have a long lifetime of about 10^7 years at 25°C, consequently bone and bone powder may be used as a natural dosimeter. Free radicals react with collagen and lead to changes in amino acid composition. For example ion radical O_2 may disrupt peptide bonds. Rupture of hydrogen bonds changes the secondary structure of the collagen macromolecule. Bowess and Moss stated in their experiment on irradiated collagen that acidic, basic amino acids and those having the ring structure were the most radiosensitive. Also, relative losses of amino acids depend on the condition of irradiation. Irradiation with doses of 500kGy led to some loss of nitrogen and an overall loss of 10 to 20% of amino acids [44]. Carboxyl- and amide groups are the most radiosensitive in amino acids. Hence along with deamination, oxidation or reduction of molecule and decarboxylation take place. The latter may be connected with free radicals, which split off the CO_2 group, as in the case of glycine and alanine-decarboxylation, or split off H_2, as in the case of glycine. Longer carbon chains may be ruptured.

Experiments carried out by Cassel stated that methionine, phenylalanine, and threonine were the most susceptible to alteration by radiation but alanine, glycine, hydroxyproline, proline and arginine were the least damaged. Relatively little damage to amino acid structure was observed at absorption dose of 20kGy [45]. Also in the case of 50kGy-irradiated bone, the group of low relative changes in the amino acid composition was formed by amino acids constituting collagen: glycine, hydroxyproline and glutamic acid [46]. Although the total amount of amino acid decreases [46-48], the content of a given amino acid may not only decrease but it may also increase [46,49]. The increase in the content is the result of transformation of one amino acid into another. Because the sensitivity of the different amino acids is not the same for different proteins [50], the same amino acid may be transformed into different amino acids depending on its surroundings.

6   EFFECT OF IRRADIATION ON HYDRATION LEVEL OF BONE AND BONE DISSOLUTION

Water is an essential bone component influencing its physical properties, such as electrical conductivity [32]. According to Nomura [13] the structural water content usually makes 0-7% of the total water content, while the bound water content makes up to 25%. The processes of water release may be observed as changes in temperature dependence of electrical conductivity σ, dielectric constant ε' and loss factor ε'' [3,37]. Irradiation leads to a decrease in the contents of the bound and structural water [3,37]. Water content is usually measured by thermo-gravimetric method. The decrease in the hydration level is explained by the decrease in the number of polar groups in the side-chains of collagen, because molecules of bound water are usually attached to polar groups [13]. A bovine bone irradiated with the dose bigger than 10kGy became brittle and dissolved in water after being powdered [51]. The degree of solubility depended on the dose of ionising radiation along with the temperature and duration of the dissolution process. Although the degree of solubility up to the dose of 100kGy was approximately constant, it rapidly increased for the doses higher than 100kGy. It is probably the effect of the domination of the degradation process [52].

7 FREE RADICALS AND ESR

The induced free radicals generated during bone irradiation are trapped for a long time and their presence can be detected by means of ESR spectroscopy. The intensity of the ESR absorption line is directly proportional to the total number of paramagnetic species in the sample, which is a basis for quantitative ESR analysis. Free radicals are generated both in the organic matrix and in the mineral component of bone. An organic radical derived from collagen was found by Stachowicz [53]. The radical is quite stable in degassed samples at room temperature but decays rapidly when heated, probably due to the inter-chain radical recombination. Admission of air to the non-heated specimen led to decay of paramagnetic species. The most stable inorganic free radical, which is known as CO_2^{-1}, has the lifetime of 10^7 year at $25°C$. The radical yield from organic matrix is much higher than from structural defects appearing in the mineral component.

However, in contrast to inorganic radicals, organic radicals recombine easily. The reason for such behaviour probably lies in the weak flexible hydrogen-bonded structure of collagen and the strong rigid crystalline structure of HAP. Free radicals can escape much easier from the collagen structure than from HAP crystals. The ESR technique is applied to the identification of irradiated food and the dating of human remains [54,55]. The method is based on the evolution of the radiation-induced free radicals in hard tissues. It means that ESR-related dose reconstructions are based on the assessment of radiation-induced radicals in HAP. The radical concentration increases linearly with the range from about 50mGy to above 100Gy, which complies well with the range of accidental doses. Theoretically, the dose in bone may be estimated in the dose range up to the dose of 100kGy [56], where the saturation of radiation-induced paramagnetic centres is obtained. The line in the ESR spectrum of long-lived singlet decreases its intensity with time and reaches a plateau. The smaller the average size of the crystals, the greater the decline in the intensity. This phenomenon can be explained in terms of the ratio of surface atoms to the overall number of atoms within the crystal. This factor increases the possibility of disappearance of free radicals localized on or near the surface in direct contact with the surrounding medium. Thus the plateau corresponds to a state of equilibrium between the surfaces of crystals and their surroundings. Therefore for the relatively large crystals the surface effect is negligible [43]. Thus the crystal size is the important factor affecting the yield of paramagnetic centres in biological materials exposed to ionizing radiation. In the tight-bone (femur diaphysis), the dose of 0.1Gy is detectable with high level of certainty, whereas in spongy bone (ribs), the certainty level of irradiation detection is about 0.3Gy [57]. In teeth, with much bigger HAP crystals, the detection level is even less than 0.1Gy. In light of the above, ESR dosimetry using mineralized tissues is a potential tool in biological dosimetry. Biophysical ESR dosimetry is becoming a tool for retrospective dosimetry, which means the evaluation of individual exposures that occurred years ago, for example the atomic bomb survivors of Hiroshima, or victims of the Chernobyl accident. ESR also seems to be a reliable method for determining whether a particular food item has been irradiated and at what dose. A linear increase in the concentration of radiation–induced free radicals in the range of 0-10kGy and in the range of 10-22kGy in the case of chicken drumstick bones and in chicken bone powder was found [58, 59]. The former dose range is the commercially used irradiation dose range

8   SUMMARY

Bone as a nanocomposite consists of two phases: organic phase – collagen, and inorganic phase – hydroxyapatite and water. Irradiation of bone due to cross-linking and degradation of collagen leads to chemical transformation of amino acids, breakdown of peptide bonds, hydrogen bridges and cross-linking of polypeptide chains, which influence the tertiary structure of collagen. The damages observed during radiation treatment of bone occur due to radiation generated inorganic radicals and free radicals derived from amino acids. Because of the long lifetime of inorganic radicals, bone and bone powder may be used as a natural dosimeter. Structural changes caused by radiation are manifested in the changes in solubility of collagen, electrical conductivity of bone, and changes in the DSC thermograms of bone. It was also reported that irradiation alters the mechanical properties of bone, particularly if, prior to irradiation, the bone had been subjected to both dehydration by freeze-drying and lipid extraction. Irradiated bone loses its mechanical resistance.

ACKNOWLEDGEMENTS

The author wishes to express appreciation to Anna Kubisz for her help in the editing of this paper.

REFERENCES

[1]   A. Bailey, J. Bendall , D. Rhodes: Int.J.Appl.Radiat.Isot. **13**, (1962)

[2]   A. Bailey, W. Thromans: Radiat-Res. **23**, (1964)

[3]   L. Kubisz, S. Mielcarek, F. Jaroszyk: Int J Biol Macromol **33**, 1-3 (2003)

[4]   E. Marzec: Int.J.Biol.Macromol. **17**, 1 (1995)

[5]   A. Bailey, D. Rhodes, C. Cater: Radiat.Res. **22**, (1964)

[6]   A. Charlesby: Poymer Journal **19**, 5 (1987)

[7]   C. Gibbs, D. Gajdusek, R. Latarjet: Proc.Natl.Acad.Sci.USA. **75**, 12 (1978)

[8]   M. Sintzel, A. Merkli, C. Tabatabay et al: Drug Development & Industrial Pharmacy **23**, 9 (1997)

[9]   W. Landis: Bone **16**, 5 (1995)

[10]   S. Weiner, H. Wagner: Annual Review of Materials Science **28**, (1998)

[11]   P. Kronick, P. Cooke: Connect Tissue Res **33**, 4 (1996)

[12]   E. Marzec, L. Kubisz, F.Jaroszyk: Int. J. Biol. Macromol. **18**, 1-2 (1996)

[13]   S. Nomura, A. Hiltner, J. Lando et al: Biopolymers **16**, (1977)

[14]   J. Currey, J. Foreman I. Laketic et al: J.Orthop.Res. **15**, 1 (1997)

[15]   A. Komender: Mater.Med.Pol. **8**, (1976)

[16]   N.Traitafyllou, E. Sotiropoulos, J. Trantafyllou: Acta Orthop.Belg **41**, Supplement (1975)

[17]   M. Anderson, J. Keyak, H. Skinner: J-Bone- Joint-Surg. **74**, 5 (1992)

[18]   P. Simonian, E. Conrad, J. Chapman et al: Clin.Orthop. **302**, (1994)

[19]   A. Hamer, J. Strachan, M. Black et al: J.Bone.Joint.Surg.Br **78**, 3 (1996)

[20]   A. Calafiori, L. Imbrogno, G. Martino et al: Boll Soc Ital Biol Sper **69**, 11 (1993)

[21]   C. Chahine: Thermochimica Acta **365**, 1-2 (2000)

[22] M. Fois, A. Lamure, M. Fauran et al: J. Polym. Sci. Part B-Polymer Physics **38**, 7 (2000)

[23] T. Sakae, H. Mishima, Y. Kozawa et al: Conn.Tiss.Res. 33, 1-3 (1995)

[24] C. Miles, M. Ghelashvili: Biophys.J. **76**, 6 (1999)

[25] A. Rochdi, L. Foucat, J. Renou: Food Chemistry, **69**, (2000)

[26] A. Bigi, A. Fichera, N. Roveri et al: Int.J.Biol.Macromol. **9**, (1987)

[27] R. Usha, T. Ramasami: Thermochimica Acta **338**, (1999)

[28] R. Usha, T. Ramasami: Thermochimica Acta **356**, 1-2 (2000)

[29] C. Miles, A.Bailey: Proc.Indian Acad.Sci. **111**, 1 (1999)

[30] C. Miles, T. Burjanadze: Biophysical Journal **80**, 3 (2001)

[31] K. Ciesla, Y. Roos, W. Gluszewski: Rad.Phys.Chem. **58**, (2000)

[32] R. Pethig: *Dielectric and electronic properties of biological materials.* (John Wiley & Sons 1979)

[33] R. Pethig: Ferroelectrics **86**, (1988)

[34] G. Ramachandran:Int.J.Peptide Res. **31**, (1988)

[35] F. Jaroszyk, E. Marzec: J.Mar.Sci. **29**, (1994)

[36] L. Kubisz: Polish J. Med. Phys. and Engineering **9**, 3 (2003)

[37] E. Marzec, L. Kubisz:J.Non-Crystalline Solids, **305**, 1-3 (2002)

[38] J. Behari , S. Guha, P. Agarwal: Connect Tissue Res **2**, 4 (1974)

[39] L. Kubisz: Int J Biol Macromol **26**, 1 (1999)

[40] W. Meyer, H. Neldel: Zeit.Tech.Phys. **18**, 588 (1937)

[41] L. Kubisz: Physica Medica **20**, Suppl. (2004)

[42] J. Vlcakova, P. Saha, V. Kreslarek et al: Synthetic Metals 113, (2000)

[43] K. Ostrowski, A. Dziedzic-Goclawska, W. Stachowicz et al: J.Ann.N.Y.Acad.Sci. 238, (1974)

[44] J. Bowes, J. Moss: Radiat-Res. 16, (1962)

[45] J. Cassel: J.Am.Leath.Chem. 54, 8 (1959)

[46] L. Kubisz, F. Jaroszyk: Sci. Proc. Riga Techn. Univ. Transport and Engineering **6**, (2002)

[47] L. Kubisz, F. Jaroszyk: Current Topics in Biophysics 17, (1993)

[48]   K. Pietrucha: Polymers in Medicine 19, 1-2 (1989)

[49]   H. Al-Khantani, H. Abu-Tarboush, M. Atia et al: Radiat.Phys.Chem. 51, 1 (1998)

[50]   P. Alexander, J. Lett: Comp.Biochem. 27, (1967)

[51]   E. Pankowski, L. Kubisz: International Conference on Medical Physics and Engineering in Health Care, Poznań, 18-20.10 2001, (2001)

[52]   E. Pankowski, L. Kubisz: IMFBE Proceedings 3, 1 (2002)

[53]   W. Stachowicz, K. Ostrowski, A. Dziedzic-Goclawska et al: Nukleonika 15, 1 (1970)

[54]   W. Gordy, W. Ard, H. Shields H.: Proc.Natl.Acad Sci U.S.A 41, (1955)

[55]   S. Mascarenhas, O. Baffa Filho, M. Ikeya: Am.J.Phys.Anthropol. 59, (1982)

[56]   W. Stachowicz, G. Burlinska, J. Michalik et al.:Nukleonika 38, 3 (1993)

[57]   W. Stachowicz, J. Michalik, G. Burlinska et al.: Appl-Radiat-Isot. 46, 10 (1995)

[58   M. Polat, M. Korkmaz, O. Korkmaz: Rad.Phys.Chem. 49, 4 (1997)

[59]   M. Polat, M. Korkmaz, B. Dulkan et al: Rad.Phys.Chem. 49, 3 (1997)

Index